Springer Series in

MATERIALS SCIENCE 129

Springer Series in

MATERIALS SCIENCE

Editors: R. Hull C. Jagadish R.M. Osgood, Jr. J. Parisi Z. Wang H. Warlimont

The Springer Series in Materials Science covers the complete spectrum of materials physics, including fundamental principles, physical properties, materials theory and design. Recognizing the increasing importance of materials science in future device technologies, the book titles in this series reflect the state-of-the-art in understanding and controlling the structure and properties of all important classes of materials.

Please view available titles in *Springer Series in Materials Science* on series homepage http://www.springer.com/series/856

Preface

Very many structural materials like metals and ceramics are of crystalline nature. Under most of the conditions, they undergo plastic deformation before they fail under load. The plastic deformation is mediated by the generation and motion of one-dimensional crystal defects, the so-called dislocations. Thus, the density and dynamic properties of the dislocations determine the plastic behavior of the respective materials and frequently also their failure. The dislocation dynamics, that is, the response of the dislocations to an external load, is controlled by the interaction between the moving dislocations with the periodic crystal lattice structure, other crystal defects from point defects over small clusters to larger precipitates, with other dislocations forming a microstructure through which the mobile dislocations have to move, and finally the grain and phase structure of the material. Considering these different interactions, the dislocation motion turns out to be quite a complex process.

In situ straining experiments in a transmission electron microscope are a powerful means for studying the microprocesses controlling the dislocation mobility. With this method, dislocations can directly be observed during their motion under load. The technique advanced when high-voltage electron microscopes became commercially available at the end of the 1960s, first in Japan and later on also in other countries. These microscopes enable the transmission of thicker specimens, which have properties similar to macroscopic bulk specimens. Besides, they offer sufficient room for elaborate deformation stages in the specimen chamber. In the last 30 years, in situ straining experiments were performed on many materials, both in conventional and in high-voltage electron microscopes. The present author and his coworkers performed such experiments over about 30 years, yielding many hours of video recordings of dislocation motion.

This book gives an introduction into the processes controlling the dislocation dynamics and its role in crystal plasticity. It is divided into two parts. Part I describes the general properties of dislocation motion in an introductory way suited also for students without prior knowledge in dislocation properties.

"Dislocations" in a mowed lawn (solarized)

It is based on lectures given in the Physics Department of the Martin Luther University in Halle (Saale). For easy understanding, the mathematical treatment is presented on a simple level. In Part II, particular materials are discussed covering semiconductors, ceramic single crystals, metals, and alloys including intermetallic alloys and quasicrystals. The whole text is, whenever possible, illustrated by electron micrographs, partly of dislocations under load taken during in situ straining experiments in a high-voltage electron microscope. Besides, video files can be downloaded from http://extras.springer.com/2010/978-3-642-03176-2 presenting characteristic video sequences of moving dislocations. The videos are commented in the chapters of Part II but are referred to also in Part I. The presentation of the different materials is not exhaustive but intended to explain the processes shown in the videos. It is a main aim of this work to make the video sequences available to a greater number of scientists and lecturers.

The experimental work frequently referred to in this book and exploited for the video clips was carried out in the Institute of Solid State Physics and Electron Microscopy of the Academy of Sciences of the GDR in Halle (Saale) and since 1992 in the Max Planck Institute of Microstructure Physics at the same place. The author is very grateful to all responsible directors, that is, the late Heinz Bethge, Volker Schmidt, and Jürgen Kirschner, for creating the experimental and collaborative environment for these studies. The continuous high-quality performance of the high-voltage electron microscope was essentially due to Christian Dietzsch. The author acknowledges the contributions of many scientists to the experiments and interpretations; first of all to Fritz Appel for the work on NaCl, MgO, and some metals, and Martin Bartsch for the studies on all other materials. He had an essential influence on the experiments and the instruction of the PhD students and post-doctoral fellows who worked in projects funded by the Deutsche Forschungsgemeinschaft and the Volkswagenstiftung. He also saved the original video recordings in digital form and helped to collect data for the book, and he discussed its contents. Many other scientists were involved in the experiments. Their names are listed below.[1] Jürgen Kirschner supported the preparation of the book after my retirement. Hans-Rainer Trebin, Boris V. Petukhov, Wolfgang Blum, Hartmut Leipner, and Ichiro Yonenaga discussed special topics or helped finding suitable references. Video clips were made available by Daniel Caillard and Volker Mohles. The author is very grateful for all these contributions. The English of the text was kindly revised by Heike Messerschmidt.

Halle (Saale) *Ulrich Messerschmidt*
December 2009

[1] Part of the experimental work was carried out by the PhD students and post-doctoral fellows Susanne Guder, Anna Wasilkowska, Dietmar Baither, Bernd Baufeld, Ralf Haushälter, Dietrich Häussler, Bert Geyer, Ludwig Junker, Lars Ledig, and Aleksander Tikhonovski. Several studies were performed in close collaboration with other laboratories. In this respect, I thank the late Peter Haasen, Manfred Rühle, Knut Urban, Toru Imura, Masaharu Yamaguchi, Eckhard Nembach, Bernd Reppich, Gerhard Sauthoff, Eduard M. Nadgornyi, Peter J. Wilbrandt, Markus Wollgarten, Michael Feuerbacher, Peter Schall, Mark Aindow, Ian Jones, Hiro Saka, Yoichi Nishino, Christina Scheu, Easo P. George, Witold Zielinski, and Michael Mills. Further names will be found in the references.

Contents

Part I

General Properties of Dislocation Motion

1 Introduction

The properties of crystalline solids can be classified into two groups. To the first one belong, for example, the elastic properties or the occurrence of X-ray diffraction patterns. These phenomena are controlled by the regular periodic structure of the crystal lattice. The second group involves properties like diffusion or the mechanical strength. Though these phenomena are also influenced by the regular crystal structure and its binding properties, they are essentially controlled by the defects in the regular arrangement of the atoms and may be called structure sensitive properties, as was introduced by Smekal [1]. The crystal defects are classified according to their extension in space into zero, one, two, and three-dimensional defects. Zero-dimensional defects or point defects include single missing atoms called vacancies, interstitial atoms, and substitutionally or interstitially incorporated foreign atoms like impurities. One-dimensional defects occur if the regular coordination of atoms is disturbed along a line. These defects are called dislocations being the topic of this book. Two-dimensional defects are grain and phase boundaries while three-dimensional ones are larger precipitates or inclusions. Since the movement of dislocations is influenced by all types of defects, it turns out to be a very complex process.

In the following section, it is shown that shearing a crystalline solid along a plane as a whole requires a high stress, which exceeds the measured strengths of materials by several orders of magnitude. Afterwards, in Sect. 1.2, dislocations are introduced. Their motion through the crystal lattice allows the shearing in small steps, yielding realistic values of the mechanical strength.

Part I of the book treats the general features of the dynamic dislocation behavior. At first, an outline of the experimental methods is given in Chap. 2. As a basis for understanding the dynamics, the geometric and elastic dislocation properties as well as the structure of dislocations in real crystals are reviewed in Chap. 3 before the dislocation motion itself is treated in Chap. 4. Chapter 5 is concerned with the kinetic processes of dislocation generation, immobilization and annihilation. Part II is devoted to the dislocation motion in particular material classes, that is, to semiconductors in Chap. 6, ceramic single crystals (Chap. 7), metals (Chap. 8), intermetallic alloys (Chap. 9), and finally to quasicrystals (Chap. 10). In these chapters, particular dislocation processes are discussed in more detail and, whenever possible, they are illustrated by video sequences.

1.1 Theoretical Yield Strength

In the 1920s, several authors tried to estimate the yield strength, that is, the critical stress necessary to plastically extend or shear a crystalline body. The following estimation of the shear strength is based on a work by Frenkel [2]. As sketched in Fig. 1.1, a shear stress τ acts on the solid and tries to shift the lattice planes A and B of distance h with respect to each other. The distance between the atoms within the planes is b. This is the periodicity distance of the lattice. Originally, the planes are in elastic equilibrium. The stress acting tries to shift the upper plane B in the x direction. After a shift by b, the body is in equilibrium again. It is assumed that the stress necessary to shift plane B is a sine function of x

$$\tau = K \sin(2\pi x/b).$$

K is an interaction strength between the planes, which has to be determined in agreement with a macroscopic quantity. In this case, it is the shear modulus μ. It is defined as

$$\mu = \left.\frac{\mathrm{d}\tau}{\mathrm{d}\gamma}\right|_{x=0},$$

where γ is the (elastic) shear strain. In Fig. 1.1, the increment in γ is given by $\mathrm{d}\gamma = \mathrm{d}x/h$. Thus,

$$\mu = h\left.\frac{\mathrm{d}\tau}{\mathrm{d}x}\right|_{x=0} = \frac{2\pi h K}{b}\cos(2\pi x/b).$$

For $x = 0$, K becomes

$$K = \frac{\mu b}{2\pi h},$$

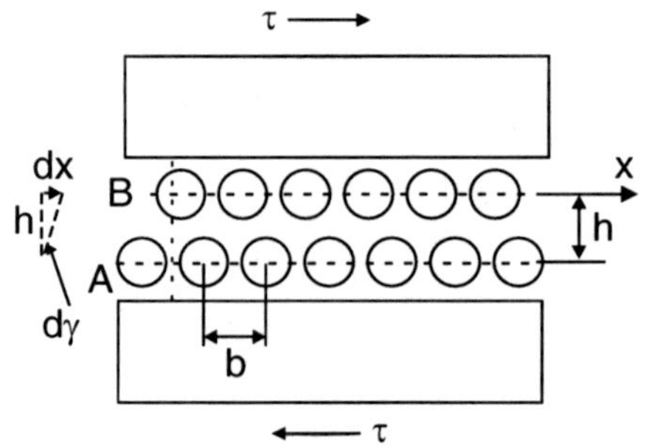

Fig. 1.1. Model for calculating the theoretical shear strength

Table 1.1. Shear modulus and real (normal) yield strength

Material	μ (MPa)	σ_y (MPa)
Al alloys	27,000	100 . . . 500
Steels	86,000	400 . . . 1,500

or

$$\tau = \frac{\mu b}{2\pi h}\,\sin(2\pi x/b).$$

The maximum value is obtained for $x = b/4$

$$\tau_{\mathrm{id}} = \frac{\mu b}{2\pi h}.$$

This is an estimate of the shear strength of an ideal crystal lattice if it is sheared along a plane as a whole. It may be noted that this strength depends on the ratio between the periodicity distance b and the spacing h between the sheared planes. With $b/h = 1/\sqrt{2}$ in the f.c.c. lattice, it follows that $\tau_{\mathrm{id}} = \mu/(\pi\sqrt{8}) \approx \mu/10$. Some values of the shear modulus and the measured technical yield stresses σ_y of materials are given in Table 1.1. The relation between the shear stress and the technical (normal) stress will be described in Sect. 2.1.

As a conclusion, the real yield stresses of materials are about two orders of magnitude smaller than the theoretical strength of an ideal crystal lattice. The difference between both is due to the existence of crystal defects, primarily of dislocations. The dynamic behavior of the latter determines the characteristics of plastic deformation.

1.2 Plastic Shear by the Motion of Dislocations

The great difference between the experimental yield stresses and the theoretical shear strength necessary to shift two lattice planes with respect to each other as a whole can be explained by the occurrence of line defects called dislocations in a crystalline solid and their motion under stress. The concept of crystal dislocations was introduced by Orowan [3], Polanyi [4], Taylor [5], and Burgers [6]. The elastic stress and strain fields of dislocations in an elastic continuum were calculated before by Volterra [7].

The process of the motion of a dislocation through a crystal is illustrated in Fig. 1.2. The undisturbed crystal is shown in Fig. 1.2a. The vertical lines inside the body mark the lattice planes. The acting shear stress τ has created a step on the left surface by breaking bonds in the left outer plane and by pushing the upper half-plane into the crystal (Fig. 1.2b). The lattice is intact everywhere except along the line D at the inner end of the inserted extra half-plane. This line is the place of the dislocation. In addition to the disturbance

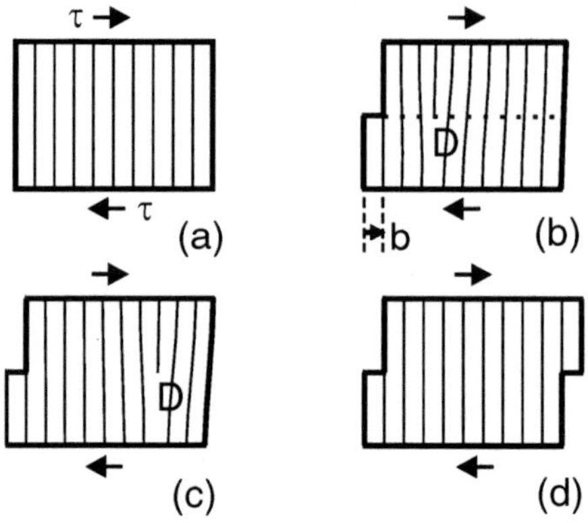

Fig. 1.2. Shearing by the motion of a dislocation through a crystal

of the coordination of the atoms near the dislocation line, the whole lattice is elastically strained, characterized by bent lattice planes. Under the action of the stress, the dislocation may move through the crystal as in Fig. 1.2c and finally emerge on the right outer plane, producing a surface step there also as in Fig. 1.2d. After the whole process, the body is sheared along the plane on which the dislocation moved by the vector $\boldsymbol{b}$, that is, it permanently changed its shape owing to a plastic strain. The shift vector $\boldsymbol{b}$ is called the Burgers vector and describes not only the shift of one part of the body with respect to the other, but also the intensity of the strain field around the dislocation. In the case of Fig. 1.2, $\boldsymbol{b}$ is perpendicular to the dislocation line.

Figure 1.3 presents a high-resolution electron micrograph of a dislocation in a real crystal, a thin PbTe foil with a (001) foil plane. The bright dots represent columns of atoms. The PbTe foil has the NaCl structure, that is, the distance between the cube planes is half the lattice constant. The dislocation in the center becomes visible when the image is viewed at a glancing angle. Viewing from the upper left corner in diagonal direction reveals that, according to the crystal structure, two extra half-planes are inserted from below and end at the location of the dislocation. The latter extends perpendicular to the image plane. As shown in the image, a dislocation involves the disturbance of the regular arrangement of atoms along a line, which causes a far-reaching elastic strain field, visible as a bending of the atom planes. In the diffraction contrast mode of transmission electron microscopy, the strong strain field near the dislocation lines is imaged and the dislocations appear as lines as in Fig. 1.4. The figure shows dislocations that have formed during plastic deformation and which are arranged on different planes.

A very important quantity for characterizing a dislocation microstructure is the dislocation density ϱ. It is defined as the length of dislocation lines per volume. Frequently, the dislocation density is measured as the number of emergence points through a surface. Both quantities are roughly equal except

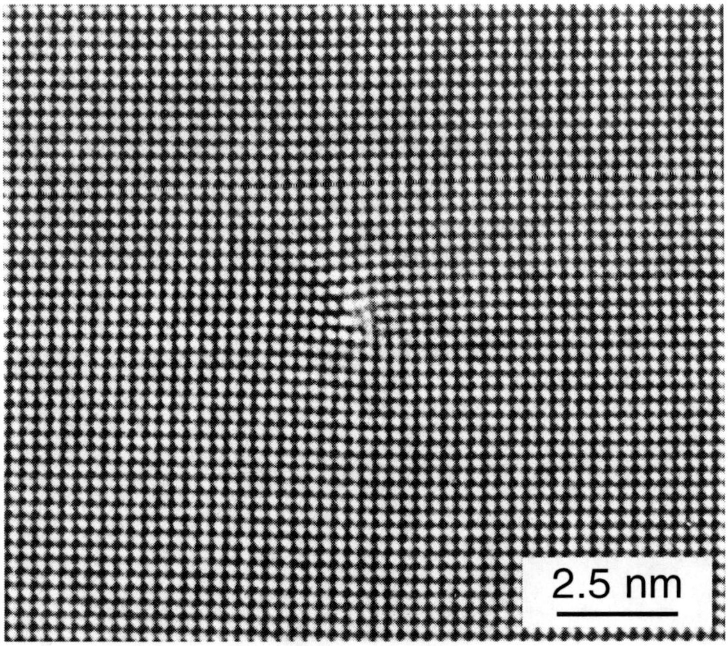

Fig. 1.3. High-resolution electron micrograph of a dislocation in a thin PbTe foil. The image plane is a (001) plane. Micrograph by Roland Scholz

for a numerical factor close to unity, which depends on the geometry of the dislocation arrangement. Characteristic densities are given in Table 1.2.

It may be noted that the dislocation densities in strongly deformed materials correspond to 1–1,000 km mm^{-3}. On the other hand, the concentration of atoms with a disturbed neighborhood is quite low, as the following estimation shows. The number of atoms with disturbed coordination is about 5 per lattice plane, that is, $5/a$ per unit length of the dislocation or $5\varrho/a$ per unit volume. Here, a is the lattice constant. Considering that the total number of atoms per unit volume is $1/a^3$, the concentration of atoms of disturbed coordination is about $5\varrho a^2$. With characteristic values of $\varrho = 10^{12}$ m^{-2} and $a = 3 \times 10^{-10}$ m, the concentration becomes about 5×10^{-7}. This concentration is not very high, and it is in the order of magnitude of that of intrinsic thermal point defects at about 70% of the absolute melting temperature.

The plastic deformation realized by the motion of dislocations is governed by their geometric properties, which will be described in Sect. 3.1. The dynamic properties are controlled by the interactions with the crystal lattice itself as well as with all the other defects in the material. The latter interactions are essentially of elastic nature. Therefore, the elastic properties of dislocations are treated in Sect. 3.2. Finally, the features of the dislocations are also influenced by the crystal lattice structure. This will be described in

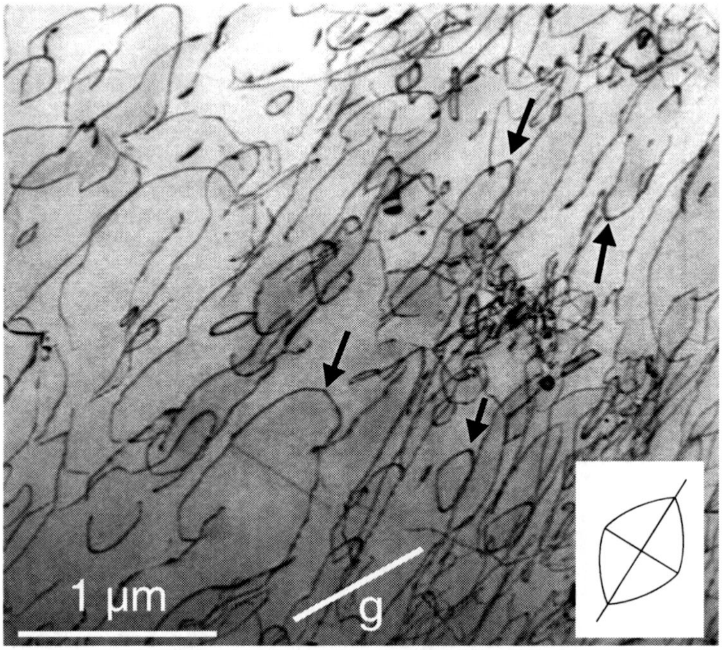

Fig. 1.4. Dislocation structure in a cross-sectional foil of an FeAl single crystal deformed about 2.5% at 533°C and 564°C. $\boldsymbol{g} = (1\bar{1}0)$ is the imaging diffraction vector near the pole $\boldsymbol{n} = [111]$. *Inset*: Equilibrium shape of dislocations with [010]. Burgers vector on a (101) plane projected onto the image plane, calculated by the line tension model using anisotropic elasticity theory as described in Sect. 3.2.7. From [8]. Copyright Elsevier Ltd. (2005)

Table 1.2. Characteristic dislocation densities in some materials

Material	ϱ (m^{-2})
Semiconductor single crystals	0 or very few
Well annealed metal crystals	$10^{7} \ldots 10^{8}$
Slightly deformed crystals	$10^{10} \ldots 10^{12}$
Strongly deformed crystals	$10^{12} \ldots 10^{15}$

Sect. 3.3. These topics are the basis of studying the dislocation motion and dynamics in Chap. 4.

For a detailed study of particular aspects of dislocation dynamics, the following textbooks are recommended. The thermodynamics of dislocation motion are treated in a fundamental way in the book *Thermodynamics and Kinetics of Slip* by Kocks, Argon and Ashby (1975) [9]. Several mechanisms of dislocation dynamics and many experimental data are given in *Dislocation Dynamics and Mechanical Properties of Crystals* by Nadgornyi (1988) [10].

The thermally activated dislocation motion is described in *Thermally Activated Mechanisms in Crystal Plasticity* by Caillard and Martin (2003) [11]. Finally, many aspects of the theory of dislocations are presented in detail in the textbook *Theory of Dislocations* by Hirth and Lothe (1982) [12], which is quoted frequently in the present book. Before the dislocation properties will be discussed in detail, the next chapter describes the experimental methods that mainly yield information on the mechanisms of dislocation dynamics.

2

Experimental Methods

Information on the dynamic dislocation behavior can be obtained by methods on different scales. This ranges from the measurement of macroscopic deformation data to microscopic techniques like transmission electron microscopy and, in particular, to in situ deformation tests in the electron microscope. Several methods have frequently been used and are described below in some detail. Others are only of supplementary character.

2.1 Macroscopic Deformation Tests

This section introduces the measurement of the macroscopic flow stress and its sensitivity to changes in the temperature and the deformation rate. These quantities will be used in Sect. 4.1 to determine the thermodynamic parameters of the dislocation motion. If a body of homogeneous cross section A and length l is loaded by a force F, it undergoes a change in length Δl. It is useful to normalize the force by the cross section and the length change by the length to obtain the (technical) tensile or compression stress σ and total strain ε_{t},

$$\sigma = F/A, \qquad \varepsilon_{\mathrm{t}} = \Delta l / l. \tag{2.1}$$

The total strain may be composed of the reversible elastic strain $\varepsilon_{\mathrm{el}}$ and the irreversible plastic strain ε

$$\varepsilon_{\mathrm{t}} = \varepsilon_{\mathrm{el}} + \varepsilon. \tag{2.2}$$

The elastic strain is determined by the laws of elasticity. Under most experimental conditions, a plastic strain in a crystalline solid is realized by shearing the material in crystallographic directions along crystallographic planes as outlined in Fig. 2.1.

The plastic shear deformation is caused by a shear stress τ resolved onto the shear plane, which results from the applied stress σ according to Schmid's law [13]

$$\tau = m_{\mathrm{s}}\sigma = \cos\varphi \cos\psi \ \sigma, \tag{2.3}$$

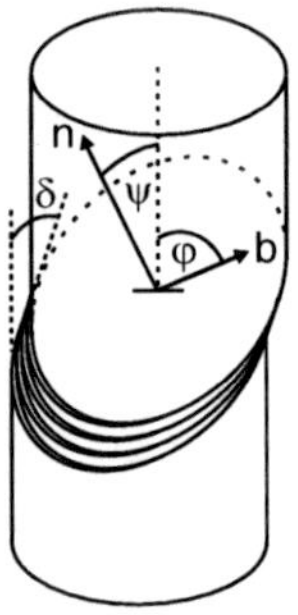

Fig. 2.1. Plastic shearing of a crystalline solid

where φ is the angle between the direction of shear $\boldsymbol{b}$ and the specimen axis, and ψ is that between the normal of the shear planes $\boldsymbol{n}$ and the specimen axis. The product of both cosines m_s is called the orientation factor (in German Schmid-Faktor). The direction of the shearing and the shear plane define a so-called slip system. Similarly, there is a relation between the (normal) plastic strain ε and the plastic shear strain $\gamma = \tan \delta$ (see Fig. 2.1)

$$\gamma = \varepsilon / m_s. \tag{2.4}$$

During the deformation, the specimen cross section and the orientation of the shearing direction and of the shear planes change. For large deformations, this should be taken into account in (2.3) and (2.4). In this book, mostly small deformations are discussed. It is therefore sufficient to consider only the starting data.

The relevant outer parameters controlling the plastic deformation are the temperature T, the plastic strain rate $\dot{\varepsilon}$, and the stress σ. In most experiments, the temperature is kept constant, at least for an interval of time. In addition, one of the other two parameters remains constant, too. If this is the stress (or in many cases simply the load F), the deformation rate will change in time, and the experiment is then called a creep test. In the other case, the deformation machine enforces a certain strain rate and the stress becomes a function of strain or time. Mostly, the total strain rate is controlled by the deformation machine so that the elastic strains in (2.2) include also the elastic deformation of the machine or, at least, parts of it. Thus, the plastic strain rate may differ more or less from the nominal one, as described later. The experiments with a constant strain rate are called quasistatic experiments. Most experiments described in this book are of this type. Special machines are designed to reach very high strain rates, for example, by using impact loading. In these dynamic experiments, neither the stress nor the strain rate and even not the temperature remain constant.

The typical result of a quasistatic deformation test is the stress–strain curve. Two examples are outlined in Fig. 2.2, where the actual applied stress σ necessary to deform the specimen is plotted vs. the strain, in the simple

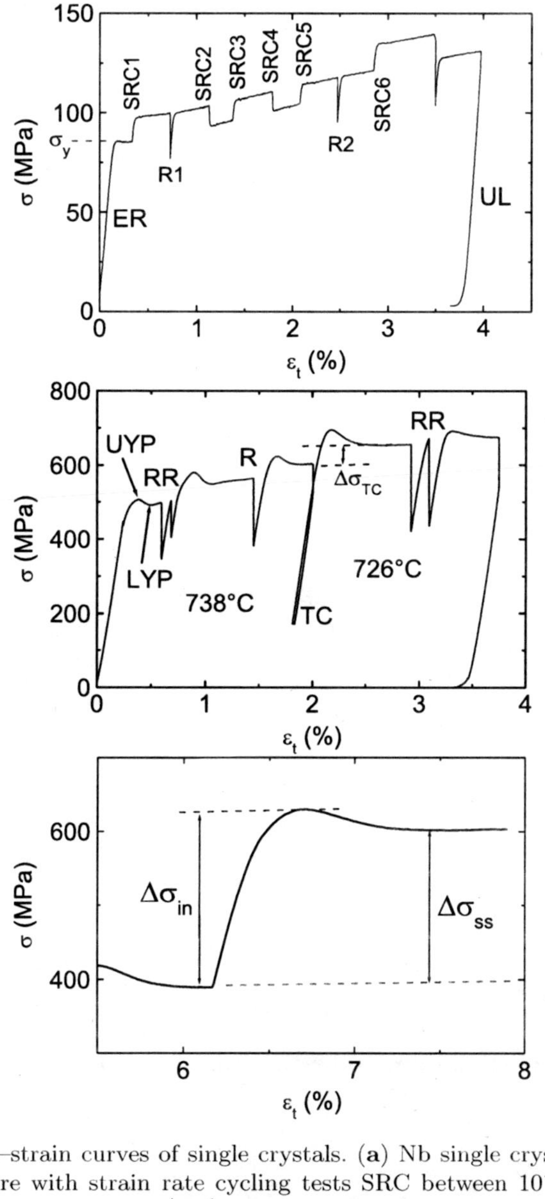

Fig. 2.2. Stress–strain curves of single crystals. (**a**) Nb single crystal deformed at room temperature with strain rate cycling tests SRC between 10^{-6} and $10^{-5}\,\mathrm{s}^{-1}$ (SRC1 to SRC4) and up to $10^{-4}\,\mathrm{s}^{-1}$ (SRC5). (**b**) $MoSi_2$ single crystal deformed at two temperatures along a $\langle 110 \rangle$ orientation at a strain rate of $10^{-5}\,\mathrm{s}^{-1}$. (**c**) Section of a stress–strain curve of an Al-Pd-Mn single quasicrystal deformed at 755°C with a strain rate change (SRC) from 10^{-4} to $10^{-3}\,\mathrm{s}^{-1}$. Data from [14–16]

case the total strain ε_t. The curve in Fig. 2.2a starts with the elastic range ER. Its slope should describe the elastic modulus (stiffness modulus) S of the specimen plus those parts of the deformation machine which are located within the strain measuring device. In practice, the connecting elements between the specimen and the machine are not ideal. For example, the grip faces of the specimen are neither fully plane nor parallel in a compression test, so that the specimen and the machine are contacted only in a small part of the specimen cross section. This part experiences a high load causing plastic deformation before the whole specimen deforms. This results in a reduced slope of the "elastic" line. It is therefore useful to determine the stiffness modulus S from the unloading curve UL at the end of the experiment where irreversible plastic deformations are strongly reduced.

At a certain stress, the specimen starts to deform plastically, leading to a decrease of the slope of the curve. Under the actual deformation conditions, the plastic strain rate depends on the stress and the "microstructure" of the specimen. As described earlier, the dislocation density is an important quantity characterizing the microstructure. The stress necessary to continue the deformation is the flow stress σ. The threshold stress for plastic deformation is called the yield stress σ_y. In curves as that in Fig. 2.2a, σ_y can be determined by linear extrapolation of the plastic range onto the elastic line. Frequently, the stress–strain curve goes through a maximum followed by a minimum called the upper and lower yield points (UYP and LYP) as in Fig. 2.2b. This behavior results from drastic changes of the microstructure, mainly a strong increase in the dislocation density. In this case, the UYP can be taken for σ_y. It is an essential aim of this book to interpret the yield stress on the basis of the dynamic dislocation behavior. After the yield point, one or several ranges of plastic deformation follow with an increase in the flow stress. This increase is called work-hardening. The slope of the flow stress vs. the plastic strain $\Theta = \mathrm{d}\sigma/\mathrm{d}\varepsilon$ is called the work-hardening coefficient.

With $S = \mathrm{d}\sigma/\mathrm{d}\varepsilon_{\mathrm{el}}$, it follows from (2.2) that

$$\dot{\varepsilon}_t = \dot{\varepsilon}_{\mathrm{el}} + \dot{\varepsilon} = \dot{\sigma}/S + \dot{\varepsilon}. \tag{2.5}$$

Thus, the plastic strain rate is different from the total or nominal strain rate of the deformation machine, namely

$$\dot{\varepsilon} = \dot{\varepsilon}_t \left(1 - \frac{1}{S}\frac{\mathrm{d}\sigma}{\mathrm{d}\varepsilon_t}\right) = \dot{\varepsilon}_t \frac{S}{S + \Theta}. \tag{2.6}$$

Both are equal only at constant stress or $\Theta = 0$, that is, at the UYP and LYP and in ranges of low work-hardening, that means during steady state deformation.

To determine thermodynamic quantities characterizing the dislocation dynamics, the sensitivity of the flow stress with respect to changes in the plastic strain rate $r = \Delta\sigma/\Delta \ln \dot{\varepsilon}$ and the temperature $\Delta\sigma/\Delta T$ have to be studied. An early review of this topic is given in [17]. On principle, this can be done by measuring stress–strain curves at different strain rates and temperatures and forming the respective difference quotients of the strain rate and temperature dependencies of the yield stress. However, this has the disadvantage that specimens deformed under different conditions may exhibit different microstructures so that the differences are not only due to the change in the outer variables. It should therefore be preferred to perform instantaneous changes of the deformation parameters during the deformation experiments and to observe the transient behavior. It is assumed that changes in the microstructure need strain or time so that the microstructure remains essentially constant during the sudden changes of the deformation conditions and that the transients can be observed. Experiments with instantaneous changes of the strain rate are called strain rate cycling tests (SRC). Such tests with increases and decreases of the strain rate by factors of 10 are included in the stress–strain curve of Fig. 2.2a. After the changes, the microstructure adjusts to a new state, leading to a yield drop effect similar to that at the beginning of the deformation. It is now possible to define two stress increments shown in Fig. 2.2c, one using the maximum stress value $\Delta\sigma_{\mathrm{in}}$ corresponding roughly to the instantaneous change in the strain rate and one describing the differences in the steady state values of the flow stress $\Delta\sigma_{\mathrm{ss}}$. In the SRC tests in Fig. 2.2a, the transient effect is very weak but is obvious at SRC6. The situation is different when the temperature has to be changed (TC). In most deformation machines, some time (in the order of 30 min) is necessary to adjust the machine to new steady state conditions. Thus, instantaneous temperature changes are not possible. Besides, the specimen and parts of the machine change their length so that the specimen has partly to be unloaded before the temperature can be changed. Accordingly, the stress increments $\Delta\sigma_{\mathrm{TC}}$ usually do not refer to instantaneous changes.

In addition to strain rate cycling experiments, the strain rate sensitivity r can well be determined by so-called stress relaxation tests, indicated by R in Fig. 2.2b. During the deformation, the testing machine is instantaneously stopped, that is, $\dot{\varepsilon}_{\mathrm{t}} = 0$ in (2.5). Since the stress is still acting, the specimen continues to deform at decreasing stress according to

$$\dot{\varepsilon} = -\dot{\sigma}/S. \tag{2.7}$$

Thus, the rate of the relaxing stress is proportional to the strain rate. The strain rate sensitivity is then given by

$$r = \frac{\partial\sigma}{\partial \ln \dot{\varepsilon}} = \frac{\partial\sigma}{\partial \ln(-\dot{\sigma})}. \tag{2.8}$$

For these experiments, it is necessary only to measure the time dependence of the relaxing stress and to plot the logarithm of the relaxation rate vs. the stress, as demonstrated by curve R in Fig. 2.3. Except for a shift along the ordinate axis, these curves describe the dependence of the strain rate on the stress. To scale this relation, the stiffness modulus S is required, which

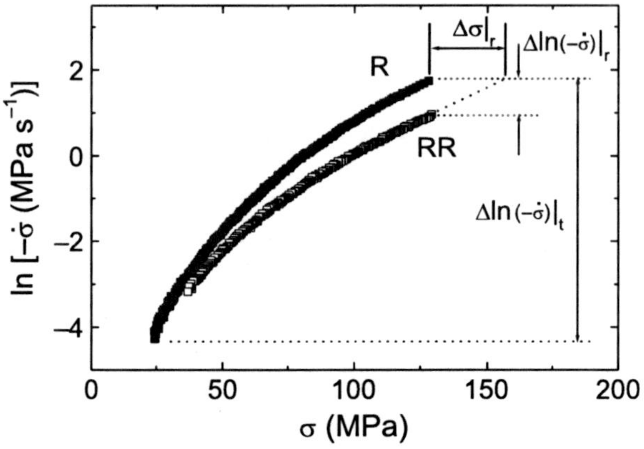

Fig. 2.3. Stress relaxation curve R and repeated stress relaxation curve RR of an Al-Pd-Mn single quasicrystal deformed at 818°C. Data from [16]

can be taken from the unloading curve, as described earlier. The inverse slope of the relaxation curves equals the strain rate sensitivity r. If the slope is determined at the beginning of the curve, the strain rate sensitivity corresponds to the strain rate just before starting the relaxation test. The dependence of r on the stress is obtained by measuring slopes at different stresses along the curve. Stress relaxation tests consume far less strain than strain rate cycling tests so that microstructure changes should be less severe. The strain and time necessary to relax a certain stress interval depend on the stiffness modulus S. It is therefore desirable to use a machine of high stiffness. This can be achieved by electronic strain control with the strain gauges attached near the grip faces of the specimen as close as possible. Nevertheless, the microstructure can change during relaxation tests. This can be studied by comparing the initial relaxation curve (R in Figs. 2.2 and 2.3) with a repeated one RR obtained after stopping the machine during reloading before the original stress and strain rate are reached again.

Alternatively, experiments with instantaneous changes of the deformation conditions can also be performed during creep tests. The counterpart to strain rate change tests are then load change tests where the strain rate goes through a transient and reaches a new value.

2.2 Stress Pulse Double Etching Technique

One of the most important topics of dislocation dynamics is the determination of the stress dependence of the dislocation velocity. Many data were collected in the 1960s and 1970s using the stress pulse double etching method. This technique was introduced by Gilman and Johnston [18,19] and consists of the following steps:

- Decreasing the dislocation density of the starting single crystals by long-time annealing at a high temperature

- Cutting suitable specimens
- Removing the surface damage from cutting by careful chemical or electrochemical polishing
- Introduction of fresh dislocations
- Marking the starting positions of the dislocations by chemical or electrochemical etching
- Moving the dislocations by applying a stress pulse
- Second etching to mark the final positions of the dislocations after the stress pulse, and calculation of the velocity from the displacement and the duration of the stress pulse.

The fresh mobile dislocations have been introduced by different methods. Johnston and Gilman used a steel ball pressed slightly onto the crystal surface. In their first study of the dislocation mobility in a metal, Stein and Low [20] scratched the surface with a steel knife. Later on, also micro-hardness indents were used to produce fresh dislocations (e.g., [21]). When many dislocations are introduced by scratching or indenting, the dislocations should be moved away from the region of high internal stresses connected with the concentrated load by a first stress pulse before marking their starting positions. This requirement was not taken into account in several measurements.

In many cases, the pyramidal shape of the etch pits reflects the symmetry of the etched crystal face, as shown in Fig. 2.4 for the $\langle 100 \rangle$ face of the LiF crystals used in the original study of Johnston and Gilman. When the dislocation leaves the starting etch pit by the stress pulse, the etch pit does not grow in depth during further etching so that the original pit gets flat at its bottom as demonstrated in the left pit and the central pit in the figure, where the dislocation moved in two steps. Instead, a new smaller pit develops in the final position of the dislocation after the stress pulses (the right pit in the figure). The etch pits are mostly imaged by usual optical or interference microscopy or by scanning electron microscopy.

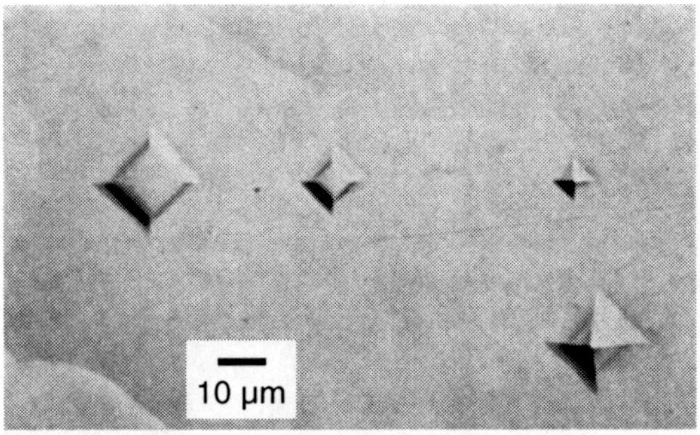

Fig. 2.4. Selective etching of a dislocation moved in two steps from *left* to *right*. From Gilman and Johnston [18]. Copyright Wiley–Blackwell (1957)

The stress pulses are usually applied in the bending geometry either in 4-point (e.g., [19]) or 3-point bending (e.g., [20]). Four-point bending has the advantage of an equal stress distribution within the inner span of the bending jig and of the opportunity to use both the tension and compression faces to study a tension–compression asymmetry. In 3-point bending, the bending moment decreases from the center outwards so that, with several scratches or indents, the stress dependence of the dislocation velocity can be studied over a certain range on a single specimen. As the plastic strains are small during the experiment, the shear stresses are calculated from the bending load applying common elasticity theory.

The duration of the stress pulses is chosen such that the displacements range between a few micrometers and about one hundred micrometers, which can easily be observed by the etching techniques. This requires, depending on the stress, durations between days and 10^{-6} s to reach dislocation velocities between 10^{-11} and 10^{3} m s^{-1}. These pulses are produced by materials testing machines down to seconds or fractions of a second, by magnetic systems similar to loudspeaker drives down to the millisecond range, and by impact loading for shorter times.

In the best case, the starting and final positions of a dislocation can individually be coordinated as in Fig. 2.4. When the dislocations move out of a scratch or an indent, this is frequently not possible. Individual dislocations may multiply on their way, or a source may emit many dislocations moving on the same or neighboring planes. As a consequence, the leading (head) dislocation moves under the action of not only the applied stress but also the internal stress field of the other dislocations moving behind. Thus, its velocity will be higher than that of an isolated single dislocation. This effect is discussed in Sect. 5.2.2, also showing that the behavior of groups of dislocations cannot unambiguously be extended to that of individual dislocations.

In spite of some limitations of the etching method, a considerable bulk of data has been measured by this technique. Almost all data available at that time have been reviewed by Nadgornyi [10].

2.3 Transmission Electron Microscopy

Transmission electron microscopy (TEM) is the most powerful method for studying the mechanisms of dislocation motion. In TEM, different modes of imaging are possible, depending on the number of diffracted beams contributing to the formation of the image. Each diffracted beam originates from the reflection of electrons at a set of lattice planes. In high-resolution electron microscopy, many reflected beams contribute to the image formation. Accordingly, individual atom columns can be resolved if the resolving power of the electron microscope is sufficient. A respective micrograph of a dislocation was shown in Fig. 1.3. Many processes controlling dislocation dynamics do not only occur on the atomic level but also on a scale between some nanometers

Fig. 2.5. Formation of diffraction contrast at a dislocation

and several micrometers. A suitable mode of imaging is then the so-called diffraction contrast. In this mode, the specimen is oriented in such a way that the diffracted intensity originates from a single set of lattice planes, as outlined in Fig. 2.5. A great part of the intensity of the incident electron beam passes the specimen as the primary beam P. The other part is reflected at the respective lattice planes forming the reflected beam R as shown for the left beam. If these planes are bent owing to the presence of a dislocation as for the right beam, the reflected beam points into another direction R′, resulting in an electron microscopy contrast, if the imaging conditions are adjusted properly. Thus, the contrast results from the elastic strain field around the core of the dislocation. A respective micrograph in bright-field contrast with the dislocations imaged as dark lines on a bright background was already shown in Fig. 1.4.

It is not possible to present the details of diffraction contrast image formation within the framework of this book. The reader is referred to standard works of TEM in materials science like [22–26]. Only the method of contrast extinction for determining the direction of the Burgers vector introduced in Sect. 1.2 is briefly mentioned here. As may be inferred from Fig. 2.5, the planes perpendicular to the Burgers vector $\boldsymbol{b}$ are most strongly distorted due to the presence of the dislocation. The planes containing the Burgers vector like the image plane of the figure are not or only weakly bent. If the image is formed by exciting these planes, the contrast of the dislocation is extinguished. The set of imaging planes is characterized by its diffraction spot and by its reciprocal lattice vector $\boldsymbol{g}$. The contrast extinction therefore requires that the vectors $\boldsymbol{g}$ and $\boldsymbol{b}$ are perpendicular to each other, or that their scalar product is zero

$$\boldsymbol{g} \cdot \boldsymbol{b} = 0. \tag{2.9}$$

To obtain the direction of the Burgers vector in space, it is necessary to find two extinction conditions with independent $\boldsymbol{g}$ vectors for the respective dislocations. An example is given in Fig. 2.6. In Fig. 2.6a, all dislocations in the specimen area are visible. In Fig. 2.6b, the set marked A in Fig. 2.6a is extinguished, in Fig. 2.6c it is set B. Finally, in Fig. 2.6d all dislocations except some small loops are invisible. Considering the indexing of the $\boldsymbol{g}$ vectors in the figure and the extinction rule (2.9), it may be concluded that the dislocations marked A have Burgers vectors parallel to [010] and those marked B have such parallel to [100]. For many TEM micrographs in this book, the $\boldsymbol{g}$ vectors are indicated. If it is written, for example, $\boldsymbol{g} = (110)$ means that planes parallel to the (110) plane are reflecting. In most cases, a reflection of higher order is excited to obtain more narrow (sharper) images of the dislocations.

2.4 In Situ Straining Experiments in the Transmission Electron Microscope

Performing complete plastic deformation experiments in situ in the specimen chamber of a transmission electron microscope provides the unique opportunity to observe the moving dislocations under stress at a resolution level that allows conclusions to be drawn on many relevant mechanisms. Such experiments were greatly facilitated by the commercial availability of high-voltage electron microscopes (HVEM) with acceleration voltages of 1 MV, mainly in Japan at the end of the 1960s. These microscopes allow the penetration of specimens of thicknesses up to about 1–2 μm compared to some 100 nm in conventional microscopes with an acceleration voltage of 200 kV. In addition, the specimen chambers of most HVEMs offer sufficient room to insert elaborate in situ stages between the pole pieces of the objective lenses. As will be mentioned below, the high voltage leads to the formation of radiation damage in most materials, which may impair the reliability of these experiments. Therefore, in situ experiments are also performed in conventional microscopes where this problem usually does not occur, for example, [27]. A number of straining stages had been designed by the pioneering work of the group of Imura and others, for example, [28–32], using different types of drive mechanisms. The adjustment of proper imaging (diffraction) conditions requires the tilting of the specimen. Most straining devices were designed for side entry specimen stages and allow tilting only around a single axis. The present author designed two straining stages for a 1 MV top entry HVEM with double tilting facilities, one for room temperature [33] and another one for temperatures up to more than 1,250°C to be able to study also ceramic materials [34]. The stages will be briefly outlined in the following.

The room temperature stage shown in Fig. 2.7a consists essentially of a single aluminium plate (1). It is supported by a screw at (2) and forms two levers (3), which are connected with the support by thin parts acting as leaf springs. The central bar (4) carries a heating coil. If the bar is heated, it pushes the levers away from each other, resulting in a deformation of the specimen (5), which is fixed to the levers from below. The hatched areas of the levers

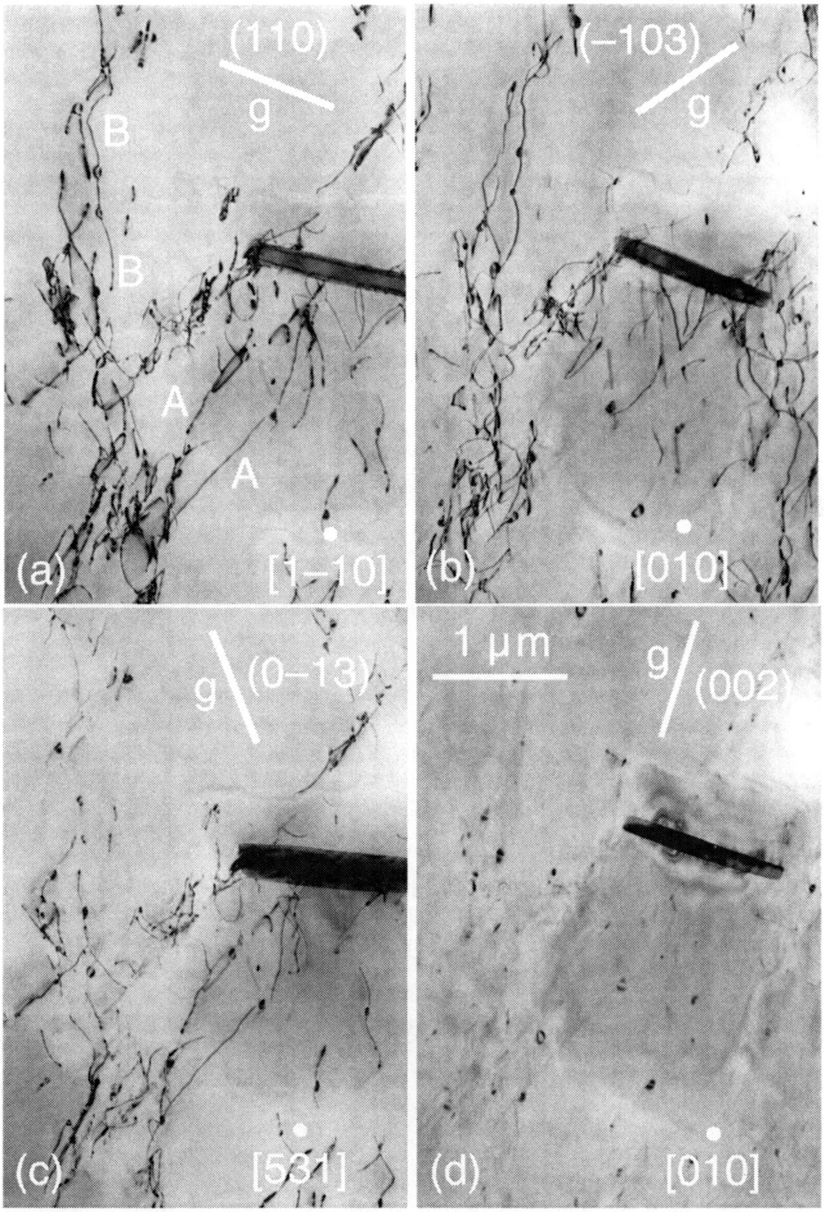

Fig. 2.6. Analysis of the direction of Burgers vectors using the $\boldsymbol{g}$ times $\boldsymbol{b}$ rule. $MoSi_2$ single crystal deformed along a $\langle 011\rangle$ direction at 380°C. The thick dark contrast near the center is a precipitate that can be used as a marker to identify individual dislocations. The directions of the respective $\boldsymbol{g}$ vectors are indicated by the indexed bars in the upper corners, the poles that are close to the viewing directions are marked by the indexed white dots in the lower edges of the images. From the work in [15]

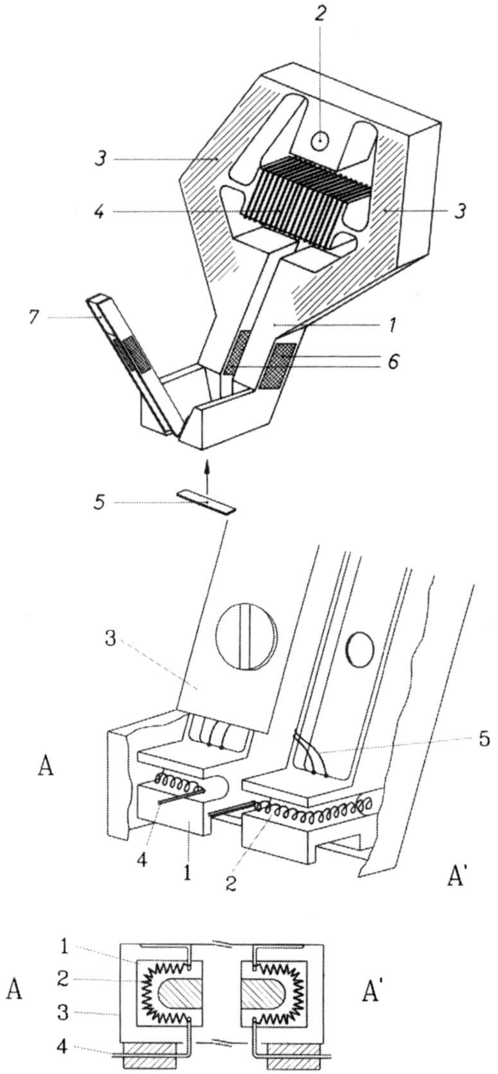

Fig. 2.7. Outline of in situ straining stages for a top entry HVEM. These stages are placed on double-tilting goniometer stages. (**a**) Room temperature deformation stage. (**b**) Hot zone of high-temperature deformation stage. From [33,34]. Copyright Elsevier Science (1976, 1994)

are water cooled. The whole equipment is inclined with respect to the electron beam so that the latter can transmit the specimen. Four semiconducting strain gauges forming a full bridge are glued to the lower parts of the levers for measuring the load. Another set of strain gauges is fixed to a pair of leaf springs (7) for measuring the specimen elongation. Thus, the stage allows the recording of the load-elongation curve during the in situ experiment.

Usually, high-temperature straining stages are heated by electrical resistance heating. However, it is difficult to generate the high heating power necessary to reach very high temperatures by this method because of the high radiation losses. A solution to the problem is heating by electron bombardment, which was introduced in the design of electron microscopy stages in [35]. The environment of the specimen in the high-temperature stage is outlined in Fig. 2.7b in perspective and in cross-section views. Two double T-shaped bars carry the specimen grips (1) at their lower ends. They are made from a W-27at% Re alloy having a high strength at high temperatures. The specimen is mounted between these grips from below by two tungsten pins fitting into bores of the specimen. As seen in the cross-sectional view, the grips have a U-shaped notch each. Tungsten filament coils (2) are situated in each notch. They are attached to the thermal shields (3) via tungsten wires (4). As the shields are at negative filament potential, they are electrically (and thermally) insulated by small Al_2O_3 spacers. For measuring the temperature, W-Re thermocouple wires with 3 and 25% Re were welded individually to the top sides of both grips. The T-shaped bars are connected to the drive mechanism. It is made of copper and stainless steel and is similar to that of the room temperature stage. For improving the heat transfer to the cooling water, heat exchangers with copper lamellae are used. Outside, they carry the semiconducting strain gauges for measuring the force acting on the specimen. The whole system is digitally controlled by a personal computer equipped with AD and DA converters. It provides the measurement of the temperatures of both grips, of the electron beam current, and of the specimen load. The system controls the electron beam current, the average temperature, the balance between both grips, and the drive voltage.

The high-temperature straining stage allows a maximum specimen elongation of 1 mm and a maximum load of 15 N. The thermal drive against water cooling allows steady state behavior and ensures a very smooth and stable operation. A number of experiments were performed at a grip temperature of 1,250°C, corresponding to a specimen temperature of 1,150°C, using a beam current of about 60 mA at 700 V. The temperature is stable within about 3 K.

During the in situ experiments, the images were recorded either as images on photographic film with a high resolution or by a video system mounted in the base of the microscope with the usual TV resolution. Figure 2.8 shows the HVEM equipped for performing straining experiments.

The specimens for in situ straining experiments have to meet the requirements of both transmission electron microscopy and of a tensile experiment. In the case of these deformation stages, they are made from thin plates about 8 mm long, 1–2 mm wide, and 0.1 mm thick. For being fixed to the stage by pins, they have two bores 5 mm apart. The thickness at the center must be small enough for electron transparency, that is, about 1 μm. Besides, the edges of the central thin area have to be thin enough to cope with the maximum load of the stage, that is, about 10 μm. Depending on the material, this shape can be produced by grinding, dimpling, and final ion milling, or by chemical

Fig. 2.8. View of the high-voltage electron microscope equipped for in situ straining experiments. In the center, the microscope column and the control desk. At the top *right* of the column, an extra turbo pump for improving the vacuum in the specimen chamber. Left of the column, the video screen for the operator of the microscope. In front, on the *right side*, the video recorder with another screen. On the *left side*, the PC for controlling all functions of the straining stage

or electrolytical polishing. In the latter case, the thinning may be performed in two steps using masks. In the first step, a shallow dimple is thinned from one side through a mask with a hole smaller than the sample width. In the second step, the whole central area is thinned, including also the edges of the sample. For brittle materials like ceramics, semiconductors, or quasicrystals, it is advantageous if there is no hole in the center of the specimen. For optically transparent materials, the thickness can be estimated by watching the occurrence of Newton fringes so that the thinning can be stopped before the sample is perforated. Figure 2.9 presents an example of the central part of a ZrO_2 specimen without a hole after in situ straining. The bright contrasts near the center show regions deformed by ferroelastic deformation. Metal specimens usually have a hole in the center of the transparent area. Because of the inhomogeneous cross section of in situ specimens, the stress distribution is not uniform. According to calculations by the finite element method [37], the stress state is complicated at the edges of the perforation lying in the tensile direction. At the transverse edges, an almost uniaxial tensile stress occurs with a stress concentration of three times the average stress in this cross section, as results from technical elasticity calculations of a plate with a cylindrical hole. Mainly these areas should be observed during in situ straining.

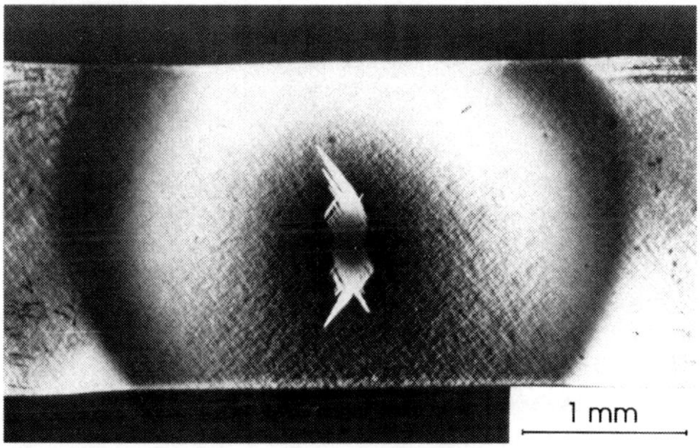

Fig. 2.9. Micro-tensile sample of a ZrO_2 single crystal deformed during an in situ straining experiment at 1,150°C. The specimen was prepared by grinding, dimpling, and ion milling. It does not have a perforation. The bright contrasts near the *center* result from ferroelastic deformation. From the work in [36]

There are two problems limiting the reliability of in situ straining experiments in an HVEM as discussed in [38,39]. The first one regards surface effects due to the relatively low specimen thickness, even in the HVEM. The surfaces close to the dislocations cause so-called image forces, which superimpose on the forces from the applied stress. These forces are estimated in Sect. 3.2.3, showing that the image forces are not of great importance in materials with high flow stresses like many contemporary materials, but they may produce artifacts in pure f.c.c. metals, which have very low flow stresses. Another effect may be called the interruption of the dislocation kinetics. During deformation, new dislocations are generated in the interior of the specimen, which may move out of the foil, but no dislocations move into the foil from outside. This results in differences in the dislocation density between the interior of the crystal and the regions near the surfaces. Therefore, in situ straining experiments are well suited to study the dynamics of individual dislocations rather than the development of complex dislocation structures.

The second problem is the occurrence of radiation damage. The imaging electrons of high energy may displace atoms from their regular places forming vacancies and interstitial atoms. The threshold voltage for this process depends mainly on the atomic number of the material. As shown in Fig. 2.10, the acceleration voltage of 1 MV exceeds the threshold voltage for most materials. Even at 200 kV, displacement damage can occur in light materials. Whether the radiation damage disturbs the deformation experiment, or not, depends on the temperature. At low temperatures, the generated primary defects are immobile. At intermediate temperatures, the defects diffuse, at first forming small clusters and later on dislocation loops. This process is described for ZrO_2 single crystals in [40]. In this temperature range, straining experiments cannot be performed. At high temperatures, the defects anneal

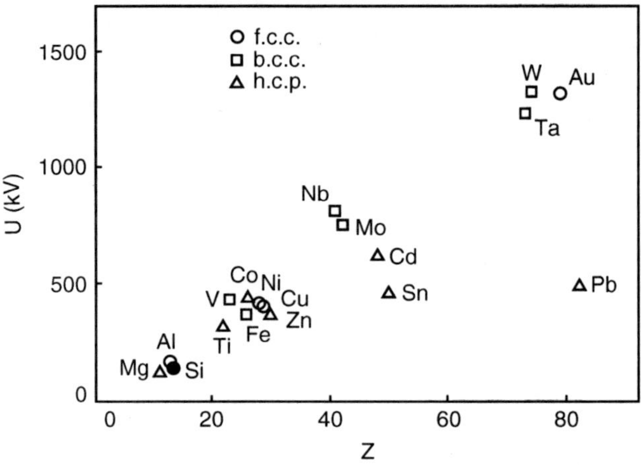

Fig. 2.10. Dependence of the threshold voltage for displacement damage by electron irradiation on the atomic number. Data from [38]

out at the near surfaces. How severe the radiation damage affects the dislocation motion in a particular experiment has to be checked individually. In most experiments described in this book, the effect is not strong so that the results should be reliable.

As mentioned in the Introduction, video sequences of dislocation motion in a number of materials can be downloaded from http://extras.springer.com/2010/978-3-642-03176-2. These video clips will be described in some detail in Part II in the chapters of the respective materials. Some videos showing general features of dislocation motion will be referred to in the following chapters. For a first glance at dislocation motion and generation, please look at Video 9.17. The video presents a rotating source which, at each revolution, produces two dislocations afterwards moving in opposite directions. They are mostly straight and aligned along crystallographic orientations. In addition to qualitative information on the dislocation behavior, in situ straining experiments yield a wealth of quantitative data, including the indexing of the planes of motion, the preferred orientations of dislocations, the density and strength of obstacles to the dislocation motion, or kinematic data like waiting times and jump distances. The evaluation of such data is discussed in the respective chapters.

2.5 Other Methods

The methods described in this section supplement the main techniques presented above to study dislocation dynamics and the processes controlling it. They are mostly limited either to particular materials or to particular processes. The contents of this section is better understood after reading Part I of this book.

2.5.1 X-Ray Topography In Situ Deformation Experiments

Using X-ray topography, dislocations can be imaged in a similar way as in the electron microscopy diffraction contrast. A recent review of the topography techniques is given in [41]. Accordingly, in situ straining experiments can also be performed in an X-ray topography arrangement to directly observe the processes controlling the dislocation motion. This requires dedicated straining stages to be positioned on the goniometer head of the X-ray topographic device. However, there are two differences with respect to TEM. The high-energy photons have a much higher penetration power compared to the electrons of usual energies, and so crystals of bulk dimensions can be observed. Besides, the width of the dislocation images is considerably greater than in TEM diffraction contrast, resulting in a poor resolution power for dislocations. This limits the application to the observation of single dislocations in crystals of a very high perfection with zero or very low dislocation densities as semiconductor single crystals, or to the observation of groups of dislocations, so-called slip bands, in otherwise perfect crystals. To record the processes by means of video systems, X-ray beams of a high intensity are necessary. This can be achieved by conventional high intensity rotating anode X-ray sources or, more recently, by synchrotron radiation.

In the first type of experiments, mainly the multiplication of dislocations in semiconductor crystals was studied [42, 43]. Microhardness indents were placed on the surface to introduce nucleation sites. An example of a dislocation source, which had emitted a number of hexagonal dislocation loops in an Si single crystal, is presented in Fig. 2.11.

The first occurrence and propagation of slip bands was studied in MgO single crystals of thicknesses between about 60 and 320 μm [44] in an equipment similar to that in [42]. First slip bands occur at about half the macroscopic yield stress. During plastic instabilities causing decreases in the applied stress, the bands propagate at high velocities of about 150 mm s^{-1}. There exists a size effect of an increasing yield stress with decreasing specimen thickness. In this respect, the X-ray topography experiments are a link between macroscopic tests and in situ straining experiments in the TEM. In optically transparent crystals, similar studies on the propagation of slip bands can be performed by optical birefringence microscopy.

2.5.2 Surface Studies of Slip Lines

The occurrence and propagation of slip bands can also be observed on the crystal surface by imaging the steps resulting from the motion of dislocations with Burgers vectors pointing out of the surface. These experiments can be performed on very different levels of spatial and temporal resolution. In

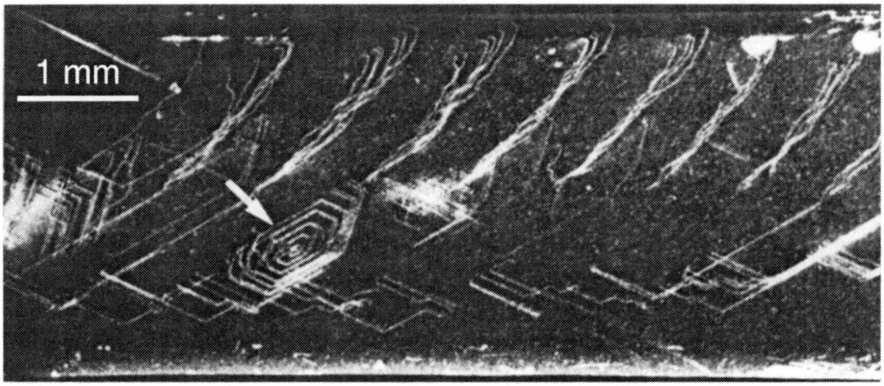

Fig. 2.11. Localized dislocation source in an Si single crystal observed during an X-ray topography in situ straining experiment. The topography device included a 30 kW X-ray generator, the in situ straining stage, and a vidicon X-ray camera. From [42]. Copyright Taylor & Francis Ltd. (1981). (http://www.informaworld.com)

the 1950s and 1960s, the surfaces of deformed crystals were imaged by optical microscopy or by surface replicas imaged in the TEM. These methods do not resolve the steps produced by moving individual dislocations. Nevertheless, important information on the distribution and length of slip bands formed by many dislocations has been obtained (e.g., the work by Seeger and coworkers [45]).

To follow the temporal development of the slip bands, the optical microscope was supplemented by a cine camera, first applied by Schwink and Neuhäuser [46], and reviewed in [47]. Later on, the cine camera was replaced by a video camera. Different methods have been used to improve the time resolution of specific processes, for example, a photo-diode was placed in the intermediate plane of the microscope to record the intensity changes of an imaged slip band (e.g., [48]). Recently, optical extensometers have been developed (e.g., [49]) where the gauge length of the specimen is coated with a pattern of dark and bright stripes, which are scanned via a special line-scan camera typically operating at a frequency of 250 Hz. The PC based evaluation of the data allows the determination of the place and the time of a deformation event. These data can be plotted in a correlation diagram.

The cinegraphic methods are mainly used to investigate the evolution of slip bands at the heterogeneous deformation as it occurs at plastic instabilities described in Sect. 5.3. The occurrence of slip bands can be correlated or uncorrelated. The heads of slip bands may move at velocities up to the $\mathrm{m\,s^{-1}}$ range. In one study [48], a second video camera was placed on the opposite side of the specimen to monitor the passage of slip bands through the crystal.

A method which is able to resolve slip steps of atomic height trailed by moving individual dislocations is the so-called heavy metal decoration technique [50,51]. It requires very smooth starting surfaces with a low density of steps. Such surfaces can be obtained by cleaving ionic crystals in a very dry atmosphere or in vacuum or by producing growth surfaces on metals. After deforming the crystals and creating the slip steps in vacuum, a small amount of gold or another heavy metal is evaporated on the surface. The gold forms small nuclei distributed randomly on the atomically smooth surfaces. Along monatomic steps, however, the nuclei are linearly arranged, thus decorating the surface step. This method allows the paths of individual dislocations with a Burgers vector component out of the surface to be imaged. The technique is well suited to study cross slip as described in Sect. 4.3 and shown in Figs. 4.9 and 4.12.

In recent years, slip steps have also been imaged by atomic force microscopy. Mostly, the resolution power is sufficient only to observe slip bands (e.g., [52]). In some studies, also the steps of a few individual dislocations were imaged showing cross slip events of superdislocations (e.g., [53]).

In general, the observations of slip bands on the crystal surfaces provide a link between the behavior of individual dislocations and the macroscopic plastic properties, in particular in the case of inhomogeneous deformation.

2.5.3 Internal Friction

If a material is deformed, the part W_{el} of the total work W_{t} expended is of elastic nature, that is, it is re-gained after unloading. The rest is partly stored in the material in the form of crystal defects produced by the deformation, the so-called stored energy W_{stor}, and partly transferred as friction W_{fr} into heat

$$W_{\mathrm{t}} = W_{\mathrm{el}} + W_{\mathrm{stor}} + W_{\mathrm{fr}}.$$

The stored energy is mostly only a small fraction (in the order of magnitude of 10%) of the nonelastic energy. Several methods have been designed to measure the stored energy, mostly calorimetric ones. On principle, the nonelastic part of the energy can be determined as the area enclosed in a stress–strain curve of a specimen loaded and unloaded cyclically. Frequently, the driving stress is harmonic

$$\sigma = \sigma_{\mathrm{a}} \exp(\mathrm{i}\omega t),$$

where σ_{a} is the oscillation amplitude, ω the circular frequency, and t the time. If the amplitudes are small so that no energy is stored, the behavior is anelastic and the relative energy loss per cycle is called internal friction. An early thorough review is given by DeBatist [54], and a later one in the proceedings [55]. In most techniques, the specimens are excited in resonance modes. Two types of measurement can be performed, either the oscillation amplitude is maintained on a constant level and the necessary power is measured, or the excitation is switched off and the decrease in the amplitude of the freely vibrating specimen is recorded. One measure of the internal friction is the ratio between the anelastic work ΔW and the total work W_{t} during one cycle.

In analogy with the electrical circuit theory, this is also designated by Q^{-1}. Another measure is the logarithmic decrement δ. It is the natural logarithm of two successive vibration amplitudes of the freely vibrating sample. Both measures are related by

$$Q^{-1} = \frac{\Delta W}{W_\mathrm{t}} = 2\,\delta.$$

Internal friction measurements can be performed over a very wide range of frequencies from less than 1 Hz up to the GHz range. The vibration modes include bending, longitudinal, and torsional oscillations. At low frequencies, an elongated (wire) specimen is excited by a torsion pendulum. Bending oscillations can be obtained, for example, by suspending the specimen in the vibration nodes on thin wires and exciting the vibrations by electrostatic forces with the specimen or a thin metal film on it being one electrode of a capacitor with another fixed electrode. At high frequencies, piezoelectric or magnetostrictive transducers are used to excite the longitudinal oscillations. At very high frequencies, the attenuation of sound waves is measured, which travel from one transducer at one end of the specimen to another one at the other end.

The energy losses in the solid are frequently due to thermally activated relaxation phenomena between different equilibrium positions of a crystal defect with a waiting time t_w for the transition given, for example, by (4.6) in Sect. 4.1. Owing to a small bias due to the applied stress, the populations of both positions change with the alternating stress. At low vibration frequencies, the defects can easily follow the vibrations. At high frequencies, they cannot follow at all. If the excitation frequency $2\pi\omega$ equals the relaxation frequency t_w^{-1}, there will be a damping maximum. Such measurements are also called acoustic spectroscopy. The variation of the internal friction is a function of the quantity $\omega\, t_\mathrm{w}^{-1}$. Thus, the variation of the internal friction for obtaining the friction maximum can either be achieved by varying the frequency ω or by varying the relaxation rate t_w^{-1} by changing the temperature. As in the resonance mode the frequency is more or less fixed, most internal friction equipments allow measurements over a wide temperature range.

Part of the internal friction studies are devoted to the properties of point defects (see the review in [56]). Anelastic effects do not occur for defects that can occupy only one position in the primitive cell or which can take two centrosymmetric positions. Point defects that produce anelastic effects may be considered elastic dipoles, which reorient under stress. Such defects are, for example, interstitials in body-centered cubic crystals. Their orientation relaxation under stress causes the Snoek effect [57–59]. The defect orientations can also relax in the stress field of dislocations producing the induced Snoek effect mentioned in Sect. 4.11. The quantities accessible by internal friction measurements are the relaxation strength giving information on the concentration of the defects and their anelastic strain contribution, the relaxation time, and the orientation dependence. The latter yields information on the symmetry of the defects.

Internal friction measurements are also used to investigate the properties of dislocations. In materials in which the intrinsic lattice friction (the Peierls mechanism, see Sect. 4.2) controls the dislocation motion, a Bordoni relaxation peak [60] is observed. The interpretation (e.g., [61]) is based on the assumption of two energetically equivalent equilibrium positions of a dislocation segment pinned at its ends with an acting internal stress. By means of thermal activation, the segment can shuttle between these equivalent positions. From the superimposed alternating stress, a bias results with one of the positions being more probable. If the excitation frequency equals the thermal vibration frequency of the dislocation segment, a relaxation peak occurs.

In materials in which the lattice friction does not control the dislocation mobility like in most metals under usual conditions, the pinned dislocation segment is treated as an elastic string having a mass per length resulting in an inertia, a line tension which keeps the segment taut, and the motion of which is viscously damped. This results in an equation of motion given in (4.75) in Sect. 4.9. This equation has to be solved under the actual boundary conditions. The model originally suggested by Koehler [62] and developed by Granato and Lücke [63] assumes that the dislocation segment is fixed at its ends by strong pinning agents like nodes in the dislocation network or precipitates, and that it is pinned in between by further weak pinning points like solute atoms or small precipitates. The pinning is described in detail in Sect. 4.5. At small amplitudes where the dislocation segments do not detach from the weak pinning points, the short segments vibrate and cause an internal friction peak with a damping independent of the exciting amplitude. The frequency of the maximum is usually in the range of tens of MHz, or higher. At frequencies below the maximum, for the logarithmic decrement the theory yields

$$\delta_{\mathrm{i}} = \alpha_\delta \mu \varrho b^2 \omega B L^4, \tag{2.10}$$

where α_δ is a constant considering the particular orientation relations and the distribution function of the segment lengths L, μ is the shear modulus, ϱ is the dislocation density (Sect. 3.1.3), b is the absolute value of the Burgers vector (Sect. 3.1.1), and B is the viscous damping constant in (4.74) in Sect. 4.9. The damping maximum is connected with a change in Young's modulus, the so-called modulus defect,

$$\frac{\Delta E}{E} = \alpha_{\mathrm{E}} \mu \varrho L^2,$$

which causes a change in the resonant frequency. α_{E} is another constant. Figure 2.12 presents the temperature dependence of the square of the resonant frequency and the internal friction of a niobium single crystal at low temperatures. Torsional oscillations at about 500 Hz were introduced by electrostatic excitation.

In a comparative study, the segment length in deformed MgO single crystals has been measured by both internal friction at different temperatures and in situ straining experiments in an HVEM [65]. For this aim, the quantity $\delta_{\mathrm{i}}/(\Delta E/E)$ has been calculated, which has the advantage that the (not

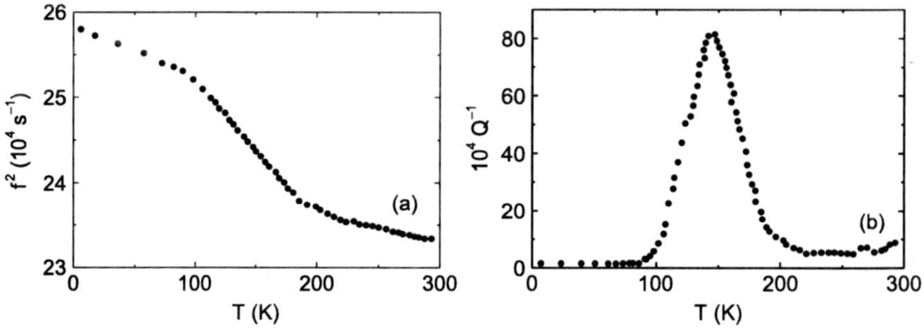

Fig. 2.12. Square of the resonant frequency (which is proportional to the respective modulus) (**a**) and internal friction (**b**) of a niobium single crystal at low temperatures. The peak is due to lattice friction. Data from [64]

well known) dislocation density is eliminated. With all the uncertainties of the internal friction method, the values of both techniques well agree.

With increasing vibration amplitude, the dislocation segments detach from the weak pinning points, leading to an increase in the internal friction. This amplitude dependent internal friction δ_{h} is additive to the amplitude independent internal friction δ_{i}. The amplitude dependent internal friction should be related to the macroscopic flow stress at small strains in materials where the flow stress is controlled by the interaction between dislocations and point-like obstacles, for example, foreign atoms or very small precipitates (Sect. 4.5). A phenomenological relation between the decrement, the stress amplitude, the strain rate at the stress amplitude, the frequency, and the stress exponent ((4.10) in Sect. 4.1) was derived in [66]. For an experimental test of the connection between both methods, it is necessary to define a comparable stress level. Different criteria have been used. As shown in [67], the stress necessary to reach a fixed plastic strain due to dislocation motion is a suitable quantity, which can be measured in both methods. The respective stresses then show the same temperature dependence.

Internal friction measurements can be performed in situ during simultaneous elastic loading or quasistatic plastic deformation as reviewed in [68]. During active plastic deformation, the internal friction increases by both a structural component δ_{s} and a dynamic one δ_{d}. The structural component arises from the increase in the dislocation density ϱ during deformation, which occurs in (2.10). The dynamic component is explained by assuming that the vibrational energy due to plastic deformation is proportional to the plastic strain γ achieved during the time of one period $2\pi/\omega$, thus $\Delta W_{\mathrm{d}} \propto \dot{\gamma}/\omega$, where $\dot{\gamma}$ is the plastic strain rate. The latter can be described by the Arrhenius relation (4.8) and (4.3) in Sect. 4.1. Superimposing the constant applied stress and the oscillating one and integrating over one period yields for the dynamic damping

$$\delta_{\mathrm{d}} = \frac{\pi M V \dot{\gamma}}{\omega\, kT},$$

if the effect of the oscillating stress on the strain rate is linearized. M is the appropriate elastic modulus and V the activation volume. This is an amplitude independent internal friction, which is proportional to the macroscopic strain rate and inversely proportional to the oscillation frequency. If the linearization is dropped, the friction becomes amplitude dependent.

Because of the nonlinearity of the Arrhenius relation, the superimposed oscillating stress causes a softening effect, the so-called acousto-plastic effect, which becomes evident either as an increase in the strain rate during creep experiments (Archbutt effect [69]) or as a decrease of the flow stress in quasistatic experiments (Blaha–Langenecker effect [70]). In most materials, the softening sets in when the oscillation amplitude reaches the amplitude dependent range.

In conclusion, internal friction measurements are versatile methods to study dynamic properties of dislocations, in particular the interaction with localized obstacles.

2.5.4 Nuclear Magnetic Resonance

Nuclear magnetic resonance (NMR) is a complementary technique to study the static dislocation structure as well as the dynamic behavior of dislocations. In contrast to many other methods, NMR measurements are only slightly influenced by the surface so that they register bulk properties. To perform dynamic measurements, the specimen chamber of a deformation machine has to be fitted between the pole pieces of the electro-magnet of the NMR spectrometer. Most experiments are performed in the quasistatic mode at a constant strain rate, which sometimes has to be quite high, that is, up to $10\,\mathrm{s}^{-1}$.

The methods are based on the interaction between nuclear electric quadrupole moments and electric field gradients of nuclei with a spin $I > 1/2$. In a cubically symmetric nuclear environment, there does not exist a static electric field gradient at the nucleus so that there will be no interaction. The strain field around a dislocation, however, destroys the symmetry resulting in an interaction. The theory of the techniques is reviewed in [71].

Dislocation density measurements are carried out in ultra-pure crystals in order to avoid lattice distortions by other defects. The quadrupole broadening of the NMR signal is then proportional to the square root of the dislocation density (e.g., [72]). For more detailed studies, the difference between edge and screw dislocations and the arrangement of the dislocations on several slip systems may be considered. Dislocation densities on different slip planes can be obtained from the orientation dependence, as shown in [73].

For studying dislocation dynamics by pulsed nuclear magnetic resonance, the Orowan equation between the strain rate, the mobile dislocation density, and the dislocation velocity (3.5) is considered assuming that the dislocation motion is not continuous but jump-like. Then, the dislocation velocity is given by $v_\mathrm{d} = \lambda/t_\mathrm{w}$, where λ is the jump distance and t_w the waiting time between the jumps. Obstacles causing the jerky motion may be precipitates and, in particular, forest dislocations. The motion of a dislocation is accompanied by movements of the surrounding atoms, resulting in a nuclear spin-lattice relaxation process with a relaxation time in a weak rotating field $t_{1\rho}$. This parameter is the suited NMR parameter as its "time windows" are in the range of t_{w}. At sufficiently high strain rates where $1/t_{\mathrm{w}}$ is large compared to the Larmor frequency of the rotating frame, there is a relation between the relaxation time and the strain rate $\dot{\varepsilon}$

$$\frac{1}{t_{1\rho}} = A\,\frac{\dot{\varepsilon}}{b\lambda},$$

where b is the absolute value of the Burgers vector and A is a factor depending on the nuclear magnetic properties of the dislocation, the mean local magnetic field in the rotating frame, and the weak rotating applied field. These parameters can be determined separately so that measurements of $t_{1\rho}$ yield the jump distance λ. The experimental values are mostly compatible with the distances between forest dislocations (e.g., [74]). More recent results on NaCl including deformation-induced point defects are given in [75].

3

Properties of Dislocations

In order to understand dislocation dynamics, first the geometric and elastic properties of dislocations have to be treated. The geometric features decide on the mode of dislocation motion. The elastic stress fields around the dislocations interact with the applied stress and stress fields of all the other defects in the crystal resulting in forces acting on the moving dislocations. These forces drive or impede the dislocation motion.

3.1 Geometric Properties

3.1.1 Burgers Vector

The dislocation introduced in Sect. 1.2 is a special case of a general crystal dislocation, the so-called edge dislocation first described by Orowan, Polanyi, and Taylor [3–5]. Such a dislocation in a primitive cubic model crystal is shown again in Fig. 3.1a. The inserted extra half-plane is marked by full circles. The edge dislocations are frequently symbolized by a T (in the present case rotated by 180°). The other special case is the screw dislocation introduced by Burgers [6] and is shown in Fig. 3.1b. In contrast to the edge dislocation, the shift vector or Burgers vector is now parallel to the dislocation line.

The Burgers vector $\boldsymbol{b}$ was introduced in Fig. 1.2 in a qualitative way. An exact definition is obtained from the so-called Burgers circuit outlined in Fig. 3.2 [6]. In this construction, the dislocation line is surrounded along a closed path by a number of defined steps as shown in Fig. 3.2a. The circuit starts and ends at A. Afterwards exactly the same sequence of steps is repeated in a perfect reference lattice demonstrated in Fig. 3.2b. Because of the presence of the dislocation in Fig. 3.2a, the circuit is not closed now. The vector of the additional step necessary to close the circuit in the reference lattice is the Burgers vector $\boldsymbol{b}$. To establish its sign, the direction of the line vector $\boldsymbol{\xi}$ and the sign of the circuit have to be considered. There exist different conventions, e.g., [76]. If the sign of the line vector is changed, the sign of the Burgers vector changes too. For a dislocation in a continuous medium, the Burgers circuit

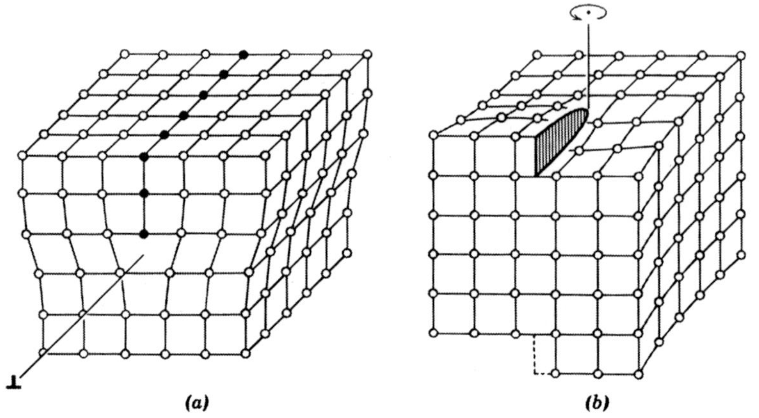

Fig. 3.1. Models of an edge (**a**) and a screw dislocation (**b**) in a cubic primitive lattice

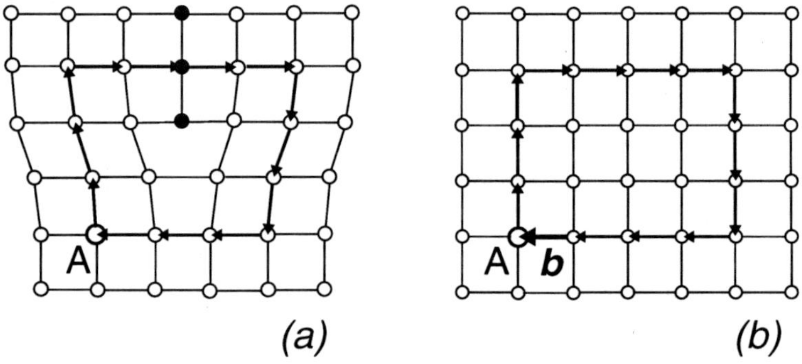

Fig. 3.2. Definition of the Burgers vector by the Burgers circuit

is replaced by a line integral over the displacements $\boldsymbol{u}$ along a closed curve C around the dislocation loop. Thus, the Burgers vector is determined by

$$\boldsymbol{b} = \oint_{\mathrm{C}} \mathrm{d}\boldsymbol{u}. \tag{3.1}$$

The Burgers vector describes both the shift of one part of the crystal with respect to the other along the face which was swept by the dislocation causing a change of the shape of the body, and the strength of the elastic strain field around the dislocation. The character of the dislocation is defined by the angle β between the Burgers vector $\boldsymbol{b}$ and the line vector $\boldsymbol{\xi}$ of the dislocation. As described earlier, $\beta = 90°$ for an edge dislocation and $\beta = 0°$ for a screw dislocation. For all the other angles, the dislocation is called a mixed dislocation. Its Burgers vector may be decomposed into an edge and a screw component according to

$$b_{\mathrm{s}} = b \cos\beta, \qquad b_{\mathrm{e}} = b \sin\beta. \tag{3.2}$$

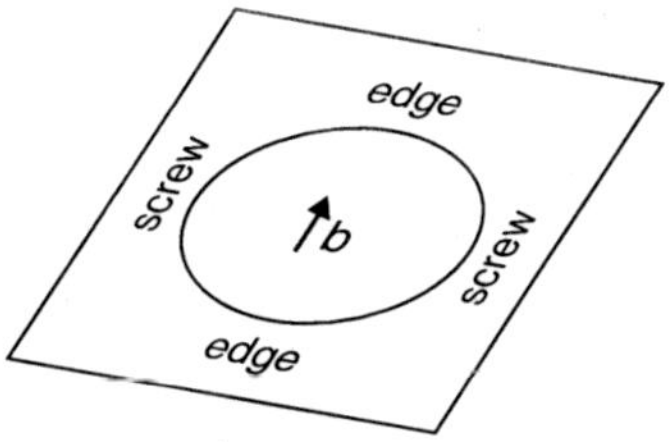

Fig. 3.3. Generation of a glide dislocation loop by shifting the faces adjoining an internal cut against each other by the Burgers vector $\boldsymbol{b}$

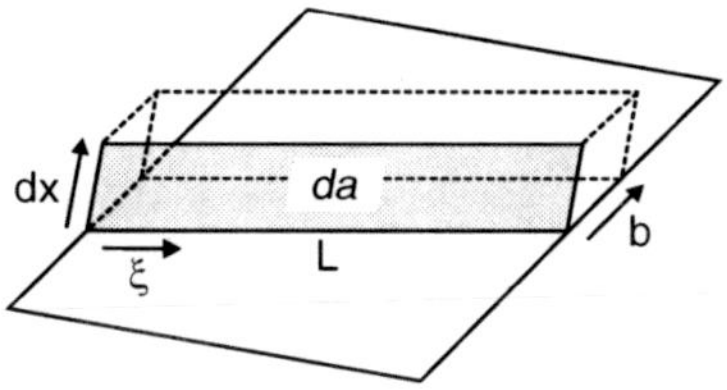

Fig. 3.4. Generation of additional crystal volume by the general motion of a dislocation segment

A dislocation loop with all characters can be generated by an internal cut ending along a closed line and by shifting the faces adjoining the cut against each other by the Burgers vector. For a glide loop, this is outlined in Fig. 3.3. Those parts of the dislocation which are parallel to $\boldsymbol{b}$ have screw character, those which are perpendicular to it are of edge character, whereas all the other parts are mixed.

There are some laws of conservation of the Burgers vector. It is constant along a dislocation line. If a line branches at a dislocation node, the sum of the Burgers vectors of all branches is zero. Besides, a dislocation cannot end inside a crystal, but only at a node, or an interior or exterior surface.

3.1.2 Glide and Climb Motion of a Dislocation

The dislocation motion can be accompanied by a change of the crystal volume. The created volume depends on the directions and magnitudes of the Burgers vector, the line vector, and the vector of displacement of the dislocation as depicted in Fig. 3.4. The figure shows a plane containing the Burgers vector $\boldsymbol{b}$ and a dislocation segment of length L and direction $\boldsymbol{\xi}$. This segment may move by a distance $\mathrm{d}\boldsymbol{x}$, sweeping an area $\mathrm{d}\boldsymbol{a}$. Since the edge component of the dislocation represents an incomplete extra lattice plane, its general motion is connected with an extension or reduction of the size of this extra plane, resulting in a change of the total crystal volume. As imaged in the figure, the additional volume is given by the product of the three vectors as

$$\mathrm{d}V = (\boldsymbol{b} \cdot \mathrm{d}\boldsymbol{a}) = (\boldsymbol{b} \cdot (L\boldsymbol{\xi}) \times \mathrm{d}\boldsymbol{x}). \tag{3.3}$$

The mode of dislocation motion is defined by the process either being conservative with respect to the crystal volume, i.e., $dV = 0$, or being nonconservative, i.e., $dV \neq 0$. In the first case, the motion is called glide or slip. This is the dominating mode of dislocation motion in most materials and at most temperatures. The nonconservative motion is called climb and is usually connected with diffusion. Details of this mode will be described in Sect. 4.10.

As seen from (3.3), the volume production is zero if the dislocation moves on a face containing the Burgers vector and the line vector. In particular, this is a plane as shown in Fig. 3.4. This plane is called the glide or slip plane of the dislocation. The glide plane is well defined for the edge and mixed dislocations. In an elastic continuum, for a screw dislocation with its Burgers vector parallel to the dislocation line, all planes containing the dislocation line are glide planes. In a crystal, however, only selected low-index crystallographic planes may act as glide planes. Usually, the glide plane of the adjoining edge part is attributed also to the screw segment of a dislocation. The motion is called cross glide or cross slip, if the screw segment changes from this plane to another glide plane.

In a real crystal, the place of a curved dislocation is not a smooth line, owing to the discrete atomic structure of the material. The dislocation then consists of straight segments parallel to crystallographic directions and transitions between these straight segments. The transitions can be of atomic height or may have the height of several atomic distances. If the transition parts are situated within the slip plane, they are called kinks or superkinks depending on their height. If the transition parts point out of the slip plane, they are called jogs. Thus, jogs connect segments of the dislocation in different slip planes. Kinks and jogs are usual parts of the dislocation. Thus, the mode of their motion is determined by the relation between the Burgers vector and the line vector, as with any other part of the dislocation. Figure 3.5 shows a dislocation loop with kinks K and jogs J. Since the kinks K extend within the slip plane, they can glide together with the dislocation. It will be described in Sect. 4.2.1 that the motion of the dislocations is realized by the formation and sidewise spreading of the kinks. In contrast to that, the slip planes of the jogs J (shaded areas) are not identical with the slip plane of the other segments of the dislocation. The slip plane of a jog in the edge part E of the dislocation loop extends perpendicular to the main course of the dislocation. Thus, the jog can glide together with the dislocation if the dislocation loop is enlarged by plastic deformation. However, the slip plane of a jog in the screw part S of the loop extends parallel to the general course of the dislocation. As a consequence, these jogs can glide along the dislocation but they cannot glide in its forward direction. These jogs represent obstacles to the dislocation motion. Their forward motion requires climb.

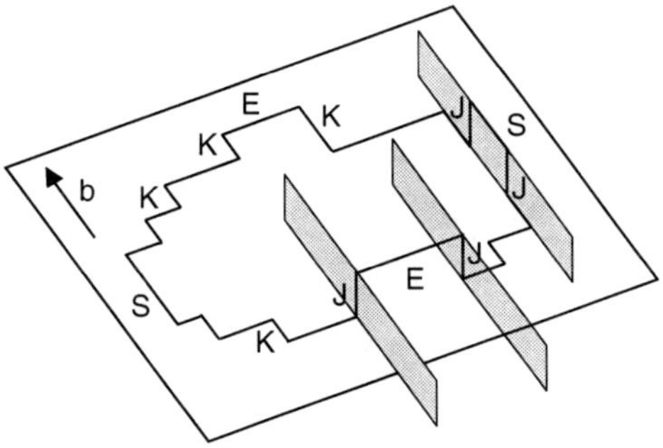

Fig. 3.5. Kinks K and jogs J in a dislocation loop with edge E and screw parts S

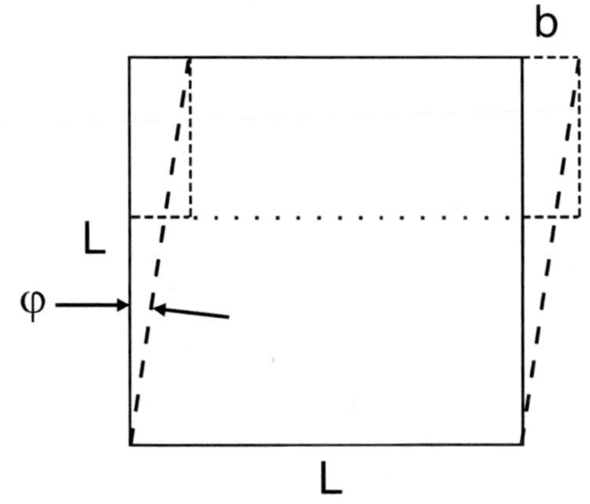

Fig. 3.6. Plastic shear strain produced by the motion of one dislocation across a cube of an edge length L

3.1.3 Relation Between Dislocation Motion and Plastic Strain and Strain Rate

The plastic strain obtained after the motion of dislocations depends on the density of dislocations, their Burgers vector, and the distance these dislocations move. Figure 3.6 shows the change of the shape of a cube of edge length L after one dislocation has moved across the cube. The body has then a surface step of length L and height b, and is sheared by the strain $\gamma = \tan\varphi = b/L$. If n dislocations glide only over a fraction λ/L of the length of the body, the strain is $\gamma = nb\lambda/L^2$. Since the dislocation density is defined by the dislocation length per volume, $\varrho = nL/L^3$, the strain is

$$\gamma = \varrho b\lambda. \tag{3.4}$$

This is the shear strain in a particular slip system. The plastic tensile strain follows from (2.4). As an example, if dislocations of the density $\varrho = 10^{12}\,\mathrm{m}^{-2}$ and the absolute value of the Burgers vector of $b = 3\times10^{-10}\,\mathrm{m}$ move over a distance of $\lambda = 10^{-4}\,\mathrm{m}$ each, the resulting shear strain is $\gamma = 3\times10^{-2} = 3\%$.

In order to obtain the strain rate $\dot{\gamma}$, (3.4) has to be differentiated with respect to the time t, $\dot{\gamma} = \mathrm{d}\gamma/\mathrm{d}t$. In general, both the dislocation density ϱ and the displacement λ may depend on time. Mostly, a limiting case is considered where a certain density of mobile dislocations ϱ_m changes slowly in time, so that only the change of the dislocation displacement is considered, i.e.,

$$\dot{\gamma} = b\varrho_\mathrm{m}\frac{\mathrm{d}\lambda}{\mathrm{d}t} = b\varrho_\mathrm{m}v_\mathrm{d}, \tag{3.5}$$

where v_d is the dislocation velocity. This is the well-known Orowan relation between the plastic strain rate, the dislocation density, and velocity [77]. The dislocation velocities range from fractions of nanometers per second to a limiting velocity, which is related to the velocity of sound. The processes controlling the dislocation velocity are the topic of this book.

Another limiting case occurs if dislocations are created at a source, move successively very quickly over a certain distance λ and are afterward annihilated or immobilized. Then,

$$\dot{\gamma} = b\lambda\frac{\mathrm{d}\varrho}{\mathrm{d}t} = b\lambda\dot{\varrho}.$$

In general, both λ and ϱ will depend on time, but the slow changes of ϱ are mostly neglected so that (3.5) can be applied.

3.2 Elastic Properties of Dislocations

The elastic strains around a dislocation result in an elastic stress field. This field reacts with the stresses originating from the external stress acting on the body, and with the stress fields of all the other crystal defects. This results in forces along the dislocation line which drive the dislocation at some segments and impede it at others.

In elasticity theory, stresses and elastic strains are tensors. Their components are denominated σ_ij and ε_ij. Most of the formulae given in this book refer to linear isotropic elasticity. Then, $i, j = 1, 2, 3$ refer to the three coordinate axes x_i. Stress or strain components with $i = j$ are normal stresses or strains. Those with $i \neq j$ are shear stresses or strains. In Sect. 3.2.7, the influence of anisotropy on the line tension of dislocations will be discussed. The strain components are derived from the elastic displacements u_i in the directions x_i according to

$$\varepsilon_\mathrm{ij} = \frac{1}{2}\left(\frac{\partial u_\mathrm{i}}{\partial x_\mathrm{j}} + \frac{\partial u_\mathrm{j}}{\partial x_\mathrm{i}}\right). \tag{3.6}$$

Thus, the definition of shear strain components is different from the technical shear strains introduced in Sect. 2.1 for the plastic shear strain

$$\gamma_\mathrm{ij} = 2\varepsilon_\mathrm{ij} \qquad \mathrm{i} \neq \mathrm{j}.$$

The symbols for stresses and strains are used in this book in the following way: σ_{ij} and $\varepsilon_{\mathrm{ij}}$ represent the components of the stress and strain tensors in formulae of elasticity theory, whereas σ denotes a tensile or compressive stress and τ denotes a shear stress resolved on a particular slip system in other chapters. The plastic tensile or compressive strains are denoted by ε and the plastic shear strains by γ as introduced in Sect. 2.1.

3.2.1 Stress Fields of Straight Dislocations

The stress fields of dislocations can be calculated by elastic straining of cylinders according to the original treatment by Volterra [7]. As an example, the model of a screw dislocation of infinite length is presented in Fig. 3.7. The three coordinate axes are denoted x, y, and z. The dislocation line extends along the z axis. The unstrained cylinders are cut along a plane, and the two faces of the cut are shifted with respect to each other by a constant displacement, by the Burgers vector $\boldsymbol{b}$. In the case of the screw dislocation, $\boldsymbol{b}$ is parallel to the dislocation line, i.e., the z axis. Then, the faces of the cut are glued together so that the body is intact again, but elastically strained. By choosing a cylinder with a hole in its center, a singularity of the stresses at the dislocation line itself is prevented.

It can be assumed that the displacements in x and y directions are zero, and that the displacement in z direction increases linearly with the cylinder coordinate angle φ

$$\begin{aligned} u_1 &= u_2 = 0 \\ u_3 &= \frac{\varphi b}{2\pi} = \frac{b}{2\pi} \arctan \frac{y}{x} . \end{aligned} \tag{3.7}$$

Equation (3.6) is used to turn from the displacements to the strains. All strains are zero except

$$\begin{aligned} \varepsilon_{13} &= \frac{1}{2}\frac{\partial u_3}{\partial x} = -\frac{b}{4\pi}\frac{y}{x^2+y^2} \\ \varepsilon_{23} &= \frac{1}{2}\frac{\partial u_3}{\partial y} = \frac{b}{4\pi}\frac{x}{x^2+y^2} . \end{aligned} \tag{3.8}$$

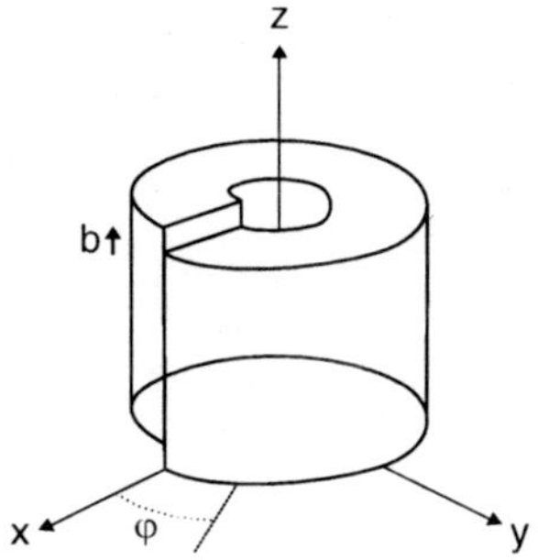

Fig. 3.7. Creation of a screw dislocation by the cut method in an elastically strained cylinder

Using Hooke's law for the relation between the shear strains and shear stresses with the shear modulus μ

$$\sigma_{\mathrm{ij}} = 2\mu\varepsilon_{\mathrm{ij}} \qquad \mathrm{i} \neq \mathrm{j} \tag{3.9}$$

yields the shear stress components

$$\begin{aligned}\sigma_{13} &= -\frac{\mu b}{2\pi}\frac{y}{x^2+y^2} = -\frac{\mu b}{2\pi}\frac{y}{r^2} = -\frac{\mu b}{2\pi}\frac{\sin\varphi}{r}\\ \sigma_{23} &= \frac{\mu b}{2\pi}\frac{x}{x^2+y^2} = \frac{\mu b}{2\pi}\frac{x}{r^2} = \frac{\mu b}{2\pi}\frac{\cos\varphi}{r}.\end{aligned} \tag{3.10}$$

Turning to cylinder coordinates with r and φ instead of x and y reveals that the stresses decay with the reciprocal value of the distance r from the dislocation. This is the slowest decay of all crystal defects. Only the stress field of cracks decays more slowly (with $1/\sqrt{r}$). Thus, dislocations are sources of long-range internal stresses. These are proportional to the absolute value of the Burgers vector b and the shear modulus μ.

For the edge dislocation, the shift of the cut faces has to be perpendicular to the line direction of the dislocation, as outlined in Fig. 3.8. In this case, the displacement in z direction is zero. The derivation of the strains and stresses is more complicated, it is not given here. For this aim, the differential equation of the *Airy* stress function has to be solved. The most general solution was given by S. Timoshenko and J.N. Goodier [78].

The components of the stress field of an edge dislocation along z with the Burgers vector parallel to x are

$$\begin{aligned}\sigma_{11} &= \frac{\mu b}{2\pi(1-\nu)}\frac{-y(3x^2+y^2)}{(x^2+y^2)^2}\\ \sigma_{22} &= \frac{\mu b}{2\pi(1-\nu)}\frac{y(x^2-y^2)}{(x^2+y^2)^2}\\ \sigma_{33} &= \nu(\sigma_{11}+\sigma_{22})\\ \sigma_{12} &= \frac{\mu b}{2\pi(1-\nu)}\frac{x(x^2-y^2)}{(x^2+y^2)^2}\\ \sigma_{13} &= \sigma_{23} = 0.\end{aligned} \tag{3.11}$$

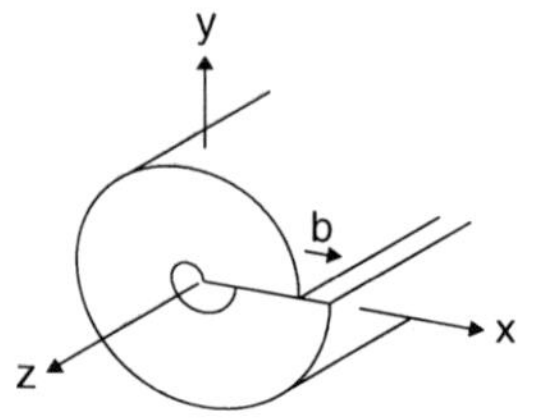

Fig. 3.8. Creation of an edge dislocation by the cut method in an elastically strained cylinder

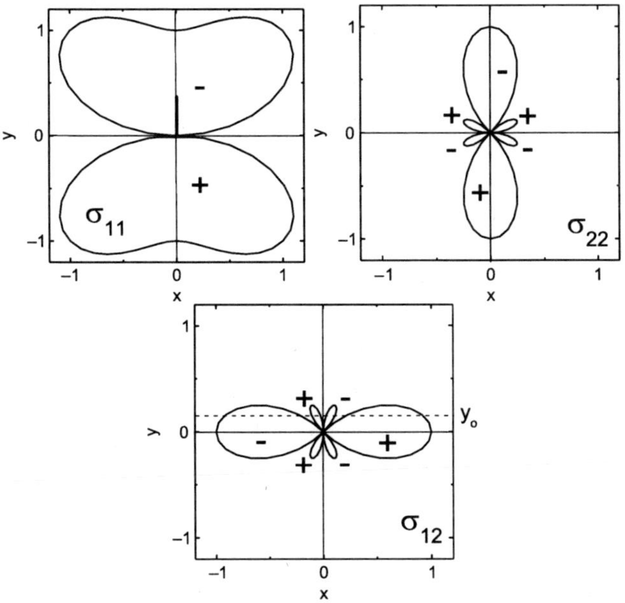

Fig. 3.9. Contours of equal stress σ_{11}, σ_{22}, and σ_{12} of an edge dislocation in z direction with the Burgers vector parallel to x. The coordinates are in units of $\mu b/(2\pi(1-\nu)\sigma_{\mathrm{ij}})$

Here, ν is Poisson's ratio. Figure 3.9 shows contours of equal stress of the edge dislocation. Characteristic features are:

- The stress field of the edge dislocation contains shear stress as well as normal stress components.
- The stress field does not show rotational symmetry as for the screw dislocation.
- The stress field is complementary (orthogonal) to that of the screw dislocation.
- The shear stresses are zero in diagonal directions and perpendicular to the Burgers vector, and
- The stress σ_{12} equals σ_{22} rotated by 90°.

In order to calculate the stress field of a mixed dislocation, the Burgers vector is decomposed in its screw and edge components according to (3.2). The stress fields of both components are superimposed linearly. Very often, the dislocations are not straight. Curved dislocations will be treated in Sect. 3.2.7.

3.2.2 Dislocation Energy

The incorporation of a dislocation into a solid increases the free energy of the latter mainly owing to the elastic stress field. The additional energy is attributed to the defect. The energy of the screw dislocation in a cylinder, as it is introduced in the preceding section, can be calculated using elementary means. The increase of the elastic energy in an elementary volume is given by

$$\mathrm{d}E = \frac{1}{2}\sum \varepsilon_{\mathrm{ij}}\sigma_{\mathrm{ij}}\mathrm{d}V, \tag{3.12}$$

where the sum has to be taken over all components of the strain and stress tensors. Considering that for the screw dislocation all normal stresses are zero and using (3.9), (3.12) reduces to

$$\mathrm{d}E = \frac{1}{2\mu}(\sigma_{12}^2 + \sigma_{23}^2 + \sigma_{31}^2)\mathrm{d}V.$$

Inserting the stress components of the screw dislocation in cylinder coordinates from (3.10 ff.) leads to the energy increase

$$\mathrm{d}E_{\mathrm{s}} = \frac{\mu b^2}{8\pi^2 r^2}(\cos^2\varphi + \sin^2\varphi)r\mathrm{d}r\mathrm{d}\varphi\mathrm{d}z.$$

The integration over the volume of the cylinder yields

$$E_{\mathrm{s}} = \frac{\mu b^2 L}{4\pi}\ln\frac{R}{r_0},$$

where L is the length of the cylinder, i.e., of the dislocation. Usually, the dislocation energy is expressed as the energy per length,

$$E_{\mathrm{ds}} = \frac{E_{\mathrm{s}}}{L} = \frac{\mu b^2}{4\pi}\ln\frac{R}{r_0} = E_{0\mathrm{s}}\ln\frac{R}{r_0}\,. \tag{3.13}$$

$E_{0\mathrm{s}}$ is the pre-logarithmic factor of the energy of the screw dislocation. The respective formula for the edge dislocation is

$$E_{\mathrm{de}} = \frac{\mu b^2}{4\pi(1-\nu)}\ln\frac{R}{r_0} = E_{0\mathrm{e}}\ln\frac{R}{r_0}. \tag{3.14}$$

Thus, the energy of the edge dislocation is slightly higher than that of the screw. For the mixed dislocation, the Burgers vector is decomposed again into its screw and edge components according to (3.2). The total energy is the sum of both parts

$$\begin{aligned} E_{\mathrm{d}} &= \frac{\mu}{4\pi}\left(b_{\mathrm{s}}^2 + \frac{1}{1-\nu}b_{\mathrm{e}}^2\right)\ln\frac{R}{r_0} \\ &= \frac{\mu b^2}{4\pi(1-\nu)}\left((1-\nu)\cos^2\beta + \sin^2\beta\right)\ln\frac{R}{r_0} \\ &= \frac{\mu b^2}{4\pi(1-\nu)}(1-\nu\cos^2\beta)\ln\frac{R}{r_0} = E_0\ln\frac{R}{r_0}. \end{aligned} \tag{3.15}$$

The radii r_o and R are called the inner and outer cut-off radii. Their definite values prevent the occurrence of singularities for both $r_0 \to 0$ and $R \to \infty$. The singularity for $r_0 \to 0$ results from the breakdown of linear elasticity theory in the core of the dislocation. Therefore, the range where the elasticity theory is not valid is cut off inside a certain radius and the energy of the dislocation core is added to the total energy. A rough estimate of the core energy is

$$E_c = \frac{1}{10}\mu b^2. \tag{3.16}$$

Alternatively, r_0 can be chosen in such a way that the elastic energy represents the energy of both the linear elastic region around the dislocation plus the energy of the dislocation core. r_0 is then called an "effective" inner cut-off radius. Its values are in the range of $0.3b < r_0 < 3b$. For straight dislocations, the outer cut-off radius R depends on the size of the crystal and of the arrangement of the dislocations in it. For a single dislocation in the body like the cylinder model in Fig. 3.7, R equals the dimension of the body. For a uniform arrangement of dislocations of opposite signs, the superimposing stress fields screen each other so that R is equal to the mutual distance between the dislocations. This is $1/\sqrt{\varrho}$ if ϱ is the dislocation density. Similarly, small dislocation loops do not cause a long-range stress field far from their place and $R = \alpha D$, where D is the diameter of the loop and $\alpha < 1$. As a consequence, the logarithmic term in the equations of the dislocation energy may take values between about 3 and 18. For a normal dislocation density of deformed crystals in the order of magnitude of $10^{13}\,\mathrm{m}^{-2}$ there follows $R \approx 3 \times 10^{-7}\,\mathrm{m}$, so that a rough estimate of the dislocation line energy is

$$E_d = \frac{1}{2}\mu b^2. \tag{3.17}$$

The energy per lattice plane amounts to about 3 eV for usual data ($b = 3 \times 10^{-10}\,\mathrm{m}$ and $\mu = 30\,\mathrm{GPa}$). Thus, the energy of even small dislocation loops equals tens of electron volts so that the probability of their thermal formation is practically zero.

In summary,

- the dislocation energy is proportional to the dislocation length
- the shear modulus and
- the square of the Burgers vector.

Due to the logarithmic factor, the dislocation energy also depends on the arrangement of the dislocations. In crystals, the Burgers vector must be a lattice vector which connects two equivalent lattice positions. Because of the strong dependence of the dislocation energy on the Burgers vector, only dislocations with the shortest possible Burgers vectors are stable.

If two dislocations of Burgers vectors $\boldsymbol{b}_1$ and $\boldsymbol{b}_2$ meet, they can react to form a third one with $\boldsymbol{b}_r$ if the reaction is associated with a gain in energy.

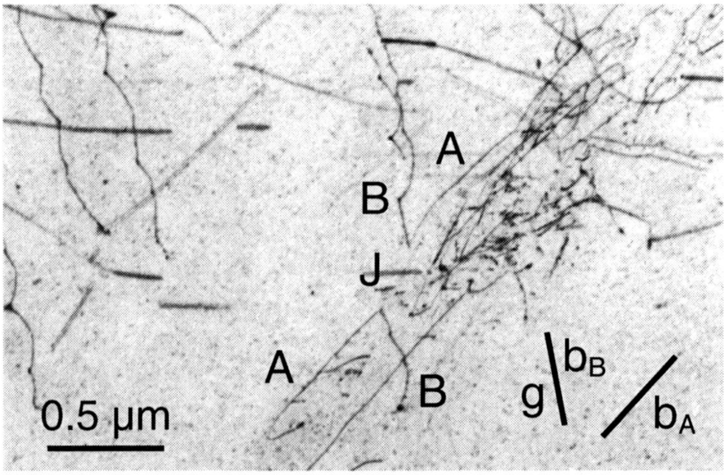

Fig. 3.10. Intersection of two slip bands with different Burgers vectors during in situ deformation of an MgO single crystal in the HVEM at room temperature with the creation of a dislocation junction marked by J. The viewing direction is turned by about 15° around [110] away from [001]. $\boldsymbol{g}$ = (010). Micrograph from [79]

Since the dislocation energy is proportional to the square of the Burgers vector, the reaction will occur if

$$b_{\mathrm{r}}^2 < b_1^2 + b_2^2.$$

This is only a very rough estimate. For a more precise estimation, also the character of the dislocations has to be considered. An example of a dislocation reaction is presented in Fig. 3.10 where two dislocations on different slip planes meet to form a dislocation junction, i.e., a common segment which is not glissile on the slip planes of the two original dislocations. In the NaCl structure of the example in Fig. 3.10, the dislocations may react according to

$$\frac{a}{2}[110](1\bar{1}0) + \frac{a}{2}[0\bar{1}1](011) = \frac{a}{2}[101](10\bar{1}).$$

a is the lattice constant. The indexes in parentheses mark the respective slip planes. It can easily be proven that the gain in energy is 50%. The original dislocations are labeled A and B. The dislocation junction J is only weakly visible since it is extinguished at the (010) $\boldsymbol{g}$ vector. Dislocation junctions harden the crystal if gliding dislocations intersect dislocations with other Burgers vectors. This was first discussed for f.c.c. crystals in [80]. In f.c.c. crystals, dislocations may form so-called Lomer–Cottrell dislocations [81,82]. These dislocations are thought to act as barriers blocking the motion of other dislocations. Then, these dislocations pile up against the barriers forming long-range stress fields as described in Sect. 5.2. The barrier strengths of several dislocation reaction products are described in [83]. At high temperatures, dislocations can form quite regular networks by dislocation reactions.

3.2.3 Forces on Dislocations

In general, a force F_η acting along a coordinate η is defined by the derivative of the free energy with respect to η. In the special case where only mechanical forces act, the free energy is replaced by the total mechanical energy W_t yielding

$$\partial W_t = -F_\eta \partial \eta.$$

The total mechanical energy is composed of the elastic energy W_{el} due to the internal stresses, e.g., by the presence of dislocations, and the potential energy W_{pot} produced by the applied forces or stresses

$$W_t = W_{el} + W_{pot}.$$

Figure 3.11 shows a body of the dimensions L_x, L_y and L_z, in which a screw dislocation S has been introduced at $x = 0$ at the outer left surface and is moved along a plane $y =$ const supported by the external stress σ_{23} into the position x near the outer right surface at L_x. σ_{23} is a shear stress in the respective plane in z direction, which is also the direction of the Burgers vector b. The total mechanical energy is then

$$W_t = \frac{\mu b^2 L_z}{4\pi} \ln \frac{L_x - x}{r_o} - L_x L_y L_z \sigma_{23} \gamma. \tag{3.18}$$

The left term is the elastic energy of the dislocation of length L_z. Since the dislocation in position x is near the surface at L_x, the outer cut-off radius R in (3.15) is suitably set equal to the distance between the dislocation and the surface $L_x - x$. The right term is the work done by an external force in z direction $F_z = \sigma_{23} L_x L_z$ shifted by a distance $\Delta z = \gamma L_y$. According to

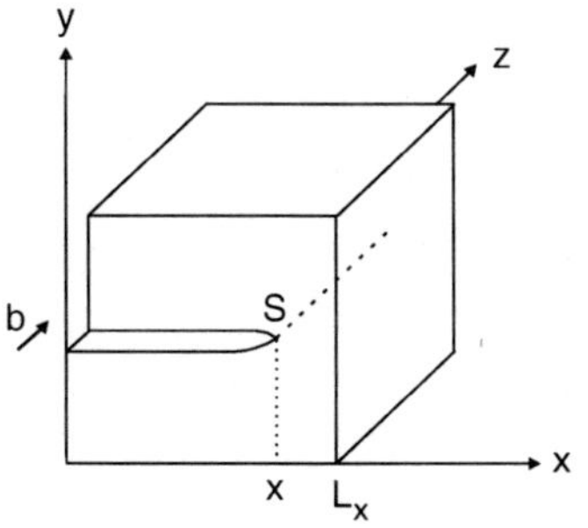

Fig. 3.11. Screw dislocation near a surface of a body

Sect. 3.1.3, one dislocation moving from $x = 0$ to x produces a plastic strain of $\gamma = bx/(L_x L_y)$, leading to

$$W_t = \frac{\mu b^2 L_z}{4\pi} \ln \frac{L_x - x}{r_o} - L_z \sigma_{23} bx.$$

The force in x direction is then

$$F_x = -\frac{\partial W_t}{\partial x} = L_z \left(\frac{\mu b^2}{4\pi(L_x - x)} + \sigma_{23} b \right). \tag{3.19}$$

The force is proportional to the length L_z of the dislocation. Usually, the force per length $f_d = F/L$ is considered. The first term in (3.19) represents an attractive force in the direction to the nearest surface. It is also called an image force as will be explained below. The second term is the force resulting from the external shear stress. In simple terminology, this yields

$$f_d = F/L = \tau b. \tag{3.20}$$

This equation is the scalar version of the Peach–Koehler formula [84], where f_d is the glide force and τ the shear stress resolved on the respective slip system.

The image force can be rationalized also by another argumentation. As a simple case, Fig. 3.12 shows a screw dislocation D at a distance l to a surface. The necessary boundary condition is that the forces perpendicular to the surface are zero. In the coordinates of Fig. 3.12, this means that

$$\sigma_{13} = 0$$

This condition can be fulfilled by extending the body to the right and placing another screw dislocation of opposite Burgers vector at the same distance l outside the original body, the so-called image dislocation ID. According to (3.10), the latter produces a stress

$$\sigma_{23} = -\frac{\mu b}{2\pi} \frac{x}{x^2 + y^2} = \frac{\mu b}{4\pi l}$$

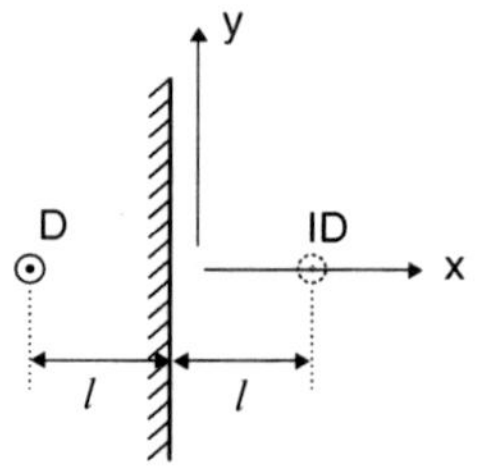

Fig. 3.12. Screw dislocation near a surface and its image dislocation

at the site of the original dislocation for $x = -2l$ and $y = 0$. The minus sign considers the negative sign of the Burgers vector of the image dislocation. This stress causes a force

$$f_{\mathrm{dx}} = \sigma_{23} b = \frac{\mu b^2}{4\pi l}, \tag{3.21}$$

which corresponds to the first term in (3.19).

As described in Sect. 2.4, image forces may limit the reliability of in situ straining experiments in a TEM. This is especially true if the image forces are of the same order of magnitude as the applied forces. In a typical in situ experiment in the HVEM (shear modulus $\mu = 30\,\mathrm{GPa}$, $b = 3 \times 10^{-10}\,\mathrm{m}$, and specimen thickness $0.5\,\mu\mathrm{m}$, i.e., $l = 0.1\,\mu\mathrm{m}$), the image force amounts to about $2 \times 10^{-3}\,\mathrm{N\,m^{-1}}$. This has to be compared with the Peach–Koehler force from the applied stress. With a characteristic flow stress of 200 MPa, i.e., a resolved shear stress of about 100 MPa, the Peach–Koehler force becomes $f_{\mathrm{d}} = \sigma b \approx 3 \times 10^{-2}\,\mathrm{N\,m^{-1}}$. Thus, the force from the applied load is more than ten times higher than the image force, with the dislocation at a typical distance to the surface. This relation will be worse for the thinner specimens in conventional electron microscopy.

In its general form, the Peach–Koehler formula reads

$$\boldsymbol{f}_{\mathrm{d}} = (\boldsymbol{b} \cdot \Sigma) \times \boldsymbol{\xi}. \tag{3.22}$$

Σ is the stress tensor and $\boldsymbol{\xi}$ again the line vector of the dislocation. This force can be decomposed into a glide component

$$f_{\mathrm{gl}} = \frac{[(\boldsymbol{b} \cdot \Sigma) \times \boldsymbol{\xi}] \cdot [\boldsymbol{\xi} \times (\boldsymbol{b} \times \boldsymbol{\xi})]}{|\boldsymbol{b} \times \boldsymbol{\xi}|}$$

and a climb component

$$f_{\mathrm{c}} = \frac{[(\boldsymbol{b} \cdot \Sigma) \times \boldsymbol{\xi}] \cdot (\boldsymbol{b} \times \boldsymbol{\xi})}{|\boldsymbol{b} \times \boldsymbol{\xi}|}. \tag{3.23}$$

Here, $\boldsymbol{b} \times \boldsymbol{\xi}$ is the normal of the slip plane and $\boldsymbol{\xi} \times (\boldsymbol{b} \times \boldsymbol{\xi})$ is the normal of $\boldsymbol{\xi}$ in the slip plane. Both force components are perpendicular to the dislocation line.

On the basis of the force concept, it is possible to describe the interaction of a moving dislocation with the external stress and the stress fields of all the other crystal defects. The moving dislocation feels the sum of all stresses, i.e., the externally applied stress and the internal stresses of the defects. These internal stresses can support or impede the action of the external stress. The obstacle action controls the glide resistance of the moving dislocations. Thus, the microscopic theory of the interaction between the moving dislocations and other dislocations as well as with other crystal defects explains the properties of the macroscopic plastic deformation. In this respect, it is necessary to distinguish between glide, cross glide, and climb motions. Accordingly, the force components in the slip plane and perpendicular to it have to be calculated. Some simple cases will be treated in the following section.

3.2.4 Interaction Between Parallel Dislocations

The elastic interactions between the moving dislocations and other dislocations in the crystal can basically be understood from the very simple case of two parallel screw dislocations with equal or opposite Burgers vectors. As illustrated in Fig. 3.13, one dislocation S1 is situated at the origin of the xy plane and extends in z direction. It is considered as the source of a field of internal stresses. The other dislocation S2 represents a probe, which is moved on an xz plane at the distance y_0 from the origin. Dislocation S1 exerts on S2 a shear stress

$$\sigma_{23} = \frac{\mu b}{2\pi} \frac{x}{x^2 + y_0^2} \tag{3.24}$$

on its xz plane of motion, according to (3.10). Considering (3.20), this stress causes a Peach–Koehler force in x direction

$$f_{\mathrm{dx}} = \sigma_{23} b = \frac{\mu b^2}{2\pi} \frac{x}{x^2 + y_0^2}. \tag{3.25}$$

In Fig. 3.14, the force is plotted as a function of x. It is zero at $x = 0$. However, there is still $\sigma_{13} \neq 0$, i.e., parallel screw dislocations always attract or

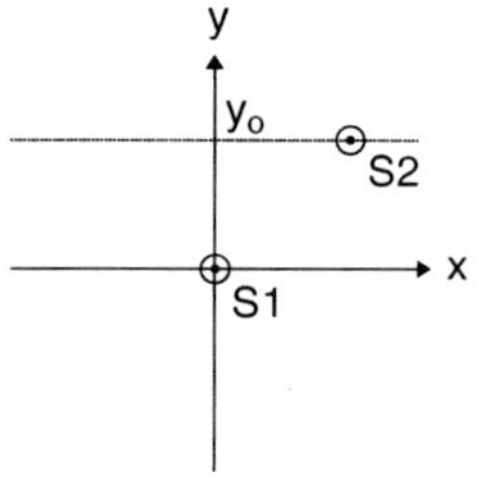

Fig. 3.13. A screw dislocation S2 moving on an xz plane in the stress field of another parallel screw dislocation S1 situated at the origin of the xy plane

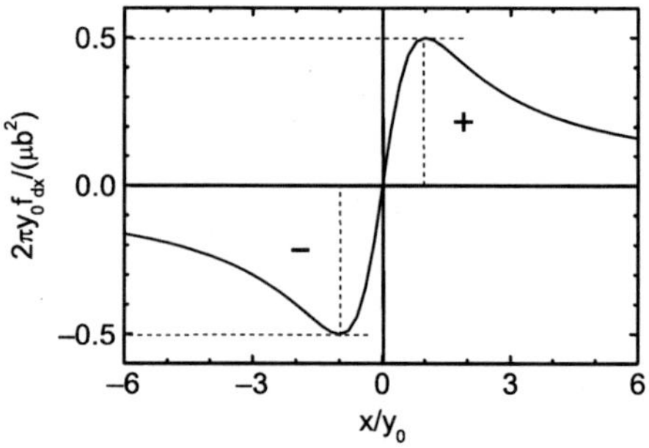

Fig. 3.14. Normalized elastic interaction force in x direction experienced by screw dislocation S2 in the field of the screw dislocation S1 of Fig. 3.13

repel each other, depending on their respective signs. For the present question of the glide motion of S2 on the xz plane, only the glide force f_{dx} is considered, no cross slip on other planes. The glide force is positive for positive x values and Burgers vectors of equal sign, but negative for negative x values. This means that screw dislocations of equal sign repel each other. For Burgers vectors of opposite sign, the dislocations attract each other on both sides and the equilibrium at $x = 0$ is stable with respect to glide on the xz plane. The site of the maximum interaction force is found by setting

$$\frac{\partial f_{\mathrm{dx}}}{\partial x} = \frac{\partial \sigma_{23} b}{\partial x} = \frac{\mu b^2}{2\pi} \frac{y_0^2 - x^2}{(x^2 + y_0^2)^2} \doteq 0.$$

The solution is $x_{\mathrm{m}} = \pm y_0$. Putting this into (3.25) yields the maximum force

$$f_{\mathrm{dxm}} = \pm \frac{\mu b^2}{4\pi y_0}. \tag{3.26}$$

If a moving dislocation is to bypass another one, the maximum interaction force has to be overcome by applying an external stress, in this case σ_{23}. Both the maximum interaction force and the necessary stress to overcome this force are proportional to $1/y_0$, i.e., the reciprocal distance between the slip plane of the gliding dislocation and the dislocation causing the internal stress field. During plastic deformation, the dislocation density increases and, consequently, y_0 decreases. Thus, to continue plastic deformation, the applied stress has to increase. This is an essential source of hardening during deformation, the so-called work-hardening.

For plastic deformation, the behavior of edge dislocations is important, too. The stress field of edge dislocations was introduced in (3.11) and Fig. 3.9. For the bypassing of two parallel edge dislocations extending in z direction with Burgers vectors parallel to x and a distance y_0 between their slip planes, the stress component σ_{12} in Fig. 3.9c is relevant. Again, one dislocation moves as a probe in the stress field of the other one located at the origin. There are positions of zero stress for $x = 0$ and $x = \pm y_0$. For parallel Burgers vectors, the signs in Fig. 3.9c indicate the directions of the force. For example, near $x = 0$ and $x > 0$ the probe dislocation is pushed to the left, and for $x < 0$ to the right, i.e., the equilibrium is stable. Table 3.1 presents the stability conditions of all equilibrium positions for equal and opposite signs of the Burgers vectors. It shows the stable position for equal signs at $x = 0$, corresponding to the tilt grain boundary, and for opposite signs at $x = \pm y_0$, corresponding to the edge dipole. Thus, the edge dislocation dipoles are stable in a position of 45° with respect to the direction of the Burgers vector.

The maximum force between two bypassing edge dislocations can be determined in the same way as described earlier for screw dislocations, yielding

$$f_{\mathrm{dm}} = \frac{\mu b^2}{8\pi(1-\nu) y_0}.$$

Table 3.1. Equilibrium positions of two edge dislocations with parallel or antiparallel Burgers vectors

	$x_0 = 0$	$x_0 = \pm y_0$
$b_2 = b_1$	Stable, grain boundary	Unstable
$b_2 = -b_1$	Unstable	Stable, dislocation dipole

The bypassing (shear) stress is given by $\tau = f_{\mathrm{dm}}/b$. In other words, an existing dislocation dipole can be decomposed at a certain stress if its height h is greater than a critical height

$$h_{\mathrm{c}} = \frac{\mu b}{8\pi(1-\nu)\tau}. \tag{3.27}$$

This is the so-called dipole opening criterion.

If dislocations capture each other in the described equilibrium configurations, the total defect energy is reduced since the stress fields annihilate far from the dipole. A reference measure is the energy of a dislocation with an outer cut-off radius R in the logarithmic term of (3.15) equal to the average distance between the dislocations. For a screw dislocation dipole, R in (3.13) has to be set equal to the mutual distance between the dislocations [85]. Since the distance between the dislocations in the dipole is mostly much smaller than the average distance between the dislocations, the energy of the dipole is remarkably lower than the above reference energy. Such a dislocation structure where the long-range stresses partially cancel each other may be called a low-energy dislocation structure (LEDS). If dislocations of equal sign arrange under stress in nonequilibrium configurations (dislocation pile-ups), where their stress fields amplify each other, the energy is higher than the reference value. Theories of work-hardening have been developed on the basis of both concepts [86, 87].

3.2.5 Interaction Between Nonparallel Dislocations

The elastic interaction between nonparallel dislocations is treated in a way similar to that between parallel ones. One dislocation is considered the origin of a field of internal stresses, and the force on the other probe dislocation is calculated by the general form of the Peach–Koehler formula (3.22). In contrast to parallel dislocations, now the force depends on the place along the dislocation. A very simple case described in detail by Nabarro [85] is that of two perpendicular straight screw dislocations S1 and S2 with the shortest distance y_0 between them as demonstrated in Fig. 3.15. The coordinate x along the probe dislocation S2 can be described by the angle θ with $\tan\theta = x/y_0$. The glide force (per dislocation length), which is the only force in this case, is then given by

$$f_{\mathrm{dy}} = \frac{\mu b^2}{2\pi y_0}\cos^2\theta. \tag{3.28}$$

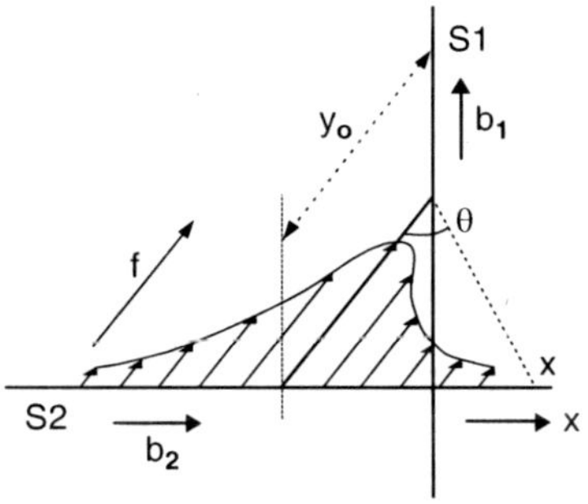

Fig. 3.15. Force between two orthogonal screw dislocations

A plot of the dependence of the force on the location along S2 is included in Fig. 3.15. The total force between dislocations of infinite length is obtained by integrating f_{dy} over x from $-\infty$ to $+\infty$ or over θ from $-\pi/2$ to $+\pi/2$, yielding

$$F_{\mathrm{y}} = \frac{\mu b^2}{2}. \qquad (3.29)$$

Thus, the total force is independent of the distance y_0 between the dislocations, as first noted by Read [88]. However, the force is increasingly concentrated in the nearest region between both dislocations if the distance becomes small. In this local region, the force is higher than that between parallel dislocations of the same distance (3.26).

A number of individual cases is reviewed by Hartley and Hirth [89] and Bullough and Sharp [90]. For a general treatment, see Hirth and Lothe [12]. In general, the forces have both glide and climb components so that the dislocations, if they comply with the forces, may assume complicated curved shapes. If the sign of the line vector or the Burgers vector of one dislocation is reversed, the sign of the force reverses, too. If two signs are changed, the sign of the force does not change. As the maximum force decreases with $1/y_0$, these interactions are of long-range character like the interaction between parallel dislocations.

If moving dislocations meet nonparallel other dislocations, they can intersect each other if the external stress is high enough to overcome the elastic interaction forces. During the intersection process, kinks or jogs are formed in both dislocations as outlined in Fig. 3.16, where a screw dislocation S1 is cut by a moving screw dislocation S2. The lattice planes around S1 are distorted as indicated in Fig. 3.16a. The dislocation line S2 moves along this distorted plane so that its right part is shifted downward and the left part is shifted upward. Finally, in Fig. 3.16b, both parts are separated by the amount of the Burgers vector $\boldsymbol{b}_1$ of S1. The connecting part is a jog J_2 in S2. Likewise, the dislocation S1 experiences a shift J_1. Whether this is a jog or a kink depends on the slip plane of S1. The jogs and kinks are subject to the rules of motion described in Sect. 3.1.2. Thus, the jog J_2 in S2 cannot glide with

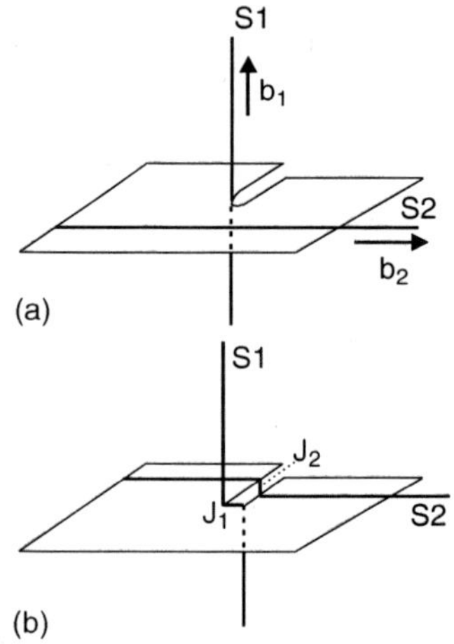

Fig. 3.16. Mutual intersection between two screw dislocations

the dislocation. It has to climb, which involves diffusion, so that the jog acts as an obstacle to the glide motion of the rest of the dislocation. But even gliding jogs may represent obstacles since they move on planes different from the main glide plane.

During the intersection process, a piece of additional dislocation line of the length of the Burgers vector of the cut dislocation is created. Since the long-range stress field of the dislocation is only slightly disturbed, the necessary energy is mainly the core energy, estimated in (3.16). While the elastic interaction between the cutting dislocations is of long-range character, the process of formation of the jog is restricted to the neighborhood of the cutting site. It is therefore of short-range nature. Long-range and short-range interactions behave differently with respect to thermal activation, as will be discussed later in Sects. 4.1 and 5.2.

3.2.6 Elastic Interaction Between Dislocations and Elastic Inclusions

While the foregoing sections treated the mutual interaction between different dislocations, the motion of dislocations is also influenced by their interactions with other crystal defects, in particular with foreign atoms solved in the crystals, or with precipitates of second phases. This is the important field of solution and precipitation hardening. To estimate the elastic interaction forces and energies in the framework of linear elasticity theory, the defects are considered as elastic inclusions with either a volume which differs from that of the regular lattice site or with a different shear or bulk modulus. Analogous to the magnetic or electric cases, the interactions are classified as parelastic or dielastic ones. Defects with a volume difference cause a compressive or dilatational elastic stress field consisting of pure shear stresses while defects with a different shear or bulk modulus do not have an own stress field. Instead, they modify the stress field of the interacting dislocation.

The hydrostatic components of the stress field of an edge dislocation produce a pressure

$$p = -\frac{1}{3}(\sigma_{11} + \sigma_{22} + \sigma_{33}). \tag{3.30}$$

Using (3.11) and polar coordinates $x = r\cos\varphi$ and $y = r\sin\varphi$, it follows that

$$p = \frac{\mu b_\mathrm{e} \sin\varphi}{3\pi r} \frac{1+\nu}{1-\nu}. \tag{3.31}$$

b_e is the edge component of the Burgers vector. The interaction energy with a spherical defect of an atomic volume difference $\Delta\Omega$ with respect to the volume Ω of the host lattice site is then given by [91]

$$\Delta W_\mathrm{i} = 3\,\frac{1-\nu}{1+\nu}\,p\,\Delta\Omega,$$

or, if the interaction constant

$$\beta = \frac{\mu b_\mathrm{e}}{3\pi} \frac{1+\nu}{1-\nu} \Delta\Omega \tag{3.32}$$

is introduced, it is given by

$$\Delta W_\mathrm{i} = 3\,\frac{1-\nu}{1+\nu}\,\beta\,\frac{\sin\varphi}{r} = \frac{\mu b_\mathrm{e}}{\pi} \Delta\Omega \frac{\sin\varphi}{r}. \tag{3.33}$$

For a dislocation moving on a plane at a certain distance to the elastic inclusion, the interaction force $F_\mathrm{i} = -\partial \Delta W_\mathrm{i} / \partial x$ in the direction of motion x has a minimum on one side of the inclusion and a maximum on the other. The signs of these extrema change if the plane of motion changes, e.g., from the compression side of the dislocation to the tension side. The maximum parelastic interaction force is given by

$$F_\mathrm{im} \approx \frac{\mu b_\mathrm{e}^2}{3} \frac{\Delta\Omega}{\Omega}, \tag{3.34}$$

if the minimum distance between the inclusion and the slip plane is set equal to half the smallest distance $b/\sqrt{6}$ between the $\{111\}$ planes in the f.c.c. structure. This force is remarkably smaller than the interaction force between nonparallel dislocations (3.29), since $|\Delta\Omega/\Omega|$ is only of the order of magnitude of 0.1. In many tables, the relative volume difference $\Delta\Omega/\Omega$ is replaced by the size misfit parameter $\delta = (1/a)(\mathrm{d}a/\mathrm{d}c)$, where a is the lattice parameter and c the concentration of the defects, with $\delta = \Delta\Omega/(3\Omega)$. A table of respective data is given in [92].

The parelastic interaction is of long-range character since it decreases with $1/r$. For mixed dislocations, the edge component of the Burgers vector enters the above equations. Screw dislocations, which do not have a hydrostatic stress field, cause only second-order effects. Dielastic interactions affect both edge and screw dislocations. While the interaction terms are weaker than for the parelastic interaction, the misfit parameters like $(1/\mu)(\mathrm{d}\mu/\mathrm{d}c)$ are greater so that both kinds of interactions may contribute to the hardening.

The models of parelastic and dielastic interactions between dislocations and spherical inclusions are mainly applied to solution hardening by substitutional foreign atoms in metals. A strong interaction may arise between dislocations and defects with a tetragonal stress field, as was first pointed out by Fleischer [93]. He calculated the interaction force between a straight screw dislocation in the cubic structure with a Burgers vector in $\langle 110 \rangle$ direction moving on a slip plane of type $\{110\}$ (as in the NaCl structure) with $y = \mathrm{const}$ at a distance b to a defect of tetragonal symmetry and $\langle 110 \rangle$ distortion. The dependence of the force F on the position x of the dislocation is given by

$$F = F_0 \frac{(x/b)^2 + \sqrt{2}(x/b) - 1}{[(x/b)^2 + 1]^2},$$

where $F_0 = \mu \Delta \epsilon b^2 / 3.86$, and $\Delta \epsilon$ is the difference between longitudinal and transverse strains of the tetragonal distortion. For the essential range of $x > b$, this function is approximated by the simpler formula [94]

$$F = F_{\mathrm{o}} \left[(x/b) + 1 \right]^{-2}, \tag{3.35}$$

which is frequently called the Fleischer approximation. As this function describes the dependence of the interaction force on the distance between the dislocation and the defect along x, it is called a *force–distance* relation or an interaction profile. Integrating this function between the applied force and the maximum force F_0 yields the interaction energy with the maximum interaction energy $\Delta W_{\mathrm{im}} = F_0 b$. This can be compared with the maximum parelastic interaction energy for spherical defects (3.33). With $r = b$ for the nearest approximation between dislocation and defect, and typical values of $\Delta \Omega = 0.1\, b^3$ on one hand and $\Delta \epsilon = 0.45$ on the other, the interaction with tetragonal defects turns out to be quite strong. In [95], the calculations of the interaction profile are extended to several orientations of the tetragonal defects and slip geometries in the NaCl, f.c.c., and b.c.c. structures as well as to elastic anisotropy. Several cases were recalculated in [96]. The results differ from the original Fleischer formula. However, the differences are too small to be verified by macroscopic measurements.

The models of the interaction of dislocations and elastic inclusions are used to interpret the hardening by solutes and small precipitates at low temperatures, where the defects do not diffuse, as described in Sect. 4.5. Spherical inclusions describe the behavior of substitutional defects, while tetragonal inclusions are applied to interstitial defects, e.g., in b.c.c. crystals and for associates between aliovalent impurities and their charge-compensating vacancies in ionic crystals, for which the model was originally derived. At high temperatures, the defects may segregate or, in the case of tetragonal defects, reorient in the stress field of the dislocations. These clouds may be dragged with the moving dislocations forming a drag stress, which will be described in Sect. 4.11.

3.2.7 Bowed-Out Dislocations

In the foregoing sections, only straight dislocations were considered. However, a dislocation segment pinned at its ends will bow out under the action of an applied stress. The bowing increases the strain energy of the segment because of the increase in length, $\Delta W_\mathrm{d} = E\Delta L$ (Sect. 3.2.2). For simplicity, the subscript d of E_d in (3.15) is dropped here. On the other hand, energy is gained from sweeping an area ΔA supported by the external resolved shear stress τ, $\Delta W_\mathrm{a} = \tau b \Delta A$. If the stress is below a critical value, the segment will find an equilibrium position, where $\Delta W_\mathrm{d} - \Delta W_\mathrm{a} = 0$. The problem of bowed-out dislocations is mostly treated within the framework of the line tension approximation derived by DeWitt and Koehler [97], analogous to the surface tension of liquids. In this approximation, the energy of the dislocation depends only on its orientation β but not on its position with respect to outer or inner surfaces and to sources of internal stresses like other dislocations, or its own self-stress, i.e., the logarithmic factor $\ln(R/r_0)$ in (3.15) is constant.

Figure 3.17 shows a finite dislocation segment of length L. On its left side, it is pinned, and it has a hypothetical node A on its right side. Shifting the node from A to B causes a change of the energy of the segment by a change in both its length and its orientation β. If $W_\mathrm{d} = E(\beta)\,L$ is the total energy of the segment, the change in energy is

$$\mathrm{d}W_\mathrm{d} = E(\beta)\partial L + \frac{\partial E(\beta)}{\partial \beta} L\partial\beta .$$

To keep the segment in equilibrium, tangential and normal forces have to act at the end point

$$F_\mathrm{t} = -\frac{\partial W_\mathrm{d}}{\partial L} = -E \qquad (3.36)$$

$$F_\mathrm{n} = -\frac{\partial W_\mathrm{d}}{\partial n} = -\frac{\partial W_\mathrm{d}}{L\partial \beta} = -E' . \qquad (3.37)$$

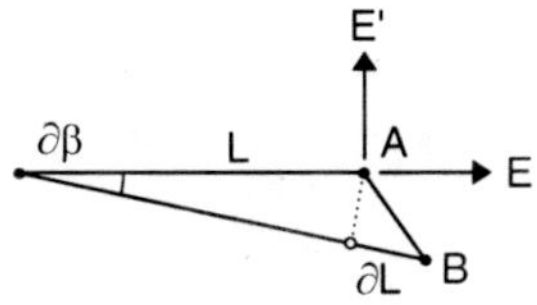

Fig. 3.17. Shifting the end point of a dislocation segment from A to B

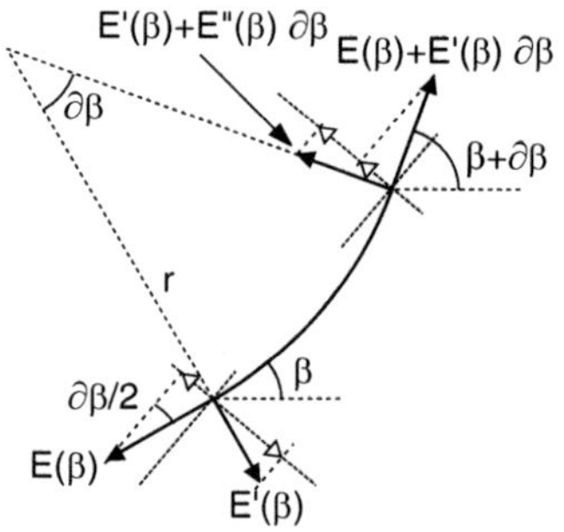

Fig. 3.18. Force components at the ends of a bent dislocation segment of infinitesimal length

n is a coordinate in normal direction at the end of the segment with $\partial n = L\partial\beta$, and E' is the derivative of E with respect to β. As the line energy tries to reduce the length of the segment, $-E$ has to act outside to balance the node. At a bent dislocation segment, these forces are in equilibrium with the Peach–Koehler force resulting from the external stress, as illustrated in Fig. 3.18. At the lower end of the segment, E and E' are taken at the orientation angle β, whereas at the upper end these quantities have to be taken at the angle $\beta + \partial\beta$. The respective quantities are indicated in the figure. Since the sign of the dislocation line vector is opposite at both ends, the sign of the force vectors is also opposite. The four forces can be projected onto the directions parallel to the main extension of the dislocation line and perpendicular to it. The latter projections are marked by the open arrow heads. The components in direction of the dislocation line cancel each other while the components perpendicular to it, i.e., in the direction of the dislocation motion, form a net force pointing at the center of the curved segment. The net force of the two tangent forces is

$$(2E + E'\partial\beta)\sin\frac{\partial\beta}{2} \approx E\partial\beta.$$

The net force of the two normal forces is

$$(E' + E''\partial\beta - E')\cos\frac{\partial\beta}{2} \approx E''\partial\beta.$$

Thus, the total force is

$$\partial F_\Gamma = (E + E'')\partial\beta.$$

This force is in equilibrium with the Peach–Koehler force (3.20) acting on the segment owing to the external stress

$$\partial F_{\mathrm{PK}} = f_{\mathrm{PK}}\partial l = \tau\, b\, r\, \partial\beta.$$

It follows that the radius of curvature of the segment is given by

$$r = \frac{E + E''}{\tau b} = \frac{\Gamma}{\tau b}. \tag{3.38}$$

$\Gamma = E + E''$ is the line tension. It is a force acting in tangential direction on a dislocation segment and describing the resistance of the segment against bowing in the case of a weak bow-out. The line tension approximation is a method to calculate the radius of curvature of a dislocation segment from the orientation dependence of the energy of a *straight* dislocation. With the line energy in the limit of isotropic elasticity (3.15), the line tension becomes

$$\begin{aligned} E &= \frac{\mu b^2}{4\pi(1-\nu)} \ln \frac{R}{r_o}(1 - \nu \cos^2 \beta) = E_e(1 - \nu \cos^2 \beta) \\ E'' &= E_e \, 2\,\nu(-\sin^2 \beta + \cos^2 \beta) \\ \Gamma &= E_e(1 + \nu - 3\nu \sin^2 \beta). \end{aligned} \tag{3.39}$$

E_e is the energy of the edge dislocation. The energies and line tensions of the pure screw and edge dislocations are given in Table 3.2. It follows that the line tension of the screw dislocation is higher than that of the edge, contrary to the relation between the line energies. Thus, the stiffness against bowing is higher for the screw dislocation.

For anisotropic crystals, the energy and its second derivative with respect to the orientation can be calculated within the framework of anisotropic elasticity theory for straight dislocations. In analogy to (3.15), the energy per dislocation length can then be written as

$$E = E_0(\beta) \ln (R/r_0) = \frac{K(\beta) b^2}{4\pi} \ln (R/r_0), \tag{3.40}$$

with $K = 4\pi E_0/b^2$ being the energy factor of the respective dislocation. It replaces the shear modulus in the respective formulae. In order to calculate $E''(\beta)$, the orientation dependence of $E(\beta)$ can be represented by a Fourier series. $E''(\beta)$ is then obtained by differentiating the individual terms with respect to β. Line tension data for a number of f.c.c., b.c.c., and h.c.p. metals are tabulated in [98].

In order to find the equilibrium shape of a long dislocation segment under stress, a parametric description of the coordinates x and y of the dislocation can be applied [97]

Table 3.2. Line energy and line tension of screw and edge dislocations in isotropic elasticity

	Screw	Edge
E	$E_e(1-\nu)$	E_e
Γ	$E_e(1+\nu)$	$E_e(1-2\nu)$

$$x = \frac{-1}{\tau b}(E \sin\beta - E' \cos\beta)$$
$$y = \frac{1}{\tau b}(E \cos\beta - E' \sin\beta). \tag{3.41}$$

For isotropic elasticity, the shape of the dislocation with its Burgers vector parallel to x is an ellipsis with the major half axis $E_e/(\tau b)$ in x direction ($\beta = 90°$) and the minor half axis $(1 - \nu)E_e/(\tau b)$ perpendicular to it.

Comparing the shape of dislocations under load with the theoretical shape offers the opportunity to determine the stress acting locally on the bowed-out dislocation segments. The first measurement of this kind was performed in [99] by fitting circles to the electron micrographs of dislocations pinned under load by neutron irradiation in order to determine the local radius of curvature. The sizes of calculated shapes of dislocation loops under stress were matched to long dislocation segments in [100] and elsewhere. Using this method, micrographs of in situ straining experiments were studied in detail in [101, 102], where the statistical data of pinned dislocations were also determined. All these methods depended on a visual comparison between the micrographs and the calculated dislocation shapes so that they are of subjective nature. The first attempt of a measurement independent of the observer was made in [103] employing an image analyzer and curve fitting techniques, analyzing, however, only a few segments. In [104], a photometer head was moved by hand over the micrographs projected in a high magnification, always in radial direction to find points of equal brightness in both flanks of the contrast profile of the dislocations. An example is shown in Fig. 3.19. The midpoint between both points of the pairs was taken as the locus of the dislocation. This method yields a constant accuracy independent of the orientation of the dislocation in contrast to simply scanning the micrographs. The local radius of curvature can be determined by polynomial regression analysis of second order over a short segment. The center of this segment is then shifted along the dislocation line to obtain the dependence of the radius of curvature on the orientation angle.

In order to determine the locally acting stress from the dislocation curvature, the line tension data are necessary. While the elastic constants are usually known very precisely, the logarithmic factor of the dislocation energy (e.g., in (3.15)) and thus of the line tension depends on the dislocation configuration. Hirth and Lothe [12] calculated the logarithmic factor of the dislocation energy for some configurations by the method of piecewise straight dislocation segments. It turns out that if the self-stress of those parts of the dislocation adjoining the segment under consideration are taken into account, the outer cut-off radius has to be replaced by a length parameter l of the dislocation configuration. The logarithmic factor can be written as

$$\left(\ln\frac{l}{r_0} + C\right). \tag{3.42}$$

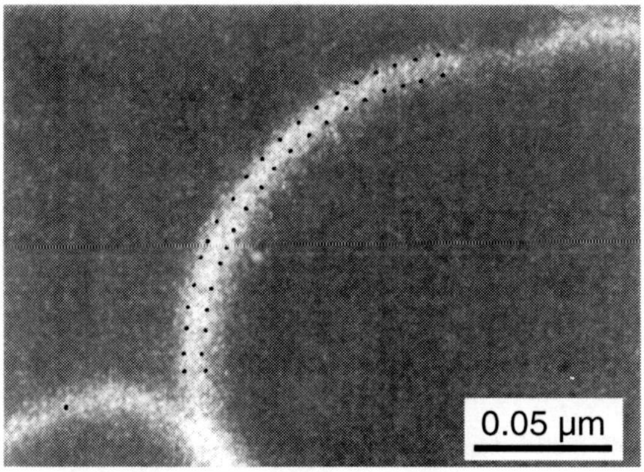

Fig. 3.19. Negative copy of a dislocation segment taken during in situ deformation of an MgO single crystal at high magnification with pairs of photometric measuring points to determine the dislocation curvature. From [104]. Copyright Taylor & Francis Ltd. (1985)

For a full hexagonal loop of the extension l Hirth and Lothe calculated the value of $C = 1.16$, for a semihexagon in a straight dislocation $C = -0.59$, and for a so-called small bow-out of length l in a straight dislocation $C = -2.39$. Thus, for a bowed-out dislocation segment, which is a very important configuration during dislocation motion, the logarithmic factor depends on the length of the segment and on the strength of the bow-out. Since $\ln(l/r_0) \approx 6$, the dislocation energy may vary by a factor of two.

The numerical method introduced in [105] offers an opportunity to determine the exact shape of a locally pinned dislocation. The dislocation is divided into small segments. One circular segment at point P is selected. All the other segments are considered straight. The segment at P is in equilibrium if the applied stress τ equals the stress on the dislocation due to its own stress field, i.e., its self-stress τ_{self}

$$\tau + \tau_{\text{self}}(\text{P}) = 0. \tag{3.43}$$

All stresses are resolved onto the glide plane. The contributions of the self-stress of all the other segments at P can be calculated by means of the Brown–Indenbom–Orlov theorem [106–108]

$$\tau_{\text{self}} = \frac{1}{b} \sum \frac{\sin(\phi - \omega)}{|\boldsymbol{r}|^2} (E_0 + E_0'') \Delta s. \tag{3.44}$$

The sum has to be taken over all elements of length Δs. $\boldsymbol{r}$ is a vector pointing from the element Δs to point P. ϕ and ω are the angles between a reference orientation and $\boldsymbol{r}$ and the tangent vector at Δs, respectively. E_0 is the prelogarithmic factor of the energy of a straight dislocation lying along $\boldsymbol{r}$, i.e., with the line direction ϕ and *not* of the dislocation orientation β at P or at Δs but having the Burgers vector of the dislocation. E_0'' is the second derivative of E_0

with respect to ϕ. Equation (3.44) is valid for arbitrary anisotropic materials. As for the line tension model, the properties of a curved dislocation are determined from the parameters of a straight dislocation of infinite length. To find the equilibrium configuration, point P is shifted along the dislocation, and the sum of the external stress and the self-stress is minimized by a numerical relaxation treatment. Equation (3.44) fails for the neighborhood of P itself since the self-stress becomes infinite. In [105], a piece of length $2r_0$ around P is cut off from the dislocation to avoid the singularity. This procedure differs from the original prescription by Brown [106] where the mean of two principal values at both sides of the dislocation has to be taken. It was pointed out in [109] that the cut-off method of [105] is incorrect, though it is used in a number of papers. The choice of r_0 replaces the inner cut-off radius of the dislocation energy.

A theoretical value for the numerical constant C for dislocations pinned at impenetrable obstacles of a distance l can be obtained from the self-stress calculations in [105]. This situation is shown in Fig. 3.20a. It is the so-called Orowan process which will be described in more detail in Sect. 4.7. The dislocation bows out between large impenetrable obstacles of the same elastic constants as the matrix, the diameter D and the inter-particle distance L. The authors compare the area ΔA (indicated in the right segment) swept by the dislocation when the stress is increased from zero to an actual value in the self-stress calculation with that obtained from the line tension model. Both areas are equal if $C = -1.61$ is chosen. This is in the middle of Hirth and Lothe's values for the semihexagon and the small bow-out,

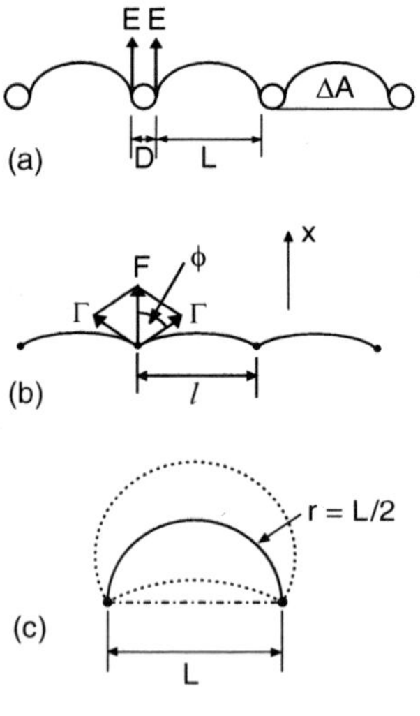

Fig. 3.20. Bow-outs of a dislocation. (**a**) Large bow-outs in a regular array of extended impenetrable obstacles. (**b**) Small bow-outs at point obstacles. (**c**) Dislocation segment of length L bowing out between two strong pinning points to the critical configuration with $r = L/2$

certainly being the best theoretical value. As mentioned earlier, both the line tension and the self-stress calculations can be performed using dislocation data from isotropic or anisotropic elasticity theory. The self-stress calculations are refined in [110, 111] by considering a possible orientation dependence of the dislocation core energy. Problems with the core cut-off in calculating the energy of dislocation loops are discussed in [112].

For small bow-outs of dislocations pinned in an array of small obstacles as outlined in Fig. 3.20b, the constant C was experimentally determined from the dependence of the dislocation curvature on the segment length in MgO single crystals [104, 113]. From the local radius of curvature at a dislocation orientation of $\beta = 0°$, $10°$, or sometimes $25°$, the major half axes e of the corresponding ellipses were calculated by $e = r(\beta) \,/\, \left(1 + \nu - 3\nu \sin^2 \beta\right)$. According to (3.15), (3.38), (3.41), and (3.42), and Table 3.2, these values depend on the segment length l as

$$\begin{aligned} e &= \frac{E_{0\mathrm{e}}}{b\tau}[\ln(l/r_0) + C] \\ &= M[\ln(l/r_0) + C]. \end{aligned} \tag{3.45}$$

Figure 3.21 shows a respective plot of experimental data from a single electron micrograph. The large scatter of the data results from the different local configurations of the evaluated dislocation segments and from the variation of the local effective stress. Nevertheless, the positive correlation is clearly proved statistically. Using the data of four similar micrographs and the effective inner cut-off radius of an edge dislocation in MgO of $r_0 = 0.4\,b$ from the atomistic calculations in [114], the intercept C equals -5.19 with 98% confidence limits of $\pm 30\%$. Similar evaluations were performed for the precipitation hardened alloys Al–4.5 wt% Zn–1.2% Mg and Al–8.36 at% Li [115], yielding the values of $C = -4.6$ and $C = -5.6$. Thus, the only available experimental data on the

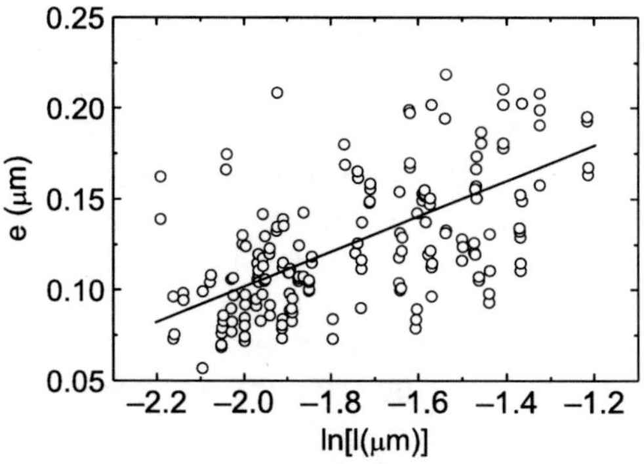

Fig. 3.21. Dependence of the major half axes of bowed-out dislocation segments on the logarithm of the segment length in an MgO single crystal during in situ deformation in an HVEM. Data from [104]

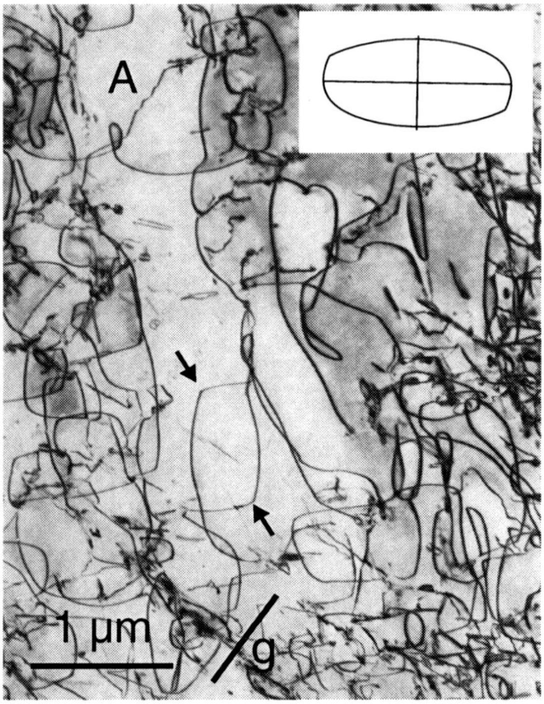

Fig. 3.22. Dislocation structure in an Fe–43 at% Al single crystal deformed at 495°C. The specimen is cut parallel to the (101) primary slip plane. $\boldsymbol{g} = (020)$ near the [101] pole. *Inset*: elastic equilibrium shape of a dislocation with $[11\bar{1}]$ Burgers vector. From [8]. Copyright Elsevier (2005)

line tension are drastically lower than the theoretical values. The problem will be followed further in Sect. 5.2.3.

In anisotropic crystals, frequently the line tension becomes negative in certain ranges of the orientation angle β, which describes the dislocation character. In these ranges, the dislocations are elastically unstable. An example is given in Fig. 3.22. The inset shows the equilibrium shape of a closed dislocation loop with its Burgers vector in horizontal direction calculated by the line tension model using anisotropic elasticity data. Between about 38° and 55° the dislocations are unstable resulting in edges in the dislocation line. In the micrograph, some edges are marked by arrows. At A, a mixed dislocation is unstable and disintegrates into small segments. The regions of elastic instability complicate the determination of the effective stress from the dislocation curvature.

A critical configuration, in particular for processes of dislocation generation (Sect. 5.1.1), is the bowing of a dislocation segment between two strong pinning agents. These may be nodes of a dislocation network or a change of the dislocation line onto planes of high slip resistance, or large particles. In the simplest approximation, the line energy is constant according to (3.17), independent of the dislocation orientation and the overall configuration. Then, the line tension equals the line energy, $\Gamma = E$, and the dislocation bow-outs have a circular shape. As shown in Fig. 3.20c, while a segment of length L bows out under stress, the half-circle with $r = L/2$ is the position of the smallest radius of curvature. Using (3.38), it follows that

$$\frac{L}{2} = \frac{\Gamma}{\tau b} \approx \frac{\mu b}{2\tau}, \tag{3.46}$$

so that the half-circle configuration is that requiring the highest stress τ

$$\tau_{\mathrm{FR}} \approx \frac{\mu b}{L}. \tag{3.47}$$

This is the critical stress to fully bow out a dislocation segment. If it bows out further, the radius of curvature increases again so that the segment becomes unstable. The half-circle is the critical configuration of the Frank–Read source [116] described in Sect. 5.1.1.

In summary of this section:

- The line tension is a tangential force which considers the change in the dislocation energy when changing the dislocation length and the orientation. For cusps in the dislocation line at relatively weak obstacles as in the configuration of Fig. 3.20b, the force on the obstacles is given by the vector sum of the line tensions of the two adjacent segments

  $$F = 2\Gamma \cos \phi. \tag{3.48}$$

  ϕ is half the angle enclosing the neighboring segments.
- For strong bow-outs like the Orowan configuration in Fig. 3.20a, the force on the obstacles equals $2E$ (or better the energy of a dislocation dipole of the respective width) and is thus not given by the line tension but by the line energy.
- The line tension approximation neglects the dependence of the local dislocation curvature on the overall dislocation configuration. The actual configuration can be considered in the line tension model by choosing appropriate values of the logarithmic factor in the dislocation energy.
- However, the only available experimental determination of the logarithmic factor yields a considerably lower line tension than the theoretical values.

3.3 Dislocations in Crystals

In the earlier sections, dislocations were considered in an elastic continuum. However, the discreteness of the crystal lattice and the particular crystal structures have a remarkable influence on the properties of the dislocations. Some general aspects will be treated in the following sections. Particular situations in specific materials like intermetallic alloys will be described in Part II of this book.

3.3.1 Selection of Burgers Vectors

In the elastic continuum, dislocations can obtain Burgers vectors of arbitrary length and orientation. However, in order to create a perfect dislocation in a crystal, the Burgers vector has to be a lattice vector, i.e., it has to connect equivalent lattice points. As expressed in (3.13)–(3.15), the energy of a dislocation is proportional to the square of its Burgers vector. Thus, according to Frank's criterion [117] a dislocation of Burgers vector b can decompose into two perfect dislocations with Burgers vectors b_1 and b_2, if

$$b^2 > b_1^2 + b_2^2. \tag{3.49}$$

As a consequence, in most crystal structures only dislocations with Burgers vectors corresponding to the shortest lattice vectors are energetically stable. In particular, dislocations with Burgers vectors of multiples of perfect Burgers vectors are unstable. Frank's criterion neglects the influence of the orientation angle β on the dislocation energy and effects of elastic anisotropy. A more exact stability criterion is therefore

$$E(\boldsymbol{b}, \beta) > E(\boldsymbol{b_1}, \beta_1) + E(\boldsymbol{b_2}, \beta_2).$$

Although the orientation of the three dislocations in space is equal, the angles β describing the dislocation characters are usually different. Table 3.3 lists the stable Burgers vectors in the most common crystal structures.

As it will be shown in Sect. 4.2.1, the lattice resistance to dislocation motion is low for short Burgers vectors. Thus, dislocations with short Burgers vectors have both a low defect energy and a low resistance to glide. In addition, the resistance stress is low for great atomic distances between the slip planes. Such planes are the planes with small Miller indices. Some of these dominating slip planes are also included in Table 3.3.

3.3.2 Stacking Faults and Partial Dislocations

Many metals crystallize in closed-packed structures. These structures can be modeled by hard spheres, which are held together by attractive forces. Both

Table 3.3. Stable Burgers vectors and dominating slip planes in simple crystal structures

Crystal structure	Stable $\boldsymbol{b}$	Slip planes
f.c.c.	$1/2\langle 110\rangle$	$\{111\}$
b.c.c.	$1/2\langle 111\rangle$, $\langle 100\rangle$	$\{112\}$, $\{110\}$, $\{123\}$
h.c.p.	$1/2\langle 11\bar{2}0\rangle$, $\langle 0001\rangle$	$\{0001\}$, $\{10\bar{1}0\}$
Diamond cubic	$1/2\langle 110\rangle$	$\{111\}$
NaCl	$1/2\langle 110\rangle$	$\{110\}$, sometimes $\{100\}$

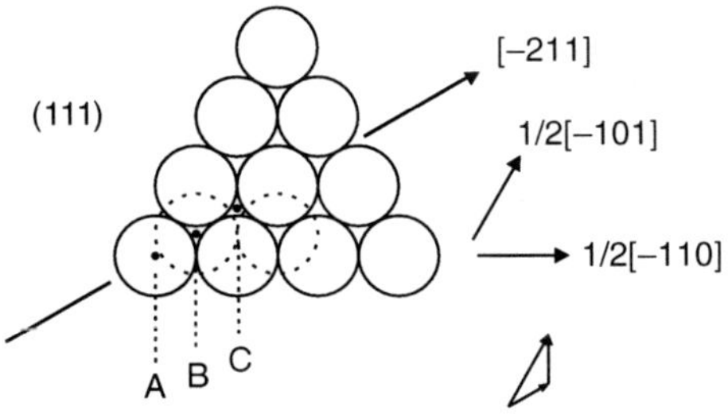

Fig. 3.23. Arrangement of atoms on a (111) plane in an f.c.c. crystal

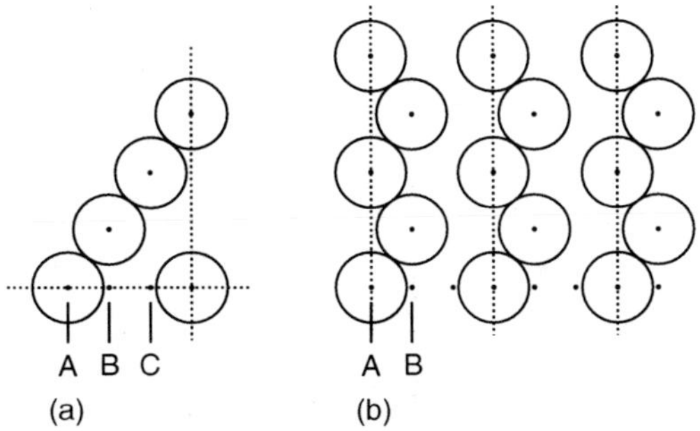

Fig. 3.24. Stacking of closed-packed planes to form (**a**) f.c.c. and (**b**) h.c.p. crystal structures. In (**a**), the $[\bar{2}11]$ direction is horizontal, the [111] direction is vertical

the f.c.c. and h.c.p. structures are based on the closed-packed plane outlined in Fig. 3.23. It is a {111} plane in the f.c.c. lattice. The vertical position of the atoms may be characterized by the center of the atom labeled A. For a close packing, the next layer has to be positioned into the holes of the A layer. The two vertical positions B and C closest to A are marked by dots. In the notation of the figure, the positions are arranged along a $[\bar{2}11]$ direction. Putting atoms into the holes B leads to a layer indicated by the two atoms with broken lines. The structures formed by stacking A, B, C planes are closed packed if two equal planes are not stacked on top of each other. As outlined in Fig. 3.24, the f.c.c. lattice forms by the

$$\ldots \mathrm{ABCABCABC} \ldots$$

stacking sequence, and the h.c.p. lattice forms by the sequence

$$\ldots \mathrm{ABABAB} \ldots \text{ or } \ldots \mathrm{BCBCBC} \ldots \text{ or } \ldots \mathrm{CACACA} \ldots$$

In this lattice, the closed-packed plane is the {0001} plane. The {111} and {0001} planes are preferred glide planes.

If the correct stacking sequence is interrupted, a planar fault is formed. According to Frank [76], the fault is called an intrinsic stacking fault if the normal stacking sequence is maintained on both sides of the fault. Such a fault can be formed by removing one plane from the correct stacking sequence, e.g.,

$$\ldots \mathrm{ABC|BCABC} \ldots$$

In the other case of extrinsic stacking faults, an additional layer is introduced into the normal stacking sequence

$$\ldots \mathrm{ABCABC}\ \overset{|}{\underset{|}{\mathrm{B}}}\ \mathrm{ABCABC} \ldots$$

The stacking faults can be characterized by the vector $\boldsymbol{R}_{\mathrm{F}}$ of the displacement in the stacking sequence perpendicular to the fault plane. It is difficult to calculate the stacking fault energy since the bonds with the nearest neighbors are not disturbed. For the same reason, the stacking fault energy is frequently small.

Stacking faults can be formed by shearing. For example, the vector $\boldsymbol{R} = 1/6[\bar{2}11]$ shifts a B plane into a C position. If all the material above that plane is also shifted, an intrinsic fault results

$$\begin{array}{l} \ldots \mathrm{ABCA\ B\ C\ A\ B\ C} \ldots \\ \qquad\qquad\ \ \downarrow\ \downarrow\ \downarrow\ \downarrow\ \downarrow \\ \qquad\qquad\ \ \mathrm{C\ A\ B\ C\ A} \ldots \end{array}$$

This shift is produced by a dislocation, the $1/6[\bar{2}11]$ Burgers vector of which is not a lattice vector. Such dislocations are called partial dislocations. The Burgers circuit cannot be performed completely in the perfect lattice. It is therefore required that the Burgers circuit starts and ends on the fault plane, as indicated in Fig. 3.25. A perfect dislocation can dissociate into partial dislocations in the f.c.c. lattice by the reaction

$$\frac{1}{2}[\bar{1}01] = \frac{1}{6}[\bar{2}11] + \frac{1}{6}[\bar{1}\bar{1}2]. \tag{3.50}$$

In the simple case, the energy balance of this reaction can be checked by Frank's criterion (3.49)

$$\frac{1}{4} \cdot 2 > \frac{1}{36}(6+6)$$
$$\frac{1}{2} > \frac{1}{3}.$$

Thus, the dissociation is energetically favorable. Of course, the stacking fault energy has to be considered. Neglecting the dependence of the dislocation energy on its orientation, the dislocation can dissociate if, at the vector addition of the Burgers vectors of the partial dislocations, there is an obtuse angle between the two vector arrows. The reaction (3.50) is sketched in the lower right corner of Fig. 3.23.

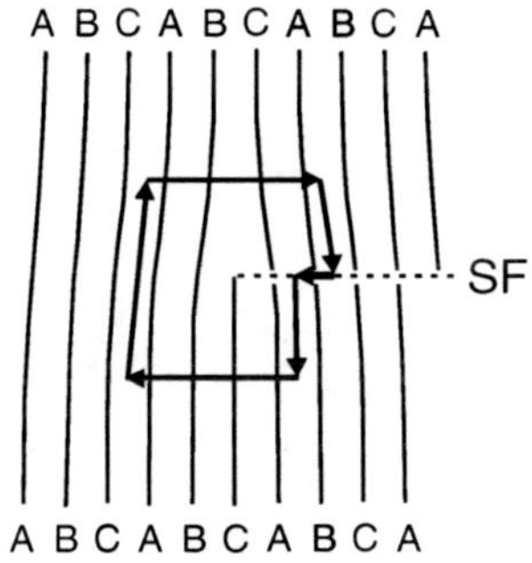

Fig. 3.25. Intrinsic stacking fault ending at a partial dislocation with Burgers circuit around the dislocation

The equilibrium width of the dissociated dislocation is obtained from the minimum of the total energy of the dislocation. This includes the energy of the two partial dislocations, the interaction energy between them and the energy of the stacking fault. Increasing the width w of the stacking fault by ∂w increases its energy per length by $\gamma_{\mathrm{sf}}\partial w$, where γ_{sf} is an energy per area. Thus,

$$\gamma_{\mathrm{sf}} = \frac{\partial(\text{energy per length})}{\partial w} = (\text{attractive force per length})$$

on the bounding partial dislocation. This force is in equilibrium with the repulsive elastic force between the partial dislocations. Considering that the Burgers vectors of the partial dislocations are not parallel as in Sect. 3.2.4, the relation between the energy of a stacking fault in the f.c.c. lattice and its equilibrium width w_0 is given by

$$\gamma_{\mathrm{sf}} = \frac{\mu b_{\mathrm{p}}^2}{8\pi w_0} \frac{2-\nu}{1-\nu} \left(1 - \frac{2\nu \cos 2\beta}{2-\nu}\right). \tag{3.51}$$

Here, b_{p} is the absolute value of the Burgers vector of the partial dislocations, and β is the orientation angle between the total Burgers vector and the orientation of the dislocation line. As a consequence, the equilibrium width of edge dislocations ($\beta = 90°$) is larger than that of screws ($\beta = 0°$).

Partial dislocations created by the dissociation of a perfect dislocation according to (3.50) are called *Shockley* partial dislocations [118]. The Burgers vectors of the total dislocation and of both partial dislocations are situated on the plane of the stacking fault. Thus, the whole dislocation can glide on this plane. An example of such a dislocation dissociated into Shockley partials in a hexagonal SiC single crystal is shown in Fig. 3.26. The dissociation reaction is given by

$$\frac{a}{3}[11\bar{2}0] = \frac{a}{3}[01\bar{1}0] + \frac{a}{3}[10\bar{1}0].$$

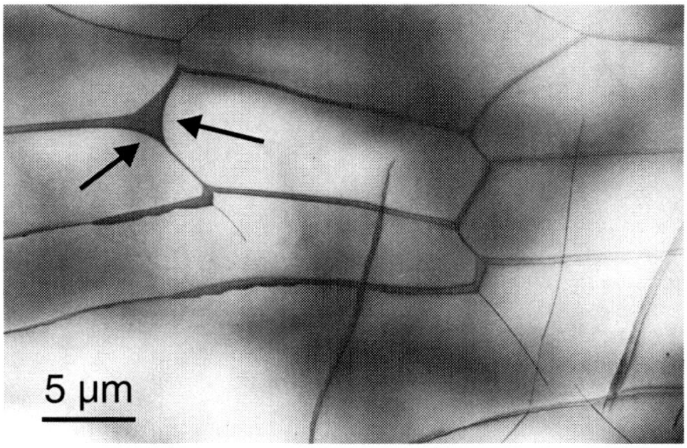

Fig. 3.26. Dislocations in an SiC single crystal plastically deformed at 1700°C dissociated into *Shockley* partials. The viewing direction is [0001]. From the work in [119]

The stacking faults cause a phase shift of the electron wave resulting in an oscillatory diffraction contrast in the TEM depending on the height of the fault in the specimen. The direction of $\boldsymbol{R}_\mathrm{F}$ can be determined by contrast extinction in analogy to (2.9) for the Burgers vector. With respect to the electron microscopy diffraction contrast, the displacement vector $\boldsymbol{R}_\mathrm{F}$ and the shearing vector $\boldsymbol{R}$ are equivalent since, in general, both vectors differ by a translation vector of the lattice. The stacking faults in Fig. 3.26 are parallel to the foil plane, thus showing a uniform area contrast. The dissociation width is very large, owing to a very low stacking fault energy in the order of magnitude of $2\,\mathrm{mJ\,m^{-2}}$. The stacking fault energy can also be determined from the radius of curvature of the partial dislocations at extended dislocation nodes. Two curved segments are indicated by arrows in Fig. 3.26. In this case, the bounding force of the stacking fault is in equilibrium with the line tension of the bowed partial dislocation. The dissociation of the dislocations essentially influences their dynamic behavior so that it is important whether the dislocations have a compact core, or are widely dissociated.

Stacking faults surrounded by a partial dislocation may also form by the collapse of condensed vacancies, e.g., after quenching a crystal from high temperature. In this case, the Burgers vector of the partial dislocation does not lie on the fault plane. Thus, these dislocations cannot glide together with the stacking fault. They are called *Frank* partial dislocations or Frank dislocation loops [120].

3.3.3 Twins

Another way to produce a planar fault in the f.c.c. lattice is to mirror the material on one side of a {111} plane, or to rotate it by 180°. Then, the stacking sequence ABC is changed into CBA, and the fault is called a twin.

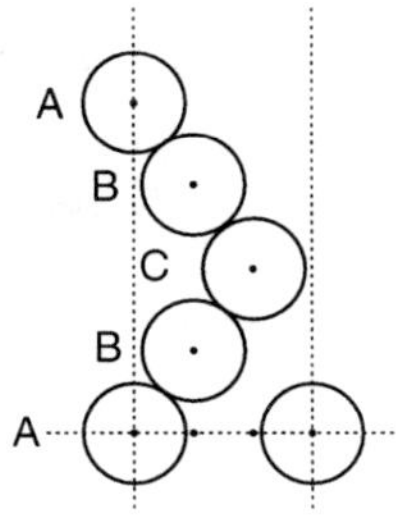

Fig. 3.27. Stacking sequence at a twin in the f.c.c. lattice. The $[\bar{2}11]$ direction is horizontal, the [111] direction is vertical

$$\ldots\text{ABCAB C BACBA}\ldots$$
$$\uparrow$$

The twin plane is labeled by the arrow. The view of the stacking sequence corresponding to Fig. 3.24 is presented in Fig. 3.27. Twins can be formed in small steps by the motion of partial dislocations on neighboring planes

$$\begin{array}{ccccccccccc}
\ldots & A & B & C & A & B & C & A & B & C & \ldots \\
 & | & | & | & | & C & A & B & C & A & \\
 & | & | & | & | & | & B & C & A & B & \\
 & | & | & | & | & | & | & A & B & C & \\
 & | & | & | & | & | & | & | & C & A & \\
 & | & | & | & | & | & | & | & | & B &
\end{array}$$

The formation of twins is an alternative way to realize plastic deformation. A twin lamella T in the intermetallic alloy TiAl will be shown in Fig. 9.8. The series of partial dislocations creating the twin lamella are clearly visible. The superdislocations S will be described in Sect. 9.3.

3.3.4 Antiphase Boundaries

Alloys consist of more than one type of atoms, e.g., atoms A and B. These alloys may crystallize in an ordered structure like the ordered intermetallic alloys. For example, the ordered alloy Ni_3Al with the $L1_2$ structure is based on the f.c.c. structure of the disordered phase. In the ordered state, however, all atoms at the edges of the cube belong to one atom type (Al), and all atoms in face centered positions belong to the other one (Ni). The Burgers vector $1/2\langle 110\rangle$ is a perfect Burgers vector in the disordered state but in the ordered one, it connects an A atom with a B atom. Dislocations with such a Burgers vector are called superpartial dislocations. The perfect Burgers vector in this direction of the superdislocation is therefore $\langle 110\rangle$. Figure 3.28 presents a piece of a crystal consisting of a primitive lattice of A atoms on the left side and an ordered region of A and B atoms on the right one. A perfect

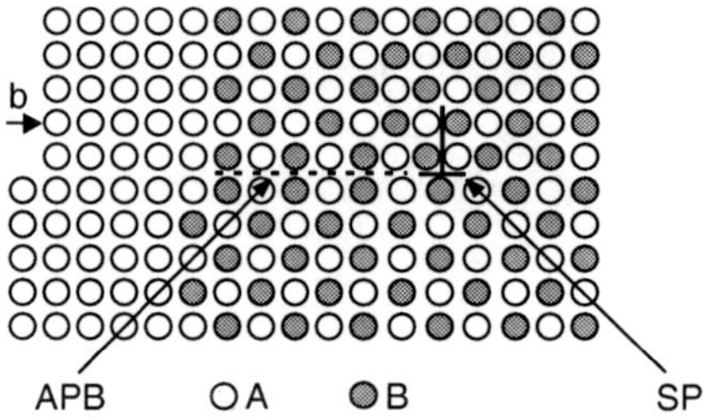

Fig. 3.28. Schematic of a dislocation having moved in an ordered region as a superpartial dislocation forming an antiphase boundary

dislocation in the primitive lattice has moved to the right into the ordered region. There, it has become a superpartial dislocation SP changing A atoms into B positions. While in the regular ordered structure, atoms are always neighbored by unlike atoms, now in the wake of the superpartial dislocation like atoms are in nearest neighbor positions. This is a fault called antiphase boundary or APB. Antiphase boundaries are two-dimensional faults like the stacking faults. In this case, the displacement vector $\boldsymbol{R}_{\mathrm{F}}$ is a lattice vector in the disordered structure, but not in the ordered one. In the latter, superdislocations frequently split into superpartials enclosing an antiphase boundary. Their width is controlled similarly to that of partial dislocations enclosing a stacking fault. The splitting of dislocations into superpartial dislocations essentially influences the properties of dislocations in ordered intermetallic alloys (Chap. 9).

4

Dislocation Motion

The character of the dislocation motion and the relation between the average velocity and the applied stress are controlled by the interplay between the forces on the dislocation segments, which vary in space and time, and the spectrum of barriers to the dislocation motion. The barriers are very different with respect to their geometric extension and to their interaction strength, ranging from arrangements of other dislocations with long-range stress fields to individual foreign atoms, which interact with the moving dislocations only in a very limited range. The character of overcoming these obstacles depends on the total energy necessary to surmount the obstacles and their local extension as well as on the external parameters like stress and temperature. If the energy is smaller than about $40kT$, where k is Boltzmann's constant and T the absolute temperature, and if the number of participating atoms is smaller than a few hundred so that collective vibrations are possible, the overcoming can be supported by thermal activation. In this case, the process strongly depends on the temperature. In the other case, the whole energy has to be expended by the external stress. Then, the parameters depend only weakly on the temperature.

In the following, thermal activation is described by the example of a dislocation segment locally pinned as in Fig. 3.20b. However, this process is far more general and plays a role in almost all situations of dislocation motion. Afterwards, the motion of a dislocation in the intrinsic periodic field of the crystal lattice is treated followed by the different kinds of obstacles to dislocation motion from single foreign atoms over precipitates to other dislocations. So far, the obstacles are regarded as fixed in the crystal. At high temperatures, dislocation motion is influenced by diffusion. Self-diffusion processes under the external stress or chemical stresses from nonequilibrium concentrations of point defects may cause the dislocations to climb. Foreign atoms may segregate to the dislocation cores and move with them. Finally, dislocation mobility laws will be discussed, which are applicable to sets of phenomenological equations describing plastic deformation.

4.1 Thermally Activated Overcoming of Barriers

The theory of the thermally activated overcoming of obstacles to dislocation motion was based on the thermodynamic treatment of viscous flow by Eyring [121]. A correct thermodynamic formulation was given by Schoeck [122]. The basic formulae are demonstrated here for the case of the interaction between gliding dislocations and localized obstacles, as described in Sect. 3.2.6, and are therefore formulated in terms of shear stresses τ and shear strain rates $\dot{\gamma}$. They can, however, analogously be applied to other situations of loading like normal stresses at problems of climb.

Figure 3.20b shows that a gliding dislocation bows out under the action of a shear stress between the obstacles resulting in a force on them, for small bow-outs given by (3.48). In general, the force a bowed-out dislocation segment of length l exerts is found by a work argument. If the point of action of the dislocation on the obstacle is shifted forward by $\mathrm{d}x$, where x is the coordinate in forward direction, the work done is $F\,\mathrm{d}x$. At the same time, the dislocation segment of length l is also shifted forward, supported by the external stress τ^*. The * indicates that an "effective" locally acting stress is meant. It will be discussed in Sect. 5.2 that

$$\tau^* = \tau - \tau_\mathrm{i}, \tag{4.1}$$

where τ is the applied shear stress and τ_i is a long-range internal stress. The gain in energy from the acting stress is

$$\mathrm{d}W = \tau^* bl\,\mathrm{d}x = F\,\mathrm{d}x,$$

which equals the energy spent on the obstacle. Thus, the force on the obstacle is

$$F = \tau^* b\, l, \tag{4.2}$$

independent of the strength and configuration of the bow-outs. If the force on the obstacle is continuously increased, the point of attack will be shifted as described by the force–distance curve of Fig. 4.1. This curve describes the elastic response to the force acting. For a certain value of F, the dislocation is in equilibrium at the position x_e on the entrance side of the obstacle. If the force reaches the maximum value of F, the obstacle strength F_max, the obstacle is spontaneously surmounted by the action of the stress τ^*. This situation holds at temperature $T = 0\,\mathrm{K}$. At all lower forces or stresses, the dislocation rests in elastic equilibrium at x_e and can surmount the obstacle only by the aid of thermal activation. If this happens, the dislocation reaches the position x_a on the exit side and is then free to move to the next obstacle. The equilibrium positions x_e (stable) and x_a (unstable) depend on the acting force. The (Helmholtz) free energy ΔF necessary to overcome the obstacle at the actual force F is the integral over the force–distance curve from x_e to

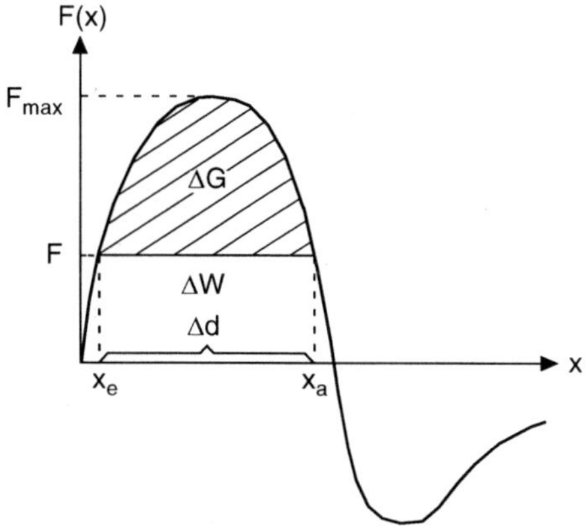

Fig. 4.1. Force–distance curve during the overcoming of a localized obstacle

x_a. (The free energy ΔF should not be mixed up with the force F.) Like the dislocation moves under the external force F, the work $\Delta W = (x_e - x_a)F$ is gained from the external force or stress. This work corresponds to the rectangle below the curve and is called the work term. The remaining part ΔG has to be supplied by thermal activation. It is the Gibbs free energy of activation for processes at constant pressure and temperature, and corresponds to the hatched area in the figure. Thus,

$$\begin{aligned}\Delta G &= \int_{x_e}^{x_a} F(x)\,\mathrm{d}x - (x_a - x_e)F, \\ &= \Delta F(F) - \Delta d\,F, \\ &= \Delta F(\tau^*) - \Delta d l b\,\tau^*, \\ &= \Delta F(\tau^*) - V\,\tau^*. \end{aligned} \tag{4.3}$$

The quantity Δd is an effective depth of the obstacle, which is called the activation distance. Frequently, the force–distance curve is expressed as $F(\Delta d)$, for example, in (3.35) for the Fleischer-type interaction between dislocations and defects with a tetragonal stress field. The area $\Delta d\,l$ is the activation area. It is the area swept by the dislocation during the activation event. Finally,

$$V = \Delta d\,lb = \Delta A\,b \tag{4.4}$$

is the activation volume. Independent of the microscopic interpretation of the introduced activation parameters of surmounting localized obstacles, these quantities have a well-defined thermodynamic meaning, valid also for other processes. The thermodynamic definition of the activation volume is

$$V = -\frac{\partial \Delta G}{\partial \tau^*}. \tag{4.5}$$

This is true although the integration limits and ΔF in (4.3) depend on the force or stress. The Gibbs free energy of activation determines the rate of overcoming the barriers by a Boltzmann factor

$$\frac{1}{t_\mathrm{w}} = \nu_0 \exp\left(-\frac{\Delta G(\tau^*)}{kT}\right), \tag{4.6}$$

where t_w is the average waiting time of the dislocation at the obstacles before their thermally activated overcoming, and ν_0 is an attempt frequency. Considering that the dislocation acts as a vibrating string, the attempt frequency depends on the vibrating segment length l as [123]

$$\nu_0 = b\nu_\mathrm{D}/l.$$

Here, ν_D is the Debye frequency of about $10^{13}\,\mathrm{s}^{-1}$. A more sophisticated treatment was given in [124] yielding typically

$$\nu_0 \approx< 10^{11}\,\mathrm{s}^{-1}.$$

For a more detailed discussion see [9]. To obtain the dislocation velocity, the activation rate of (4.6) has to be multiplied by the distance λ of forward motion after successful activation. Accordingly,

$$v_\mathrm{d} = \frac{\lambda}{t_\mathrm{w}} = \lambda\nu_0 \exp\left(-\frac{\Delta G(\tau^*)}{kT}\right) = v_{\mathrm{d}0} \exp\left(-\frac{\Delta G(\tau^*)}{kT}\right). \tag{4.7}$$

This is the Arrhenius equation of the dislocation velocity. For the considered array of localized obstacles, the jump distance λ is approximately equal to the obstacle distance or segment length l. Combining (4.7) with the Orowan equation for the strain rate (3.5) yields the Arrhenius equation of the strain rate

$$\begin{aligned}\dot\gamma &= b\varrho_\mathrm{m}\lambda\nu_0 \exp\left(-\frac{\Delta G(\tau^*)}{kT}\right) \\ &= \dot\gamma_0 \exp\left(-\frac{\Delta G(\tau^*)}{kT}\right).\end{aligned} \tag{4.8}$$

$\dot\gamma_0$ is the pre-exponential factor combining the parameters in front of the exponential function.

Arrhenius relations are applied to many problems of dislocation dynamics. The thermodynamic meaning of the activation parameters ΔF, ΔG, and V is always the same, whereas the microscopic structural interpretation of, for example, the activation volume as $V = \Delta d\,lb$ is valid only for the model of overcoming localized obstacles. The latter is further developed in Sect. 4.5 considering the statistical problems related to the nonregular arrangement of the obstacles.

The activation parameters can be measured by macroscopic deformation tests. The experimental techniques are reviewed in Sect. 2.1, pointing out that the most convenient methods are differential tests with instantaneous changes of the deformation parameters and/or stress relaxation tests. The relation between the (normal) stress σ measured in the tests and the resolved shear stress used above is given by the Schmid relation $\tau = m_s\sigma$ (2.3) with the orientation factor m_s. For the instantaneous changes, the internal stress τ_i is considered to remain constant so that the applied stress changes are equal to the changes of the locally acting stress, $\partial\tau = \partial\tau^*$. These relations will be discussed in Sect. 5.2.2. Differentiating (4.8) with respect to τ and taking (2.8) into account yields the activation volume

$$\frac{1}{m_s r} = \left.\frac{\partial \ln(\dot{\gamma}/\dot{\gamma}_0)}{\partial \tau}\right|_{T=\mathrm{const}} = -\frac{1}{kT}\left.\frac{\partial \Delta G}{\partial \tau^*}\right|_{T} = \frac{V}{kT}, \tag{4.9}$$

where r is the strain rate sensitivity. Considering the differentiation rules for integrals shows that this relation is valid although ΔF and the integration limits in (4.3) depend on τ^* or τ, respectively.

Frequently, the so-called stress exponent m is used to characterize the dynamic dislocation behavior. It is based on a phenomenological power law relation between the effective stress and the strain rate

$$v_\mathrm{d} = \mathrm{const} \times \tau^{*m} \quad \text{or} \quad \dot{\gamma} = \mathrm{const} \times \tau^{*m}. \tag{4.10}$$

Thus,

$$m = \frac{\partial \ln v_\mathrm{d}}{\partial \ln \tau^*} = \frac{\partial \ln \dot{\gamma}}{\partial \ln \tau^*} = \frac{\tau^* \partial \ln \dot{\gamma}}{\partial \tau^*} = \frac{\tau^*}{m_s r} = \frac{\sigma^*}{r} = \frac{V\tau^*}{kT}. \tag{4.11}$$

The first equality sign is valid only if the preexponential factor $\dot{\gamma}_0$ remains constant during the measurement. The work term $V\tau^*$ is the work by which the activation energy is reduced owing to the action of the stress. As usually $\tau^* \neq \tau$, the true stress exponent m has to be distinguished from the experimental one

$$m' = \frac{\partial \ln \dot{\gamma}}{\partial \ln \tau} = \frac{\tau}{m_s\, r} = \frac{\sigma}{r}. \tag{4.12}$$

In contrast to the activation volume, the relevant energy parameters are not obtained directly. Differentiating (4.8) with respect to T leads to

$$\left.\frac{\partial \ln(\dot{\gamma}/\dot{\gamma}_0)}{\partial T}\right|_{\tau} = -\frac{1}{kT}\left.\frac{\partial \Delta G}{\partial T}\right|_{\tau} + \frac{\Delta G}{kT^2}$$

or

$$Q_\mathrm{e} = kT^2 \left.\frac{\partial \ln(\dot{\gamma}/\dot{\gamma}_0)}{\partial T}\right|_{\tau} = \Delta G + T\Delta S = \Delta H, \tag{4.13}$$

since $\partial\Delta G/\partial T = -\Delta S$ is the activation entropy. Thus, the experimental activation energy Q_e, which is obtained from a temperature change test in a creep experiment with constant stress, equals the activation enthalpy ΔH. In a quasistatic test with constant strain rate,

$$Q_{\mathrm{e}} = \Delta H = -kT^2 \left.\frac{\partial \tau}{\partial T}\right|_{\dot{\gamma}} \left.\frac{\partial \ln(\dot{\gamma}/\dot{\gamma}_0)}{\partial \tau}\right|_T = -\frac{kT^2}{r} \left.\frac{\partial \sigma}{\partial T}\right|_{\dot{\gamma}} . \tag{4.14}$$

The contributions to the activation entropy arise mainly from the fact that all elastic interactions, that is, both the long-range interactions causing τ_{i} as well as the short-range interactions determining the activation energy, are proportional to a relevant elastic constant (mostly the shear modulus μ), which depends on the temperature. If this is taken into account in (4.3) and it is considered that the derivatives of the integration limits are zero because of the equilibrium, it follows that

$$\begin{aligned}\frac{\partial \Delta G}{\partial T} &= \frac{1}{\mu}\frac{\partial \mu}{\partial T}\left(\Delta F + \tau_{\mathrm{i}} V\right) \\ &= \frac{1}{\mu}\frac{\partial \mu}{\partial T}\left(\Delta G + \tau V\right).\end{aligned}$$

Thus,

$$Q_{\mathrm{e}} = \Delta G - \frac{T}{\mu}\frac{\partial \mu}{\partial T}\left(\Delta G + \tau V\right)$$

or

$$\Delta G = \frac{Q_{\mathrm{e}} + \frac{T}{\mu}\frac{\partial \mu}{\partial T}\tau V}{1 - \frac{T}{\mu}\frac{\partial \mu}{\partial T}} . \tag{4.15}$$

This equation, derived by Schoeck [122], contains only measurable quantities, not, e.g., τ^*. As $\partial\mu/\partial T < 0$, it follows that the Gibbs free energy is smaller than the experimental one, $\Delta G < Q_{\mathrm{e}}$. To determine the interaction potential of a special process controlling the dislocation mobility, it is necessary to measure V and ΔG over a range of the stress τ as wide as possible. This can be done by selecting wide ranges of the strain rate $\dot{\gamma}$ and the temperature T. However, it is then mostly questionable whether the preexponential factor $\dot{\gamma}_0$ remains constant as supposed in the derivations of the formulae. According to (4.8), ΔG describes the preexponential factor for certain values of $\dot{\gamma}$ and T. At constant $\dot{\gamma}$, ΔG should be proportional to T. This may be used as a test on the constancy of $\dot{\gamma}_0$ and the dominance of a single process controlling the dislocation motion. As will be shown later, the different mechanisms imply different values of the activation volume and the activation energies. Thus, measuring these parameters may help identify the mechanisms of dislocation mobility.

4.2 Lattice Friction

Owing to the periodic lattice potential, a dislocation moving in a crystal experiences an intrinsic friction stress. As outlined in Fig. 4.2a, without external stress, an edge dislocation is in a symmetric configuration with the minimum energy W_0. When the dislocation moves in x direction, the configuration

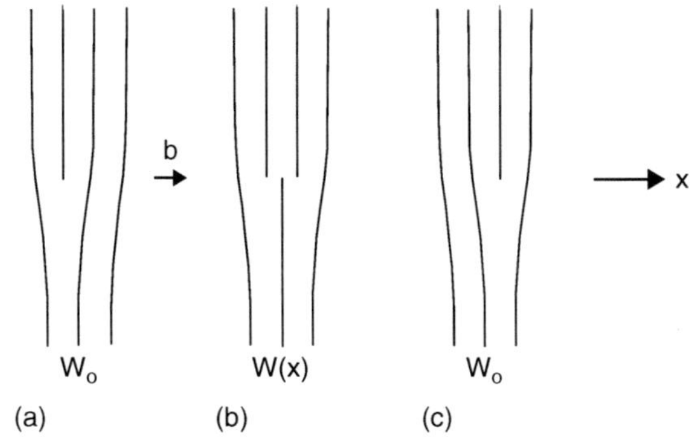

Fig. 4.2. Shifting a dislocation through the lattice with two symmetric configurations

becomes asymmetric. Accordingly, the energy increases and is a function of the dislocation position. To increase the dislocation energy requires a stress. In the middle of the way to the next equivalent minimum energy position, at a distance of the Burgers vector b, another symmetric position occurs, as sketched in Fig. 4.2b. Finally, the dislocation reaches the next minimum energy position as in Fig. 4.2c.

Thus, in a crystal, a straight dislocation moves in a periodic lattice potential, which causes a frictional stress. This stress is called the Peierls–Nabarro stress. However, if a straight dislocation moves from one minimum energy position, a so-called Peierls valley, to the next one as a whole, many bonds have to be broken and restored simultaneously, still requiring a relatively high energy. This energy can be expended at small steps by the sidewise motion of kinks. Starting with a straight dislocation, as a first step, a kink pair has to be formed, which requires the kink pair formation energy. Afterwards, the kinks of opposite sign of the pair spread in opposite directions slowly shifting the whole dislocation to the next Peierls valley. As the kinks also move in a periodic lattice potential, each step of motion needs a respective energy. Up to this point, the necessary energy was supplied by the acting mechanical stress. However, the kink formation and migration energies are usually not very high. Besides, only a small number of atoms is involved in an elementary step of the kink motion. Consequently, the kink formation and migration can be supported by thermal activation, resulting in a strongly temperature-dependent friction stress. In the following, these processes will be outlined mainly following the detailed treatment in [12].

4.2.1 Peierls–Nabarro Model

To model the motion of a straight dislocation in the periodic lattice potential, Dehlinger and Kochendörfer [125] suggested to apply the calculations of Frenkel and Kontorova [126] of a one-dimensional array of balls connected

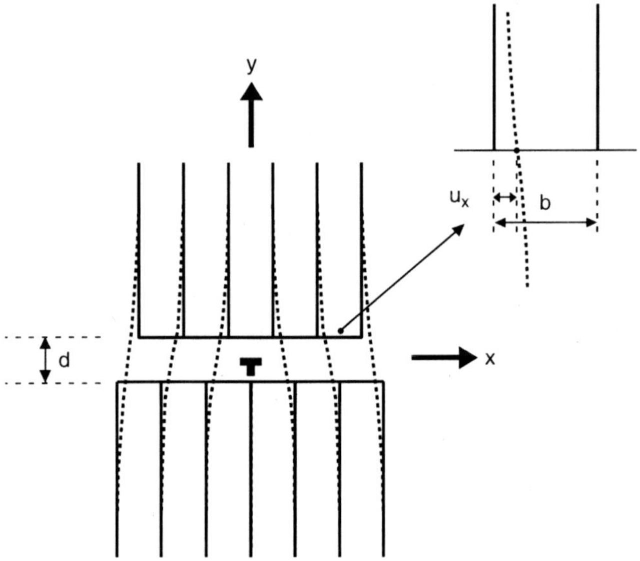

Fig. 4.3. Joining two semi-infinite elastically strained crystals to form an edge dislocation

by springs and moving on a periodic substrate. At the same time, Orowan suggested the problem of dislocation motion in a periodic potential relief to Peierls who solved it in analytical form [127]. Later on it was extended by Nabarro [128].

In the Peierls–Nabarro model, an edge dislocation is simulated by two elastic half-crystals, which are shifted with respect to each other by half the Burgers vector b as outlined in Fig. 4.3 by the two sets of parallel lattice planes. The half-crystals are positioned at a distance d. To join the lattice planes of the two half-crystals, these planes have to be displaced on the glide plane in the middle between the two half-crystals ($y = 0$) by

$$\begin{aligned} 2u_x(x) + b/2 \qquad & \text{for} \qquad x > 0 \\ 2u_x(x) - b/2 \qquad & \text{for} \qquad x < 0 \end{aligned}$$

with respect to the bottom half-crystal. The displaced and joined lattice planes are represented by dotted lines in Fig. 4.3. Far from the dislocation, the displacements have to fulfil the boundary condition $u_x(\infty) = -u_x(-\infty) = -b/4$. To realize the displacements, internal stresses have to be applied. Similar to the Frenkel treatment of the theoretical shear strength of Sect. 1.1, it is assumed that the stresses depend on the displacements via a sinusoidal function and that the displacements in y direction are comparatively small

$$\sigma_{xy}(x, 0) = -K \sin \frac{4\pi u_x}{b}.$$

K is a constant, which is determined by fulfilling Hooke's law for small strains

$$\mu = \frac{1}{2}\frac{\partial\sigma_{xy}}{\partial\varepsilon_{xy}} = \frac{d}{2}\frac{\partial\sigma_{xy}}{\partial u_x},$$

where $\varepsilon_{xy}(x,0) = (2u_x + b/2)/(2d)$. For $u_x = 0$, it follows that

$$\sigma_{xy}(x,0) = -\frac{\mu b}{2\pi d}\sin\frac{4\pi u_x}{b}. \tag{4.16}$$

As found out by Eshelby [129], this shear stress distribution can be realized by a continuous distribution of infinitesimal edge dislocations along the x axis, leading to an integral equation with the solution

$$u_x = -\frac{b}{2\pi}\arctan\frac{2x(1-\nu)}{d}. \tag{4.17}$$

This relation fulfils the boundary condition stated above. Besides, $u_x(d/[2(1-\nu)]) = -b/8 = (1/2)u_x(\infty)$. Accordingly, in the range $-d/[2(1-\nu)] < x < d/[2(1-\nu)]$, the disregistry is greater than half its maximum value of $-b/4$ at $x = \infty$. This is a possibility to define a width of the dislocation within which linear elasticity theory is certainly not valid. The width replaces the core region of the Volterra dislocation, which is cut out by the inner cut-off radius. The method of a continuous distribution of infinitesimal edge dislocations also allows the calculation of the whole stress field of the Peierls dislocation. The relations reduce to the stress field of the Volterra dislocation (3.11) for distances great with respect to $d/[2(1-\nu)]$. The Peierls dislocation does not show the divergence of the stresses in the core of the Volterra dislocation.

The energy of the Peierls dislocation is composed of the elastic energy of the two half-crystals and the misfit energy of the atom rows between them. For comparing the result with that of the Volterra dislocation, the elastic energy can be calculated in two half-cylinders, which have to be deformed by surface forces to yield the displacements u_x. Following the original treatments of Peierls [127] and Nabarro [128], the result is

$$\begin{aligned} W(x) &= \frac{\mu b^2}{4\pi(1-\nu)} + \frac{\mu b^2}{2\pi(1-\nu)}\exp -\frac{2\pi d}{b(1-\nu)}\cos\frac{4\pi x}{b} \\ &= \frac{\mu b^2}{4\pi(1-\nu)} + \frac{W_\mathrm{P}}{2}\cos\frac{4\pi x}{b}. \end{aligned} \tag{4.18}$$

W_p is called the Peierls energy and represents the magnitude of the periodic displacement potential. The dislocation energy varies with double the lattice periodicity b because of the two symmetrical positions in Fig. 4.2. However, (4.18) does not correspond to the expectation from Fig. 4.2 as the symmetrical positions turn out to be positions of maximum energy. This result originates from the method of summing up the energies of the individual lattice rows in the two original treatments. The method was improved by Hirth and Lothe

[130] omitting this problem. The maximum of the derivative of $W(x)$ with respect to the position x of the dislocation yields the stress necessary to move the dislocation over the potential barrier

$$\tau_{\mathrm{P}} = \frac{1}{b}\left[\frac{\partial W(x)}{\partial x}\right]_{\max} = \frac{2\pi W_{\mathrm{P}}}{b^2} = \frac{2\mu}{1-\nu}\exp -\frac{2\pi d}{b(1-\nu)}. \tag{4.19}$$

This is the Peierls–Nabarro stress or shortly Peierls stress to move a straight dislocation lying in a Peierls valley only by mechanical stress, that is, at very low temperatures. For $d \approx b$, it is in the range of $10^{-4} \ldots 10^{-2}\mu$, that means in the range of the real shear strengths of materials. The lattice resistance is controlled by the structure of the dislocation cores. Therefore, the Peierls–Nabarro model is too coarse for detailed predictions of the flow stress of crystals. Similar to the theoretical yield stress of perfect crystals in Sect. 1.1, the Peierls stress is small for small Burgers vectors b and large distances d between the lattice planes. According to the experimental results, the Peierls stress is small for close-packed structures like f.c.c. metals, where it does not play a role except at low temperatures. It is high for materials with directional bonds like the covalent semiconductor and ceramic crystals, which become plastic only at high temperatures. The b.c.c. metals show an intermediate behavior.

The Peierls–Nabarro model is refined in many ways: by relaxing the severe assumptions made and by using more realistic potentials than the sinusoidal one (e.g. [131]). More detailed information on the dislocation core configurations and energies as well as the Peierls stress can be obtained from atomistic calculations. The crystal is divided into an outer zone, which is treated by continuum elasticity theory and a core region where the interactions between the atoms are calculated discretely. The positions that are first determined by the elastic solution are then varied to find the equilibrium configurations. The results of these calculations depend very sensitively on the chosen interaction potentials between the atoms. The situation is comparatively simple in ionic crystals with Coulomb attraction, Born–Mayer repulsion, and van-der-Waals potentials. Early calculations of this kind were made in [132] for NaCl and in [133] for MgO. For metallic materials, the semi-empirical embedded-atom method (EAM) is quite successful (for a review see [134]). In recent studies of dislocation cores and the Peierls stress, details of the dislocation behavior were interpreted, for example, for intermetallic alloys [135] or on the influence of shear stresses in more than one slip plane and of nonglide stresses in b.c.c. Mo [136]. In a model cubic primitive lattice, the Peierls stress was calculated by molecular dynamics simulation [137]. Applying a sinusoidal interaction between the atoms, the Peierls potential also turns out to be close to sinusoidal, as assumed in most continuum models. As a general result of the atomistic calculations, it may be stated that the effective core radius introduced in Sect. 3.2.2, which is chosen to include the core energy within the elastic solution, is slightly less than $0.5b$ in ionic crystals, and $0.5b$ to a few b in metals.

Hirth and Lothe [12] suggest to use a phenomenological equation for the Peierls potential

$$W(x) = \frac{W_p}{2}\left(1 - \cos\frac{2\pi x}{h}\right) = W_p \sin^2\frac{\pi x}{h}, \tag{4.20}$$

where h either equals $b/2$ or b, depending on the degree of equivalence of the two equilibrium configurations. Accordingly, the Peierls stress becomes

$$\tau_p = \frac{\pi W_p}{bh}. \tag{4.21}$$

4.2.2 Double-Kink Model

In the same way in which the high energy necessary to shear two lattice planes with respect to each other as a whole is expended in small steps by the motion of a straight dislocation from one Peierls valley to the next, this shift can again be performed in small steps by the generation and sidewise spreading of kinks. At low stresses, the kinks may occur in their thermal equilibrium concentration. At higher stresses, they form under the action of the stress. For dislocations not lying in a Peierls valley, that is, not being parallel to a low-index crystallographic direction, kinks are necessary for geometric reasons. As shown in Fig. 4.4a, kinks may appear as single kinks K or kink pairs K^+K^-. Depending on the binding properties of the material, the kinks can be abrupt as in Fig. 4.4a or may have a defined width w as in Fig. 4.4b. If the kinks are treated within the framework of the line tension model, the dislocation changes in a smooth way from one Peierls valley to the next (Fig. 4.4c). The sidewise motion of the kinks along the dislocation as indicated by the arrows in Fig. 4.4c shifts the dislocation forward (upwards) in small steps.

Elastic Properties of Kinks

The properties of the kink formation depend on their energy, which can be estimated from continuum elasticity theory. Kink pairs or double-kinks of height h can be formed by putting a segment of length L of a straight dislocation into the neighboring Peierls valley. For kinks of small h with respect to L, the energy of the kink pair is given by

$$W_{fdk} = 2W_{fk} + W_{int}, \tag{4.22}$$

where W_{fk} is the formation energy of a single kink and W_{int} is the interaction energy between both kinks. Using the configuration of Fig. 4.4a of sharp kinks, the kink formation energy can be determined for kinks of height h large compared to the inner cut-off radius r_0. This condition is not fulfilled for elementary kinks. Since the long-range stress field of a dislocation is not disturbed remarkably by introducing a kink, the core energy of (3.16) of a

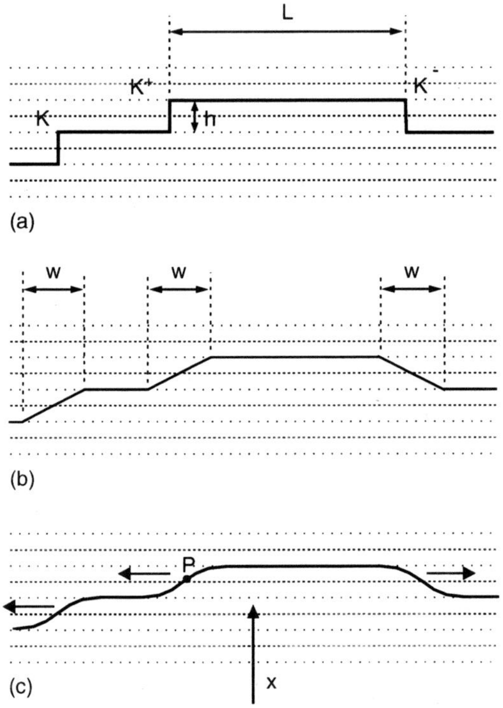

Fig. 4.4. Configurations of single kinks K and kink pairs K^+K^-. (**a**) Abrupt kinks. (**b**) Kinks of width w. (**c**) Smooth kinks. *Dotted lines*: Peierls valleys, *dashed lines*: Peierls hills

piece of dislocation of length h can be taken for the kink formation energy, $W_{\text{fk}} \approx (1/10)\mu b^2 h$. This problem does not occur for the interaction energy. For a kink pair in a screw dislocation, it is given by

$$W_{\text{int}} = -\frac{\mu b^2 h^2}{8\pi L}\frac{1+\nu}{1-\nu}. \tag{4.23}$$

There follows the attractive interaction force

$$F_{\text{int}} = \frac{\partial W_{\text{int}}}{\partial L} = \frac{\mu b^2 h^2}{8\pi L^2}\frac{1+\nu}{1-\nu}. \tag{4.24}$$

This expression can be verified by applying Brown's theorem (3.44). The same equation holds for the extended kink configuration in Fig. 4.4b as long as $L > w$. For a kink pair in an edge dislocation, the factor $1 + \nu$ has to be replaced by $1 - 2\nu$ (see Table 3.2 for the line tensions of screw and edge dislocations). It should be noted that the interaction force decreases quickly with $1/L^2$ in contrast to long dislocations where the interaction force decreases with the reciprocal distance (e.g. Sect. 3.2.4). The calculation of the energy of individual kinks is more complicated as they have to form at a surface so that image terms have to be considered.

The kink energy can also be determined using the line tension model in its simple version of constant line tension equal to the line energy E_d, for instance (3.17). The dislocation is assumed to have the smooth kink configuration of Fig. 4.4c. At some point P of the kink, the line tension acts as a force in tangential direction of the dislocation line. This force has a component in forward x direction. It is balanced at any point along the kink by the force $\partial W(x)/\partial x$ resulting from the Peierls potential, taken, for example, from (4.20), and by the force from the applied stress τ^*. Thus, the equilibrium condition is [138]

$$\frac{\mathrm{d}E}{\mathrm{d}x} = -E_\mathrm{d}\frac{\mathrm{d}^2x}{\mathrm{d}z^2} + \frac{\mathrm{d}W_\mathrm{P}}{\mathrm{d}x} - \tau^* b = 0. \tag{4.25}$$

The solution of the problem for zero stress yields

$$W_\mathrm{fk} = \frac{2h}{\pi}\sqrt{2W_\mathrm{p}E_\mathrm{d}}. \tag{4.26}$$

This formula, which expresses the kink formation energy in terms of the Peierls energy and the dislocation line energy, links the Peierls–Nabarro model for the motion of a straight dislocation at low temperature, that is, a quantity describing the periodic lattice potential, with the double-kink model of realistic dislocation motion. For the formation of the kinks at constant temperature, the free energy of kink formation ΔF_fk has to be used instead of the energy. However, the uncertainties of the elastic estimates above are greater than the entropy contribution so that $\Delta F_\mathrm{fk} \approx W_\mathrm{fk}$. Like the dislocation energy $W(x)$ in (4.18), the kink energy, too, shows a lattice periodicity resulting in an energy barrier to the motion of kinks. The respective potential is sometimes called the Peierls potential of second order.

Kinks in Thermal Equilibrium

At a finite temperature, the equilibrium configuration of a dislocation is not a straight line lying in a Peierls valley but a dislocation containing a certain concentration of kinks. Inside the crystal, the kinks can only be formed as kink pairs with equal numbers of positive and negative kinks and a well-defined free energy of formation $2\Delta F_\mathrm{fk}$ of the kink pairs. Their concentration is governed by a minimum in the free energy of the system, that is,

$$\frac{\partial \Delta F}{\partial n} \doteq 0, \tag{4.27}$$

if n is the number of kinks of one sign along a dislocation of length L. Each kink increases the free energy of the system by ΔF_fk, yielding the total increase

$$n\Delta F_\mathrm{fk}.$$

On the other hand, a configurational entropy occurs

$$\Delta S = k \ln P_{\mathrm{n}},$$

where P_{n} is the number of possibilities to arrange the n kinks of one sign at the respective number of possible kink sites $N = L/a$ inside L. For a single kink, $P_1 = N$. For two kinks, $P_2 = \frac{1}{2}N(N-1)$. For three kinks, $P_3 = \frac{1}{6}N(N-1)(N-2)$. The factors 1/2, 1/6, etc. consider that the individual kinks cannot be distinguished from each other. Thus, for n kinks of one sign

$$P_n = \frac{1}{n!}N(N-1)\ldots(N-n+1) = \frac{N!}{(N-n)!n!}. \tag{4.28}$$

For the change in the free energy by introducing n kink pairs, this yields

$$\Delta F = 2\left(n\Delta F_{\mathrm{fk}} - kT\ln P_{\mathrm{n}}\right).$$

Inserting (4.28) and using Stirling's formula for large n,

$$\ln x! = x\ln x - x$$

yields

$$\frac{\partial \Delta F}{\partial n} = \Delta F_{\mathrm{fk}} - kT \ln\frac{N-n}{n} \doteq 0$$

for the equilibrium condition (4.27). Introducing the kink concentrations of positive and negative kinks $c_+ = c_- = n/L$ and considering that $n \ll N$ finally results in an equilibrium concentration of kinks

$$c_+ = c_- = \frac{1}{a}\exp\left(-\frac{\Delta F_{\mathrm{fk}}}{kT}\right). \tag{4.29}$$

This derivation is typical also for a number of other processes where defects exist in thermal equilibrium. A very prominent case, which will be discussed in Sect. 4.10, is the occurrence of thermal point defects, vacancies, and interstitial atoms.

Thermally Activated Motion of Kinks

The kink formation energy, which frequently is of the order of magnitude of the dislocation core energy, is not very high, for simple metals it is only a fraction of one electron-volt. The kink migration energy is due to the periodic variation of the kink energy. This variation is sometimes called the secondary Peierls potential. It is mostly small so that the kink motion can be aided by thermal activation. Since the height of the elementary kinks is also small (in the order of b), the force acting from the applied stress is small, too. Therefore, the kink motion can be treated by the theory of diffusion. For a general introduction into diffusion see [139, 140]. The diffusion coefficient for the diffusion of kinks along the dislocation line is defined by

$$D_{\mathrm{k}} = a^2 \omega, \tag{4.30}$$

where a is the jump distance of the kink and ω the jump frequency. When a small force F acts on the kink, the jump frequency in forward direction becomes higher than that in backward direction, resulting in a drift velocity, which is given by the Einstein relation

$$v_{\mathrm{k}} = \frac{D_{\mathrm{k}}}{kT} F. \tag{4.31}$$

In general, the force F can originate from a mechanical stress, but as thermodynamic force also from a chemical potential $\mu = kT \ln(c/c_0)$, where c is the concentration of some species and c_0 its equilibrium value. The force is then

$$F = -\frac{\partial \mu}{\partial x} = -\frac{kT}{c}\frac{\partial c}{\partial x}. \tag{4.32}$$

Equations (4.31) and (4.32) together yield Fick's first law

$$J = c v_{\mathrm{k}} = -D_{\mathrm{k}} \frac{\partial c}{\partial x}. \tag{4.33}$$

Here, J is the flux or current density. The frequency ω to move a kink over the activation barrier is given by

$$\omega \approx \nu_0 \exp\left(-\frac{\Delta F_{\mathrm{mk}}}{kT}\right). \tag{4.34}$$

In this case, the attempt frequency ν_0 is approximately equal to the Debye frequency $\nu_{\mathrm{D}} = 10^{11} \ldots 10^{12}\,\mathrm{s}^{-1}$. ΔF_{mk} is the free energy of kink migration. For kink motion, the force in (4.31) results from a shear stress τ acting on the slip system of the dislocation, that is,

$$F = \tau b h.$$

Correctly speaking, τ is the locally acting stress. The superscript * is dropped in the following considerations. Together with (4.30), (4.31), and (4.34), this relation yields the velocity of the kink along the dislocation

$$v_{\mathrm{k}} = D_{\mathrm{k}} \frac{\tau b h}{kT} \approx \nu_{\mathrm{D}} a^2 \exp\left(-\frac{\Delta F_{\mathrm{mk}}}{kT}\right) \frac{\tau b h}{kT}. \tag{4.35}$$

Except in crystals with strong directional bonds, the kink migration energy is quite small and much smaller than the kink formation energy. Therefore, kinks often move very fast so that their velocity is controlled by high speed damping processes, which will briefly be described in Sect. 4.12. Under certain conditions, the diffusion coefficient of kinks is then constant, approximately equal to the limiting case of $\Delta F_{\mathrm{mk}} \ll kT$, but this coincidence is fortuitous.

To conclude the dislocation velocity v_{d} from the kink velocity, the concentration of kinks c_{k} must be known

$$v_{\mathrm{d}} = hc_{\mathrm{k}}v_{\mathrm{k}}. \tag{4.36}$$

There are different situations depending on the magnitude of the force acting on the kinks, or on the applied stress. For small stresses, the kink concentration does not differ remarkably from the thermal equilibrium concentration in (4.29). For high stresses, kink pairs are formed by stress-assisted thermal activation so that the concentration of kinks is much higher than in the small stress regime. In the latter case, (4.29) and (4.35) can be inserted into (4.36). Considering further that $c_{\mathrm{k}} = c_+ + c_-$ yields an expression for the dislocation velocity in the small stress region,

$$v_{\mathrm{d}} = \frac{2\tau bh^2}{akT} D_{\mathrm{k}} \exp\left(-\frac{\Delta F_{\mathrm{fk}}}{kT}\right). \tag{4.37}$$

Double-Kink Nucleation at High Stresses

At high stresses, kink pairs are formed by stress-assisted thermal activation. Only kink pairs moving the dislocation forward, which are favored by the applied stress, need to be considered as kink pairs moving the dislocation backwards will immediately shrink and annihilate. Because of the mutual elastic interaction energy of the kinks of the pair (4.23), which decreases with increasing kink separation L, and the increasing energy gained from the external stress, the formation energy of a double-kink ΔF_{fdk} will show a maximum at a critical separation L^* of the kinks, as schematically shown in Fig. 4.5. At the critical separation, the attractive kink interaction force is balanced by the force τbh due to the external stress, which drives the kinks apart. Kink pairs with separations smaller than L^* collapse with a gain in energy. Those with $L > L^*$ tend to separate further. However, kink pairs close to L^* can still annihilate so that only those which are wider than L' are really stable. The energy difference between the states at L^* and L' is in the order of magnitude of kT. This behavior of the free energy of double-kink formation and the related nucleation and annihilation properties very much resemble the steady state nucleation theory [141–145] as it was developed, for example, for crystal growth [146]. With some approximations, the result of this theory yields the double-kink nucleation rate

$$J_{\mathrm{dk}} = \frac{\tau bh}{a^2kT} D_{\mathrm{k}} \exp\left(-\frac{\Delta F^*_{\mathrm{fdk}}}{kT}\right). \tag{4.38}$$

$\Delta F^*_{\mathrm{fdk}}$ is the maximum value of the double-kink formation energy at the critical separation L^*. To obtain the dependence of the double-kink formation

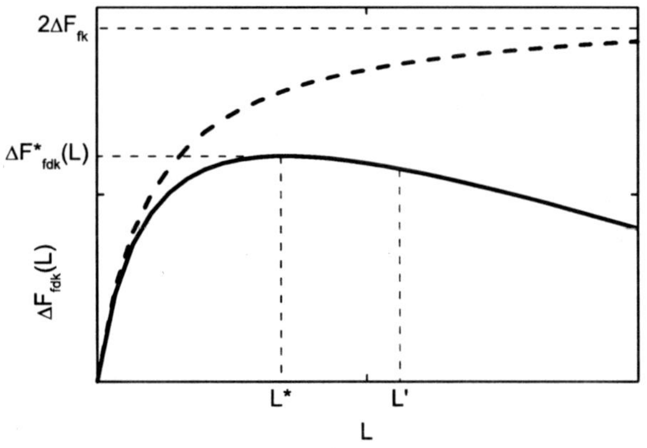

Fig. 4.5. Schematic plot of the dependence of the double-kink formation energy on the separation of the kinks. *Dashed line*: without external stress. *Solid line*: with external stress. See text below

energy $\Delta F_{\mathrm{fdk}}(L)$ on the kink separation as depicted in Fig. 4.5, both the contribution of the kink interaction energy (4.23) introduced by Seeger and Schiller [147, 148] and the work done by the external stress have to be considered. With average data of screw and edge dislocations, the formation free energy for kink pairs of a separation larger than the kink width w can be written as

$$\Delta F_{\mathrm{fdk}}(L) = 2\Delta F_{\mathrm{fk}} - \frac{\mu b^2 h^2}{8\pi L} - \tau b h L. \tag{4.39}$$

The critical separation L^* of the kink pairs can be obtained by

$$\left[\frac{\partial \Delta F_{\mathrm{fdk}}}{\partial L}\right]_{L=L^*} = 0,$$

yielding

$$L^* = \sqrt{\frac{\mu b h}{8\pi\tau}}.$$

Inserting this into (4.39) results in the critical double-kink formation energy

$$\Delta F^*_{\mathrm{fdk}} = 2\Delta F_{\mathrm{fk}} - \sqrt{\frac{\mu\tau b^3 h^3}{2\pi}}. \tag{4.40}$$

Evaluations of experimental data are often based on this formula, c.f. [149]. For a relatively low stress of $\tau = 10^{-4}\mu$ and $h = b$, the critical kink separation is $L^* = 20b$. Thus, the kinks are well separated. Using (4.19) and (4.26), a Peierls stress of $\tau_{\mathrm{p}} = 10^{-2}\mu$, and a dislocation energy of $E_{\mathrm{d}} = \mu b^2/2$ (3.17), the formation energy of two separated kinks becomes $2\Delta F_{\mathrm{fk}} = 10\sqrt{8}b^3\mu/\pi^{3/2}$. Considering that $\mu b^3 = 5\,\mathrm{eV}$, $2\Delta F_{\mathrm{fk}} \approx 0.25\,\mathrm{eV}$. Compared to this value, the last term in (4.40) amounts to only 10% of the total energy. Thus, for this relatively low stress, the Seeger–Schiller term in (4.40) is not very important so that $\Delta F^*_{\mathrm{fdk}} = 2\Delta F_{\mathrm{fk}}$ is a sufficient approximation, as suggested in [150],

$$J_{\mathrm{dk}} = \frac{\tau b h}{a^2 kT} D_{\mathrm{k}} \exp\left(-\frac{2\Delta F_{\mathrm{fk}}}{kT}\right). \tag{4.41}$$

For high stresses, the kinks of the pair are not clearly separated so that the elastic formalism with the separation of kink formation and interaction energies cannot be applied. Then the kink pair has to be treated as a small bow-out considering the elastic energy and the potential energy of the Peierls relief. Such theories [151, 152] lead to different stress dependencies of the kink-pair formation energy. In the high-stress limit, with τ being in the vicinity of the Peierls stress τ_{p}, the stress dependence of the activation energy becomes independent of the particular form of the Peierls potential, that is, $\Delta F_{\mathrm{fdk}} \propto (\tau_{\mathrm{p}} - \tau)^{5/4}$ [153, 154]. This effect was confirmed experimentally, for example, in [155, 156]. For a review see [157]. In the usual double-kink model, the dislocation velocity is constant. However, at the presence of foreign atoms, the traveling distance Δx may not be proportional to the time Δt. Instead, $\Delta x \propto \Delta t^{\delta}$ with $\delta < 1$ [158]. This kind of kinetics is now being discussed in different fields of physics, chemistry, and biology, see for example [159].

To summarize the above, double-kink formation is a process operating on a level between the atomic one and that of elasticity theory. The elastic description applied is therefore an approximation. In particular, the free energies of activation can be replaced by the energies as discussed above in the section of elastic properties of kinks. For low stresses, the double-kink formation energy can conveniently be replaced by twice the formation energy of isolated kinks so that the formation rate is given by (4.41). For higher stresses, the stress dependence of the formation energy according to the Seeger–Schiller approach (4.40) should be used. In the high-stress range, other potentials after [151,152] have to be applied. However, the experimental decision between these different cases is quite difficult.

Dislocation Velocity in the Range of Double-Kink Nucleation

A straight dislocation in a Peierls valley is considered, which moves forward by double-kink nucleation, by the spreading of the kinks of the pair in opposite directions, followed by the mutual annihilation with kinks of the neighboring kink pairs after sweeping a distance λ, or by the storage of kinks at obstacles, which limit a dislocation segment of length λ_0. The dislocation velocity is controlled by the shorter distance of both. A suitable average between both cases is given by

$$v_{\mathrm{d}} = h \frac{\lambda \lambda_0}{\lambda + \lambda_0} J_{\mathrm{dk}}. \tag{4.42}$$

To determine the sweeping distance λ of the kink pairs, their lifetime t^* is considered. It is given by

$$t^* = \frac{\lambda}{2v_{\mathrm{k}}}. \tag{4.43}$$

v_{k} again is the kink velocity of (4.35). To ensure a steady state motion of the dislocation, a new kink pair has to be created during the lifetime t^* along a length $\lambda/2$ of the dislocation. Thus,

$$t^* = \frac{2}{\lambda J_{\mathrm{dk}}}. \tag{4.44}$$

Solving for λ and using (4.35) and (4.38) yields

$$\lambda = 2\sqrt{\frac{v_{\mathrm{k}}}{J_{\mathrm{dk}}}} = 2a \exp\left(\frac{\Delta F_{\mathrm{fk}}}{kT}\right). \tag{4.45}$$

After (4.42), the dislocation velocity is given by

$$v_{\mathrm{d}} = 2h\sqrt{v_{\mathrm{k}} J_{\mathrm{dk}}} = \frac{2\tau b h^2}{akT} D_{\mathrm{k}} \exp\left(-\frac{\Delta F_{\mathrm{fk}}}{kT}\right) \tag{4.46}$$

for the case of long dislocations ($\lambda_0 \gg \lambda$). This equation is identical with (4.37) for the drift of thermal kinks at low stresses, and therefore universal over a wide range of stresses. For short dislocations, that is, $\lambda_0 \ll \lambda$, the dislocation velocity becomes

$$v_{\mathrm{d}} = h\lambda_0 J_{\mathrm{dk}} = \frac{\tau b h^2 \lambda_0}{a^2 kT} D_{\mathrm{k}} \exp\left(-\frac{2\Delta F_{\mathrm{fk}}}{kT}\right). \tag{4.47}$$

There are some remarkable differences between the two cases of (4.46) and (4.47). At long dislocations, kinks store at the obstacles limiting the segments so that these segments bow out while short dislocations maintain straight segments. For short dislocations, the activation energy is higher by ΔF_{fk}, and the dislocation velocity is proportional to the segment length. The latter effect was proved by in situ straining experiments on semiconductor single crystals in the TEM, as mentioned in Chap. 6.

There are many refinements of the double-kink model. As mentioned earlier, some of them aim at more realistic Peierls potentials. In several metals and intermetallic alloys, the dislocation cores are not planar (e.g. [160]) and may have different stable configurations. As an example, in b.c.c. transition metals, the dislocations with $1/2\langle 111 \rangle$ Burgers vectors have cores of threefold symmetry. This can be considered by a two-dimensional Peierls potential on the $\{111\}$ plane [161]. The plane on which the kink pairs will be formed then depends on the orientation of the stress field. The calculations reproduce the plastic anisotropy of these materials. Mathematical methods to simulate the dynamic behavior of dislocations are summarized in [162].

4.2.3 Characteristics and Experimental Evidence of the Double-Kink Model

The double-kink mechanism, often also designated as the Peierls mechanism, describes the lattice friction of a dislocation in an otherwise undisturbed crystal lattice at a finite temperature. The decisive quantity is the Peierls stress τ_{p}, which is the flow stress at temperatures approaching 0 K. In addition, it determines the kink formation energy according to (4.26). The latter controls the dislocation mobility in the range of the operation of the model. Frequently, the contribution of the lattice friction to the total flow stress at a finite temperature is also called the Peierls stress.

Dislocation motion controlled by the lattice friction exhibits the following features:

- The dislocations show straight segments oriented along the Peierls valleys, that is, low-index crystallographic directions, as demonstrated in Fig. 4.6 for solar silicon. The corners between the straight segments may be rounded.
- The dislocations move in a smooth viscous mode. Both features are shown in Video 6.1.
- The macroscopic deformation behavior is characterized by the (at least approximate) proportionality between the dislocation velocity and the effective stress after (4.46), which is valid for both low stresses with the

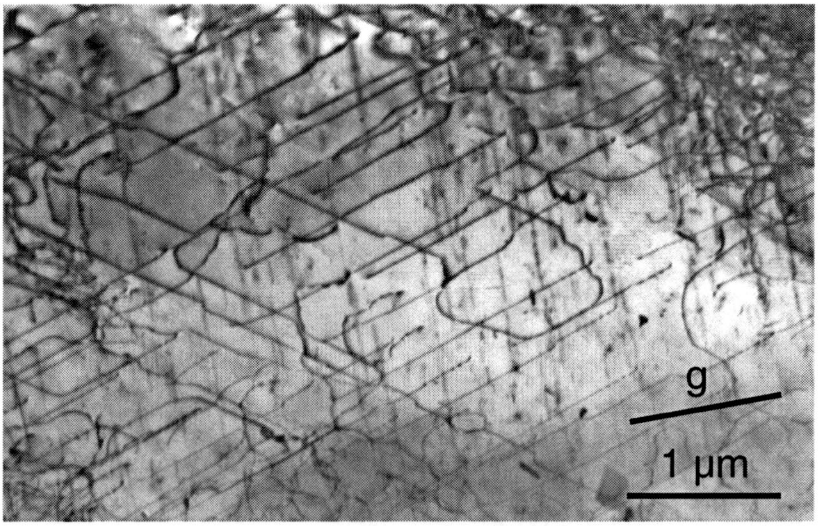

Fig. 4.6. Dislocation structure formed during in situ deformation of a solar silicon specimen in the HVEM. The dislocations are arranged parallel to the three $\langle 110 \rangle$ directions on the $(1\bar{1}\bar{1})$ slip plane. The viewing direction is $[011]$, $\boldsymbol{g} = (1\bar{1}1)$. Micrograph from the work in [163]

concentration of thermal kinks and high stresses with stress-assisted kink nucleation. Considering the Orowan relation between the dislocation velocity and the strain rate (3.5) shows that also the strain rate is proportional to the stress. It follows that the stress exponent m of (4.11) is equal or close to unity. The consequence is a high strain rate sensitivity r approximately equal to the effective stress τ^* and a small activation volume V of the order of magnitude of only a few b^3. The small activation volume is related to the size of the critical kink pair at kink nucleation.

The temperature range in which the double-kink mechanism operates is determined by its activation energy, which, for long dislocations, is the energy of kink formation ΔF_{fk} plus, for slowly moving kinks, the energy of kink migration ΔF_{mk}. Usually, the double-kink mechanism controls the dislocation mobility at the low-temperature end of the thermally activated range. In closed-packed metals where atoms can easily be shifted with respect to each other, the activation energy of the double-kink mechanism is small, which restricts this mechanism to temperatures close to 0 K. On the other hand, materials with directional bonds as semiconductors and ceramics have high kink energies so that the lattice friction controls the deformation behavior also at higher temperatures. As the double-kink mechanism shows also a strong dependence on the temperature, the respective flow stresses at low temperatures may be higher than the fracture stress, resulting in a brittle behavior with a well defined brittle to ductile transition. An intermediate behavior is observed in some oxide and alkali halide crystals and in b.c.c. metals.

4.3 Slip and Cross Slip

In crystalline materials, slip systems are selected having a low stress necessary to overcome the lattice friction. As discussed earlier, these are the systems with the shortest lattice vectors as Burgers vectors and low-index planes with large spacings ((4.20) and (4.21)). For a certain orientation of the crystal or the crystal grain, those possible slip systems are being activated, which have a high orientation factor (2.3). If several slip systems of low friction stress have high orientation factors, more than one system will operate leading to multiple slip. In particular orientations, the orientation factor of one slip system is remarkably higher than that of the others so that the crystal deforms in single slip.

In materials with a low stacking fault energy, the dislocations may dissociate into partial dislocations, including a stacking fault as described for f.c.c. crystals by (3.50). The dissociation occurs by glide on the plane of the stacking fault so that this plane contains the directions of the Burgers vectors of both partial dislocations as well as that of the total Burgers vector. Thus, dislocation glide will be restricted to this plane. Consequently, several criteria decide on the activation of the particular slip systems. For an undissociated screw

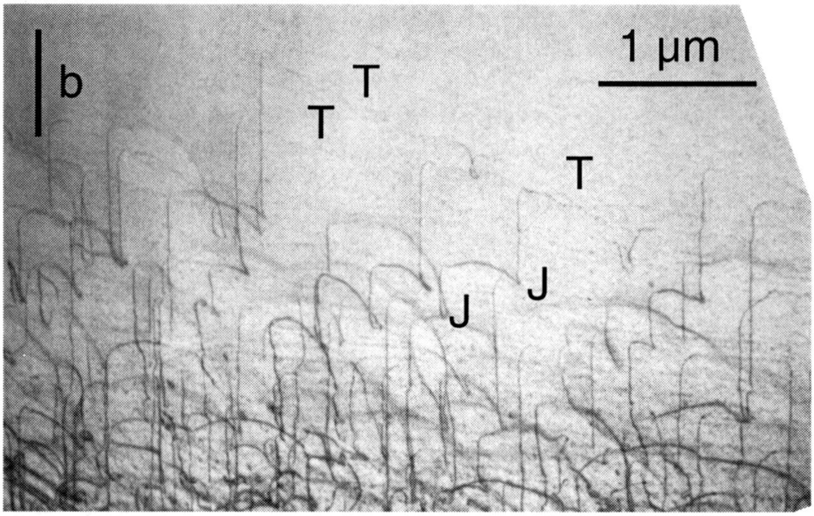

Fig. 4.7. Dislocation structure formed during in situ deformation of an MgO single crystal in the HVEM showing extended cross slip of screw dislocation segments. The viewing direction is ⟨100⟩. The trace of the {011} slip plane runs in horizontal direction. ***b*** projection of the Burgers vector, T slip trails of moving dislocations. The slip trails are imaged in weak contrast owing to elastic strains of the slip steps in surface contamination layers. J jogs. From the work in [164]

dislocation with its Burgers vector parallel to its line vector, also other low-index planes are possible slip planes. A change of a screw dislocation from the preferred slip plane onto another one is called cross slip or cross glide. Cross slip may occur as a single event on long screw dislocation segments, or with frequent changes of the slip plane, or even in a non-crystallographic way. In the latter cases, the slip traces on the surface are not straight but consist of short straight segments, or they are even wavy (wavy slip). In most cases, cross slip is a local phenomenon where only a relatively short screw segment changes to another slip plane. An intermediate case is demonstrated in Fig. 4.7, showing a slip band in an MgO single crystal during in situ deformation. Straight screw segments leave the original slip plane with a horizontal surface trace. The trails T of the dislocation motion run along non-crystallographic paths. The cross slip is possibly initiated by the internal stresses of the slip band and by image forces near the surface. As only the screw segments cross slip, they are connected to the rest of the dislocations by jogs J, which pin the dislocations and cause the neighboring segments to bow out.

Cross slip plays an important role in many phenomena of plastic deformation. It may be responsible for both dislocation multiplication as well as recovery by mutual annihilation of screw dislocations (Sect. 5.1). In intermetallic alloys, dissociated dislocations can get locked by cross slip of one superpartial. On the other hand, cross slip can help to bypass glide obstacles like large precipitates. The ability to cross slip depends on the structure of the dislocation core. Undissociated dislocations can easily cross slip onto other planes of low lattice resistance. However, the partial dislocations of a dissociated screw dislocation usually do not have screw character so that they cannot leave the original slip plane. Either a dislocation reaction is necessary resulting in a partial dislocation with a Burgers vector on the cross slip plane [165], or the stacking fault has to constrict on the original plane before the dislocation may extend on the cross slip plane [166].

Cross slip of dissociated dislocations in the f.c.c. structure is treated in the model of Escaig [167] based on the suggestions of Friedel [168] and the calculations by Stroh [169]. In this model described in detail in [11], the dislocation has to form a constriction of the stacking fault in the original glide plane of a certain length before dissociating again on the cross slip plane. The final configuration is sketched in Fig. 4.8. Initial nuclei of constrictions may pre-exist at jogs or dislocation nodes. Extending the constriction increases the energy of the system. The energy difference is calculated considering the stacking fault energies on both planes, the energies and line tensions of all dislocation segments involved as well as their interaction energies. The total energy difference depends on the widths of the dissociated dislocations and the local stresses on both planes. Estimates for copper yield an energy, which decreases with increasing stress and which amounts to about 1.1 eV at zero stress for the case of pre-existing constrictions. Thus, the cross slip should be a thermally activated process. The activation volume decreases from about $300b^3$ at zero stress to some tens of b^3 at high stresses. Refined calculations of the activation parameters of cross slip are reviewed in [170]. Recently, cross slip of dissociated dislocations has been modeled on an atomic scale by the embedded atom method (EAM) [171] and by molecular dynamics simulations [172, 173], essentially confirming the results of the continuum models. Schöck [174] criticized several models of cross slip and derived the stress conditions for the spreading of cross-slip.

The parameters of cross slip were studied experimentally in Cu applying a particular technique [175, 176]. The crystals were first deformed in a symmetrical multiple slip orientation into a stage of strong work-hardening. Afterwards, smaller specimens were cut out of the predeformed samples and deformed in

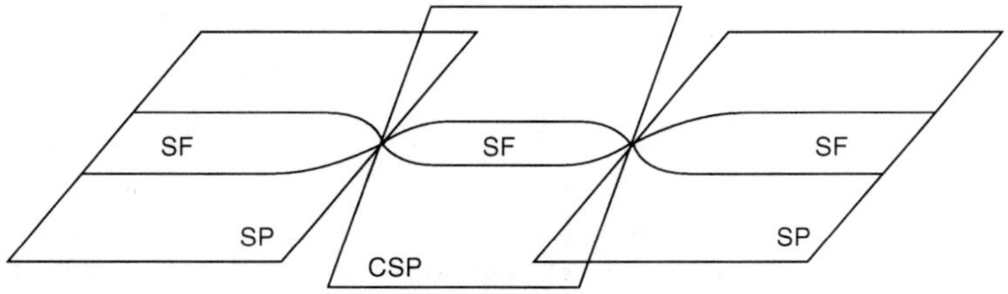

Fig. 4.8. Cross slip of a segment of a dissociated dislocation. SP original slip plane, CSP cross slip plane, SF stacking faults

a single slip orientation, where the dislocations of the predeformation have a high orientation factor on their cross slip plane but where the original slip systems are out of stress. It is assumed that the critical flow stress of the second deformation represents the stress necessary to activate cross slip of the dislocations of the predeformation. In a narrow temperature range, the activation energy agrees with the theoretical value quoted earlier. The described measurements probably present the only experimental estimation of the activation parameters of cross slip.

Similarly, also statistical data on the occurrence of cross slip are quite rare. A method well suited to image cross slip events is the heavy metal decoration technique of slip lines of individual dislocations on alkali halide cleavage surfaces described in Sect. 2.5.2, where the paths of individual dislocations with a Burgers vector component out of the surface are imaged. A corresponding example is presented in Fig. 4.9. During cleavage, some grown-in dislocations are cut. At the emergence points S of these dislocations with a screw component out of the surface, cleavage steps originate. The straight slip lines mark the paths of individual dislocations, which had moved during plastic deformation on {011} planes inclined with respect to the surface. When these dislocations stop moving as indicated by E in the figure, they may shorten

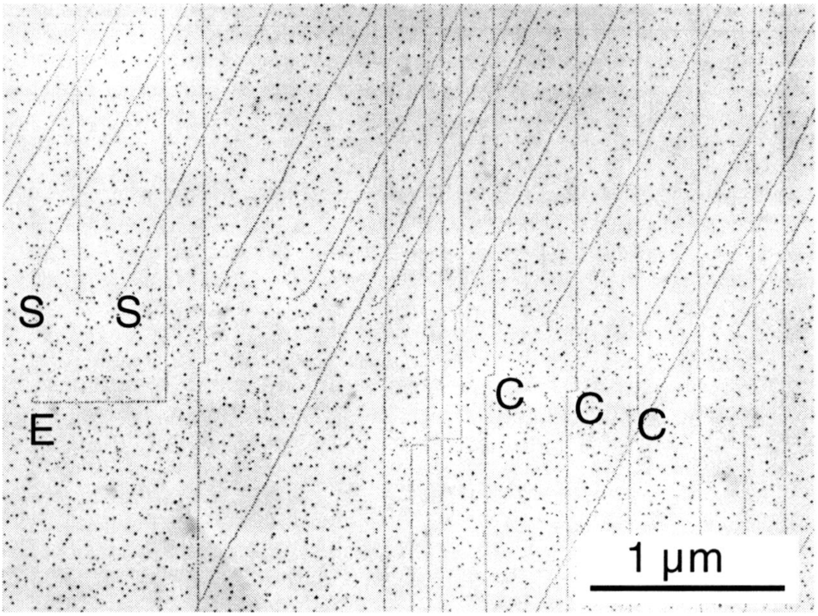

Fig. 4.9. Gold decoration surface replica of an NaCl single crystal deformed at room temperature showing cleavage steps and slip steps of dislocations with $1/2\langle 011\rangle$ Burgers vectors that had moved on {011} planes. ⟨100⟩ surface. Micrograph from [177]. Copyright Elsevier Sequoia (1982)

their length by cross slipping on $\{010\}$ planes under the action of image forces. Using line tension arguments, the flow stress on the cross slip plane can be estimated from the cross slip distance on the surface. As a result, the flow stresses on the cross slip plane are approximately equal to those on the primary slip plane, indicating that both are controlled by the same mechanism [178]. Segments of a certain length often cross slip during the dislocation motion from the primary slip plane onto the cross slip plane and, after some motion there, they move back to a plane parallel to the original slip plane. Several of such events are marked by C. It should be noted that under the applied stress acting in $\langle 001 \rangle$ direction there is no shear stress component on the cross slip plane. For the respective geometry with $\{011\}$ slip planes intersecting the $\{100\}$ surface, the cross slip height h, that is, the distance between the original and final slip planes, is given by the displacement of the slip step divided by $\sqrt{2}$.

Figure 4.10 presents the frequency distribution $H(h)$ of the cross slip heights h in an NaCl single crystal with 32 ppm divalent foreign atoms deformed at room temperature. In alkali halide crystals, the concentration of divalent additions significantly determines the yield stress. The histogram shows frequencies strongly decreasing with increasing cross slip height. A very simple theoretical model of the distribution of the cross slip heights is given in [180], assuming that there is first a certain fixed probability per swept area of the primary slip plane to leave this plane by cross slip, and second another probability per swept area of the cross slip plane to return to a plane parallel to the primary plane. Under these assumptions, the frequency distribution of the cross slip heights becomes

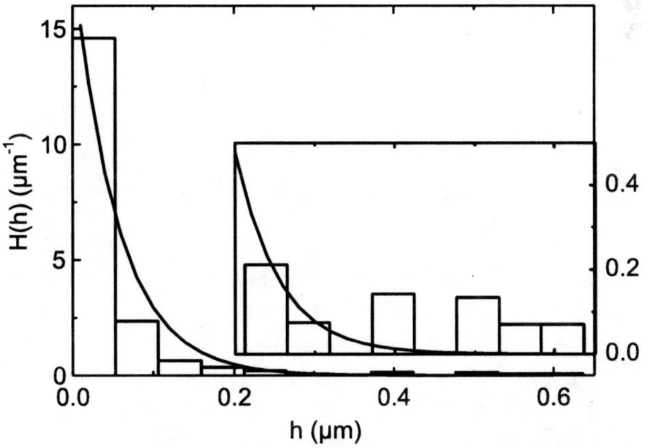

Fig. 4.10. Frequency distribution of the cross slip heights of an NaCl single crystal with 32 ppm divalent foreign atoms deformed along $\langle 001 \rangle$ at room temperature. Total of 262 cross slip events. Data from [179]

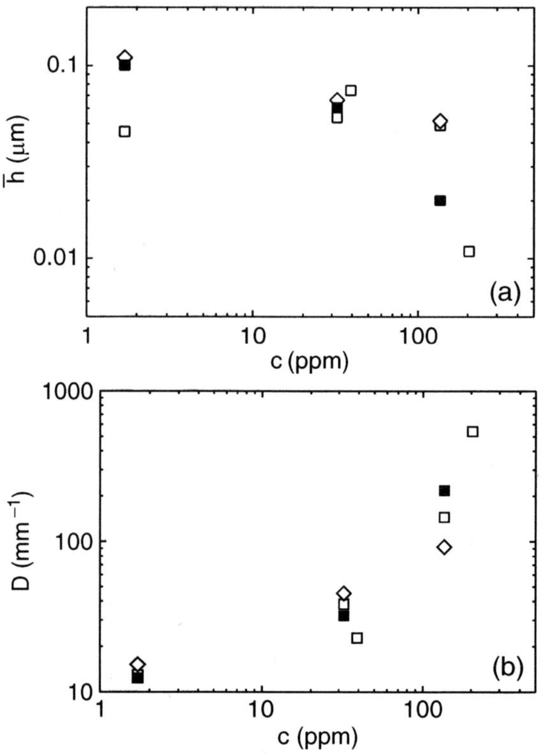

Fig. 4.11. Dependence of cross slip height (**a**) and total density of cross slip events along the slip line length (**b**) on the concentration of divalent cationic impurities in NaCl single crystals. *Open squares*: deformation at room temperature along ⟨100⟩. *Full squares*: liquid nitrogen temperature along ⟨100⟩. *Open diamonds*: room temperature along ⟨110⟩. Data from [177]

$$H(h) = \frac{1}{\bar{h}} \exp\left(-\frac{h}{\bar{h}}\right) ,$$

where $\bar{h}$ is the average cross slip height. The theoretical distribution is included in Fig. 4.10 as a solid line. It agrees roughly with the experimental distribution. The main difference consists in the long tail for large heights in the experimental values. These properties are characteristic also of the crystals with other concentrations of divalent additions. Figure 4.11a shows the dependence of the average cross slip heights on the divalent impurity concentration c for two orientations of the loading axis at room temperature, and for the ⟨001⟩ axis at liquid nitrogen temperature. There is almost no dependence on the loading axis and the deformation temperature. The values decrease slightly for small impurity concentrations and more strongly for higher ones. In Fig. 4.11b, the total density D of cross slip events per slip line length is plotted as a function of c. There is a strong increase of D with increasing c but again no dependence on the specimen orientation and temperature. As mentioned earlier, the concentration of divalent additions essentially determines the yield stress of the alkali halide crystals. It also influences the dislocation density, which builds up at the first stage of deformation, and thus the intensity of the internal stresses. As both the mean cross slip height and the total density depend on the impurity concentration but not on the temperature, the cross slip processes seem to be of athermal character and be influenced by the internal stresses rather than by the total flow stress.

A direct correlation between the local density of forest dislocations and the density D of cross slip events was observed in the latent hardening study in [181]. NaCl crystals were deformed consecutively along two different compression axes so that the dislocations having formed during the first deformation were out of stress during the second one. As demonstrated in Fig. 4.12, a first surface decoration with the large gold nuclei revealed the emergence points F of the dislocations generated during the first deformation which, during the second deformation, act as forest dislocations. The slip steps in horizontal direction created during the second loading are marked by a second decoration with smaller palladium nuclei. They show cross slip C and immobilization events I. Individual micrographs reveal a local correlation between the forest dislocation and cross slip densities, proving the role of internal stress fields as a source of cross slip.

In summary, cross slip may show quite different characteristics. On the one hand, in materials of low stacking fault energy and with dissociated dislocations, the initiation of cross slip seems to be a thermally activated process. In contrast to that, in some materials with compact dislocation cores, cross slip seems to be easily initiated and may be of athermal character. The driving forces result from the internal stress field of other dislocations and from components of the applied stress. Further motion on the cross slip plane is governed by these stresses and the usual mechanisms controlling the dislocation mobility on the respective planes. Collective cross slip of many dislocations is supposed to cause the decrease of the work-hardening rate after heavy plastic deformation [182]. As will be discussed in Sect. 5.1.1, individual cross slip events are a prerequisite to dislocation multiplication.

4.4 The Locking–Unlocking Mechanism

As mentioned earlier, the cores of the dislocations need not be planar. In particular, a dislocation with a certain total Burgers vector may have several stable configurations extended on different planes with different energies. These configurations may be glissile on a certain plane, or totally sessile, or sessile with respect to the glide plane. To move the dislocation in a sessile

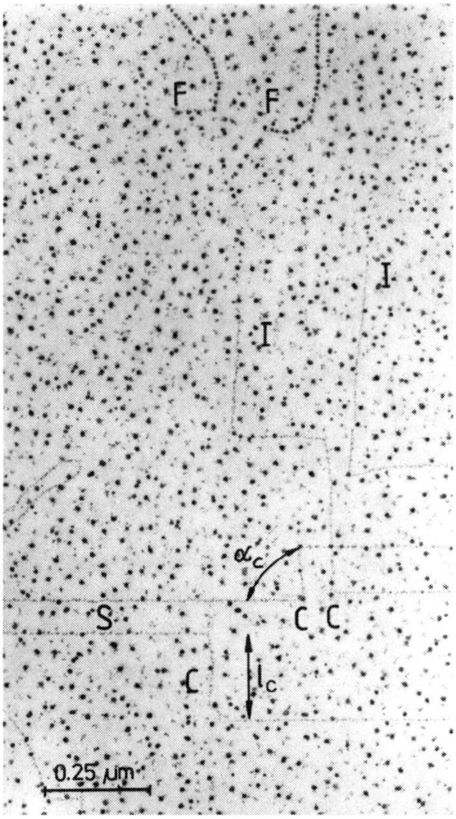

Fig. 4.12. Double decoration micrograph of a latent hardening experiment with forest dislocations F marked by beginning cleavage steps decorated with large gold nuclei and slip lines of the second deformation decorated with small palladium nuclei and running horizontally, with cross slip C and immobilization events I. From [181]

configuration, it has to be unlocked, that is, to be transformed into a glissile state. The core transformations can be assisted by thermal activation, leading to a state of higher or lower energy. Unlocking transformations into a state of higher energy are only possible under the action of an external stress. The theory is similar to that of the formation of a critical kink pair in the double-kink mechanism (Sect. 4.2.2). The probability of unlocking or forming a critical bulge is given by

$$P_{\mathrm{ul}} = \frac{\nu_{\mathrm{D}} b}{L^*} \exp\left(-\frac{\Delta F_{\mathrm{cb}}}{kT}\right),$$

where ν_{D} is again the Debey frequency, L^* the width, and ΔF_{cb} the formation energy of the critical bulge.

The transformation into a core of lower energy is similar to the process of cross slip of an extended dislocation as described, for example, in [183] with an activation energy similar to that of a constriction. The transformation may lead to a glissile configuration as for cross slip, or to a sessile one. In the latter case, the dislocation becomes locked. The probability of locking is given by a formula similar to that above for unlocking with an activation energy ΔF_{lock}.

A dislocation can move by a succession of locking and unlocking events. This process is similar to the kink pair mechanism as noted in [184]. A detailed theory was developed in [185]. The activation energy is the difference $\Delta F_{\text{cb}} - \Delta F_{\text{lock}}$ between the energies of unlocking and locking. The properties of the mechanism resemble in some aspects the double-kink mechanism, in particular with respect to the occurrence of long straight dislocation segments oriented along crystallographic directions. However, in contrast to the Peierls mechanism, the dislocations move jerkily over distances remarkably larger than those between the Peierls valleys. Besides, if dislocation segments are pinned between obstacles, macro-kinks pile up against these obstacles in contrast to smoothly curved bows for the Peierls model. Dislocation motion by locking–unlocking is observed in some intermetallic materials. In Fe single crystals, a transition occurs below about $-20°\text{C}$ from the double-kink mechanism to the locking-unlocking mechanism connected with a change in the activation parameters [186].

4.5 Overcoming of Localized Obstacles

A way to increase the flow stress of a material consists in alloying it with foreign atoms. Depending on the concentration of the alloying element and the respective phase diagram, the foreign atoms can either be solved in the host lattice (substitutional or interstitial solutes) or they may precipitate as a second phase. Usually, the interaction of these defects with the dislocations is of short-range character. If the Peierls stress is low, the dislocation segments bow out freely under stress, reaching an equilibrium configuration as described in Sect. 3.2.7. The way of overcoming the obstacles depends on the strength and size of the latter. With respect to the size, the dimensions of the obstacles parallel to the main direction along the dislocation line and in forward direction have to be distinguished. Obstacles the dimension of which along the dislocation line is not small compared to the mutual distance between the obstacles are called extended obstacles. They will be treated in the following section. Obstacles pinning the dislocation only along a segment that is short with respect to the distance are designated in this book as localized obstacles. They may still have a finite size along the forward direction of the dislocation motion. Extended and localized obstacles were illustrated above in Fig. 3.20a, b.

The problem of overcoming localized obstacles can be divided into two parts:

The overcoming of individual obstacles
The statistical problem originating from a nonregular arrangement obstacles on the glide plane

The process of surmounting the individual obstacles has been treated in the foregoing sections, in particular in Sect. 4.1. In a regular arrangement, the local effective stress causes a force $F = \tau^* bl$ (4.2) on each obstacle. It is opposed by an interaction force, which is described by the force–distance curve introduced in Sect. 3.2.6. Since the localized obstacles are mostly small with low interaction energies, their overcoming is supported by thermal activation.

The statistical problem is mostly treated for a random arrangement of the obstacles. Owing to the nonregular arrangement, the forces on the individual obstacles are not equal and can be obtained from (3.48). To reduce the number of parameters, the relevant quantities can be normalized. A characteristic length is the so-called square lattice distance l_{sq}, where l_{sq}^2 is the average area of the slip plane per obstacle. Thus,

$$l_{\mathrm{sq}} \approx \frac{b}{\sqrt{c}}. \tag{4.48}$$

Here, c is the atomic concentration of the obstacles. The forces are normalized using the line tension Γ

$$f = F/(2\Gamma) \tag{4.49}$$

and the obstacle strength

$$f_0 = F_0/(2\Gamma). \tag{4.50}$$

For the approximation of constant line tension, $f = \cos\phi$, where ϕ is half the angle of attack. The normalized stress is given by

$$\tau' = \frac{\tau^* b l_{\mathrm{sq}}}{2\Gamma}. \tag{4.51}$$

Analytical solutions of the statistical problem are available for some situations of the athermal surmounting of the obstacle array. In this case, the applied stress must be high enough so that the dislocation can overcome the strongest configuration in the array. It was shown by Schwarz and Labusch [187] that the solutions depend on the relation between the extension x_0 of the obstacle in x direction and its strength f_0 as described by the parameter

$$\xi_0 = (x_0/l_{\mathrm{sq}}) f_0^{-1/2}. \tag{4.52}$$

In the limit of $\xi_0 \ll 1$, the obstacles have no extension and are called point obstacles. All obstacles oppose the dislocation motion and the dislocation bows out between them. This case was first treated by Friedel [188], it is designated "Friedel" statistics. If $\xi_0 \gg 1$, the dislocation is in touch with obstacles on both the entrance and the exit sides of their force–distance curves. Thus, the obstacles cause forces on the dislocation in the forward and backward directions. The respective theories follow "Mott" statistics [189]. In the following, both cases are described.

4.5.1 Friedel Statistics

As pointed out earlier, Friedel statistics is valid for small isolated obstacles of relatively high strength. An example is given in Fig. 4.13. The obstacles are small precipitates in an MgO single crystal. The dislocations form cusps at the obstacles, resulting in forces acting on them. Owing to the nonregular arrangement of the obstacles, the forces are not equal. Obstacles with a high force, that is, an acute angle of attack, have a high probability to be quickly surmounted. This is shown in the figure, where always the obstacles marked by arrows are overcome. In Fig. 4.13c, the moving dislocation segment is imaged twice due to its motion during the exposure of the film. Thus, the advanced dislocation segment spreads sidewise (upwards in the figure), similar to a kink in the double-kink model. The jerky motion of the dislocations overcoming localized obstacles is also presented in the Video 7.1. For simple geometrical reasons, the average obstacle distance along the dislocation line is not equal to the square lattice distance as observed by Friedel [188] and outlined in Fig. 4.14. The figure presents two dislocation segments of length l in position 1, bowing out between three point obstacles to the equilibrium radius r. If the central obstacle is overcome, the dislocation moves to position 2 with the same radius of curvature, sweeping the hatched area. According to Friedel statistics, on average this area contains one additional obstacle, that is, it equals l_{sq}^2.

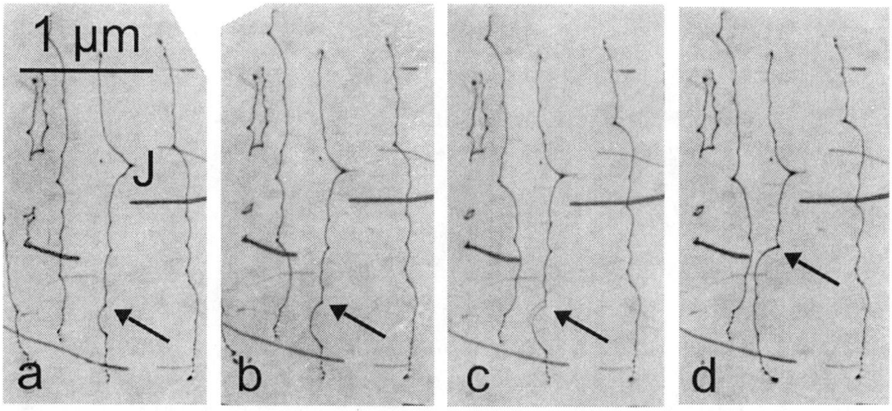

Fig. 4.13. Image sequence of dislocations in an MgO single crystal overcoming localized obstacles during in situ deformation in an HVEM at room temperature. Micrographs from the work in [190]

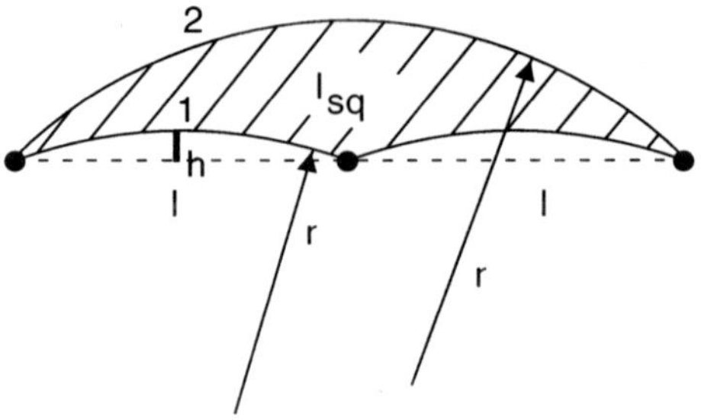

Fig. 4.14. Geometrical representation of the Friedel relation between the average segment length (obstacle distance) and the square lattice distance at overcoming localized obstacles

With the approximations that the area of the segment of a circle is about $2/3\,lh$, with h being the height of the bowed segment, and $\sqrt{1-\epsilon} \approx 1 - \epsilon/2$, it follows for the average distance that

$$\bar{l} = \mathrm{k}_1 l_{\mathrm{sq}}^{2/3} (2r)^{1/3} \tag{4.53}$$

or, with (3.38) $r = \Gamma/(\tau b) = l_{\mathrm{sq}}/(2\tau')$

$$\bar{l} = \mathrm{k}_1 l_{\mathrm{sq}} \tau'^{-1/3}. \tag{4.54}$$

Consequently, the average obstacle distance depends on the stress. In general, $\bar{l}$ is greater than l_{sq} and approximates it for high stresses ($\tau' \to 1$). Considering the stress dependence of the obstacle distance in the equation of the Gibbs free energy of activation (4.3) leads to the introduction of a numerical factor of 3/2 into the relation for the experimental determination of the activation volume (4.9), which then reads [191]

$$V = \frac{3}{2} \frac{kT}{m_{\mathrm{s}} r}. \tag{4.55}$$

If the Friedel length is inserted into (4.2), the stress of the athermal overcoming of a random array of obstacles of the strength F_0 becomes

$$\tau_0 = \mathrm{k}_2 \frac{F_0^{3/2}}{b\, l_{\mathrm{sq}} \sqrt{2\Gamma}}, \tag{4.56}$$

or, in the dimensionless form,

$$\tau_0' = \mathrm{k}_2 f_0^{3/2}. \tag{4.57}$$

In the simple derivation according to Fig. 4.14, the numerical constants k_1 and k_2 are equal to unity. The functional form of the Friedel relations were checked first by computer simulation [192] and later by both computer simulation [193–197] and analytical theories [198, 199]. These calculations confirm the Friedel relations and yield values close to unity for the numerical constants. Suitable data are $k_1 = 0.73$ [200] and $k_2 = 0.95$ [199]. An extension of the statistical theories including also inertia effects of the dislocations was given in [201]. These effects become important only at very low temperatures.

In addition to the macroscopic flow stress at 0 K and the average obstacle distance, the statistical theories yield information on the microscopic shape of the moving dislocations. When a stress is applied to an originally straight dislocation, it contacts a number of obstacles and bows out between them. If no obstacle is surmounted, certain distributions of the obstacle distances and forces on the obstacles result. These so-called starting distributions were calculated in [202]. If the stress is increased successively, some obstacles will be overcome, with the dislocation proceeding forward. After reaching a certain threshold stress, the dislocation experiences its strongest configuration, and after a further increment of the stress it passes the remaining part of the obstacle array. The strongest configuration determines the athermal flow stress of (4.56). In this configuration, both the obstacle distances and the forces acting on the obstacles exhibit characteristic frequency distributions presented in Fig. 4.15. The distribution of the segment lengths in Fig. 4.15a differs from the exponential distribution of points randomly arranged on a line, showing an asymmetric maximum. In the force distribution of Fig. 4.15b, the frequencies increase with increasing forces up to the maximum force f_0. If the lengths are expressed in units of $2 f_0\, r$ and the forces in units of f_0, the curves can be normalized as shown in the figure.

In a real material, the obstacles are usually not of equal strength. The concentrations of obstacles of different types α can be characterized by their individual square lattice distances $l_{\mathrm{sq}\alpha}$. As in an array composed of obstacles of similar concentrations and strengths the concentrations add, there follows

$$\frac{1}{l_{\mathrm{sq}}^2} = \sum \frac{1}{l_{\mathrm{sq}\alpha}^2} .$$

For the simplest case of two types of obstacles of not very different strength, that is, stronger ones with $\alpha = \mathrm{s}$ and weaker ones with $\alpha = \mathrm{w}$, the relative fractions on the slip plane are

$$x_{\mathrm{w}} = \frac{l_{\mathrm{sqs}}^2}{l_{\mathrm{sqw}}^2 + l_{\mathrm{sqs}}^2} \quad \text{and} \quad x_{\mathrm{s}} = \frac{l_{\mathrm{sqw}}^2}{l_{\mathrm{sqw}}^2 + l_{\mathrm{sqs}}^2} .$$

Each type of obstacles produces an athermal flow stress according to (4.57). If small terms are neglected, the combined flow stress is then [203]

$$\tau_0'^2 = x_{\mathrm{w}} \tau_{0\mathrm{w}}'^2 + x_{\mathrm{s}} \tau_{0\mathrm{s}}'^2 , \tag{4.58}$$

or, in un-normalized form

$$\tau_0^2 = \tau_{0\mathrm{w}}^2 + \tau_{0\mathrm{s}}^2 . \tag{4.59}$$

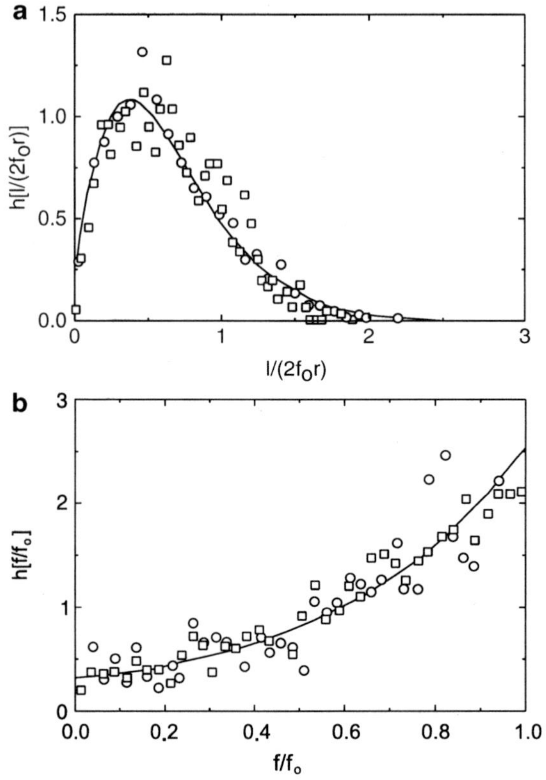

Fig. 4.15. Theoretical distribution of normalized obstacle distances and forces on the obstacles. *Curves*: according to the analytical theory in [199]. *Squares*: according to the numerical simulation in [198]

The composed average obstacle distance is given by

$$\bar{l} = x_{\mathrm{w}}\bar{l}_{\mathrm{w}}(\tau_0', f_{0\mathrm{w}}) + x_{\mathrm{s}}\bar{l}_{\mathrm{s}}(\tau_0', f_{0\mathrm{s}}).$$

Here, the mean values $\bar{l}_{\mathrm{w}}$ and $\bar{l}_{\mathrm{s}}$ are not defined by (4.54). This equation presupposes that the stress is given by (4.57) and the respective strength $f_{0\alpha}$. In a mixture of obstacles, however, the flow stress is not equal to any of the individual flow stresses.

The relative concentrations of the obstacles of different types along the dislocation in the strength-determining configuration do not correspond to their fractions x_α. As the composed flow stress is high with respect to the weak obstacles, the force acting on many of them is larger than their strength ($f > f_{0\alpha}$) so that they are spontaneously overcome. The density of weak obstacles is therefore lower than expected from their density on the slip plane. For the cases treated in [203], a sufficient approximation of the concentration of weak obstacles along the dislocation line is [204]

$$c_{\text{w}} = e^{-3.14} \left(f_{0\text{s}}/f_{0\text{w}}\right)^{-1.28} e^{4.28x_{\text{w}}}.$$

The composed force distribution follows from the distribution $h(f, \tau_0')$ for a single type of obstacles at the stress τ_0' by the mixing rule

$$h(f) = h(f, \tau_0') \sum_{\alpha} x_{\alpha} K_{\alpha}(f)$$

with the weight factors $K_{\alpha} = 1$ for $f \leq f_{0\alpha}$ and $K_{\alpha} = 0$ for $f > f_{0\alpha}$. An example is presented in Fig. 4.16 for $f_{0\text{w}} = 0.2$ and $f_{0\text{s}} = 0.5$. The composite distribution abruptly decreases always at the strengths of the individual obstacle types.

According to (4.57), for thermal activation the applied stress τ' is lower than the athermal flow stress. The statistical theories mostly consider a single obstacle type only. It is assumed that the dislocation velocity is controlled by the obstacles, which the highest force in the strongest configuration is acted on[194]. The waiting time at this obstacle is defined by (4.6) and the simple box potential $\Delta G = \Delta d(f_0 - f)$. This approximation is valid for high stresses and low temperatures. In the opposite case, the probability of activation is approximately equal at all obstacles. In computer experiments, the thermally activated motion of dislocations was simulated using also other potentials, for example, the Fleischer approximation [197]

$$\Delta G = \Delta G_{\text{o}} \left[1 - (f/f_{\text{o}})^{1/2}\right]^2 \tag{4.60}$$

following from (3.35). The effect of the thermal activation on the configuration of the dislocation line primarily consists in the rapid overcoming of obstacles, which are subject to high applied forces f/f_0 so that these obstacles disappear

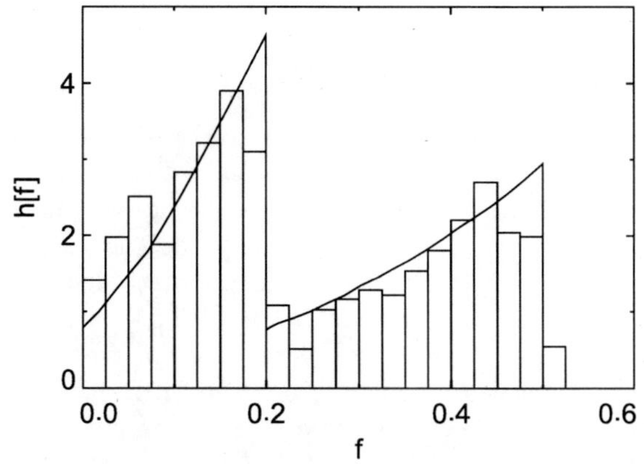

Fig. 4.16. Composite force distribution for the athermal overcoming of obstacles of strengths $f_{0\text{w}} = 0.2$ and $f_{0\text{s}} = 0.5$. *Curve*: analytical theory. *Histogram*: computer simulation. Data from [203]

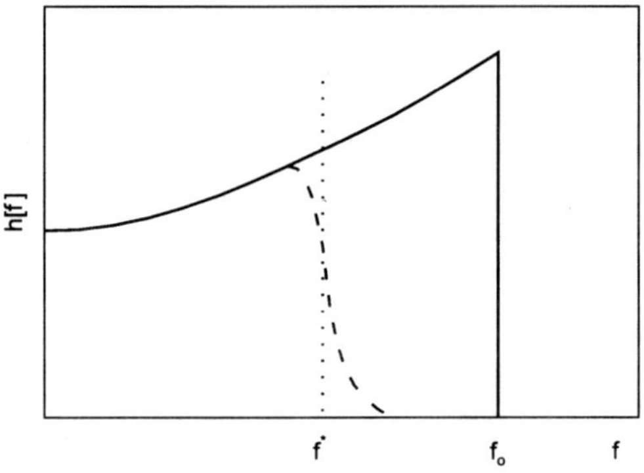

Fig. 4.17. Schematic representation of the effect of thermal activation on the force distribution. The force f^* is defined by the applied stress via (4.61)

from the force distribution. In the range of small forces, the distribution is affected only little. This behavior results in a strong decrease of the frequencies above a characteristic force f^* as indicated in Fig. 4.17. The force f^* correlates with the applied stress by a Friedel relation analogous to (4.57)

$$\tau' = \mathrm{k}_2 (f^*)^{3/2} . \tag{4.61}$$

The dislocation velocity in an array of obstacles of Fleischer type of equal strength can well be represented by an activation energy resulting from inserting the effective force f^* from (4.61) into the Fleischer potential (4.60) [197]

$$\Delta G = \Delta G_0 \left[1 - \left(\frac{(\tau'/\mathrm{k}_2)^{2/3}}{f_0} \right)^{1/2} \right]^2 . \tag{4.62}$$

The numerical factor of $\mathrm{k}_2^{-2/3} = 1.08$ is close to that cited earlier. It is important to note that the force f^* where the distribution breaks off enters (4.62) and not the average force. The latter is essentially smaller than f^*.

Only at low stresses τ'/τ'_0, all obstacles have to be surmounted by thermal activation. At higher stresses, part of the configurations is unstable with $f > f_0$. In this connection, the term "unzipping" is used as after a successful activation at one obstacle, several neighbored obstacles are overcome spontaneously. The dislocation motion becomes jerky on a mesoscopic scale. The effect does not depend on the temperature. The fraction j of configurations that is overcome spontaneously is obtained from computer simulation. In Fig. 4.18, it is plotted vs. the relative stress. The curve is represented by the formula [201]

$$j = (\tau'/\tau'_0)^{0.454} \exp\left[-10.38 \left(1 - \tau'/\tau'_0\right)^{3.78} \right] .$$

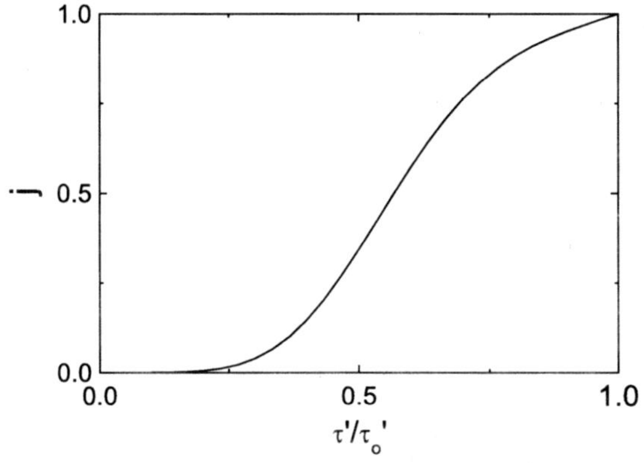

Fig. 4.18. Dependence of the fraction of unstable dislocation configurations during thermally activated dislocation motion on the relative stress. Data from [201]

The fraction j is small up to about $\tau' = 0.3\,\tau_0'$ where almost all obstacles have to be overcome by thermal activation. At high stresses, almost all configurations are surmounted spontaneously and the dislocation stops only at a few very strong configurations so that its motion is most jerky and the character is almost athermal.

In the case of thermal activation, the distributions of the obstacle distances and forces depend not only on the stress τ'/τ_0' but also on the ratio between ΔG_0 and kT. The limiting cases for low and high temperatures were calculated in [205] by computer simulation. Accordingly, the segment length distribution depends only little on the special conditions. It is always approximately equal to the athermal case (Fig. 4.15a). In the limit of high temperatures, the typical maximum (Figs. 4.15b and 4.17) disappears. After constant frequencies at low forces, the distribution slowly tends to zero.

No comprehensive theory is available of the thermally activated surmounting of an array of obstacles of different strengths. It may be assumed that the mixing rules for the athermal case of obstacles of similar concentrations and strengths described earlier, in particular (4.58) and (4.59), are still valid. As discussed by Kocks [206], other arguments apply to the thermally activated surmounting of an array of many weak and a few strong obstacles like localized obstacles and dislocations to be cut. These obstacles have to be overcome simultaneously so that the Gibbs free energies ΔG of both processes have to be equal. Instead of the quadratic superposition rule for similar obstacles, a linear superposition rule should be used for both the stress components and the reciprocal activation volumes,

$$\tau^* = \tau_{\mathrm{w}}^* + \tau_{\mathrm{s}}^* \tag{4.63}$$

$$\frac{1}{V} = \frac{1}{V_{\mathrm{w}}} + \frac{1}{V_{\mathrm{s}}}. \tag{4.64}$$

It was shown by Schoeck [207] that the linear superposition rule of the stress contributions holds for low temperatures. At high temperatures, however, the waiting times at the weak obstacles are very short so that only the strong obstacles control the flow stress. Computer simulation studies [208] of the thermally activated motion through an array of obstacles of very different strengths suggest that under most conditions the dislocation motion is jerky as shown in Fig. 4.18, that the strain rate sensitivity is controlled by the strong obstacles and is exclusively controlled by τ'/τ'_0.

The transition from a regular arrangement of obstacles to a random one was treated in [209]. The flow stress decreases rapidly already for small deviations from regularity.

In summary:

- The macroscopic properties of dislocation motion in an array of localized obstacles obeying Friedel statistics are well described by the Friedel length (4.53) or, in normalized form, (4.54).
- The relation between the athermal flow stress and the obstacle strength is obtained by replacing the square lattice distance in a regular arrangement of the obstacles by the Friedel length (4.56) or, in normalized form, (4.57).
- The same relation holds also for thermal activation. However, instead of the obstacle strength, the force where the force distribution breaks off has to be used.
- The distribution of the obstacle distances and its mean value are a measure of the density of localized obstacles actually impeding the dislocation motion.
- The kinematic character of the dislocation motion allows conclusions to be drawn on the relation between the actual stress and the athermal flow stress of the obstacle array. At low relative stresses, the dislocation motion is jerky on the scale of the square lattice distance, and the obstacles are surmounted with the aid of thermal activation. At high relative stresses, the dislocations move very jerkily on the scale of several hundreds of the square lattice distance.

The range of application of Friedel statistics is hardening by solute atoms and small precipitates of low concentration but high strength. Experimental data on the microscopic aspects of dislocations overcoming localized obstacles will be described in the sections on ceramic crystals and metallic alloys in Part II. Details of the dislocation configurations and the kinematic behavior of dislocations from in situ straining experiments will be given in Sect. 7.2.

4.5.2 Mott Statistics

If ξ_0 in (4.52) is not small with respect to unity, the extensions of the obstacles cannot be neglected and Friedel statistics is not valid. This holds in particular

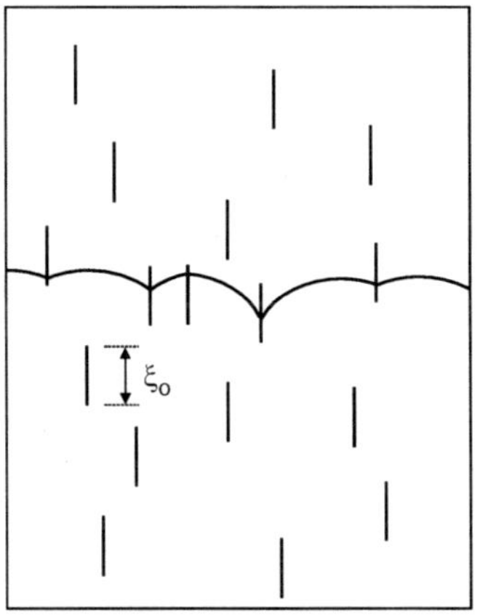

Fig. 4.19. A dislocation interacting with localized obstacles of width ξ_0 in x direction which obey Mott statistics

for weak obstacles with small f_0. Applying a suitable transformation, Schwarz and Labusch [187] show that the width in z direction along the dislocation line can be neglected for weak obstacles so that they act like ribbons extended only in x direction, as demonstrated in Fig. 4.19.

Schwarz and Labusch treat the problem by establishing the differential equation of the motion of the dislocation (see Sect. 4.9) in the field of obstacles in the approximation of constant line tension. For the interaction profile between the dislocation and a single obstacle, a force–distance curve according to

$$f(\xi) = \mathrm{A}_0(\xi/\xi_0)\,(1-(\xi/\xi_0)^n)^2$$

is used in the interval $-1 \leq (\xi/\xi_0) \leq 1$. This interaction profile includes a repulsive and an attractive branch. The results are almost independent of the exponent n. For the simulation of thermal activation described below, $n = 2$ was used. The constant A_0 is chosen as to obtain $f_0 = 1$. The equation of motion contains also a damping term. For the athermal case, it was solved by numerical simulation. For $\xi_0 < 0.8$ and $n = 2$, the athermal flow stress is

$$\tau_0' = \mathrm{k}_2 f_0^{3/2}\,(1 + 0.55\xi_0),$$

confirming the result of Friedel statistics (4.57) for $\xi_0 = 0$. The constant k_2 is also almost equal to the value quoted above. For $0.8 < \xi_0 \leq 4$, the simulation results suggest

$$\tau_0' = \mathrm{k}_2 f_0^{3/2}\,(1 + 2.5\xi_0)^{1/3}. \tag{4.65}$$

Thus, the athermal flow stress depends on the width of the obstacles. It is higher than that for Friedel statistics. The motion of the dislocation supported by thermal activation is simulated by adding random point forces to the equation of motion [210]. The results of the simulations are summarized in empirical relations. All Arrhenius plots meet a reasonable straight line of

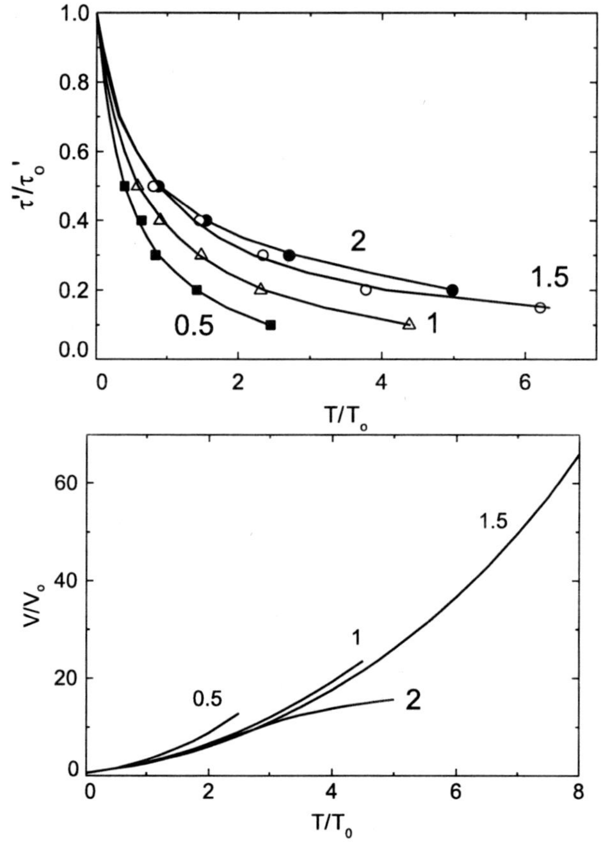

Fig. 4.20. Results of the computer simulations in [210] of dislocations moving through obstacles extended in the x direction. (**a**) Dependence of the relative stress τ'/τ'_0 on the relative temperature T/T_0. The parameter is ξ_0. (**b**) Dependence of the effective activation volume normalized by $V_0 = g\xi_0 b l_{sq}^2$ on the relative temperature

the slope -1 if $\ln v_d$ is plotted vs. a quantity $F_s(\tau', \xi_0)\,\Delta E_0/(kT)$. Here, v_d is the dislocation velocity, F_s a scaling factor depending on the stress and the width of the obstacles, and $\Delta E_0 = gF_0x_0$ is the activation energy at zero force for overcoming the individual obstacles. g is a geometrical factor close to unity depending on the shape of the interaction profile. The scaling factor F_s represents the ratio between an effective activation energy ΔE_{eff} and the activation energy ΔE_0 to overcome the individual obstacles. Using the modified Arrhenius equation (4.7) for a constant dislocation velocity $v_d = v_{d0}\exp(-F_s\Delta E_0/(kT))$, one can write

$$F_s(\tau'/\tau'_0, \xi_0) = \frac{kT}{\Delta E_0}\ln(v_{d0}/v_d) = \frac{T}{T_0}$$

with $T_0 = \Delta E_0/(k\ln(v_{d0}/v_d))$. Inversion of this relation yields the relative flow stress τ'/τ'_0 as a function of the temperature with ξ_0 as a parameter. This is plotted in Fig. 4.20a. Thus, the flow stress normalized by the athermal flow stress decreases very slowly with increasing temperature, the more slowly the larger is ξ_0. For point obstacles, the flow stress contribution is already small at T_0.

The other quantity that can be compared with experimental results is the activation volume

$$V = -\frac{\partial \Delta E_{\text{eff}}}{\partial \tau} = -\frac{\Delta E_{\text{o}}}{\tau_{\text{o}}} \frac{\partial(\Delta E_{\text{eff}}/\Delta E_{\text{o}})}{\partial \tau'} .$$

Using the Friedel relation (4.56) for τ_0 shows that the first factor in the equation becomes $V_0 = \Delta E_0/\tau_0 = g\xi_0 b l_{\text{sq}}^2$. The second factor V/V_0 is obtained numerically from Fig. 4.20a, and it is plotted in Fig. 4.20b. In the Friedel case, the activation volume is of the order of $b^2 l_{\text{sq}}$. In the Mott–Labusch situation, even V_0 is greater than that as g and ξ_0 are close to unity but l_{sq} is greater than b. In addition, Fig. 4.20b indicates activation volumes being larger than V_0 by a factor of several tens.

In conclusion, the theory of Labusch and Schwarz shows that for weak obstacles extended in the x direction of forward motion

- The athermal flow stress depends on the normalized obstacle width.
- The obstacles are not overcome individually by thermal activation but collectively in groups.
- As a consequence, the effective activation energy and the activation volume are far greater than the values for overcoming the single obstacles.
- The obstacles contribute to the flow stress at high temperatures where individual obstacles are no longer active.

Mott–Labusch statistics is applied to solution hardening, especially of substitutional alloys with a high concentration of weak obstacles. In addition to a relatively steep low-temperature increase, the model explains the plateau-like weak decrease of the flow stress at higher temperatures. In Sect. 9.4, the model will be applied to NiAl containing 0.5 at% Ti.

4.6 Transition from the Double-Kink Mechanism to the Overcoming of Localized Obstacles

The activation parameters of the double-kink (Peierls) model and the models of overcoming localized obstacles are quite different so that these mechanisms act in different ranges of temperature and dislocation velocity or strain rate. In the transition region, both mechanisms act simultaneously and the net dislocation velocity is controlled by both the waiting time t_{w} of the dislocation pinned at the obstacles, (4.3) and (4.6), and the traveling time t_{t} between the stable configurations determined by the double-kink mechanism. In the limiting case of localized obstacles acting alone, the traveling time is neglected,

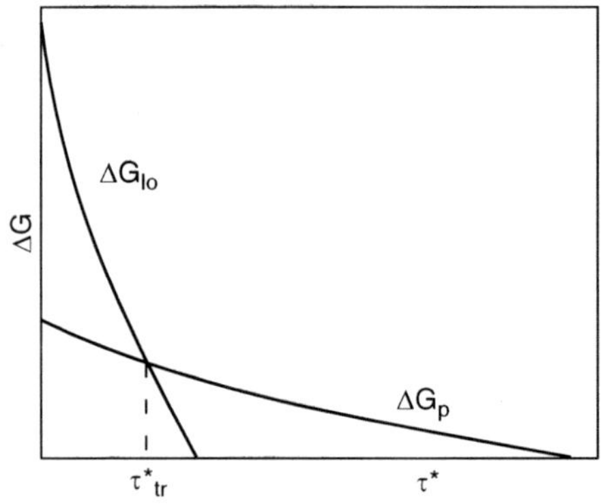

Fig. 4.21. Schematic plot of the dependence of the Gibbs free energy of activation on the effective stress for overcoming localized obstacles and the double-kink mechanism

$t_w \gg t_t$. If only the Peierls mechanism acts, $t_w \ll t_t$. In a simplified treatment where the difference in the pre-exponential factors is neglected, the transition between both mechanisms occurs when the respective Gibbs free energies of activation are equal, as outlined in Fig. 4.21. At high temperatures corresponding to low stresses, where localized obstacles control the deformation, ΔG_{lo} is large. The activation volume V is large, too, as it contains the obstacle distance l or multiples of it. This leads to a strong dependence of the activation energy ΔG_{lo} on the effective stress τ^*. The double-kink mechanism has a low activation energy $\Delta G_{\mathrm{p}} \approx \Delta F_{\mathrm{mk}} + \Delta F_{\mathrm{fk}}$ or $\approx \Delta F_{\mathrm{mk}} + 2\Delta F_{\mathrm{fk}}$, (4.46) or (4.47), and a small activation volume resulting in a very weak dependence of ΔG_{p} on τ^*. Both curves intersect at the transition stress τ^*_{tr} or the corresponding transition temperature.

In a more detailed picture, it turns out that both mechanisms influence each other. A theory of dislocation motion in the combined relief of the extended Peierls barriers and localized obstacles was developed in [211]. If the Peierls mechanism is neglected, the dislocation segments bowing out under stress acquire only a single equilibrium configuration, which can be described, for example, by the line tension approximation. In the more general case of the combined action of the double-kink mechanism and localized obstacles, owing to the periodic Peierls potential, many locally stable or metastable configurations appear, which can be characterized by their bowing heights h in the center of the segments. Different positions are shown in Fig. 4.22.

The equilibrium states are revealed in the line tension approximation by minimizing the energy of a curved dislocation in the field of the Peierls potential as discussed already in (4.25). For the segments pinned at their ends at $z = 0, x = h; z = \pm l/2l, x = 0$ the integration yields [212]

$$E = 2\int_0^h \sqrt{2\Gamma\left[W_{\mathrm{p}}(x) - W_{\mathrm{p}}(h) + \tau^* b(h-x)\right]}\,\mathrm{d}x + \left[W_{\mathrm{p}}(h) - \tau^* bh\right] l .$$

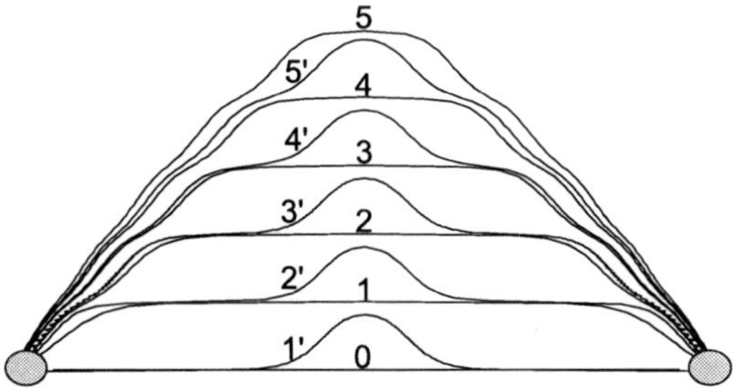

Fig. 4.22. Stable and metastable configurations of a dislocation segment bowing out between localized obstacles under the simultaneous action of the Peierls mechanism. From [212]. Copyright EDP Sciences (2000)

The solutions of the problem are presented in Fig. 4.22. The configurations with an extended flat top designated by numbers n are stable solutions, whereas the solutions with a tip designated by n' are unstable and correspond to ordinary kink pairs on a straight dislocation. In the intermediate stable positions, the angles of attack at the obstacles are larger than in the limiting configuration without the Peierls mechanism. Correspondingly, the forces on the obstacles are smaller. To study the kinematics of the dislocation motion, the activation energies for the transitions between the stable positions have to be calculated. For small bowing, i.e., for dislocation segments with an extended straight top, the activation energy corresponds to the usual double-kink mechanism, which can be approximated for the harmonic Peierls potential (4.20) by

$$\Delta G_{\mathrm{po}}(\tau^*) \approx 2\Delta F_{\mathrm{fk}} \left[1 - \frac{\pi \tau^*}{8\tau_{\mathrm{p}}} \left(\ln \left(\frac{16\tau_{\mathrm{p}}}{\pi \tau^*} \right) + 1 \right) \right] , \tag{4.66}$$

where

$$\Delta F_{\mathrm{fk}} = \left(\frac{2}{\pi} \right)^{3/2} \sqrt{\Gamma \tau_{\mathrm{p}} b h^3} \, .$$

The stress dependence of the activation energy with both the Peierls potential and localized obstacles being present is characterized by several branches corresponding to the numbers n of the different Peierls valleys. For high stresses, all these branches form the curve of the usual Peierls mechanism (4.66). However, with decreasing stress, the curves of the different configurations n reach a limiting stress where the activation energy strongly increases. Below this limiting stress, there does not exist an equilibrium configuration with the respective n value. These abnormal increases of the respective activation energies result in either a strong increase in the activation volume with respect to the usual Peierls mechanism at low stresses $\tau^* < 0.1\,\tau_{\mathrm{p}}$, or in a very steep decrease of the strain rate sensitivity during the transition between the action of the Peierls mechanism at low temperatures and localized obstacles at higher ones, as shown in Fig. 4.23 in comparison with experimental data

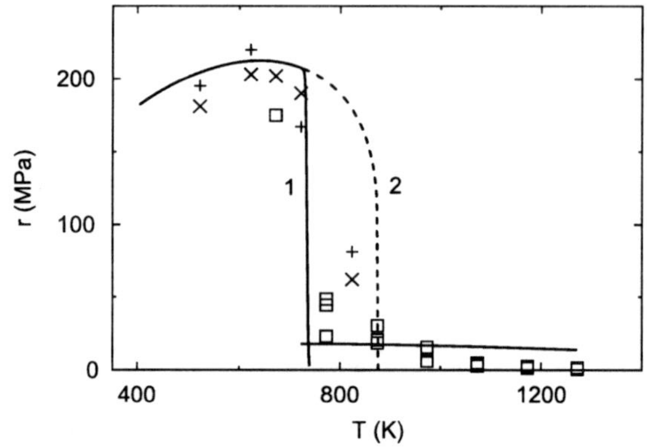

Fig. 4.23. Strain rate sensitivity of cubic ZrO_2 as a function of temperature. Experimental data from [213] (*open squares*) and from deformation under confining hydrostatic pressure [214] (*upright and tilted crosses*). Theoretical curve below transition temperature after transition model (*full line 1*) and after simple theory (*dashed line 2*). Theoretical curve above transition temperature for precipitation hardening, see Sect. 7.3.5. Data from [212]

from ZrO_2 single crystals. The properties of this material are described in more detail in Sect. 7.3.

If the double-kink mechanism acts, localized obstacles cannot only impede the dislocation motion, but they can also advance it by facilitating double-kink nucleation, thus leading to softening. For a recent review, see [215].

4.7 Overcoming of Extended Obstacles

Large precipitates can frequently not be treated by the model of localized obstacles which extend only in forward (x) direction of the dislocation motion. Kocks et al. [9] described the interaction between penetrable particles with dislocations in a simple line tension model. Impenetrable obstacles can be passed only by forming a dislocation loop around the obstacle. This process will be described below.

The following line tension model is restricted to isotropic line tension, where the dislocation segments bow out to form circular arcs. It can also be extended to anisotropic line tension. The resisting forces of the particles are in equilibrium with the line tension forces of the bowed-out dislocation segments between the particles. The balance is attained at the border line of the cutting face of the obstacles as will be discussed later. The model allows a unified description of the different resistance mechanisms and the calculation of the force–distance profile of extended obstacles. These may cause the following resistance effects on the dislocations.

1. Friction stresses in the interior of the particles. They result from all mechanisms requiring energy to be spent being proportional to the area swept by the dislocation like the formation of an antiphase boundary in intermetallic alloys (Sect. 3.3.4). The friction stress is given by $\tau_\mathrm{f} = \gamma/b$, where γ is the fault energy per area. The resistance effect can be represented by the dislocation being bowed-out inside the particle in backward direction [216]. The radius of curvature is defined by (3.38) with the line tension equal to the line energy $\varGamma = E_\mathrm{p}$ for the isotropic line tension
$$r_\mathrm{p} = \frac{E_\mathrm{p}}{b\,(\tau_\mathrm{f} - \tau^*)}. \tag{4.67}$$
The index p indicates that these quantities are related to the interior of the particle. The stress that determines the curvature in the particle is the difference between the friction stress τ_f and the local effective stress τ^*, which supports the dislocation motion also inside the particle.
2. Forces acting at the border line of the particles. Their displacement along the border line may create, for example, the energy necessary to form additional interface area at the surface steps trailed by the dislocations. The forces may be taken into account by their components parallel to the edge and screw directions of the dislocations. They cause a knee in the dislocation line at the particle border.
3. Different line energies in the particle and the matrix, for example, owing to different elastic constants. This effect also causes a knee in the dislocation line.

During the forward motion of the dislocation, the contact point between the dislocation and the particle moves along the border line. According to [9, 216], the components of the acting forces and the line tension forces along the tangent to the particle border always have to be in equilibrium, as outlined in Fig. 4.24. The figure shows part of the cutting plane of a particle with the border B. The dislocation line L extends outside and inside the particle. At the intersection between the dislocation and the border line, the tangent to the border has the angle θ with respect to the x coordinate. The dislocation has the tangent angle ψ inside the particle and ϕ outside of it. At the intersection point, the forces K_x and K_z act on the dislocation. Only K_x is plotted in the figure. In addition, line tensions (or energies) E_p act in the particle and E_m in the matrix. At the intersection point, the line tensions act as forces in tangential directions. A variational procedure shows that these forces are in equilibrium if the sum of their projections onto the tangent direction of the particle border vanishes [9, 216]

$$K_x \cos\theta + K_z \sin\theta + E_\mathrm{p} \cos(\psi - \theta) - E_\mathrm{m} \cos(\phi - \theta) = 0. \tag{4.68}$$

The angle ϕ determines the obstacle force according to (3.48)

$$F = 2E_\mathrm{m} \cos\phi. \tag{4.69}$$

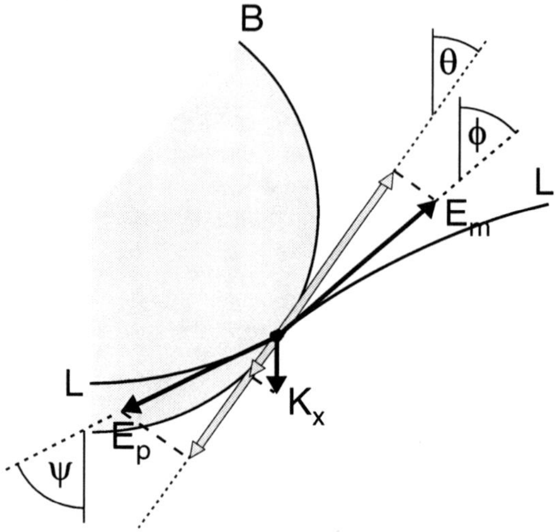

Fig. 4.24. Force balance along the tangent to the border line of the cutting face of an extended obstacle

Using $\cot\theta = x'$ and

$$\cos\psi = zb\left(\tau_{\mathrm{f}} - \tau^*\right)/E_{\mathrm{p}}$$
$$\cos\phi = \left(l/2 - z\right)b\tau^*/E_{\mathrm{m}},$$

the equilibrium condition (4.68) can be written as

$$x'K_x + K_z + x'\tau_{\mathrm{f}}bz - \frac{x'\tau^* bl}{2} + E_{\mathrm{p}}\sqrt{1 - \left[\frac{zb\left(\tau_{\mathrm{f}} - \tau^*\right)}{E_{\mathrm{p}}}\right]^2}$$
$$- E_{\mathrm{m}}\sqrt{1 - \left[\frac{\left(l/2 - z\right)b\tau^*}{E_{\mathrm{m}}}\right]^2} = 0. \quad (4.70)$$

This equation establishes a relation between the different resistance mechanisms and the stress τ^*. It can be used to calculate the equilibrium position x_{e} on the entrance side of the dislocation in the particle, which is dependent on τ^*. Even in the simple case that the obstacle resistance consists only of a force K_x at the border of the cutting plane, the obstacle resistance force $F = \tau^* lb$ is not equal to $2K_{\mathrm{x}}$. This led Kocks, Argon and Ashby [9] to the conclusion that the respective activation distance Δd need not be equal to the difference between the equilibrium x coordinates at the exit and entrance sides of the particle, $x_{\mathrm{a}} - x_{\mathrm{e}}$. To solve the problem, from geometric considerations, they suggested a straight extrapolation of the dislocation outside the particle to the symmetry line (x axis) leading to the reaction coordinate

$$d_{\mathrm{KAA}} = x - z\cot\phi.$$

However, it can be shown that simple interaction profiles constructed by using this formula and (4.70) violate the thermodynamic relation (4.5). To calculate the force–distance laws of overcoming the extended obstacles, the Gibbs free energy of activation ΔG can be established considering the "chemical" terms, that is, the work against the forces K_x and K_z and against the friction stress τ_f, as well as the energy changes due to the changing length of the dislocation line inside and outside the particles [217]. According to (4.3), the work of the external stress τ^* by sweeping the area V/b has to be subtracted. All terms have to be taken between the stable equilibrium position $x_\mathrm{e}, z_\mathrm{e}$ of the dislocation on the entrance side of the particle and the labile equilibrium position $x_\mathrm{a}, z_\mathrm{a}$ on the exit side, as outlined in Fig. 4.25. These positions depend on τ^* and are obtained by (4.70). Differentiating ΔG with respect to τ^* yields the activation volume V after (4.5). For the extended obstacles, a description in terms of V and τ^* is more meaningful than that in terms of Δd and F as the relation $F = \tau^* b\, l$ (4.2) is valid only for $z \ll l$.

The result of the calculation is very simple [217]. The activation area V/b equals the area swept by the dislocation between the entrance and exit positions as indicated by the hatched area in Fig. 4.25. This is independent of the acting resistance mechanisms. The formulae show that, in general, the activation distance $\Delta d = V/(b\, l)$ is not equal to the difference of the x coordinates $x_\mathrm{a} - x_\mathrm{e}$. Exceptions are only cases with $z_\mathrm{a} = z_\mathrm{e}$.

To illustrate the results, several cases of particles of octahedral shape with a rhombic cutting plane are calculated in [217]. Such particles act as obstacles in crystals with the NaCl structure like MgO [218]. Figure 4.26 shows the central and a noncentral cutting plane. At the borders, the slope is piecewise constant (for edge dislocations $x' = 0.707$ on the entrance side and $x' = -0.707$ on the exit side). Edge and screw dislocations intersect the particles in different ways. Figure 4.27 compares the V vs. τ^* curve with the simple $x_\mathrm{a} - x_\mathrm{e}$ vs. τ^* curve in a normalized way for a particle with a friction stress τ_f inside the particle and a difference between the dislocation energies in the particle E_p and the matrix E_m. Except for $\tau^* = 0$, the normalized activation

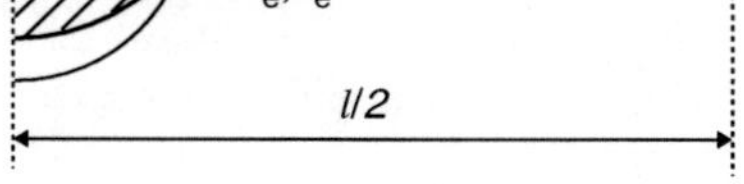

Fig. 4.25. Half of an extended obstacle and the dislocation line in the entrance and exit positions. The hatched area is the activation area

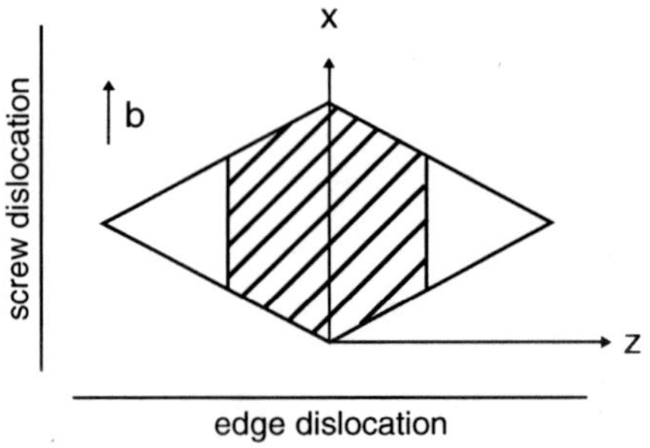

Fig. 4.26. Shape of the {110} cutting plane of an octahedral particle in the NaCl structure with edges parallel to ⟨110⟩ directions. The outer border marks a plane for a cut through the particle center, the hatched area is for a noncentral cut

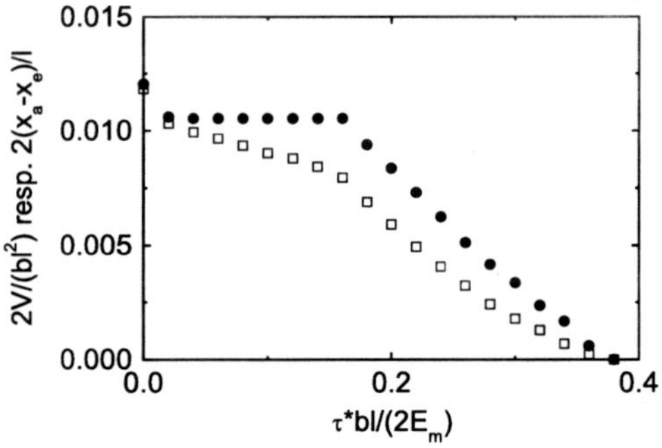

Fig. 4.27. Dependence of the normalized activation volume $2V/(bl^2)$ (*open squares*) and the normalized x difference $2\,(x_\mathrm{a} - x_\mathrm{e})\,/l$ (*full circles*) on the normalized stress for the interaction between edge dislocations and particles with $K_\mathrm{x} = K_\mathrm{z} = 0$, $\tau_\mathrm{f} bl/\,(2E_\mathrm{m}) = 10$, and $E_\mathrm{p}/E_\mathrm{m} = 1.1$. Data from [217]

volume is smaller than that following from the simple x difference. Thus, the curves of stress vs. activation volume (force–distance curves) obtain a narrow tip.

In particles with polygonal cutting planes, the dislocations can usually assume equilibrium positions only within restricted ranges of stress. These depend on x' and are different on the entrance and exit sides. For the particles of Fig. 4.27, stable positions are possible only for $0.16 \leq \tau^* bl/(2E_\mathrm{m}) \leq 0.382$ in the entrance half and for $\tau^* bl/(2E_\mathrm{m}) \leq 0.023$ in the exit half of the particle. As long as $\tau^* bl/(2E_\mathrm{m}) < 0.16$, the dislocation cannot enter the obstacle. It is pinned at the tip and bows out in the matrix with increasing stress so that $z_\mathrm{e} = 0$. Between $\tau^* bl/(2E_\mathrm{m}) = 0.16$ and 0.382, the dislocation enters the particle and forms a stable equilibrium position with z_e depending on the stress. On the exit side, the stress $\tau^* = 0$ corresponds to a certain z_e value. Regions of the particle with smaller z are overcome spontaneously. The different stress ranges are clearly visible in Fig. 4.27. The calculations were performed for the central cut. If the dislocation cuts the particle off-center, the cutting face is smaller. Correspondingly, also the activation energy is lower so that the particles cut at different heights produce a whole spectrum of obstacle strengths.

For particles with a smoothly curved border of the cutting plane exerting a friction stress τ_f, the dislocation can enter the obstacle only if the normalized maximum force is less than unity, $f_0 = \cos\phi_{min} < 1$. In view of Fig. 4.24, this means that the radius of curvature of the dislocation inside the particle (4.67) is greater than the radius $D/2$ of the particle

$$D < \frac{2E_p}{b(\tau_f - \tau^*)},$$

or, with $\tau_f = \gamma/b$ and $\tau^* \ll \tau_f$,

$$\gamma D < 2E_p.$$

Thus, while small particles can be cut, the dislocations cannot enter larger particles. These can be overcome only by bowing out the dislocation in the matrix to the critical Frank–Read configuration (3.47) shown before in Fig. 3.20a and c. This process was first studied by Orowan [219]. The critical stress to overcome a regular array of impenetrable obstacles is called the Orowan stress τ_{OR}. In the simple model, the Frank–Read stress is applied with the inter-particle distance L as the segment length. As the particles cannot be penetrated, the dislocations have to form loops around them, which is outlined in Fig. 4.28.

A complex analysis of the Orowan problem in a row of equidistant particles was given by Bacon et al. [105, 220], including elastic anisotropy as

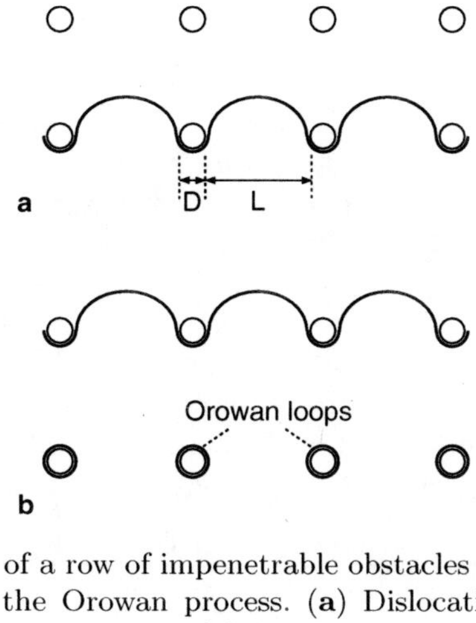

Fig. 4.28. Overcoming of a row of impenetrable obstacles of diameter D and inter-particle distance L by the Orowan process. (**a**) Dislocation bowing out between particles in the critical configuration. (**b**) The dislocation has passed the lower row of particles and has formed Orowan loops around them

well as the self-interaction along the dislocation line. The particles of circular shape on the slip plane of diameter D and interparticle distance L (Fig. 3.20a) are considered impenetrable obstacles of the same elastic constants as the matrix so that the dislocation has to bend around the particle border. For finding the equilibrium shape of the segments bowing out between the particles, numerical calculations were performed using the Brown–Indenbom–Orlov theorem as described in Sect. 3.2.7. The balance between the applied stress and the self-stress of the dislocation (3.43) was found by an iteration treatment. The applied stress was increased by small steps until no balance was reached. The respective stress is the Orowan stress. In the last stable configuration, the dislocation branches adjoining the particle on both sides were always nearly parallel. At higher stresses, the configuration collapses, and forms a closed loop around the particle before the remaining dislocation moves away.

The calculations yielded the following results.

- The numerical data on the Orowan stress can be summarized by using line tension arguments. As pointed out at the end of Sect. 3.2.7, the effective line tension of strongly bowed segments like those in the Orowan configuration is given by the line energy $E(\phi)$ (energy per dislocation length) of the segments adjacent to the pinning agents having an orientation angle ϕ. If the starting dislocation is a screw dislocation, this is the energy of an edge dislocation, and vice versa, as noticed already in [221]. In the general case of a mixed dislocation, the adjacent segments are not always perpendicular to the direction of the starting dislocation as sketched in Fig. 4.29a. To shift the dislocation forward, the length of these branches has to be increased. Thus, the Orowan stress can be written as

$$\tau_{OR} = \frac{2E_0(\phi)}{bL}\left(\ln\frac{X}{r_0} + B\right),$$

where $E_0(\phi)$ is the prelogarithmic factor of the energy of a straight dislocation of orientation ϕ, and r_0 is the inner cut-off radius as in (3.15). B is a numerical constant with an average of 0.65. The outer cut-off or screening distance X has to be chosen according to the distance to the dislocation closest to the considered one. Thus, when L is small compared to D, $X \approx L$ should be an appropriate value corresponding to the line tension solution. On the other hand, for small obstacles with L being large compared to D, the dislocation segments near the obstacles interact most strongly with the branches on the opposite sides of the obstacles so that the screening distance should be $X \approx D$. The simplest average between D and L with the described limiting behavior is the harmonic mean. Thus

$$\tau_{OR} = \frac{2E_0(\phi)}{bL}\left(\ln\frac{DL}{r_0(D+L)} + 0.65\right). \tag{4.71}$$

It is shown in [105, 220] that the numerical calculations fit this relation well.

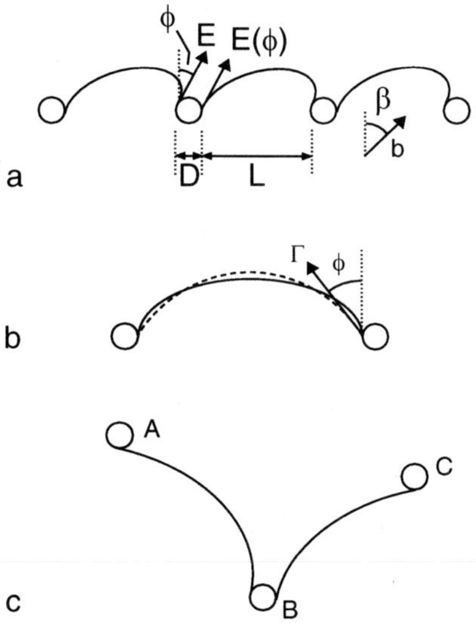

Fig. 4.29. Orowan process of a mixed dislocation in an anisotropic crystal. (**a**) Dislocation bowing out between particles. (**b**) Schematic of the comparison between the shape of a bowed segment calculated by the self-stress method (*solid line*) and the respective line tension analogue (*dashed line*). (**c**) Triplet of particles for calculating the breaking condition of the dislocation at the central obstacle B for the simulation of the Orowan stress in a random array of obstacles after [222]

- Elastic anisotropy may be considered by calculating $E_0(\phi)$ by means of anisotropic elasticity theory. For a number of metals, the respective line energy and line tension data are compiled in the form of Fourier coefficients in [98]. The angle ϕ depends on the character of the dislocation β and the crystal system. It is zero for screw and edge dislocations but may amount up to 30° for $\beta = 60°$. For details, see [105]. To express the results in terms of parameters of isotropic elasticity, it is possible to define a shear modulus by $\mu_{\mathrm{a}} = 4\pi E_{0\mathrm{s}}/b^2$, where $E_{0\mathrm{s}}$ is the prelogarithmic term of the energy of a screw dislocation, and a Poisson ratio ν_{a} by $1 - \nu_{\mathrm{a}} = E_{0\mathrm{s}}/E_{0\mathrm{e}}$, with $E_{0\mathrm{e}}$ being the prelogarithmic term of the energy of an edge dislocation. Both energies are calculated by anisotropic elasticity theory.
- The influence of the dislocation self-interaction can be observed by comparing the shape of the bowed-out dislocation segments in the critical configuration determining the Orowan stress calculated by the self-stress method (solid line in Fig. 4.29b) with that following from the line tension model (dashed line). The line calculated by the self-stress method touches the obstacles in perpendicular direction (in the figure) as described earlier. The dislocation line in the line tension solution, that is, the solution with L appearing in the logarithmic factor exhibiting approximately the same swept area, touches the particle at an angle ϕ. Thus, the effect of

the self-interaction is to pull both branches attached to the obstacle closer together. Accordingly, in the line tension approximation, the normalized force on the obstacle according to (3.48) and (4.49) is smaller than unity. Consequently, in the line tension model impenetrable obstacles appear as penetrable ones with a breaking force $f_0 < 1$.

To include randomness of the distribution of the particles into the self-stress calculations of the Orowan stress, computer simulations similar to those in [192] were performed for small particles with $D \leq 0.1\,L$ in the limit of isotropic elasticity and $\nu = 0$ [222]. The stability of the dislocation near an individual obstacle like B in the triplet ABC of Fig. 4.29c is primarily determined by the interaction between the two bowing segments AB and BC. Therefore, the breaking conditions of such triplets were calculated over a wide range of parameters using the self-stress relaxation technique outlined earlier. These results were represented by an empirical formula to be incorporated into the computer simulation code of the motion of a dislocation through a random array of particles. For this simulation, the dislocation segments neighboring the triplets were represented by two straight lines and all farther segments by single straight lines. The self-stress of all these straight segments at the position of the triplet was calculated and added to the applied stress. Thus, the motion of the dislocation was simulated without relaxation or iteration treatments. The result can be written again in a line tension analogue with the effect of the self-interaction expressed by a normalized breaking force f smaller than unity or a breaking angle ϕ different from zero

$$\tau_{\mathrm{OR}} = \frac{2\Gamma}{bL}\cos\phi\left[\left(0.8 + \frac{2\phi}{5\pi}\right)\cos\phi\right], \tag{4.72}$$

with

$$\Gamma = E_{\mathrm{d}} = \frac{\mu b^2}{4\pi}\ln\frac{1.2\,L}{r_0}$$

for $\nu = 0$ and

$$\cos\phi = \frac{\ln(3.3D/r_0)}{\ln(0.8L/r_0)}.$$

In these formulae, the line tension is that of a freely bowing segment disregarding the self-interaction. The normalized breaking force $\cos\phi$ considers the self-interaction, and the bracketed term represents the influence of the random distribution of the particles, which reduces the flow stress as in [192].

The topics of this section are important for the understanding of precipitation hardening and particle strengthening. Relevant reviews are found in [223–225]. If a metal is alloyed with some other substance and subjected to a sufficiently high temperature, the additions may solve in the matrix. These solutes can be treated as localized obstacles, giving rise to solution hardening. Depending on the obstacle strength, extension, and concentration, Friedel or Mott statistics has to be applied. This state can be frozen in by rapid quenching to a low temperature. Controlled by the phase diagram and the ability

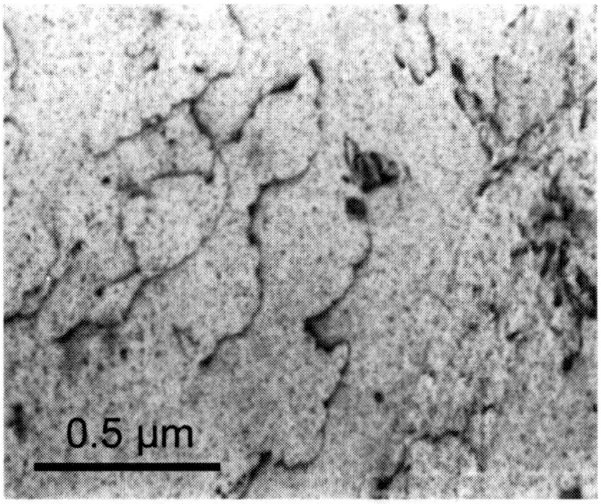

Fig. 4.30. Precipitation hardening in the technical alloy Al-4.54Zn-1.18Mg (wt%) homogenized for 2 h at 400° C, quenched and aged for 7 days at room temperature. The precipitates of the η' phase are visible as *dark dots*. Micrograph from [226]

of diffusion, that is, by annealing at a sufficiently high temperature, a second phase may precipitate. In any case, the flow stress is proportional to the reciprocal value of either the square lattice distance l_{sq} or the inter-particle distance L. Thus, the flow stress is approximately proportional to the square root of the concentration of the obstacles on the slip plane. If the precipitates are small, they are often coherent and may be treated as elastic inclusions as outlined in Sect. 3.2.6. During further annealing, the defect structure coarsens, with larger particles growing at the expense of smaller ones. This process is called ageing. For penetrable particles with a friction stress τ_f, the obstacle strength increases with increasing particle size. This leads to an increasing flow stress or age hardening. Above a critical size, the dislocations cannot enter the particles any longer. They bow out around the particles and overcome them by forming Orowan loops. At this stage, the obstacle strength no longer increases and the decrease of the obstacle concentration leads to a decrease of the flow stress. The alloy with the maximum flow stress is in the peak aged state. Afterwards, it is overaged. A typical example of these precipitation hardened alloys are the aluminium alloys discussed in Sect. 8.1. Figure 4.30 shows dislocations during in situ deformation in an HVEM of an Al-Mg-Zn alloy hardened by precipitates of the η' phase. The particles are visible as small dark dots. Because of their high obstacle strength, the dislocations are strongly bowed-out. Usually, incoherent particles cannot be intersected. Such particles are, for example, oxide dispersoids in metal matrices, forming the so-called oxide dispersion strengthened (ODS) alloys described in Sect. 8.3. At low temperatures, the oxide particles are overcome by the Orowan mechanism. They are very stable also at high temperatures being then surmounted by climb.

As long as the activation energy to overcome the obstacles and the number of involved atoms are small, the particles are passed by the aid of thermal activation, and the flow stress depends on the temperature and the deformation rate. The dislocation motion is then jerky on the scale of the particle distance. This is a small scale for solution hardening with Mott statistics, and a larger one for precipitation hardening with Friedel statistics. The force–distance curve of even relatively large particles may have a narrow tip so that thermal activation may still play some role. The Orowan mechanism for large impenetrable particles is of fully athermal character as the energies to form even small dislocation loops are too high for thermal activation, as mentioned in Sect. 3.2.2. In this case, the stress has to be high enough to overcome the strongest configurations. The dislocation motion is then very jerky on a large scale.

4.8 Dislocation Intersections

In most cases of plastic deformation, dislocation motion is not restricted to a single slip system, instead several slip systems are activated simultaneously (multiple slip). Then, the moving dislocations have to intersect dislocations of other slip systems, usually having different Burgers vectors. These dislocations are called forest dislocations. The process of dislocation cutting is very complex. The crossing dislocations experience a mutual long-range elastic interaction, as outlined in Sect. 3.2.5. In general, the interaction includes components of the interaction force out of the glide plane, which may induce cross glide or climb. Within the glide plane, the interaction leads to pinning either by attractive or repulsive forces with the dislocations bowing out between the forest dislocations. Figure 4.31 presents an example of an edge dislocation pinned by a screw dislocation with perpendicular Burgers vector (arrow). The bow-outs may be strong so that the moving dislocations have to form the critical half-circle (Frank–Read) configuration (3.47) to overcome the forest dislocations. Then, the segment length L equals the average distance between the forest dislocations $\varrho_\mathrm{f}^{-1/2}$, if ϱ_f is the forest dislocation density. Thus, the contribution of the dislocation forest to the flow stress can be written as

$$\tau_\mathrm{forest} = \alpha_\mathrm{f} \frac{\mu b}{2\pi} \sqrt{\varrho_\mathrm{f}} \,. \tag{4.73}$$

The factor α_f describes the strength of the actual interaction force and should be less than 2π. An alternative process is the formation of dislocation junctions if the interaction is attractive and reduces the dislocation energies (see Sect. 3.2.2 and Fig. 3.10). For b.c.c. crystals, values of $\alpha_\mathrm{f} \approx 0.4\,\pi$ were computed by virtual displacement of the triple nodes bounding the junctions [227]. In addition to these long-range interactions, there is a short-range contribution due to the formation of jogs or kinks in both dislocations as already shown in Fig. 3.16. As described in Sect. 3.2.5, the formation energy of the jogs

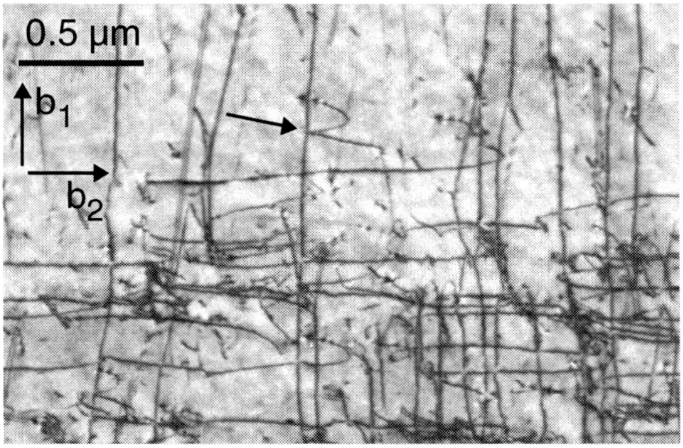

Fig. 4.31. Edge dislocation that is strongly pinned by a screw dislocation of perpendicular Burgers vector during in situ deformation of an MgO single crystal at room temperature. From the work in [164]

is approximately equal to the core energy $E_c = \mu b^2/10$ which, being in the order of magnitude of 0.5 eV, is well within the range of thermal activation at room temperature. If the dislocations are dissociated, the jog energy involves also the energy for the constriction (see Sect. 4.3). In ionic crystals, where the inserted half-plane consists of more than one lattice plane, jogs of different heights are possible. Then, jogs of small height (e.g., half-jogs in the NaCl structure) are electrically charged.

As the discussion has shown, the interaction between the dislocations during the intersection has both long-range and short-range components. Therefore, it is difficult to determine the interaction profile and the total activation energy. The activation distance is certainly less than about $10b$, and the segment length equals the distance between the forest dislocations $\approx \varrho_f^{-1/2}$. As the forest density for multiple slip increases with increasing plastic strain, the contribution of dislocation intersections to both the flow stress and the activation parameters depends on the strain. Unfortunately, also other processes depend on the strain so that it is difficult to clearly verify the contribution of dislocation cutting.

The influence of the forest dislocation density on the flow stress can quite directly be studied by so-called latent hardening experiments as mentioned in Sect. 4.3. Figure 4.32 shows a plot according to (4.73) of the secondary flow stress in NaCl single crystals as a function of the square root of the forest screw dislocation density generated during a primary deformation up to different plastic strains [228]. Linear regression analysis yields a straight line with $\alpha_f \approx 1.4\pi$, where the anisotropic shear modulus $\mu = \sqrt{c_{44}\left(c_{11} - c_{12}\right)/2}$ was taken. Considering that in addition to the screw dislocations counted in Fig. 4.32 also edge dislocations contribute to the forest, it follows that $\alpha_f \approx \pi$, which meets the theoretical expectation. The small intercept on the τ_s axis

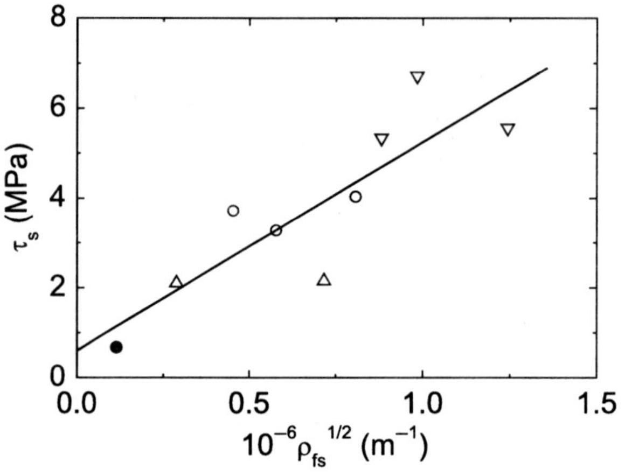

Fig. 4.32. Latent hardening experiments on high-purity NaCl single crystals deformed along two different ⟨100⟩ axes. Dependence of the secondary flow stress on the square root of the density of screw dislocations produced during a primary deformation on a slip system that is stress-free during the secondary deformation. 0% primary plastic shear strain (*full circle*), 3% (*upward triangle*), 6% (*open circles*), 12% (*downward triangles*). Data from [228]

indicates a flow stress contribution from the interaction between dislocations and aliovalent impurities. According to measurements of the strain rate sensitivity, the contribution of forest cutting is essentially of athermal nature.

After their formation, the jogs have to move together with the dislocation. As a jog is a piece of dislocation line that is not contained in the slip plane, it has its own slip plane defined by its Burgers vector and line vector as mentioned above in Sect. 3.1.2 and sketched in Fig. 3.5. According to the latter, jogs can glide (i.e., move conservatively without the production or annihilation of point defects) in the direction of the Burgers vector of the dislocation. This is the forward direction in edge components of dislocation loops but the direction along the dislocation in the screw parts. In mixed dislocations, jogs glide in a direction inclined to that of the dislocation line. As the jogs do not glide on the main glide plane of the dislocation, which is usually a plane of low glide resistance, they may experience additional glide resistance so that the dislocation has to form a cusp at them. This is shown in Fig. 4.33. The dislocation moving on the slip system with Burgers vector $\boldsymbol{b}_1$ mainly has an edge orientation. It has cut dislocations with the projection of the Burgers vector $\boldsymbol{b}_2$ of the slip system with a horizontal trace and obtained several jogs, which cause cusps in the dislocation line (arrow). As the curvature of the dislocation between the cusps indicates, the jogs cause a considerable resistance to the motion of the dislocation. This resistance should be higher in dislocations with large screw components as there the jogs have to move faster

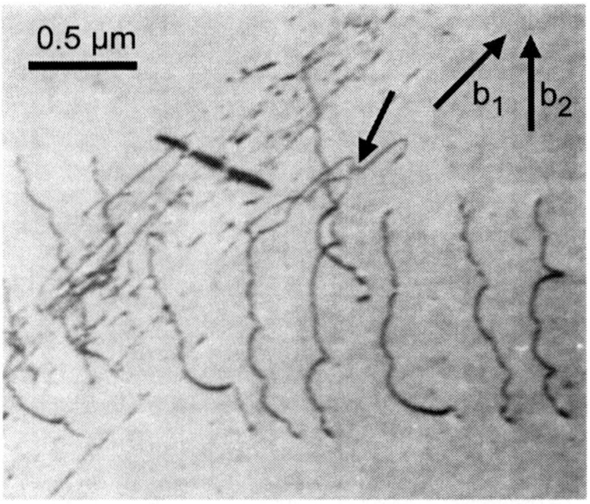

Fig. 4.33. Intersection between oblique slip bands during in situ deformation of an MgO single crystal with conservative motion of jogs in an edge dislocation (arrow). From the work in [79]

than the rest of the dislocations. Jogs in pure screw dislocations can move with the dislocations only by climb, that is, by the production or absorption of point defects. This process is called jog dragging and will be described in Sect. 4.10.5.

4.9 Dislocation Motion at High Velocities and Low Temperatures

The main topic of this book is dislocation dynamics under normal conditions, that is, at dislocation velocities up to the centimeter-per-second range and at temperatures between about liquid nitrogen temperature and about 90% of the melting temperature. At high velocities and low temperatures, special effects have to be considered, which will only briefly be outlined here, mainly following Hirth and Lothe [12]. See also the review by Alshits and Indenbom [229].

The elastic properties of dislocations described in Sect. 3.2 are those of resting dislocations. These solutions are valid up to velocities of some fractions of the velocity of sound. To treat dislocation motion at high speeds, inertial terms have to be included into the elastic equilibrium condition as body forces $-\rho_0 \left(\partial^2 u_{\mathrm{i}}/\partial t^2\right)$ caused by the acceleration of the volume element. Here, ρ_0 is the mass density of the material and t the time. For a screw dislocation in an elastically isotropic medium, the differential equation defining the stress field of the dislocation takes the form of the static one if a "relativistic" transformation is applied in analogy to the electro-magnetic case. The displacement field of the screw dislocation can then be written as

$$u_3(x,y,t) = \frac{b}{2\pi} \arctan \frac{\eta y}{x - v_\mathrm{d} t},$$

with

$$\eta = \sqrt{1 - \frac{v_\mathrm{d}^2}{c_\mathrm{t}^2}}.$$

Here, v_d is the dislocation velocity and c_t is the transverse sound wave velocity. For $v_\mathrm{d} = 0$, this formula equals (3.7). For v_d approaching c_t, that is, $\eta \rightarrow 0$, the stress field is distorted in such a way that, compared to the static case, the stresses are lower in x direction and higher in y direction. In the limit of $v_\mathrm{d} = c_\mathrm{t}$, the stresses are zero in x direction and infinite along y. Of course, the linear elasticity theory applied breaks down for these high stresses. The energy per unit length of the moving screw dislocation can be expressed as

$$E_\mathrm{ms} = E_\mathrm{ds}/\eta,$$

where E_ds is the respective energy of the resting screw dislocation (3.13). In relativistic mechanics, there is the same relation between energy and rest energy of a moving body. If $1/\eta$ is expanded to the first power of $v_\mathrm{d}^2/c_\mathrm{t}^2$, for certain boundary conditions, the energy of the moving screw dislocation can be written as

$$E_\mathrm{ms} = E_\mathrm{ds} + \frac{E_\mathrm{ds}}{2c_\mathrm{t}^2} v_\mathrm{d}^2.$$

The factor $m_\mathrm{d} = E_\mathrm{ds}/c_\mathrm{t}^2$ can be called the effective mass of the dislocation per length, in analogy to the respective formula in mechanics.

The described properties of dislocations together with effects of elastic nonlinearity lead to the conclusion that the respective sound velocity is a limiting velocity of the dislocation motion. Nevertheless, supersonic motion is possible if the dislocation gains energy by transformation of the range close to the slip plane into a state of lower energy, for example, by the motion of a single partial dislocation removing the attached stacking fault. Atomistic simulations have shown that a dislocation subjected to high stresses and created in the supersonic state at a strong stress concentration can continue to move at velocities higher than the respective sound velocity [230].

In the limit of linear elasticity theory, dislocations move at subsonic velocities without radiation of energy and without friction. However, elastic waves, that is, phonons, are scattered at the dislocations and transfer momentum to them. This scattering of thermal elastic waves results in a viscous damping of the dislocation motion at finite temperatures [231, 232]. In addition, the discreteness of the crystal lattice expressed by the Peierls barriers has to be considered, leading to the radiation of phonons from the dislocation cores. A further contribution to the drag stems from the scattering of conduction electrons in metals [233]. The viscous drag force due to phonon and electron damping is proportional to the dislocation velocity, that is,

$$f_{\mathrm{d}} = \tau^* b = B\, v_{\mathrm{d}}, \tag{4.74}$$

where f_{d} is the force per unit length, and B is the viscous drag coefficient. According to Leibfried's theory [231], the temperature dependence of the drag coefficient is given by

$$B = \frac{3kTz}{20c_{\mathrm{t}}b^2}.$$

Here, z is the number of atoms per unit cell. The linear temperature dependence was confirmed in [234] by molecular dynamics simulations using an embedded atom potential. The electron scattering contribution to the damping is supposed to be independent of the temperature. Turning to the superconducting state, it freezes out, resulting in a softening of the materials as first observed in [235–237], reviewed in [238] and recently in [239]. Other quantum effects at low temperatures are quantum-mechanical tunneling and the zero point vibrations of the dislocation segments adjoining localized obstacles. The first quantitative theory of these effects was given in [240]. These effects also result in a softening with respect to the Arrhenius relation. A dislocation moving in the field of the Peierls potential periodically accelerates and decelerates, resulting in the emission of lattice waves. Simulations in [241] proved that this process results in an energy loss, which is not proportional to the dislocation velocity and which is about one order of magnitude larger than the theoretical value of the phonon scattering mechanism.

To summarize the results on the dynamic behavior of dislocations at high speed and/or low temperature, the equation of motion of a dislocation segment can be established as

$$m_{\mathrm{d}} \frac{\partial^2 x}{\partial t^2} - \Gamma \frac{\partial^2 x}{\partial z^2} + B \frac{\partial x}{\partial t} = b\tau^* + \sum f_{\mathrm{di}}. \tag{4.75}$$

The first term stems from the dislocation inertia. The second term results from the line tension of the dislocation curvature. The third term represents the dynamic dislocation drag. On the right side of the formula are the forces due to the external stress and the sum of the internal forces due to other defects in the lattice. To model thermally activated dislocation motion, random forces can be added as done, for example, in [210].

In summary,

- Dislocations move viscously at high speeds and sufficiently high temperatures, and their mobility is controlled by the dynamic drag coefficient B.
- At (very) low temperatures, the dislocation motion is underdamped and inertial effects become important. They cause a short dynamic overshooting of the force acting on obstacles with respect to the overdamped case and result in a softening of the material.

- A softening is also observed after transition to the superconducting state. It is explained by both the direct consequence of the reduction of the damping coefficient B and the resulting increase of the importance of inertial effects.

4.10 Dislocation Climb

Up to this point, dislocation motion has been treated at temperatures where diffusion jumps are slow compared to the elementary steps of dislocation motion. Thus, the obstacles to glide were considered immobile. In the following sections, dislocation motion at high temperatures will be discussed where diffusion cannot be neglected. The first case is dislocation climb introduced in Sect. 3.1.2 with a point defect flux to or from the dislocations. This process can be driven by the applied stress or by chemical forces resulting from non-equilibrium point defect concentrations. Climb is an intrinsic mode of dislocation motion in cases where glide is suppressed, for example, in hexagonal metals loaded along the c axis, where the glide systems with the basal and prism planes have zero orientation factors for glide [242, 243]. Climb also is the dominating mode of dislocation motion during plastic deformation of quasicrystals as described in Chap. 10. In most cases, climb occurs in addition to glide, playing an important role in recovery processes where the dislocations stored during the deformation form low energy structures, which cannot be reached by glide alone, or where the dislocations even annihilate each other (Sect. 5.1.3). Besides climb, diffusion-controlled processes without a diffusion flux to or from the dislocations may influence the dislocation mobility. Point defects have elastic (and other) interactions with the dislocations, which result in point defect atmospheres around the dislocations consisting of both intrinsic (vacancies and interstitial atoms) and extrinsic point defects (foreign atoms), or of clouds of oriented states of defects with nonspherical stress fields. These atmospheres are dragged with the moving dislocations causing a frictional force. These processes are described in Sect. 4.11. To understand diffusive processes during dislocation motion, first point defect equilibria will be discussed.

4.10.1 Point Defect Equilibrium Concentrations

In a similar way, as a straight dislocation is not the state of lowest Gibbs free energy but contains an equilibrium concentration of kinks at a finite temperature (Sect. 4.2.2), the equilibrium state of a crystal contains certain equilibrium concentrations of intrinsic point defects. The derivation of the equilibrium concentrations is based on the same ideas as that for kinks. The increase in the free energy by the formation energy of the defects is compensated by the configurational entropy arising from the many different possibilities to arrange the defects. An intrinsic point defect is introduced into a crystal if an atom from a ledge in a surface step is removed and inserted into the interior of the crystal to form an interstitial, or it is removed from the interior and placed at a ledge of a surface step to form a vacancy. The process is connected with a change in the external volume of the crystals ΔV_{ext}. In analogy with (4.29), the atomic equilibrium concentration of interstitials or vacancies, that is, the concentration related to the number of possible interstitial or vacancy sites, is given by

$$c_0 = \exp\left(-\frac{\Delta G_{\mathrm{f}}}{kT}\right), \tag{4.76}$$

where ΔG_{f} is the Gibbs free energy of the formation of the point defects. The formation energy ΔG_{f} is calculated in isotropic elasticity theory by inserting a small sphere into a hole, which is either too small or too large for it. As in closed-packed structures there is no space for the interstitials, they are strong centers of repulsion, with their formation energies being high (several electron-volt). Consequently, in most materials the concentration of vacancies is much higher than that of the interstitials. Frequently, interstitials form a so-called dumbbell configuration where two atoms share one lattice site. Instead of spherical symmetry, these defects have a stress field of tretragonal symmetry.

The situation is more complicated in ionic crystals because charge neutrality has to be maintained. Therefore, in monovalent crystals like those of the NaCl structure, either equal numbers of vacancies in both sublattices (Schottky disorder) have to be created, or equal numbers of vacancies and interstitials in one sublattice (Frenkel disorder). Even mixtures of both kinds of disorder are possible.

Point defects as centers of compression or dilatation react with the hydrostatic components of external and internal stress fields

$$p = -(1/3)\left(\sigma_{11} + \sigma_{22} + \sigma_{33}\right).$$

The interaction results in a change in the formation energy by the reversible work $p\Delta V^*$. The thermodynamic formation volume ΔV^* is different for vacancies and interstitials. It is approximately equal to the atomic volume, $\Delta V^* \approx \Omega$. Hirth and Lothe [12] discuss the differences in the formation volume of vacancies and interstitials in detail. Owing to the reversible work of introducing a point defect, the equilibrium concentration with external and internal pressures being present is given by

$$c = \exp\left(-\frac{\Delta G_{\mathrm{fi}} - p\Omega}{kT}\right) = c_0 \exp\left(\frac{p\Omega}{kT}\right). \tag{4.77}$$

Formulae of this type are valid for all sources or sinks for point defects, that is, outer or inner surfaces and dislocations. The latter are sources of internal hydrostatic stresses. Accordingly, the point defect equilibrium concentrations differ near the dislocations with respect to the rest of the crystal. As long as the equilibrium concentrations are not established, the dislocations emit or absorb point defects and they climb. In addition, point defects can be emitted or absorbed at the surfaces. These processes take place until the equilibria with all sources and sinks of point defects are attained. On the other hand, external and internal stresses may force the dislocations to climb, resulting in nonequilibrium concentrations of point defects. Both situations are treated in the sections below following the presentation by Friedel [123].

4.10.2 Climb Forces

Force Required for Athermal Climb

As introduced in Sect. 3.1.2, climb motions of dislocations change the volume of the crystal. The volume change due to a displacement of a dislocation segment of length L and orientation $\boldsymbol{\xi}$ by $\mathrm{d}\boldsymbol{x}$ is given by (3.3). For athermal climb, that is, for climb caused only by the action of mechanical stresses, the volume change is realized by the generation of vacancies or interstitials. The force necessary follows from the number of defects to be created and their formation energy. For pure climb, the Burgers vector $\boldsymbol{b}$ and the line vector $\boldsymbol{\xi}$ of the dislocation are perpendicular to the displacement vector $\mathrm{d}\boldsymbol{x}$ in Fig. 3.4. Thus, both the cross products $\boldsymbol{\xi} \times \mathrm{d}\boldsymbol{x}$ and $\boldsymbol{b} \times \mathrm{d}\boldsymbol{x}$ lie in the slip plane. In addition, $\boldsymbol{b} \times \mathrm{d}\boldsymbol{x}$ is perpendicular to $\boldsymbol{b}$ so that the volume change in scalar notation is

$$\mathrm{d}V = bL \sin\beta \, \mathrm{d}x. \tag{4.78}$$

β is again the angle between the Burgers vector and the line direction. If β is small, the volume change is realized by the formation of individual point defects. Then, the energy per dislocation length necessary for producing a number $\mathrm{d}N$ of point defects is given by

$$\frac{\mathrm{d}W}{L} = \Delta G_\mathrm{f} \frac{\mathrm{d}N}{L} = \Delta G_\mathrm{f} \frac{\mathrm{d}V}{L\Omega}.$$

ΔG_f is the formation energy of the respective point defects, and the atomic volume Ω is the approximate formation volume. Introducing the volume change from (4.78) yields

$$\frac{\mathrm{d}W}{L} = \Delta G_\mathrm{f} \frac{b \sin\beta}{\Omega} \, \mathrm{d}x.$$

The necessary climb force per dislocation length is then

$$f_\mathrm{cl,max} = \frac{\mathrm{d}W}{L \, \mathrm{d}x} = \Delta G_\mathrm{f} \frac{b \sin\beta}{\Omega}. \tag{4.79}$$

A rough estimation of the formation energy of a vacancy is $\Delta G_\mathrm{fv} \approx \mu\Omega/4 \approx \mu b^3/4$. Accordingly, for pure climb of a pure edge dislocation ($\sin\beta = 1$), the climb force is

$$f_\mathrm{cl,max} = \frac{\mu b}{4}$$

or the required stress $\mu/4$. This is a stress in the order of magnitude of the theoretical yield stress of Sect. 1.1, which real crystals cannot sustain. Consequently, climb usually does not take place in an athermal way as assumed here, but in a thermally activated mode in small steps by climb of jogs connected with diffusion.

Applied Climb Forces

The climb force estimated above is the force required for pure athermal climb. The applied forces driving climb essentially have three origins.

1. The climb component of the Peach–Koehler force resulting from a mechanical stress Σ. It is given by (3.23) comprising a normal stress acting on the edge component and a shear stress acting on the screw component of the dislocation. The climb force per dislocation length resulting from a uniaxial stress on a dislocation with its Burgers vector parallel to the x axis is given by
$$f_c = \sigma_{11} b \sin\beta, \tag{4.80}$$
or $f_c = \sigma_{11} b$ for the pure edge dislocation. The reversible work in creating a vacancy is $f_c ha = f_c \Omega/b$. Here, h is the atomic distance perpendicular to the slip plane, and a is the atomic distance along the dislocation line. Two different situations are illustrated in Fig. 4.34. In the first one (a), an edge dislocation is located in a crystal subject to an external pressure p. The pressure reduces the energy of the formation of the point defects, usually vacancies, at the surfaces by $p\Omega$ so that the equilibrium concentration of the defects is given by (4.77). However, the energy is reduced by the same amount also at the dislocation so that the equilibrium concentrations on both sources or sinks are equal. Thus, after the equilibrium concentrations have been established, the dislocation will no longer climb. A different situation holds in Fig. 4.34b, where a uniaxial stress σ_{11} acts on the crystal. The uniaxial stress reduces the formation energy by $\sigma_{11}\Omega = -p\Omega$. Under

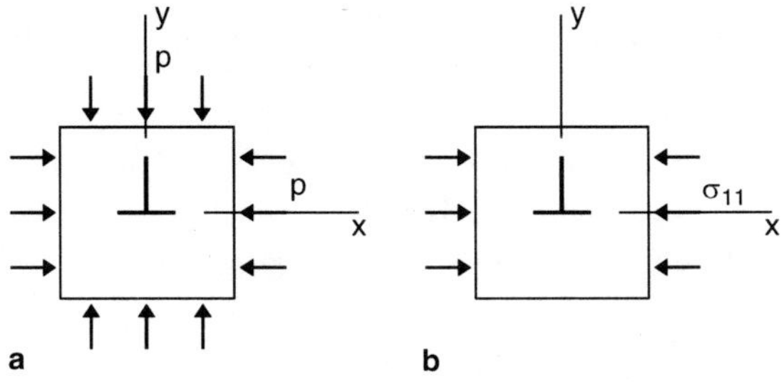

Fig. 4.34. An edge dislocation in a crystal under external pressure p (**a**) and uniaxial stress σ_{11} (**b**)

these conditions, the equilibrium point defect concentrations c are equal along the dislocation and at the two outer surfaces perpendicular to the x axis. They are given by (4.77). The equilibrium concentrations at the other four surfaces, however, equal the concentration c_0 without external stress (4.76). Consequently, there appears a point defect flux between these four surfaces and the dislocation, the dislocation climbs under the action of the external stress σ_{11}. This is a mechanism of high-temperature creep (Nabarro–Herring creep [244, 245]). It is observed in decagonal quasicrystals as described in Sect. 10.3.2 and illustrated in the Videos 10.6 and 10.9.

Usually, climb does not occur by the displacement of the dislocation line as a whole but by the motion of jogs along it. If a climb force acts on the dislocation, the force on the jog of height h is given by

$$F_{\text{jog}} = f_{\text{c}} h. \tag{4.81}$$

The elastic work done in shifting the jog along the dislocation and emitting a vacancy is then $F_{\text{jog}} a = f_{\text{c}} h a = f_{\text{c}} \Omega / b$. This work, which reduces the formation energy of the vacancy, is the same as that discussed earlier. Accordingly, the equilibrium point defect concentration near the dislocation does not depend on the structure of the dislocation.

During high-temperature deformation, not only climb processes induced by external stressees are important but also those due to internal stresses. They enable the formation of low-energy dislocation structures and the annihilation of dislocations if this is not possible by glide or cross glide alone, thus contributing to recovery.

2. The line tension of prismatic dislocation loops. Pure prismatic loops are loops with their Burgers vector being perpendicular to the loop plane. Thus, they are loops of pure edge dislocations, which can glide on a cylinder defined by the dislocation line and the Burgers vector. The force follows from (3.38) as

$$f_{\text{lt}} = \frac{\Gamma}{r}, \tag{4.82}$$

with the line tension Γ and the radius of curvature r. The line tension drives the loops to shrink, which for prismatic loops is only possible by climb.

3. The osmotic or chemical force. Osmotic climb forces result from sub- or supersaturations of intrinsic point defects. To derive the osmotic force, (4.79) is re-interpreted in the following way. If a force f acts on the dislocation, it may reduce or enhance the energy of formation of a point defect by an amount

$$\Delta\left(\Delta G_{\text{f}}\right) = f \frac{\Omega}{b \sin \beta}.$$

Then, the equilibrium concentration of vacancies near the dislocation, for instance, will be

$$c_{\mathrm{v}} = \exp\left(-\frac{\Delta G_{\mathrm{fv}} - f\Omega/(b\,\sin\beta)}{kT}\right)$$
$$= c_{\mathrm{v0}}\,\exp\left(\frac{f\Omega}{b\,\sin\beta\,kT}\right) = c_{\mathrm{v0}}\,\exp\left(\frac{\Delta\,(\Delta G_{\mathrm{f}})}{kT}\right). \qquad (4.83)$$

c_{v0} is again the equilibrium concentration of vacancies without a force acting. The force f can be considered an osmotic force due to the sub- or supersaturation of vacancies $c_{\mathrm{v}}/c_{\mathrm{v0}}$

$$f_{\mathrm{os}} = \frac{b\,\sin\beta}{\Omega}kT\ln\frac{c_{\mathrm{v}}}{c_{\mathrm{v0}}} = \frac{b\,\sin\beta}{\Omega}\,\mu_{\mathrm{cv}}. \qquad (4.84)$$

The osmotic force is often denominated a chemical force resulting from the chemical potential of nonequilibrium vacancies $\mu_{\mathrm{cv}} = kT\,\ln\,(c_{\mathrm{v}}/c_{\mathrm{v0}})$.

Large supersaturations of vacancies can result from radiation damage, from plastic deformation, and from quenching the crystal from a temperature near the melting point. The latter supersaturations can yield very high climb forces. If a dislocation sweeps and absorbs a complete layer of voids, the actual defect concentration is $c_{\mathrm{v}} = 1$. Inserting this into (4.84) and using (4.76) leads to the maximum force of (4.79)

$$f_{\mathrm{os}} = \Delta G_{\mathrm{fv}}\frac{b\,\sin\beta}{\Omega} = f_{\mathrm{cl,max}}.$$

Real quenching experiments do not result in such high climb forces. Nevertheless, quenching from a temperature T_1 to a temperature T_2 causes a supersaturation

$$\frac{c_1}{c_2} = \exp\left[-\frac{\Delta G_{\mathrm{fv}}}{k}\left(\frac{1}{T_1} - \frac{1}{T_2}\right)\right]$$

and a respective osmotic climb force

$$f_{\mathrm{os}} = \frac{b\,\sin\beta}{\Omega}\Delta G_{\mathrm{fv}}\frac{T_2 - T_1}{T_1} = f_{\mathrm{cl,max}}\frac{T_2 - T_1}{T_1}.$$

Thus, quenching the specimen from a temperature near the melting point may lead to quite high osmotic climb forces. Point defect supersaturations are generated also during plastic deformation. The osmotic force may then support climb under an external stress.

4.10.3 Emission- or Absorption-controlled Climb

Similar to a gliding dislocation, which does not shift a long straight segment from one Peierls valley to the next one but which realizes this shift by many small steps by shifting kinks, a climbing dislocation does not move a long straight segment in a single step. Instead, the motion takes place by emission or absorption of individual point defects at ledges J of the inserted half-plane

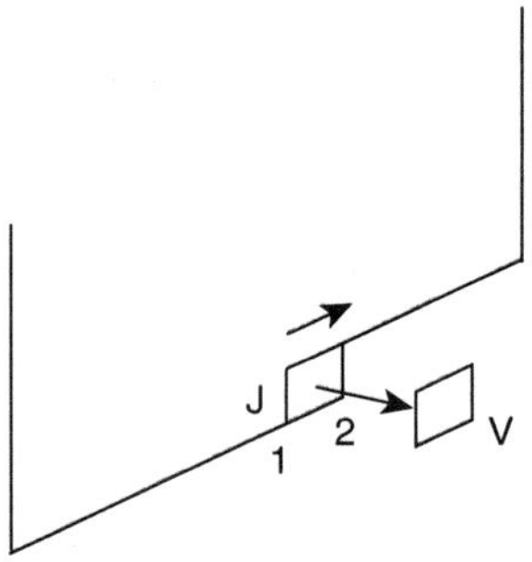

Fig. 4.35. Edge of the inserted extra half-plane of an edge dislocation with a jog J moving from position 1 to position 2 by the emission of a vacancy V

of a dislocation as illustrated in Fig. 4.35 for a pure edge dislocation. The ledges are identical with the jogs introduced in Sect. 3.1.2. To extend the extra half-plane, the jog has to be shifted to the right from position 1 to position 2. This is possible by the absorption of an interstitial or by the emission of a vacancy V. In most cases, the latter will happen because of the usually much higher formation energy of interstitials. Therefore, in the following only climb via vacancies will be treated. In deriving the climb velocity of a dislocation, two cases have to be distinguished. In the first one, the concentration of vacancies c_{v} is uniform in the crystal, also directly near the dislocation line. The climb velocity is then controlled by the emission or absorption of the vacancies. They diffuse to or from the dislocation so quickly that a concentration gradient does not form. This case is being treated in the present section. In the second case, vacancies are emitted or absorbed so quickly that a concentration gradient develops around the dislocation. Then, the climb velocity is not controlled by the emission or absorption of the point defects but by the diffusion problem around the dislocation. This will be discussed in the next section.

The emission rate of a vacancy at a jog under the action of an external force can be written as

$$\nu_{\mathrm{forw}} = n\nu_0 \exp\left(-\frac{\Delta G_{\mathrm{fv}} + \Delta G_{\mathrm{mv}} - f_{\mathrm{c}}\Omega/b}{kT}\right).$$

ν_{forw} is the rate of forward motion under the force f_{c}. The activation energy implies the formation energy of the vacancy ΔG_{fv}, as the vacancy has to be created, as well as the migration energy ΔG_{mv}, since the vacancy has to be separated from the jog. The sum of these energies is reduced by the reversible work done by the applied force. ν_0 is again an attempt frequency (vibration frequency of the jog). n is the coordination number, which is the number of lattice sites neighboring the jog. In the f.c.c. lattice, $n = 11$. The backward jumps are controlled by the concentration of vacancies that might jump to the jog being annihilated. By analogy with (4.77), the concentration near the jog is

$$c_{\mathrm{v}} = \exp\left(-\frac{\Delta G_{\mathrm{fv}} - f_{\mathrm{os}}\Omega/b}{kT}\right).$$

A vacancy can perform a diffusional jump to the jog from each of the n neighboring lattice sites so that the rate of backward jumps is

$$\begin{aligned}\nu_{\mathrm{backw}} &= n\nu_0 c \exp\left(-\frac{\Delta G_{\mathrm{mv}}}{kT}\right)\\ &= n\nu_0 \exp\left(-\frac{\Delta G_{\mathrm{fv}} + \Delta G_{\mathrm{mv}} - f_{\mathrm{os}}\Omega/b}{kT}\right).\end{aligned}$$

The velocity of the jogs along the dislocation is

$$\begin{aligned}v_{\mathrm{j}} &= \frac{a}{\sin\beta}(\nu_{\mathrm{forw}} - \nu_{\mathrm{backw}})\\ &= \frac{n\nu_0 a}{\sin\beta}\exp\left(-\frac{\Delta G_{\mathrm{fv}} + \Delta G_{\mathrm{mv}}}{kT}\right)\left[\exp\frac{f_{\mathrm{c}}\Omega}{b\,kT} - \exp\frac{f_{\mathrm{os}}\Omega}{b\,kT}\right]\\ &\approx \frac{D_{\mathrm{sd}}}{b\,\sin\beta}\left[\exp\frac{f_{\mathrm{c}}\Omega}{b\,kT} - \exp\frac{f_{\mathrm{os}}\Omega}{b\,kT}\right].\end{aligned} \tag{4.85}$$

In this equation,

$$D_{\mathrm{sd}} \approx n\nu_0 ab \exp\left(-\frac{\Delta G_{\mathrm{fv}} + \Delta G_{\mathrm{mv}}}{kT}\right) \tag{4.86}$$

is the self-diffusion coefficient via vacancies. The migration of the jogs results in a dislocation velocity

$$v_{\mathrm{cl}} = v_{\mathrm{j}}\, c_{\mathrm{j}}\, \sin\beta, \tag{4.87}$$

where

$$c_{\mathrm{j}} = \frac{a}{l_{\mathrm{j}}\,\sin\beta} \tag{4.88}$$

is the jog concentration along the dislocation, and l_{j} is the distance between the jogs. $\sin\beta$ takes into account that only lattice sites of the edge component of the dislocation are considered. The $\sin\beta$ terms in (4.85) and (4.87) cancel each other. If the jogs are in thermal equilibrium, their concentration is

$$c_{\mathrm{j}} = \exp\left(-\frac{\Delta G_{\mathrm{fj}}}{kT}\right), \tag{4.89}$$

with ΔG_{fj} being the formation energy of jogs. As discussed in Sect. 3.2.5, the jog energy is approximately equal to the core energy (3.16) $\Delta G_{\mathrm{fj}} \approx \mu b^3/10$, which is not a high energy so that the thermal jog density is high at high temperatures. At low temperatures and high climb stresses, jogs can nucleate in a way similar to that described for kinks (Sect. 4.2.2). For small forces, the bracketed terms in (4.85) can be approximated by

$$\frac{(f_c - f_{os})\,\Omega}{b\,kT} = \frac{f_{cl}\Omega}{b\,kT}.$$

Then, (4.85), (4.87), and (4.88) yield the climb velocity of the dislocation as

$$v_{cl} \approx D_{sd}c_j \frac{f_{cl}\Omega}{b^2\,kT} \approx D_{sd}c_j \frac{f_{cl}b}{kT}. \tag{4.90}$$

This is the classical Einstein relation between the diffusion rate and the applied force for climb via the motion of jogs. The activation energy is the self-diffusion energy plus the jog formation energy, $\Delta G = \Delta G_{fv} + \Delta G_{mv} + \Delta G_{fj}$.

4.10.4 Diffusion-controlled Climb

The preceding derivation of the climb velocity of dislocations is based on the condition that the applied forces are not very small and that there is no concentration gradient of point defects around the dislocations. This means that the emitted defects diffuse fast enough away from the dislocations or that the defects to be absorbed diffuse fast enough towards them. If this is not the case, the dislocations become saturated with vacancies and a concentration gradient develops around them. Then, the climb velocity is no longer determined by the process of emission or absorption of the point defects at the dislocations but by the diffusion flux owing to the concentration gradient. It was shown in [246,247] that this situation holds in many experiments. During diffusion-controlled climb, the flux of vacancies emitted or absorbed by the dislocation has to be equal to the flux through some border face around the dislocation [248]. If the dislocation climbs at a velocity v_{cl}, the flux per dislocation length is $J = v_{cl}b\,\sin\beta/\Omega$. It is determined by a diffusion problem around the dislocation applying Fick's law

$$J = \frac{v_{cl}b\,\sin\beta}{\Omega} = \frac{b\,D_{sd}}{\Omega}\operatorname{grad} c. \tag{4.91}$$

The solution depends on the proper boundary conditions and contains a characteristic length l_c. For a climbing straight dislocation in an environment of other parallel dislocations of both signs, suitable boundary conditions are given by a cylinder of radius R around the dislocation. Outside this radius, the concentration of point defects equals the equilibrium concentration without stress, $c \approx c_0$. The value of R is about half the distance between the dislocations, similar to the outer cut-off radius for determining the dislocation energy (Sect. 3.2.2). The characteristic length is then $l_c \approx b\,\ln(R/b)$. The climb velocity (i.e., the velocity in the direction perpendicular to $\boldsymbol{b}$ and the line vector $\boldsymbol{\xi}$) is [12]

$$v_{cl} = \frac{2\pi D_{sd}f_{cl}\Omega}{b^2\,\sin\beta\,kT\ln(R/b)}, \tag{4.92}$$

or, neglecting the weak dependence on R, it becomes

$$v_{\mathrm{cl}} \approx \frac{D_{\mathrm{sd}} f_{\mathrm{cl}} \Omega}{b^2 \sin\beta \, kT}. \tag{4.93}$$

This equation is very similar to that for emission or absorption-controlled climb (4.90), except that the jog concentration is not contained. The transition to saturation of the dislocation with vacancies begins when the distance between the jogs l_{j} in (4.88) becomes smaller than the characteristic diffusion length l_{c}. For thermal jogs, this occurs at high temperatures. In an exact way, the climbing dislocation should be treated by a diffusion problem in a moving reference system. However, under usual climb conditions, the above static solution is sufficient. Other geometric configurations of the dislocations lead to different estimates of the characteristic length l_{c}. An important case is the climb under the attractive force between the dislocations of an edge dislocation dipole. It is a mechanism of recovery of the dislocation density. Both dislocations are located near the center of the cylinder and both emit or absorb vacancies simultaneously into the neighborhood. The climb velocity of each of the two dislocations is then approximately half of that of a single dislocation under the same force.

The activation energy for diffusion-controlled climb is the self-diffusion energy $\Delta G_{\mathrm{sd}} = \Delta G_{\mathrm{fv}} + \Delta G_{\mathrm{mv}}$. For emission or absorption-controlled climb, the jog formation energy ΔG_{fj} has to be added. As for low climb forces in all formulae the climb velocity is proportional to the climb force and accordingly also to the climb component of the stress, the stress exponent m of (4.10) is equal or close to unity. The activation volume is very small, approximately the atomic volume Ω.

In several diffusion problems, diffusion along preferred paths is important. Such preferred paths are the cores of the dislocations. The diffusion along the dislocation cores is termed pipe diffusion. Its activation energy is considerably lower than the self-diffusion energy, down to half of it. Pipe diffusion plays a role in climb processes driven by the line tension of the dislocations, for example, in changes of the shape of prismatic dislocation loops due to nonconservative dislocation motion.

4.10.5 Jog Dragging

For climb of a dislocation as a whole, as it was treated above, the work term like $f_{\mathrm{c}}\Omega/b = \sigma_{11}\Omega$ in (4.85) is small so that almost the whole diffusion energy has to be expended. Usually, this is possible only by the help of thermal activation at high temperatures. An exception is a dislocation with a high screw component and a relatively low jog density subjected to a glide stress. Figure 4.36a shows such a dislocation with two jogs J of atomic height and opposite sign. In principle, the jogs can glide in the direction of the Burgers vector, that is, along the shaded areas. However, different processes may

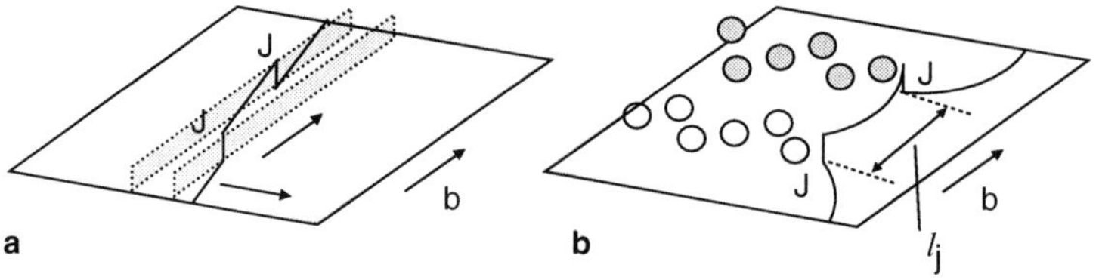

Fig. 4.36. Jog dragging. (**a**) Dislocation with near-screw character containing two jogs J and slip planes of the jogs (*shaded areas*). (**b**) Emission of different types of point defects by jog dragging

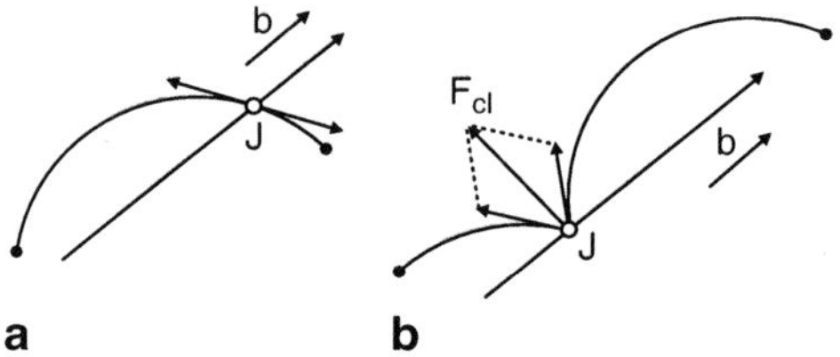

Fig. 4.37. Equilibrium configurations of jogs with respect to glide. (**a**) Glide configuration. (**b**) Configuration with climb force

prevent glide. The first one is of dynamic nature. Because of the large screw component, the jogs have to glide much faster than the other parts of the dislocation. Besides, usually the jogs have to glide on a crystallographic plane different from the main slip plane of the dislocation. Both processes lead to a higher glide resistance of the jogs so that they cannot follow the main parts of the dislocation by glide alone. In pure screw dislocations, there is no glide force at all acting on the jogs. Localized obstacles pinning the dislocation influence the motion of the jogs. As pointed out in [249, 250] and outlined in Fig. 4.37, a jog (open circle) situated between two localized obstacles (solid dots) can have two equilibrium configurations with respect to glide. The arrow indicates the line parallel to the Burgers vector along which the jog can glide. In Fig. 4.37a, the line tensions of the segments adjoining the jog cancel each other so that there is neither a glide nor a climb force resulting. In Fig. 4.37b, the glide components cancel each other but there remains a force F_{cl} in climb direction. When a jog is formed by dislocation intersection, a deep cusp develops in the dislocation line because of the strong elastic interaction between the cutting dislocations. Consequently, intersection jogs will be generated in the configuration of Fig. 4.37b. Thus, the presence of localized obstacles stabilizes jogs in the their nonconservative configurations. A model of stage I deformation was based on these ideas [251].

If the jogs in near-screw dislocations cannot glide sidewise, the segments of length l_j between the jogs bow out under stress as in Fig. 4.36b. While for climb of a straight dislocation, the climb force acting on a jog is $F_{jog} = f_c h = \sigma_{11} h b \sin\beta$ ((4.80) and (4.81)), it now becomes $F_{jog} = \tau^* b\, l_j$ (4.2).

Accordingly, the glide velocity of the dislocation controlled by the climb of the jogs is given by

$$v_{\mathrm{d}} = b\nu_{\mathrm{o}} \exp\left(-\frac{\Delta G_{\mathrm{f}} + \Delta G_{\mathrm{m}} - \tau^* b^2 l_{\mathrm{j}}}{kT}\right). \tag{4.94}$$

This process is called jog dragging (see, e.g., [252,253]). Here, the activation distance was set to b. The activation energy comprises the energy of formation of the point defects ΔG_{f} and their migration ΔG_{m}, as the defects have to be separated from the jog. Separation could also occur by sidewise glide of the jog. However, if sidewise glide were easily accessible, the jog had glided away before the point defect had formed. Thus, if the jog moves by the creation of point defects, the migration energy has to be taken into account. In Fig. 4.36, the two jogs are of different sign. Accordingly, one jog is a vacancy-producing jog while the other one is an interstitial-producing jog. At low velocities and small stresses, the interstitial jog moves by absorbing vacancies. However, the concentration of available vacancies limits this process so that at high stresses the jog has to produce interstitials. As their formation energy is mostly much higher than that of vacancies, the interstitial jogs represent stronger obstacles controlling the total dislocation velocity.

The jog distance l_{j} can be several orders of magnitude larger than h. As a consequence, the work term $F_{\mathrm{jog}} b = \tau^* b^2 l_{\mathrm{j}}$ can now assume a large fraction of the self-diffusion energy, thus enabling jog dragging even at quite low temperatures like room temperature. Accordingly, the activation energy depends strongly on the stress. The activation volume is given by $V = b^2 l_{\mathrm{j}}$, which is comparable to that of localized obstacles. However, there are clear differences between both processes. Jog dragging is not subject to Friedel statistics, and the forward slip distance after successful thermal activation is b but not l_{sq}. The jogs are created by dislocation intersection, they are certainly not in thermal equilibrium. The segment length l_{j} can therefore vary during plastic deformation.

Jog dragging, that is, climb of jogs driven by the glide component of the applied stress field including the creation of isolated point defects, is restricted to jogs of only very few atomic distances in height. Higher jogs drag dislocation dipoles, so-called debris. This process is discussed in Sects. 5.1 and 5.2. Jog dragging is thought to contribute to the thermal part of the flow stress and to be a main process of the generation of point defects during plastic deformation. However, the parameters of the model may vary to a large extent, impeding the identification by macroscopic tests.

4.11 Drag Forces due to Point Defect Atmospheres

The hydrostatic stress field of dislocations changes the equilibrium concentrations of intrinsic and extrinsic point defects near the dislocation cores. Therefore, atmospheres form with point defect concentrations deviating from

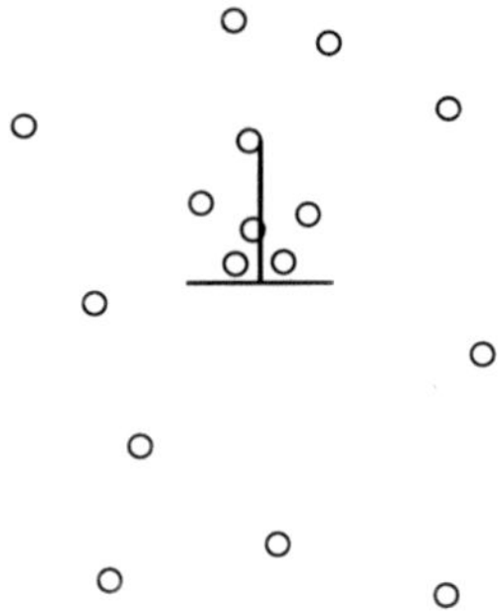

Fig. 4.38. Vacancy atmosphere around an edge dislocation

the stress-free equilibrium concentration thus reducing the total energy of the system of dislocations and point defects. While moving, the dislocations are shifted away from the centers of the atmospheres, which are dragged behind. This is connected with an increase in energy and a respective friction force. These processes will be discussed below.

The edge components of dislocations involve a hydrostatic stress field plotted in Fig. 3.9a, b and given by (3.30) and (3.31). φ is the coordinate angle around the dislocation. The latter equation describes a region of hydrostatic compression on the side of the inserted half-plane of the dislocation, and a region of tension on the opposite side. Inserting the distribution of the pressure around the dislocation into (4.76) reveals an increased equilibrium concentration of vacancies and a decreased concentration of interstitials in the region of compression. The opposite behavior is observed for the tension side as illustrated for the vacancies in Fig. 4.38. The above treatment is valid for the edge component of straight rigid dislocations. Straight screw dislocations do not react with the centers of expansion or dilatation. However, as the stress field of such an elastic inclusion is of pure shear character, the dislocations will not remain straight. Instead, they develop a wavy shape where an initially straight screw dislocation also contains parts with an edge component so that both edge and screw dislocations interact with the centers of dilatation. Besides, higher-order effects in the pressure become important at the high stresses close to the dislocation cores. These effects result in an attractive interaction between vacancies and all types of dislocations.

In alloys, foreign atoms solved in the matrix essentially determine the dislocation mobility. The solute atoms can either be placed on regular lattice sites to form a substitutional alloy, or they are situated on interstitial sites. In both cases, the defect can be approximated by a spherical elastic inclusion as done for intrinsic point defects and outlined in Sect. 3.2.6. If the solute is inserted near a dislocation with its field of hydrostatic pressure p around it, the reversible work of insertion is given by $p\,(\Omega_{\mathrm{s}} - \Omega) = p\Delta\Omega$, where Ω is the atomic volume, and Ω_{s} is the volume of the solute. This work changes the concentration of solutes around the dislocation from the average (equilibrium) concentration c_{s0} to

$$c_s = c_{s0} \exp\left(-\frac{p\Delta\Omega}{kT}\right). \tag{4.95}$$

Inserting the interaction constant β from (3.32), there follows a distribution of the solutes around the dislocation as

$$c_s = c_{s0} \exp\left(-\frac{\beta \sin\varphi}{rkT}\right). \tag{4.96}$$

The distribution is very similar to that of intrinsic point defects as presented in Fig. 4.38. Solutes with $\Omega_s < \Omega$ accumulate on the compression side of the dislocation, while the dilatational region is depleted. Again, the energy of the dislocation is reduced by the formation of the solute atmosphere. Integrating the solute distribution in a cylinder around the dislocation with an inner and outer cut-off radius shows that the dislocation accumulates solute atoms. With characteristic values of the solutes, in a 2% alloy about one solute atom per atomic plane segregates in the elastic stress field of the dislocation at about 300°C [12]. Additional solute atoms can segregate in the dislocation cores. This process is described by a binding energy between solute and dislocation. The possibilities of arranging the solutes in the dislocation core depend on the crystallographic direction of the dislocation line. This effect may promote particular orientations of the dislocations, which need not correspond with the directions of the Peierls valleys. Further elastic interactions between dislocations and solute atoms were pointed out in Sect. 3.2.6.

For dislocations dissociated into partial dislocations, solute segregation effects at the stacking faults may be important. As at the stacking fault, the regular stacking sequence is violated, the energy of a solute at the fault plane will, in general, differ from that in the undisturbed matrix. Therefore, also the equilibrium concentration at the fault will differ from that in the bulk. The problem is treated in the framework of adsorption theory. Solute atoms can be attracted or repelled by the fault, leading to the accumulation or depletion of solutes. In both cases, the fault energy is reduced, yielding an increased dissociation width of the dislocation (e.g. [254]). During the motion, the dislocation is dragged out of its equilibrium configuration, resulting in a friction force. This mechanism was first suggested by Suzuki [255, 256] and is called Suzuki effect. For a review see [257].

If the dislocation is moved under the action of a shear stress τ, it is dragged out off its equilibrium solute atmosphere described by (4.96). Here, c_{s0} is the solute concentration at $r = \infty$. With the displacement x of the dislocation with respect to the frozen-in atmosphere, the necessary force per unit length of the dislocation is

$$F = \tau b = \frac{\partial W(x)}{L \partial x}.$$

L is the dislocation length and $W(x)$ is the reversible work to drag the dislocation out of the cloud. It is given by $W(x) = \beta \sin\varphi / r$ or, in cartesian coordinates,

$$W(x) = \frac{\beta y}{x^2 + y^2}.$$

The force the displaced solute atmosphere exerts on the dislocation is equal to that the dislocation exerts on the atmosphere. Thus, the atmosphere is driven to move with the dislocation. The respective diffusion problem is a very complex two-dimensional one. It was first attacked by Cottrell and Jawson [258] and Cottrell [259]. The effect is therefore called Cottrell effect, and the solute cloud Cottrell atmosphere. The treatment was revised by Hirth and Lothe [12]. It is reduced to a one-dimensional problem of the diffusion in a moving potential well in layers parallel to the slip plane. The solutions depend on the dislocation velocity. At low velocities, the distribution of the solute concentration around the dislocation is close to the equilibrium distribution (4.96), and this distribution diffuses with the dislocation. Its velocity is given by

$$v_{\mathrm{d}} = \frac{\tau b^4 DkT}{c_{\mathrm{s0}}\beta^2 \ln (y_2/r_1)}. \tag{4.97}$$

D is the diffusion coefficient and y_2 and r_1 are cut-off parameters. The outer cut-off radius is given by $y_2 = D/v_{\mathrm{d}}$ as long as it is smaller than the distance R between the dislocations or the size of the crystal. Otherwise, $y_2 = R$. In the first case, the dependence of the velocity on the force or stress is not linear. The inner cut-off radius $r_1 = \beta/(kT)$ limits the range in which the many assumptions of the calculation are valid. If r_1 turns out to be smaller than the usual inner cut-off radius r_0, r_1 can be replaced by the latter.

At high dislocation velocities, the Cottrell atmosphere is distorted, reduced in size, and the concentration of the solutes tends towards the equilibrium concentration c_{s0}. Above a critical velocity v_{dc}, the dragging stress even starts to decrease with increasing velocity. In the high-velocity range, the extension of the Cottrell atmosphere is no longer controlled by diffusion, that is, by a random motion of the solutes, but by the directional drift in the stress field of the dislocation. In the upper range with no solute cloud left, the dislocation velocity shows a reciprocal dependence on the stress

$$v_{\mathrm{d}} = \frac{\pi c_{\mathrm{s0}} D\beta}{\tau b^6 kT}. \tag{4.98}$$

In the intermediate range, no analytical treatment is available but the limiting behavior suggests the existence of a maximum drag stress. Figure 4.39 presents a schematic plot of the velocity dependence of the drag stress. The abscissa is a logarithmic axis either of the dislocation velocity or, via the Orowan equation (3.5), of the plastic strain rate $\dot{\gamma}$. As in thermal activation theory, an increase in the logarithm of the rate corresponds to a decrease in temperature, the abscissa can also be read in opposite direction as a temperature axis. Three ranges of the curve can be distinguished, the low-velocity region A with an increasing drag stress and a diffusing Cottrell atmosphere,

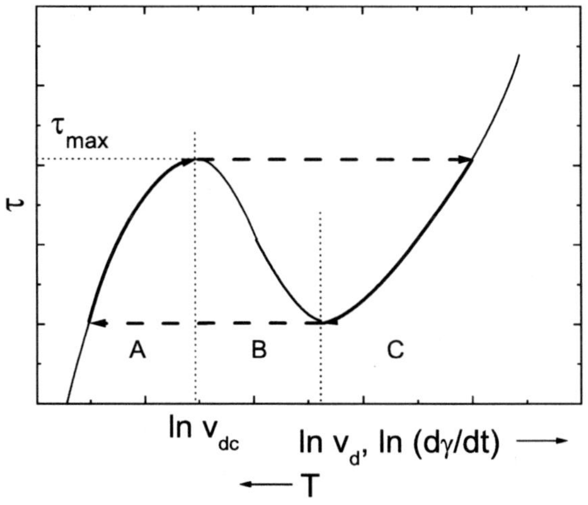

Fig. 4.39. Schematic representation of the dependence of the drag stress due to the interaction between dislocations and solute atoms on the dislocation velocity. Ranges A and B are controlled by the Cottrell atmosphere, in range C the solutes act as localized obstacles

the high-velocity range B with a decreasing drag stress and a reduced atmosphere moving by drift, and finally region C at very high velocities where the solutes can be considered fixed and acting as localized obstacles.

The more exact treatment of the Cottrell effect in [258] yields estimates of the critical dislocation velocity v_{dc} and the maximum drag stress $\tau_{\max}$

$$v_{\mathrm{dc}} = \frac{\dot{\gamma}_{\mathrm{c}}}{b\varrho_{\mathrm{m}}} \approx \frac{4\,DkT}{\beta} \qquad \text{and} \tag{4.99}$$

$$\tau_{\max} \approx \frac{17\,c_{\mathrm{s0}}\beta}{b^4}. \tag{4.100}$$

Several more refined calculations have shown that the numerical factors especially in (4.100) are uncertain up to factors of about 5.

The rate or temperature dependence of the drag stress of Fig. 4.39 and the respective dislocation behavior show some remarkable features. The most important one is the anomalous increase of the drag stress with increasing temperature in range B. This is one possibility to explain the so-called flow stress anomalies in some alloys and particularly in several intermetallic alloys. Figure 4.40 presents an example of the intermetallic alloy FeAl, where the flow stress anomalously increases between about 250 and 550°C. The mechanisms of the flow stress anomaly will be described in more detail in Sect. 9. Irrespective of the active mechanism in FeAl, (4.100) can be used for a rough estimate of the concentration of solutes necessary to explain a significant contribution of the Cottrell effect to the flow stress. In Fig. 4.40, the anomalous increase of the flow stress amounts to about $\tau_{\max} = 2 \times 10^{-3}\,\mu$. With a

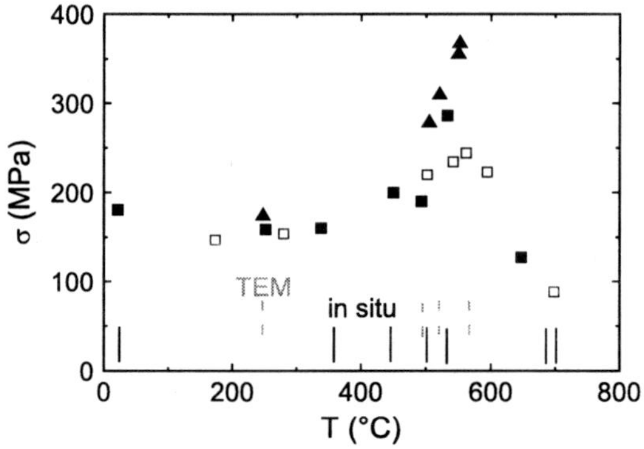

Fig. 4.40. Temperature dependence of the flow stress in Fe-43 at% Al single crystals loaded along a [753] axis. *Solid symbols*: initial flow stress of individual experiments. *Open symbols*: flow stress after temperature changes corrected for work-hardening. *Triangles*: $\dot{\varepsilon} = 10^{-4}\,\mathrm{s}^{-1}$. *Squares*: $\dot{\varepsilon} = 10^{-5}\,\mathrm{s}^{-1}$. *Dashed grey vertical lines* indicate temperatures of TEM studies of the dislocation structure, and *black vertical lines* those of in situ straining experiments. Data from [8]

difference in the atomic volumes of the solute and the matrix of $\Delta\Omega = 0.1\,b^3$ and $\nu = 0.3$ in (3.32), the necessary solute concentration is $c_{s0} \approx 5 \times 10^{-3}$. Thus, solute concentrations of the order of magnitude of several 1,000 ppm are necessary to reach a remarkable contribution of the Cottrell effect to the flow stress.

In a plot like Fig. 4.39, with the stress plotted as a function of the logarithm of the strain rate, the slope of the curve $\partial\tau/\partial\ln\dot{\gamma}$ is directly related to the strain rate sensitivity

$$r = \frac{\partial\sigma}{\partial\ln\dot{\varepsilon}} = \frac{\partial\tau}{m_s\partial\ln\dot{\gamma}} .$$

m_s is the orientation factor. In the same figure, in range C of the action of localized obstacles, the curve is bent upwards, that is, the strain rate sensitivity r increases with increasing strain rate $\dot{\gamma}$ or stress τ. This is the normal behavior of thermally activated processes. Contrary to that, in the low velocity range A of the Cottrell effect, the curve tends to the maximum at v_{dc}. That means that the strain rate sensitivity decreases with increasing strain rate or stress towards zero. In the present book, this behavior is called an inverse dependence of the strain rate sensitivity r on the strain rate $\dot{\gamma}$ or the stress τ. Finally, in range B of the curve, the strain rate sensitivity is negative. Then, the deformation is unstable as the flow stress decreases with increasing strain rate. Such plastic instabilities will be described in Sect. 5.3.

In the above three ranges, the kinematic behavior of the dislocations is different. In the normal range C, the behavior depends on the mechanism controlling the dislocation mobility. When the action of the solutes as localized obstacles is significant, the dislocation motion is jerky on a microscopic scale. In the stable range A of the Cottrell effect, the dislocation motion is smooth and viscous, while in the unstable range B dislocations frequently are generated in avalanches, moving very fast until the stress is relaxed to a value where only a slow steady motion is possible.

The segregation of solute atoms at dislocations is frequently designated as strain ageing. In static strain ageing, the resting dislocations acquire their equilibrium solute clouds. To break the dislocations away from their solute clouds requires a stress much higher than that for steady state motion. This leads to a temporary increase of the stress followed by a decrease, that is, a yield point or yield drop effect. In dynamic strain ageing, the mobility of the dislocations is controlled by the Cottrell effect or a similar mechanism. This is often associated with a very low or even negative strain rate sensitivity and with plastic instabilities. Point defect atmospheres may arise not only from extrinsic point defects but also from intrinsic ones as described above and illustrated in Fig. 4.38. This topic will be discussed in more detail in Sect. 9. In materials with a high Peierls stress, the segregation of solutes to dislocations is treated in [260].

Point defects with a tetragonal stress field may orient in the stress field of the dislocations and give rise to the induced Snoek effect. Such defects are, for example, interstitials in the dumbbell configuration or divacancies or associates between an aliovalent impurity atom and a charge-compensating vacancy in ionic crystals. In stress-free cubic crystals, the orientations of these defects are equally distributed over the three equilibrium orientations. In the stress field of a dislocation, the defects may orient so that the three equilibrium orientations are no longer equally occupied. These so-called Snoek atmospheres may be dragged together with the moving dislocations, similar to the Cottrell atmospheres. Here, only the orientational state of the solutes moves with the dislocations without a net solute diffusion. After the dislocations have moved away, the oriented state relaxes so that the equilibrium orientations are again equally occupied. The problem has been attacked by Schoeck and Seeger [261]. They show that the maximum drag stress of the induced Snoek effect occurs when the dislocation moves during the relaxation time $t_{\mathrm{rel}} = 1/\omega$ just over the width r_{s} of the potential well. Thus, the critical velocity at the flow stress maximum is given by

$$v_{\mathrm{dc}} = \frac{r_{\mathrm{s}}}{t_{\mathrm{rel}}} = r_{\mathrm{s}}\omega .$$

The diffusion jump frequency ω is given by a formula similar to that for the motion of kinks (4.34), with the migration energy of the moving point defect. Hirth and Lothe [12] solve the problem similarly to the Cottrell problem, which for the dislocation velocity in the limiting low-velocity range yields

$$v_{\mathrm{d}} = \frac{3\tau b^4 \omega}{2\pi c_{\mathrm{s}} kT},$$

with c_{s} being the concentration of the tetragonal defects. The drag stress in the high-velocity drift region reads

$$\tau = \frac{4\pi c_s \omega \beta^2}{3 v_d b^4 kT} \ln \frac{v_d}{\omega b}.$$

As a consequence, the induced Snoek effect shows a behavior similar to that of the Cottrell effect with a positive strain rate sensitivity at low velocities, and a negative one at high velocities. In both effects, the drag stress is proportional to the solute concentration, in contrast to the action of the solutes as localized obstacles where the flow stress contribution is proportional to the square root of the concentration of defects. The temperature dependence of the Cottrell effect is given by the diffusion coefficient of the solutes, which mostly includes the activation energy of the formation of vacancies in addition to the activation energy of migration, as most diffusion mechanisms are vacancy mechanisms. The activation energy of the Snoek effect is only the migration energy.

4.12 Dynamic Laws of Dislocation Mobility

In many materials, dislocation velocities v_d have been measured as a function of the applied (effective) stress τ^* and temperature T, mainly by selective etching described in Sect. 2.2. The stress dependence of screw and edge dislocations in LiF single crystals is demonstrated in Fig. 4.41 as a double-logarithmic plot taken from the pioneering work of Johnston and Gilman [19]. The dislocation velocities increase strongly with increasing stress. The velocities of edge dislocations are higher than those of screws. At high velocities, the slope of the curves decreases in agreement with the asymptotic behavior near the velocity of sound. The curves in Fig. 4.41 are typical also for other materials in

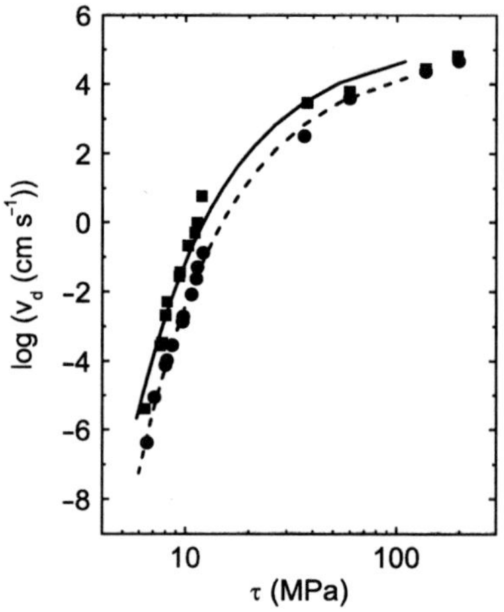

Fig. 4.41. Velocity of edge (*squares*) and screw dislocations (*circles*) in an as-grown LiF single crystal as a function of stress. Regression curves according to (4.104). Data from Johnston and Gilman [19]

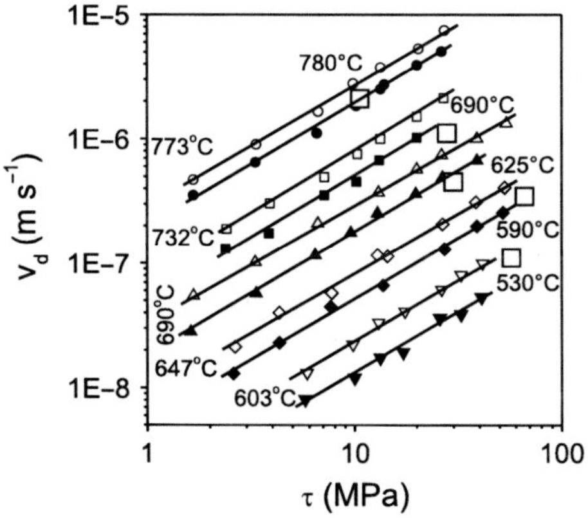

Fig. 4.42. Shear stress dependence of the velocity of screw (*full symbols*) and 60° dislocations (*open symbols*) in high-purity silicon crystals grown by the float zone technique. The dislocation motion was monitored by X-ray topography. Temperatures are indicated on the left side. Data from Imai and Sumino [262]. For comparison, single data from in situ straining experiments in an HVEM from [163] shown by *large open squares*. These data will be discussed in Sect. 6.2. Temperatures are indicated on the right side

which the dislocation mobility is not controlled by the double-kink mechanism. Characteristic curves for materials with the double-kink mechanism are presented in Fig. 4.42. Here, dislocations of 60° character move faster than screw dislocations. For these materials, the high-speed range has not been accessible. Some further examples will be presented in Part II. For incorporation into other theories, it is useful to describe the relation $v_{\mathrm{d}}(\tau^*, T)$ over a wide range of the parameters by phenomenological laws. In a variety of relevant deformation processes, the dislocation motion is supported by thermal activation so that the Arrhenius equation (4.7) with the Gibbs free energy of activation from (4.3) represents the natural, physics-based law

$$v_{\mathrm{d}} = v_{\mathrm{d0}} \exp\left(-\frac{\Delta G(\tau^*)}{kT}\right) = v_{\mathrm{d0}} \exp\left(-\frac{\Delta F(\tau^*) - V(\tau^*)\,\tau^*}{kT}\right). \quad (4.101)$$

The activation volume is then given by

$$V = kT\,\frac{\partial \ln v_{\mathrm{d}}}{\partial \tau^*}. \quad (4.102)$$

The disadvantage of (4.101) consists in the fact that the activation free energy ΔF and the activation volume V are, in general, still functions of τ^*. If a particular process is suggested in the range where diffusion can be neglected, the respective stress dependence of ΔG can be inserted, for example, (4.62) for the combination of the Fleischer potential and Friedel statistics. Kocks et al. [9] propose a general formula

$$\Delta G = \Delta G_0 \left[1 - \left(\frac{\tau^*}{\tau_0} \right)^p \right]^q ,$$

where p primarily describes the tail of the interaction profile, and q, the top. The possible ranges are $0 < p \leq 1$ and $1 \leq q \leq 2$. The respective shapes of the potentials are given in [9]. For practical use, this formula contains too many variables so that the exponents have to be chosen for theoretical reasons. If the dislocation velocities are analyzed over a wide range of the temperature, it has to be considered that ΔG_0 and τ_0 are proportional to the shear modulus, which depends on the temperature. Therefore, these quantities should be multiplied by a factor $\mu(T)/\mu_0$, with μ_0 being the shear modulus at zero temperature. Respective analyses of experimental data are discussed, for instance in Sect. 7.2.5, Fig. 7.21, for a simple box potential, and in Sect. 7.3.5 for a phenomenological potential suggested in [9]. A further disadvantage of the Arrhenius relation (4.101) is the incorrect limiting behavior for both very low and very high velocities. The correct behavior of $v_d \to 0$ for $\tau^* \to 0$ is obtained by including back jumps with a changed sign of the work term $V\tau^*$, which corresponds to the exchange of the exponential function by two times the hyperbolic sine function. At high stresses where the dislocation velocities are reduced by damping, (4.101) can no longer be used.

A phenomenological dynamic law, which approximates the Arrhenius relation but shows correct behavior at low velocities, is the power law similar to (4.10)

$$v_d = v_{d0} \exp\left(-\frac{\Delta G_0}{kT} \right) \left(\frac{\tau^*}{\mu} \right)^m . \tag{4.103}$$

For a fixed temperature, the Boltzmann factor with the activation energy can be included into v_{do} so that there are only two parameters, the pre-factor v_{d0} and the exponent m. Except for high velocities, this relation gives a good fit to experimental data, thus being frequently used. In Fig. 4.41, the low-velocity range exhibits a relatively great slope of $m = \mathrm{d} \ln v_d / \mathrm{d} \ln \tau \approx 20$, typical of the obstacle mechanisms. The slope in the high-velocity range is $m \approx 1.8$, close to $m = 1$ characteristic of the viscous range. In Fig. 4.42, the slope m is close to unity as expected for the Peierls mechanism.

Gilman [263, 264] has suggested an equation of the form

$$v_d = v_{d0} \exp\left(-\frac{\Delta G_0}{kT} \right) \exp\left(-D/\tau^* \right) \tag{4.104}$$

to interpret the measurements in [19]. Again, the Boltzmann factor can be included in v_{d0} for a fixed temperature. This relation describes the correct limiting behavior at both low and high velocities and fits the measured velocities over a range of more than ten orders of magnitude as the regression lines in Fig. 4.41 demonstrate. This equation has little physical justification as the processes are very different in the different velocity ranges. In spite of its simplicity, it is rarely used in further evaluations.

To determine the parameters of the phenomenological laws, dislocation velocities have to be plotted vs. the stress at constant temperature in suitable coordinates to obtain straight lines. Plots of $\ln v_d$ vs $1/T$ at a fixed stress, so-called Arrhenius plots, yield the activation energy. Many data on the dislocation mobility are concluded indirectly from macroscopic deformation tests described in Sect. 2.1 using the formalism of thermal activation of Sect. 4.1. The dislocation velocity is related to the plastic strain rate $\dot{\gamma}$ by the Orowan relation (3.5) via the mobile dislocation density ϱ_m. In the special tests of determining the dislocation velocity, the dislocation density is mostly low so that τ_i in (4.1) is also low, and τ^* occurring in the dislocation mobility laws is equal to the applied stress τ. This does not hold for the macroscopic deformation experiments. Therefore, incremental tests are performed with sudden changes of the deformation parameters: the strain rate by strain rate cycling (SRC) or stress relaxation tests, and the temperature by temperature change tests. It is assumed and mostly fulfilled only approximately that then ϱ_m and τ_i remain constant so that

$$\partial \ln (v_d/v_{d0}) = \partial \ln (\dot{\gamma}/\dot{\gamma}_0) \qquad \text{and} \qquad \partial \tau^* = \partial \tau .$$

This problem will further be discussed in Sect. 5.2.2. Changes in ϱ_m and τ_i after the sudden change of the deformation parameters will result in transient effects. Some of them will be described in Part II. An example of determining the parameters of a dislocation mobility law will be given in Sect. 7.2.

The parameters of the Arrhenius law (4.101) may help identifying the processes controlling the dislocation mobility. Table 4.1 lists the main mechanisms discussed in this chapter, giving characteristic values of the activation parameters.

Table 4.1. Characteristic values of activation parameters of dislocation mobility

Mechanism	Act. vol. (V/b^3)	m	Activation energy	Comments
Peierls mechanism	1–20	1–3	Related to line energy	V only weakly dependent on stress
Loc. obstacles, Friedel case	100–10^3	30–10^3	Fraction to several eV	V approx. $\propto c^{-1/2}$
Loc. obst., Mott-Labusch case	10–10^4	10–10^3	Mostly <1 eV	
Dislocation intersection	$>10^3$ for $\varrho < 10^{13}\,\mathrm{m}^{-2}$			
Climb	≈ 1	≈ 1	Energy of self-diffusion	
Jog dragging	10–10^3	>1	Energy of self-diffusion minus work term	

Most activation energies depend on the type of binding and the elastic constants. They can therefore vary between a few tenths of an electron-volt up to several electron-volt. It should be kept in mind that for processes taking place within time intervals of seconds up to some days, reasonable rates, for example, in (4.6), are necessary. For atomic jumps in 100 s, for instance, with an attempt frequency of $\nu_0 = 10^{13}\,\mathrm{s}^{-1}$, the Gibbs free energy of the process should be $\Delta G \approx 35\,kT$ or approximately 0.9 eV at room temperature. Thus, at low temperatures the Gibbs free energies of activation will amount to fractions of an electron-volt and to several ones at high temperatures. However, the activation energies at zero stress are higher than ΔG by the work term $V\tau^*$, which is large at low temperatures, and small at high ones.

To perform simulations of the motion of many dislocations, the dynamic laws have to be inserted into the computer codes. This can be done by defining discretization nodes along the dislocation lines and physical nodes where the dislocations branch. Then, the simulation consists in integrating the nodal equations of motion [265].

5

Dislocation Kinetics, Work-Hardening, and Recovery

In the preceding chapter, the motion of individual dislocations was described as a function of the locally acting stress. During plastic deformation, however, the dislocations have to move in an evolving dislocation structure. The main effect of this structure is a spatially and temporarily varying field of internal stresses, which are superimposed onto the external stress. The effective stress τ^* controlling the velocity of the moving dislocations is influenced by these internal stress fields. The most important parameter determining the strength of internal stresses is the total dislocation density. The next section on dislocation kinetics will describe the processes of the evolution of the dislocation density and structure. The formation of the dislocation substructure with an increasing dislocation density results in a hardening of the crystals. This influence of the internal stress fields on the dislocation dynamics and the flow stress will be treated in Sect. 5.2. Finally, Sect. 5.3 will discuss plastic instabilities.

5.1 Dislocation Kinetics

In most crystalline materials, the dislocation density increases drastically during plastic deformation leading to work-hardening. This process may be described by an evolution law of the dislocation density containing a rate of dislocation generation minus rates of dislocation immobilization and annihilation. The dependencies of the dislocation density itself on the experimental parameters like stress, strain, and temperature have been extensively studied for numerous materials. However, the processes of dislocation generation and recovery are much less understood, at least on a quantitative level, although the basic models are as old as about 50 years. The present section describes these models and summarizes the information on dislocation kinetics obtained from in situ straining experiments in an HVEM.

5.1.1 Models of Dislocation Generation

As described in Sect. 3.2.2, the energy to form even small dislocation loops is very high so that the homogeneous nucleation of dislocations requires stresses of about one tenth of the shear modulus. The generation of dislocations during plastic deformation has to take place at far lower stresses. An exception is the concentrated loading of the crystals using micro-indentation, where the acting force during elastic loading suddenly decreases down to a lower level. During this pop-in effect, dislocations are nucleated at the high stresses under the indenter tip (e.g., [266]). Most materials, especially those for structural applications, contain a certain dislocation density originating from the production process. These grown-in dislocations form a three-dimensional network with certain distributions of the segment lengths between the nodes [267–269]. The first mobile dislocations are formed by elongation, that is, bow-out, of such existing dislocation segments. For this process, a segment of a certain length has to lie on a slip plane, and the stress has to be high enough to fulfil the same criterion as that for dislocation sources described below. If the grown-in dislocations are pinned by impurities, an additional stress component is necessary to unpin the dislocations. The dislocation density further grows when the moving dislocations increase their length, mainly by one of the following mechanisms.

The best-known model of dislocation generation is the Frank–Read source [116] mentioned in Sect. 3.2.7, which may be characterized as a localized source. A dislocation segment lying on a slip plane is pinned at its ends. The pinning may be caused by some extrinsic pinning agents A and B like precipitates as shown in Fig. 5.1a, when the dislocation changes onto another plane where it is not mobile (b), or by nodes of the dislocation network (c). The case of Fig. 5.1b is sketched again in Fig. 5.2. Under the applied stress, the mobile segment moves through different stages marked a to d. At stage d, the arcs marked by arrows have opposite signs of the line direction. When they approach each other, they can annihilate and the original source dislocation e is restored. This localized source can operate repeatedly to emit a greater number of dislocations.

In the cases of Fig. 5.1a, b, only one dislocation is involved so that all dislocations are generated on the same slip plane leading to planar slip. In Fig. 5.1c, where the dislocation nodes act as pinning agents, three cases are possible depending on whether the Burgers vectors b_{A2} and b_{A3} of node A

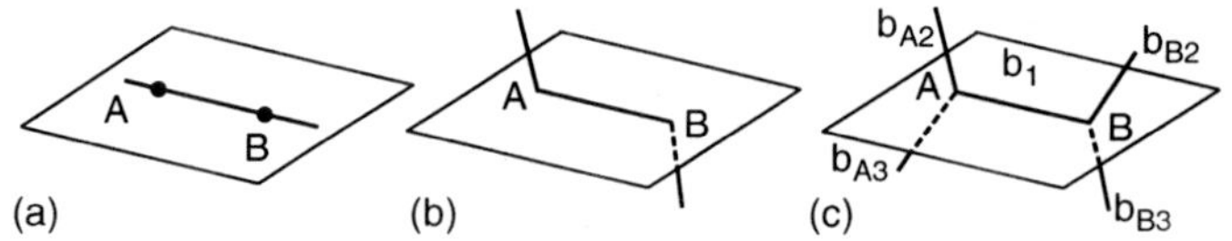

Fig. 5.1. Different types of pinning agents for localized dislocation sources

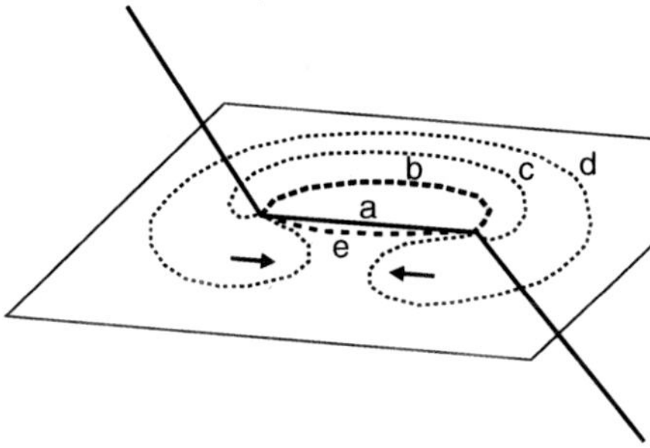

Fig. 5.2. Localized (Frank–Read) source for the generation of many dislocations on a single slip plane

and the Burgers vectors b_{B2} and b_{B3} have components perpendicular to the glide plane of the source dislocation with Burgers vector b_1, or not.

1. These Burgers vectors do not have a component perpendicular to the slip plane. Then, the dislocations are all emitted on the same plane as in Fig. 5.1b.
2. If the dislocations with Burgers vectors b_{A2} and b_{B2} have equal components perpendicular to the plane of the source, each subsequent dislocation loop enters a new parallel plane at the distance of the normal component of the Burgers vectors. Such a source is called a spatial source [87].
3. If the normal components are not equal, the source assumes a spiral character. These so-called pole mechanisms are important for the creation of twins. The different types of sources are reviewed in more detail in [270].

The dislocations emitted from the source cause a back stress on the source dislocation. The dynamics of the emission of dislocations into an otherwise undisturbed crystal was modeled in [271] by the motion of straight screw dislocations emitted from the source in one direction. The dependence of the velocity of the ith dislocation on the effective stress is described by an exponential law like the Arrhenius equation (4.101), in the form of

$$v_{\mathrm{d}i} = A \exp\left(B\tau^*\right),$$

where A and B are constants. The effective stress on the ith dislocation is given by the applied stress τ plus the sum of the internal stresses of all the other dislocations

$$\tau^* = \tau + \tau_{\mathrm{d}i}.$$

The shear stress exerted by the jth dislocation on the probe dislocation i is given by (3.10) with $y = 0$ (all dislocations move on the same plane) and the mutual distance $x_j - x_i$

$$\frac{\mu b}{2\pi} \frac{1}{x_j - x_i},$$

or that of all dislocations j

$$\tau_{\mathrm{d}i} = \frac{\mu b}{2\pi} \sum_{j=0, i \neq j}^{N} \frac{1}{x_j - x_i} = \frac{\mu b}{2\pi} \xi_i .$$

The index $j = 0$ represents the source. Thus, the velocity of the ith dislocation is

$$v_{\mathrm{d}i} = A \exp\left(B\tau\right) \exp\left(\frac{B\mu b \xi_i}{2\pi}\right),$$

or in the abbreviated form

$$v_{\mathrm{d}i} = \frac{\mathrm{d}x_i}{\mathrm{d}t} = v_{\mathrm{a}} \exp\left(\alpha \xi_i\right),$$

where $\alpha = B\mu b/(2\pi)$ and v_{a} is the velocity of an isolated dislocation under the uniform stress τ. The latter equation forms a set of differential equations which were solved numerically. The dislocation source was represented by a source dislocation fixed at $x = 0$. After reaching an effective stress higher than the initiation stress $\tau^* = \tau_{\mathrm{s}}$ at a site $x \geq 0$, the source dislocation was mobilized with a new source dislocation being introduced.

The above study yielded the following main results:

- After initial transients, all dislocations reach approximately the same constant velocity, as demonstrated in the displacement vs. time plot of Fig. 5.3 for four selected dislocations. Thus, during their steady state motion, all dislocations experience the same effective stress τ^*. In other terms

$$\tau^* = \tau + \Delta\tau,$$

with a fixed $\Delta\tau$.

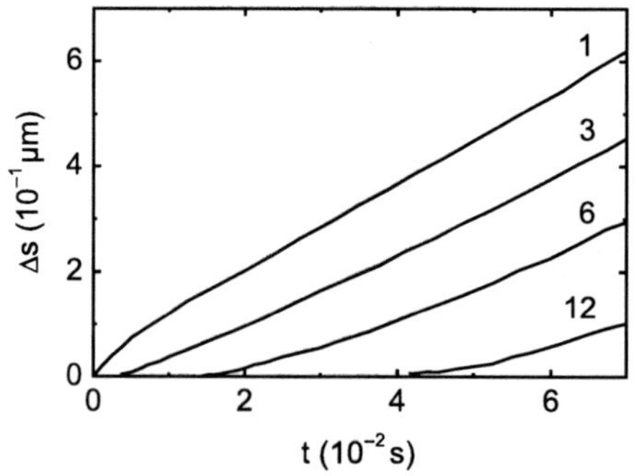

Fig. 5.3. Collective movement of dislocations emitted from a source. Displacement vs. time plot for dislocations number 1, 3, 6, and 12. Parameters: $\partial \ln v_{\mathrm{d}}/\partial(\tau^*/\mu) = \mu B = 1000$, applied stress $\tau/\mu = 5 \times 10^{-3}$, source stress $\tau_{\mathrm{s}}/\mu = 2 \times 10^{-3}$. Velocity of an individual dislocation at the same stress $v_{\mathrm{a}} = 1\,\mu\mathrm{m\,s^{-1}}$, velocity of leading dislocation $v_1 = 9.1\,\mu\mathrm{m\,s^{-1}}$. Data from [271]

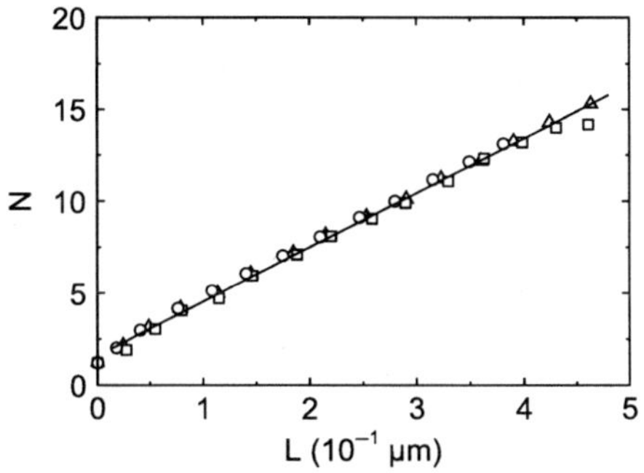

Fig. 5.4. Linear relation between the number N of dislocations emitted from a source and the distance of the leading dislocation to the source L. Parameters: normalized activation volume parameter $\mu B = 500$ (*circles*), 750 (*triangles*), 1,000 (*squares*), applied stress $\tau/\mu = 5 \times 10^{-3}$, source stress $\tau_s/\mu = 2 \times 10^{-3}$. Data from [271]

- The velocity of the leading dislocation v_1, and accordingly also that of all the others, is higher than the velocity of an individual dislocation v_a under the same applied stress τ. Thus, $\Delta\tau > 0$ with

$$\frac{v_1}{v_a} = \exp(B\Delta\tau)\,. \tag{5.1}$$

- As demonstrated in Fig. 5.4, the number of emitted dislocations is proportional to the distance of the first dislocation from the source, that is,

$$\frac{\mathrm{d}N}{\mathrm{d}L(t)} = \frac{\pi\Delta\tau}{\mu b}\,.$$

  This relation is independent of the normalized activation volume parameter $\partial \ln v_d/\partial(\tau^*/\mu) = \mu B = \mu V/(kT)$.
- The positions of the dislocations resemble those in a pile-up (see Sect. 5.2) in front of the source dislocation.

The described model of a Frank–Read source corresponds to the properties of slip localized in a narrow slip band propagating into a crystal of low dislocation density. Such a situation holds for some measurements of the velocity–stress relation on dislocations generated at a scratch. The development of a constant velocity of the leading dislocation, that is, of the head of the slip band, is a prerequisite to such kind of measurements. On the other hand, the velocity of the head dislocation is much higher than that of the individual dislocations, thus making the interpretation difficult. According to (5.1), the velocity ratio depends on the stress difference $\Delta\tau$, which, in turn, depends on the applied stress τ and weakly on the source initiation stress τ_s. Velocity ratios were found up to 200. In another study [272], the behavior of a double-ended array was analyzed, that is, of a source that emits dislocations in opposite directions. The parameters were chosen to fit the experimentally observed differences between the velocity of individual dislocations and head dislocations in slip bands in NaCl single crystals [273]. According to this work, the head dislocations do not reach a constant velocity. Instead, the velocity decreases with increasing time. This underlines the problem of deducing velocity–stress relations from measurements on head dislocations in dislocation groups.

The dynamic behavior of localized dislocation sources during plastic deformation in regions with a high dislocation density or in polycrystalline materials with slip distances limited by grain boundaries differs from that just described. Related experimental observations will be mentioned in the next section.

The second mechanism of dislocation generation, which creates individual isolated dislocation loops, is the double-cross slip mechanism suggested by Koehler [274] and Orowan [275] and experimentally first observed indirectly by Johnston and Gilman [19, 276]. It is shown schematically in Fig. 5.5. A screw dislocation (a) moves on its glide plane, which is identical with the image plane. Cross slip of a segment of length L results in two superjogs J (b). They can move conservatively along the dislocation line but not in the forward direction of the dislocation so that they act as pinning agents, and the segments near the jogs bow out. The segment between the jogs then multiplies similarly to the Frank–Read source (c).

The sources can act only if certain criteria are fulfilled. The length of the bowing segment has to be larger than the critical length L_c to overcome the half-circle configuration of (3.47). Inserting characteristic values of $\mu = 30$–$100\,\mathrm{GPa}$, $b = 3 \times 10^{-10}\,\mathrm{m}$, and $\tau = 100$–$1000\,\mathrm{MPa}$ results in L_c to be in the order of magnitude of 100 nm. In stage (c) of the double-cross slip mechanism, the branches adjoining the jogs have to pass each other on their parallel planes. This is only possible if the height of the jogs, that is, the distance between the parallel glide planes, is larger than a critical value h_c given by the dipole opening criterion (3.27). It is obvious that h_c is about 20 times smaller than L_c. Thus, both criteria involve a dependence of the dislocation generation processes on the stress.

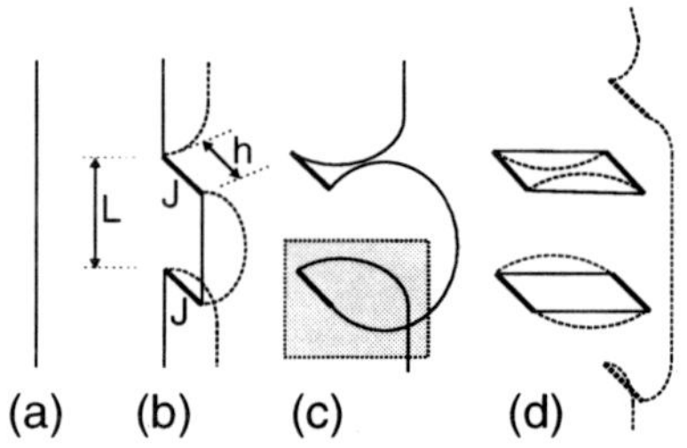

Fig. 5.5. Double-cross slip mechanism of the generation of individual dislocation loops and of the production of dislocation debris

If the jogs are not high enough to fulfil the dipole opening criterion, dislocation dipoles are trailed by the moving dislocation as outlined in Fig. 5.5d. The dipoles can be terminated by conservative motion of the jogs away from the dipoles. The dislocation dipoles formed during the motion of screw dislocations are also called dislocation debris. The accumulation of the debris during plastic deformation hardens the crystals. After the flow stress has increased, originally stable dipoles can fulfil the dipole opening criterion and open to form new dislocation loops.

Experimental values of the frequency of multiplication events can be concluded from the statistical data on cross slip processes in NaCl single crystals. The frequency distribution $H(h)$ of the cross slip heights h was given in Sect. 4.3 (Fig. 4.10). Only those cross slip events with $h > h_c$ are relevant for multiplication. The latter fraction of the total cross slip density D (Fig. 4.11) determines the density of cross slip events with sufficient height for a multiplication

$$D_c = D\Delta h \sum_{h_c}^{\infty} H(h) .$$

$\Delta h = 0.053\,\mu\text{m}$ is the width of the columns in Fig. 4.10. Using (3.27) and the corresponding data of NaCl single crystals, $\mu = 18\,\text{GPa}$, $b = 3.96 \times 10^{-10}\,\text{m}$, the flow stress $\tau = 1\,\text{MPa}$, and $\nu = 0.3$, the critical cross slip height for multiplication becomes $h_c = 0.4\,\mu\text{m}$. The total density of cross slip events along the path of the moving dislocation is $D = 38\,\text{mm}^{-1}$. The tail of the histogram with $h > h_c$ (Δh times the sum) corresponds to a fraction of 0.022. Thus, $D_c = 0.8\,\text{mm}^{-1}$, or a dislocation has to travel an average distance of about 1.2 mm before a multiplication occurs. This large distance is in accordance with the very long slip lines of individual dislocations on the NaCl single crystals. As all parameters involved, that is, the frequency distribution of cross slip heights $H(h)$, the critical height for multiplication h_c, and the total density of cross slip events D, depend on the stress, the multiplication rate may be assumed to be proportional to the stress as supposed below.

L_c and h_c are well below the foil thickness of about 500 nm in an HVEM in situ experiment, so that the mechanisms of dislocation generation can well be observed. The intermediate configuration of stage (c) is often metastable. It is characterized by the highlighted α-like configuration in Fig. 5.5c, which is frequently observed in dislocation structures under stress. The double-cross slip mechanism usually emits only a single new dislocation loop. As the generated dislocation moves on a plane parallel to the original one, slip may spread leading to the width of the slip bands to grow, which is in contrast to the Frank–Read source, which emits many dislocations on the same or a neighboring plane.

Cross slip events occur during the motion of dislocations. Accordingly, the increase in the dislocation density $\mathrm{d}\varrho_+$ should be proportional to the area $\mathrm{d}A$ swept by all dislocations as suggested in [277]. Considering the dependence of the creation rate of dislocations on the stress, discussed above, the creation rate may be written as

$$d\varrho_+ = w\tau^* dA = w\tau^* \varrho ds = (w/b)\tau^* d\gamma, \tag{5.2}$$

where the multiplication coefficient w is a constant, ds is the displacement of all dislocations, and dγ is the increment in shear strain. The multiplication is driven by the local effective stress τ^* rather than by the total stress τ. Such an expression was used by Haasen and Alexander [278, 279] to explain both the initial increase of the dislocation density in semiconductor single crystals and the related yield drop effect.

5.1.2 Experimental Evidence of Dislocation Generation

As mentioned earlier, dislocation generation can well be studied during in situ straining experiments in the TEM. An overview is given in [280]. In several materials, new dislocations are generated in localized (Frank–Read) sources. Mostly only half the operating segment of the sources in Figs. 5.2 and 5.5 is visible, as in Fig. 5.6 taken in duplex steel. The other half is cut away by the specimen surface. Then, the operating segment rotates around the single pinning agent, forming a new dislocation loop during each revolution. The figure presents three stages where always one dislocation is emitted between the stable stages. A video sequence of the same source is given in Video 8.8. Video 9.16 shows a similar source in an $MoSi_2$ single crystal.

In contrast to the dynamic behavior of the Frank–Read source described in the preceding section, which emits dislocations into the undisturbed crystal volume, the dislocations emitted from the present sources pile up either against grain or phase boundaries or against groups of other dislocations. As will be discussed in Sect. 5.2, the pile-ups produce a back stress on the source, which opposes the applied stress. As a consequence, long-range the source stops operating and starts again only after one of the outer dislocations is released from the pile-up, thus reducing the back stress.

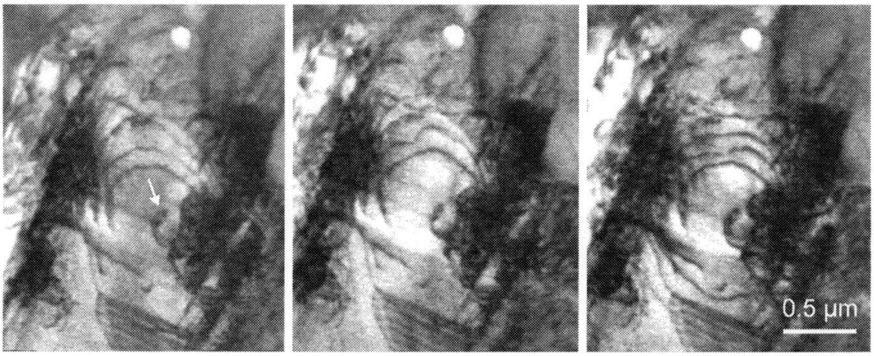

Fig. 5.6. Three stages of the operation of a localized dislocation source (Frank–Read source) in an austenite grain during the in situ deformation of duplex steel taken from a video recording. Frames from [280]

In many materials, the new dislocations are generated by the double-cross slip mechanism. One example showing all the aspects of this mechanism is the deformation of MgO single crystals at room temperature [281]. Figure 5.7 presents several stages of the opening of a new dislocation loop at a jog in a screw dislocation. If the jog is not high enough for the loop to open ($h < h_c$), a dislocation dipole is trailed behind the jog. Figure 5.8a shows a dislocation with a jog, which is indicated by a small trailed dipole (arrow). In Fig. 5.8b, the dislocation has proceeded further making the trailed dipole well visible. The dipole and the upper part of the dislocation are moving during the exposure of the film. In Fig. 5.8c, the dipole has achieved a stable position now marked by strong contrast. Finally, the dipole has terminated in d, that is, the jog again had glided along the dislocation, thus separating the dipole from the dislocation. In some materials, many dipoles are created during plastic deformation. These processes were first observed by Johnston and Gilman [276] and may be responsible for part of work-hardening.

While the deformation is going on, the stress may increase because of work-hardening. As a consequence, dipoles having formed during earlier stages of the deformation may now open as the bypassing criterion (3.27) is now

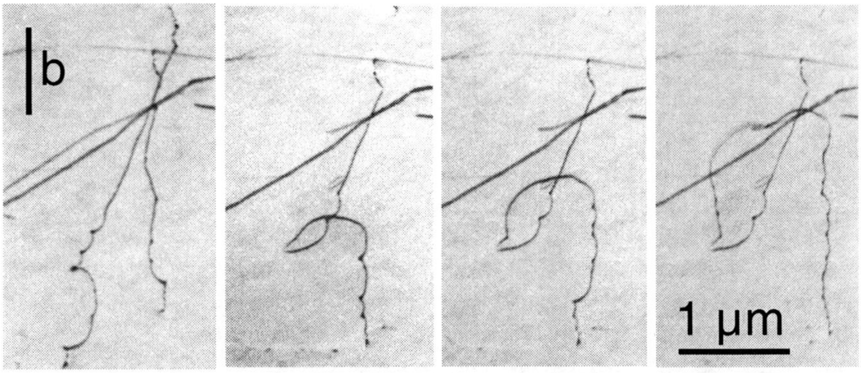

Fig. 5.7. Formation of a new dislocation by the opening of a loop at a jog during in situ deformation of an MgO single crystal at room temperature. b is the projection of the Burgers vector. Micrographs from [281]

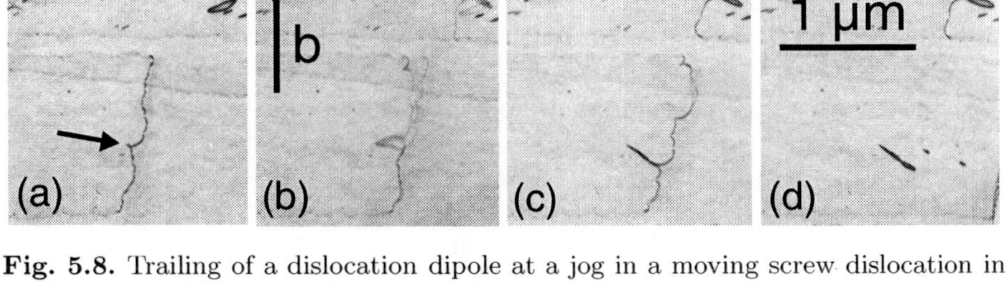

Fig. 5.8. Trailing of a dislocation dipole at a jog in a moving screw dislocation in MgO. Micrographs from [281]

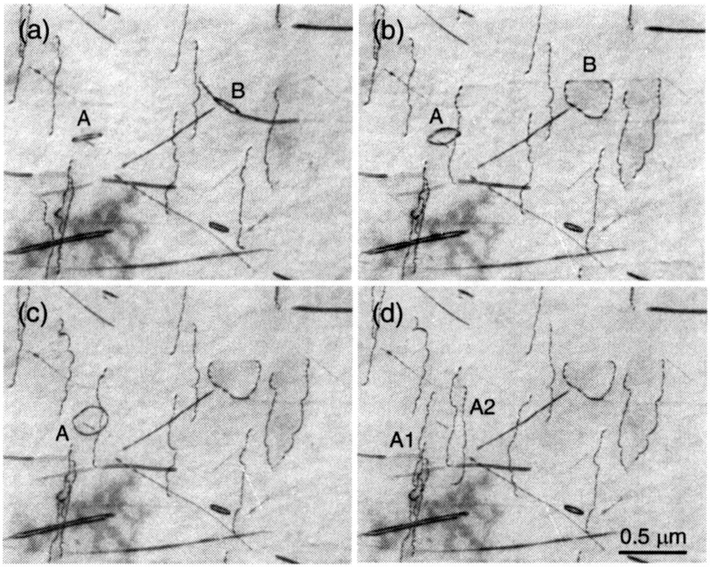

Fig. 5.9. Opening of dipoles to form glide loops during in situ deformation of an MgO single crystal. Micrographs from [281]

fulfilled. This is shown by two examples in the sequence of Fig. 5.9. In Fig. 5.9b, dipole A opens to a loop. The loop of dipole B emerges partly through the surface so that a half-loop remains. In Fig. 5.9c, loop A opens further and emerges through both surfaces in d leaving a pair of new dislocations A1 and A2. Several dislocation multiplication processes are demonstrated in Video 7.1.

A survey of the data on dislocation generation from the in situ deformation studies performed in Halle is presented in Table 5.1. Although these experiments reveal the different mechanisms of dislocation generation, the understanding on a quantitative level is still poor. In contrast to the localized Frank–Read sources, dislocation multiplication requires cross slip. While in the f.c.c. metals the cross slip planes are of the same type as the primary slip planes resulting in an equal dislocation mobility, in most other materials the mobility on the cross slip planes is lower than on the primary ones. In ZrO_2-10 mol% Y_2O_3, for example, the flow stress on the cross slip planes is about 30% higher than that on the primary ones [289]. The flow stress ratio between both planes certainly influences the width of the cross slip events. As described earlier, the double-cross slip mechanism leads to the growth of slip bands by the sidewise spreading of slip. Since double-cross slip events of smaller height may lead to multiplication at higher stresses, the slip bands become narrower in strong materials and at low temperatures.

In NaCl single crystals, cross slip is initiated by long-range internal stress fields. Only in materials where extended dislocations have to constrict before cross slip, the initiation process is thermally activated. The multiplication

Table 5.1. Burgers vectors, slip and cross slip planes as well as type of dislocation generation in different materials studied at the Halle HVEM facility and summarized in [280]

Material, load axis	Burgers vector	Slip planes	Cross slip planes	FR	DCS
Si, Ge	$1/2\langle 110\rangle$	$\{111\}$	$\{111\}$	[163]	
NaCl, $\langle 100\rangle$	$1/2\langle 110\rangle$	$\{110\}$	$\{100\}$		[179]
MgO, $\langle 100\rangle$	$1/2\langle 110\rangle$	$\{110\}$	$\{122\}$, $\{211\}$		[281]
ZrO_2, $\langle 112\rangle$	$1/2\langle 110\rangle$	$\{100\}$	$\{110\}$, $\{111\}$		[282]
Al–Zn–Mg	$1/2\langle 110\rangle$	$\{111\}$	$\{111\}$		[226]
Al–Ag	$1/2\langle 110\rangle$	$\{111\}$	$\{111\}$		[283]
Al–8Li	$1/2\langle 110\rangle$	$\{111\}$	$\{111\}$		[284]
Duplex steel austenite	$1/2\langle 110\rangle$	$\{111\}$		[280]	
Duplex steel ferrite	$1/2\langle 111\rangle$	$\{110\}$, $\{112\}$, $\{123\}$	$\{110\}$, $\{112\}$, $\{123\}$		[280]
Ti–6Al	$a/3\langle 11\bar{2}0\rangle$	$\{0001\}$, $\{1\bar{1}00\}$	$\{0001\}$, $\{1\bar{1}00\}$	[285]	[285]
γ–Ti–52Al	$1/2\langle 110\rangle$	$\{111\}$	$\{111\}$		[286]
NiAl, $\langle 100\rangle$	$\langle 100\rangle$	$\{100\}$, $\{110\}$, $\{210\}$	$\{100\}$, $\{110\}$, $\{210\}$		[287]
$MoSi_2$, $\langle 210\rangle$	$1/2\langle 111\rangle$	$\{110\}$		[288]	

FR localized Frank–Read source, DCS double-cross slip mechanism

event, that is, the bowing after (3.47) and the bypassing after (3.27) are also athermal processes. This is confirmed by the very instantaneous character of multiplication in ZrO_2 where the flow stress is also of athermal nature at the respective temperature as described in Sect. 7.3. As a consequence, the multiplication rate of (5.2) contains the applied stress but not explicitly the temperature. However, the latter enters the rate of dislocation annihilation as climb is involved in this process so that the evolution of the dislocation density depends strongly on temperature, as will be discussed below.

While cross slip is a prerequisite to dislocation multiplication, localized sources should operate if cross slip is limited. This is obvious for dislocations with $1/2\langle 111\rangle$ Burgers vectors in $MoSi_2$, where there exist only $\{110\}$ planes as easy slip planes. In the other cases, the situation is not so clear, for example, in Ti-6 at% Al, where slip and cross slip planes are of different type, with both mechanisms of dislocation generation occurring as demonstrated in Videos 8.4–8.6. The type of dislocation generation certainly influences the spreading of slip. Planar slip as it is observed, for example, in $MoSi_2$ and Ti–6Al requires the operation of localized sources. Slip localization is frequently accompanied with plastic instabilities (Sect. 5.3). Both phenomena are usually discussed on the basis of processes that soften the material, for example, by destroying the long-range or short-range order (as in Ti–6Al), thus facilitating dislocation motion in a narrow region. At the same time, the mechanism of dislocation creation must generate the new dislocations within this narrow region. This may occur either by the localized Frank–Read sources or by multiplication at high stresses where cross slip events of small height are sufficient for the emission of a new dislocation.

5.1.3 Dislocation Immobilization and Annihilation

After their formation, the dislocations move through the crystal. If there are no other obstacles like grain or phase boundaries and if the dislocation density is low, the dislocations can travel until they reach the crystal surface, produce a slip step by shifting the two sides of the slip plane against each other by the Burgers vector, and disappear. Otherwise, the traveling distance is limited by a number of processes. In sufficiently large grains or in single crystals the dislocations may become immobile by being captured in a region of high internal stress caused by the presence of other dislocations. The basic mechanism is the formation of dislocation dipoles owing to the mutual attraction of dislocations with Burgers vectors of opposite sign. As described in Sect. 3.2.4, dislocations on parallel slip planes have equilibrium positions with respect to glide. If the applied stress is lower than the critical stress defined by the dipole opening criterion (3.27), these dislocations cannot pass each other.

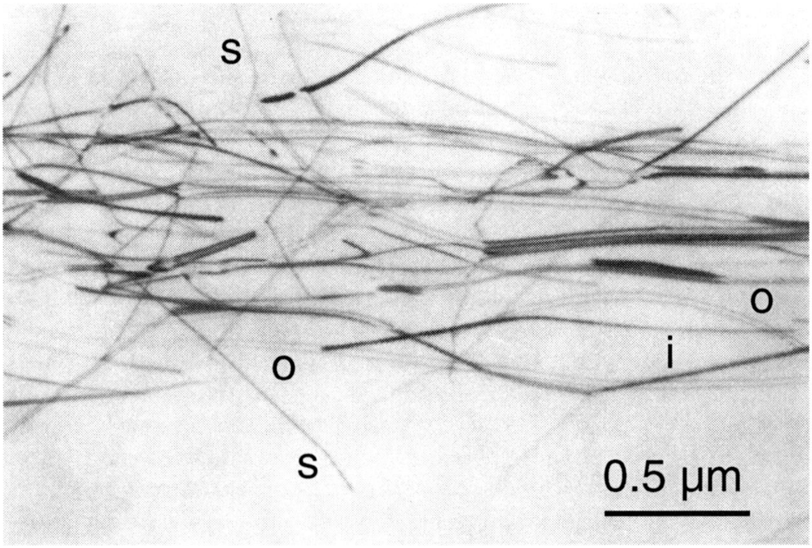

Fig. 5.10. Region of a high density of dislocation dipoles in a deformed MgO single crystal. Dipoles imaged in inside contrast (i) or outside contrast (o). Single dislocations (s). Micrograph from the work in [281]

According to Tetelman [290], the dislocations with line directions not being parallel can reorient to arrange themselves parallel and to form the dipole. As discussed in Sect. 3.2.4, the energy of the dipole is lower than that of the two isolated dislocations. In contrast to the short dipoles trailed behind high jogs (e.g., Fig. 5.8), the dipoles formed by the mutual capturing of moving dislocations are mostly very long as demonstrated by the dislocation structure in a deformed MgO single crystal in Fig. 5.10. The dipoles are imaged as thin double lines (o) or as thick lines of high contrast (i). In transmission electron microscopy, the image of a dislocation is shifted with respect to its geometrical projection. Direction and sense of the shift depend on the direction and sign of the Burgers vector and the imaging $\boldsymbol{g}$ vector. Therefore, the images of the two dislocations constituting the dipole are shifted in opposite directions. The contrast appears either inside or outside the geometrical projection of the dipole. Dipoles of small height imaged in inside contrast (i) often appear as thick single lines, while dipoles in outside contrast (o) appear as two thin separated lines.

Further dislocations can be captured in the vicinity of the dipoles. The static equilibrium situations between an infinitely long edge dislocation dipole and an approaching individual edge dislocation are investigated in [291]. The dipole can trap another dislocation in several configurations. Under stress, the formed free tripole moves like a single dislocation as the dipole has no long-range stress field. If one dislocation of the dipole is fixed, the approaching single dislocation may decompose the dipole. The dynamics of encounters between an edge dislocation dipole and a third edge dislocation (tripole configuration) or another widely spaced dipole (quadrupole) was studied in [292], assuming a linear dependence of the dislocation velocity on the local stress as in the Peierls mechanism. There are four situations possible.

1. The moving dislocation passes the dipole without changing it.
2. All dislocations form a stable configuration, that is, the moving dislocations are trapped.
3. The dipole is decomposed so that all dislocations become mobile. The stress necessary for this decomposition with the help of another dislocation is only half the value of the dipole opening criterion (3.27). Accordingly, this process is very frequent.
4. The dislocations rearrange to form a new dipole.

As a consequence, dislocation structures of a low density are very dynamic. Nevertheless, further dislocations are trapped in the regions of an increased dislocation density, yielding complex dislocation structures with the separation of regions of a high dislocation density from those of a low one. In these structures, the total elastic strain energy is lower than that of the same number of dislocations arranged randomly. The structures limit the traveling distances of the moving dislocations.

In addition to the 45° dipole configuration of two edge dislocations of opposite sign, two edge dislocations of equal sign form a stable equilibrium configuration when they arrange themselves along the direction perpendicular to their Burgers vector, as listed in Table 3.1. Further dislocations on parallel planes can be trapped to form an irregular tilt boundary. Climb can result in a regular boundary structure if the temperature is high enough. The formation of tilt boundaries during recovery is called polygonization.

The kinetics of the mobile dislocation density can be described by combining the formation rate of dislocations (5.2) with an expression for the immobilization. The basic process of the immobilization is the meeting of two dislocations. The respective rate is therefore proportional to the square of the dislocation density [293]

$$\mathrm{d}\varrho_- = -\mathrm{q}\varrho^2\,\mathrm{d}s .$$

The process coordinate is the slip displacement s as the dislocations meet while they are traveling. As an elastic process, the temperature does not enter the immobilization rate. Together with the Orowan equation (3.5), the differential equation for the dislocation density may then read as

$$\mathrm{d}\varrho = (\mathrm{w}\tau^*\varrho - \mathrm{q}\varrho^2)\mathrm{d}s = \frac{1}{b}(\mathrm{w}\tau^* - \mathrm{q}\varrho)\mathrm{d}\gamma . \tag{5.3}$$

In [293], the dependence of the creation rate on the stress is not considered. The equation can then be integrated to yield a mobile dislocation density approaching exponentially a steady state value. A linearly increasing total (mobile and immobile) dislocation density is obtained by neglecting the immobilization rate. The outlined dislocation kinetics is typical of low temperatures where dislocations cannot disappear.

As the dislocations in the dipole-like configurations attract each other, they tend to mutually annihilate. For screw dislocations, this is possible by cross slip if the attractive force is higher than the glide resistance on the cross slip plane. Figure 5.11 presents an example of the annihilation of segments of screw dislocations of opposite Burgers vectors during in situ deformation in MgO. The annihilated segments are marked by white triangles. The instantaneous annihilation of the whole dislocations is prevented by the bowing of the dislocation segments between localized obstacles on the main slip plane so that parts of the dislocations cannot orient in pure screw direction. The figure shows also the reorientation of the dipole D1 and the disappearance of the dipole D2 between Fig. 5.11b and c. The latter might have disappeared by the annihilation of one of its constituting dislocations by another gliding dislocation, while the remaining dislocation moved away and stopped as dislocation d in Fig. 5.11c. As annihilation by cross slip is restricted to pure screw dislocations, it is probably not a very frequent process.

Non-screw dislocation dipole configurations can annihilate only by climb. The velocity of dislocations of a dipole climbing against each other was briefly outlined in Sect. 4.10.4. As in other cases of diffusion-controlled climb, the climb velocity is proportional to the self-diffusion coefficient, for example, (4.93). If the climbing velocity along the dislocations of the dipole is not uniform, the dipole can be pinched-off via pipe diffusion into small prismatic

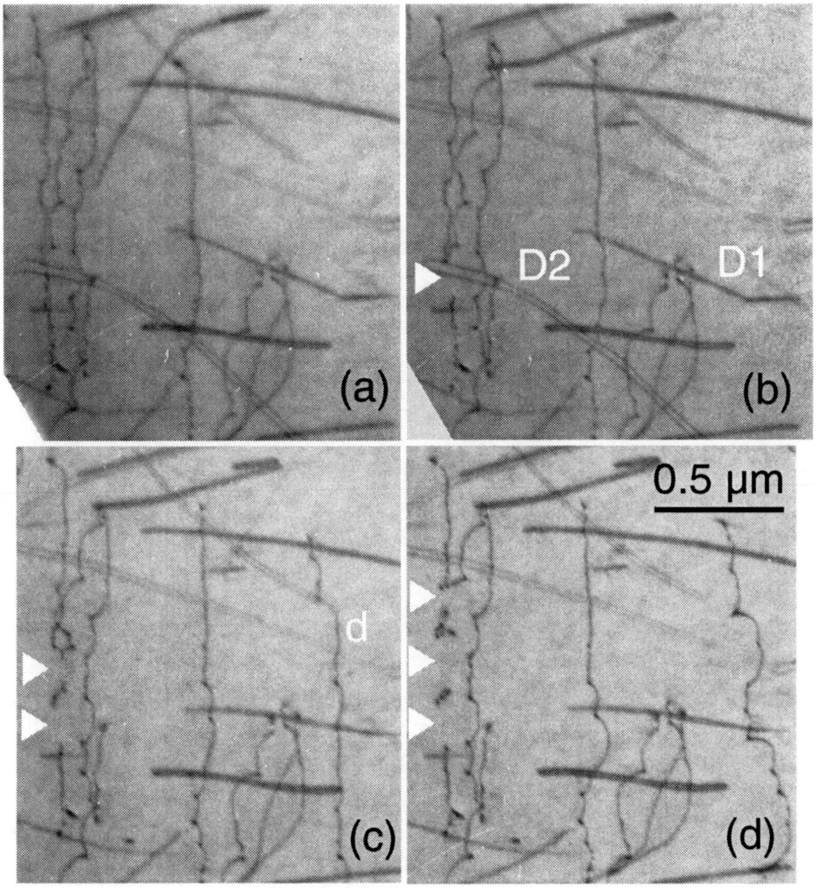

Fig. 5.11. Mutual annihilation of segments of two screw dislocations as well as annihilation and reorientation of dipoles during in situ deformation of an MgO single crystal. Micrographs from [79]

dislocation loops. Further climb is then driven by the line tension. TEM observations of the loop shrinkage during annealing yield information on the self-diffusion coefficient. As there is some similarity to the annealing of dislocations, the method will briefly be outlined following Seidman and Baluffi [294].

The prismatic dislocation loop is considered to be of toroidal shape of ring radius r_{loop} and radius of the tube r_{t}. The flux of the point defects emitted from or absorbed at the surface of the toroid is controlled by diffusion into the lattice having the equilibrium concentration of point defects. The flux per unit length of the dislocation is then given by

$$J = \left(\frac{2\pi r_{\mathrm{t}}}{b^2}\right) c_{\mathrm{v0}} \left(\frac{12 D_{\mathrm{v}}}{3a^2}\right) \left[\exp\frac{\mu_{\mathrm{cv}}}{kT} - 1\right] .$$

The first factor is the number of atomic sites per unit length of the tube around the dislocation. Fraction c_{v0} of these sites is occupied by vacancies. The third term is the jump frequency of the vacancies having a diffusion coefficient D_v across the tube surface. The factor 12/3 considers the coordination number and the fact that only one third of the jumps are forward jumps. Finally, the bracketed term is the fractional difference between the inward and outward fluxes. μ_{cv} is the chemical potential of the vacancies at the dislocation. The product of the diffusivity of vacancies and their concentration is the self-diffusion coefficient via vacancies $D_{sd} = c_{v0} D_v$. To obtain the total flux, i.e., the number of vacancies emitted or absorbed per unit time, J has to be multiplied by the length of the dislocation

$$\Phi = 2\pi r_{loop} \, J = \frac{(2\pi r_{loop})\,(2\pi r_t)}{b^2 a^2} D_{sd} \left[\exp\frac{\mu_{cv}}{kT} - 1\right].$$

In its area, the prismatic dislocation loop contains $n = \pi r_{loop}^2 a/\Omega$ extra atoms or vacancies. Thus, the shrinkage rate is given by

$$\Phi = \frac{\partial n}{\partial t} = -\frac{2\pi r_{loop} a}{\Omega}\frac{\partial r_{loop}}{\partial t},$$

resulting in

$$-\frac{\partial r_{loop}}{\partial t} = \frac{\Omega}{2\pi r_{loop} a}\Phi \approx \frac{8\pi r_t}{a^2} D_{sd}\left[\exp\frac{\mu_{cv}}{kT} - 1\right]. \tag{5.4}$$

Hirth and Lothe [130] treat the problem by assuming that point sources are located on the dislocation at a distance x_0 and that the flux is controlled by diffusion into a shell having a diameter much larger than the loop radius. The shrinkage rate of the loop is then

$$-\frac{\partial r_{loop}}{\partial t} = \frac{2\pi D_{sd}}{b\ln(8r_{loop}/x_0)}\left[\exp\frac{\mu_{cv}}{kT} - 1\right]. \tag{5.5}$$

The driving force of the climb is the line tension Γ of the dislocation loop. With (4.82) and (4.84), the chemical potential becomes

$$\mu_{cv} = \frac{\Omega}{b} f_{lt} = \frac{\Omega\Gamma}{b r_{loop}}.$$

The prismatic loop is everywhere of edge character so that $\sin\beta = 1$ and the line tension equals the line energy E_{de}. For the dislocation loop, the outer cut-off radius is equal to some fraction of the loop diameter, or $R = \alpha\, r_{loop}$ with α near 1. Thus, with (3.14)

$$\mu_{cv} = \frac{\mu b\Omega}{4\pi(1-\nu)r_{loop}} \ln\frac{\alpha r_{loop}}{r_0}. \tag{5.6}$$

This has to be inserted into (5.4) or (5.5). For small forces, that is, $0 < \mu_{cv}/(kT) \ll 1$, the exponential $\exp(\mu_{cv}/(kT))$ can be replaced by $1+\mu_{cv}/(kT)$.

Equation (5.5) has the advantage that then the logarithmic terms cancel if the inner cut-off radius of the dislocation is set to $r_0 = \alpha x_0/8$, yielding

$$-\frac{\partial r_{\text{loop}}}{\partial t} = \frac{\mu b \Omega D_{\text{sd}}}{2(1-\nu)kTr_{\text{loop}}} \,. \tag{5.7}$$

This equation can easily be integrated. For larger stresses, (5.4) or (5.5) with (5.6) have to be integrated numerically. As mentioned earlier, the observation of loop shrinkage during annealing opens the possibility of measuring the self-diffusion coefficient via the velocity of climb. This method can be used at temperatures of deformation tests which are considerably lower than those of conventional diffusion experiments. Besides, in more-component materials like ionic crystals or intermetallic alloys, the climb rate is controlled by the diffusion of the slower diffusing component. In the loop-annealing studies, this rate is measured directly. In [295], the method is extended to loops of dissociated dislocations of nonpure prismatic character.

Dislocation annihilation can be introduced into the kinetic equation of the dislocation density in a way similar to immobilization. Again, two dislocations are involved in the process leading to second-order kinetics. In contrast to immobilization, the process coordinate is now the time t, and the rate depends on the temperature so that the kinetic equation may read

$$\mathrm{d}\varrho = w\tau^* \varrho \mathrm{d}s - q\,\varrho^2\,\mathrm{d}t = \left(w\tau^* \varrho v_{\mathrm{d}} - q\,\varrho^2\right)\,\mathrm{d}t, \tag{5.8}$$

with v_{d} being the dislocation velocity. The temperature depending annihilation rate coefficient can be written as

$$q = q_0 \exp\left(-\Delta G_{\text{ann}}/kT\right). \tag{5.9}$$

ΔG_{ann} is the activation energy of annihilation by climb. After the discussion above, it should equal the activation energy of self-diffusion. Kinetic equations of the dislocation density similar to (5.8) and (5.9) are part of a system of constitutive equations describing the plastic behavior of materials. Such a set of equations will be discussed in Sect. 5.2.4. Kinetic equations with annihilation by climb can be applied to deformation at high temperatures. An example is given in Video 9.15 showing the dynamic formation and annihilation of dislocations during the deformation of an FeAl single crystal. Dislocation length disappears also by shrinkage of relatively large loops.

5.2 Work-Hardening and Recovery

In the preceding sections, the processes of the generation of dislocations and their immobilization and annihilation were described, leading to an evolution of the density of mobile and stored dislocations. This section discusses the dislocation structures forming during deformation, their internal stress fields, and their influence on the dislocation motion.

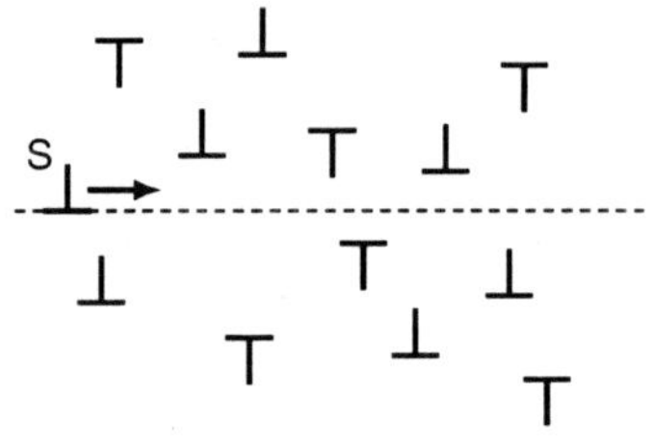

Fig. 5.12. Motion of a probe dislocation S through a random arrangement of parallel dislocations of opposite signs

5.2.1 Work-Hardening Models

In the early stages of plastic deformation of materials generating the new dislocations by multiplication, the dislocation density is not very high, and the dislocations are distributed in the developing slip bands quite uniformly. The mobile dislocations have to move in the field of the long-range internal stresses of all the other dislocations as sketched in Fig. 5.12 and first treated by Taylor [5]. Parallel dislocations of both signs are considered to be distributed randomly. The mutual elastic interactions between two dislocations passing each other on parallel planes of distance y_0 were discussed in Sect. 3.2.4. According to this, the bypassing stress is

$$\tau = \frac{\mu b}{4\pi y_0}$$

for screw dislocations, and

$$\tau = \frac{\mu b}{8\pi(1-\nu)y_0}$$

for edge dislocations. The average distance between the dislocations in the random array is given approximately by $1/\sqrt{\varrho_\mathrm{p}}$, where ϱ_p is the density of parallel dislocations. It may be assumed that

$$\bar{y}_0 = \frac{1}{2\pi\sqrt{\varrho_\mathrm{p}}} \tag{5.10}$$

is a suitable effective distance between the slip planes to represent the bypassing stress in the random array of dislocations. Thus, the latter can be described by

$$\tau_\mathrm{i} = \pi\,\frac{\mu b}{2\pi}\,\sqrt{\varrho_\mathrm{p}} = \alpha_\mathrm{p}\frac{\mu b}{2\pi}\,\sqrt{\varrho_\mathrm{p}}\;. \tag{5.11}$$

The factor α_p for parallel dislocations assumes values between 1 and 2π, depending on the details of the particular model. A frequently used value is $\alpha_\mathrm{p} = \pi$ corresponding to screw dislocations and the estimate in (5.10). The bypassing stress was designated τ_i to characterize it as some average internal

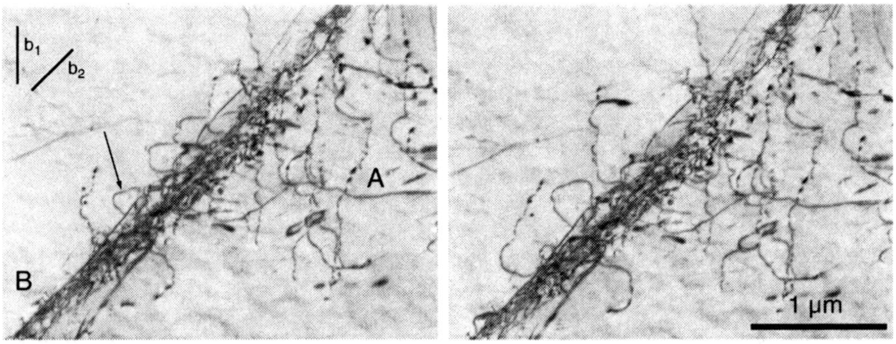

Fig. 5.13. Cutting of dislocations of slip system A with the projection of the Burgers vector $\boldsymbol{b}_1$ through dislocations of the oblique system B with Burgers vector $\boldsymbol{b}_2$ during in situ deformation of an MgO single crystal. Micrographs from the work in [281]

stress. Its meaning will further be discussed in the next section. The important properties are its proportionality to the shear modulus and the square root of the dislocation density. The increase of the internal stress component due to the increasing dislocation density is termed work-hardening or strain hardening, and Taylor hardening in the special case of randomly distributed parallel dislocations.

In general, slip is not restricted to a single slip system as considered in the Taylor hardening model. In multiple slip, dislocations of the primary slip system have to intersect dislocations of the other (secondary) slip systems. These processes have been described in Sect. 4.8. Figure 5.13 shows an example of dislocations of a slip band A in an MgO single crystal which intersect an oblique slip band B. The dislocations of the band B act as forest dislocations between which the gliding dislocations A have to bulge. In the empty region behind band B, the dislocations can freely expand like the loop indicated by an arrow. Interactions between dislocations of different slip systems include also the formation of junctions as discussed in Sect. 4.8. According to [296], so-called collinear interactions form strong pinning agents. In this interaction, two dislocations AA′ and BB′ in Fig. 5.14 of the same Burgers vector glide on non-coplanar planes, for example, on the slip and cross slip planes in an f.c.c. crystal. When they meet at C, parts of the dislocations between D and E can annihilate. In elastic equilibrium, the dashed segments form right angles with the intersection line between the slip planes at the transition points D and E. In the case of junctions, this angle is smaller than 90° so that the re-mobilization stress is higher for the collinear interaction.

The formula for the contribution τ_{forest} of the intersection of a dislocation forest of density ϱ_f to the flow stress (4.73) is exactly equal to (5.11) with another interaction constant α_f. On a local scale, the square root dependence was illustrated above in Fig. 4.32. Thus, this dependence of the internal stress contribution to the flow stress on the dislocation density is of very general

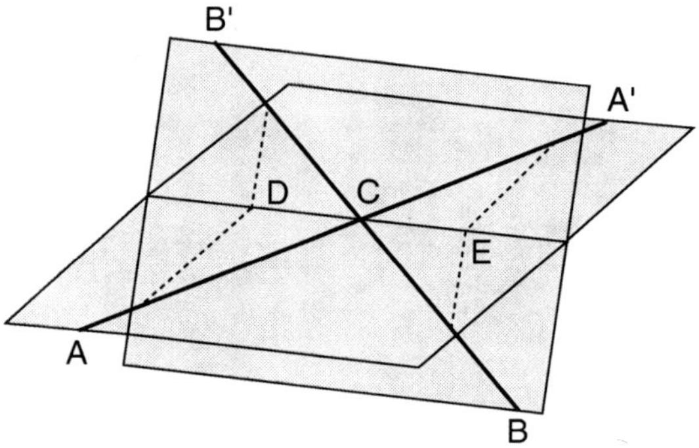

Fig. 5.14. Collinear interaction after [296] between two dislocations of the same Burgers vector on non-coplanar slip planes. *Dashed lines* show segments after the reaction. The lines between D and E are annihilated

character. In this respect, the total dislocation density is the most important parameter to describe a dislocation structure. The square root dependence of the internal stress component on the dislocation density has been proved by many experiments (see the review in [297]) and by three-dimensional computer simulations of the evolution of the dislocation structure during plastic deformation [298], reviewed in [299]. Recently, the results of three-dimensional simulations of the development of dislocation structures and the mean free path of dislocations are incorporated into macroscopic hardening models on the basis of Taylor hardening predicting the orientation dependence of the stress-strain curves of f.c.c. metals [300]. A typical experimental value is the interaction constant in f.c.c. Cu of $\alpha = 0.66\,\pi$ for the total dislocation density, and of $\alpha = 0.52\,\pi$ for dislocations breaking through the forest in cell walls (see the next section) [301]. Equation (5.11) with a suitable constant α is a further equation in the set of constitutive equations describing plastic deformation.

However, as (5.11) is explained using very different models, the numerical factor α_{p} does not represent a universal constant. This is shown by a study of deformed NaCl single crystals [302] presented in Fig. 5.15. The dislocation density was varied either by work-hardening up to different plastic strains in a material of constant composition (squares), or by a change in composition but with measurements at a constant (low) plastic strain (triangles). The dislocation density was determined by electron microscopy of replicas of etched cross-sectional faces. As the figure shows, the square root dependence of the flow stress on the dislocation density is observed in both sets of measurements. The slope of the curves, however, is most different.

The random distribution of dislocations assumed in the Taylor hardening model is a structure of intermediate dislocation strain energy where the outer cut-off radius R in (3.15) equals the mutual distance between the dislocations or $\sqrt{\rho}$. Such a uniform distribution of dislocations may occur in the early stages of deformation. With increasing strain, the dislocation structure becomes heterogeneous with regions of low dislocation density and others

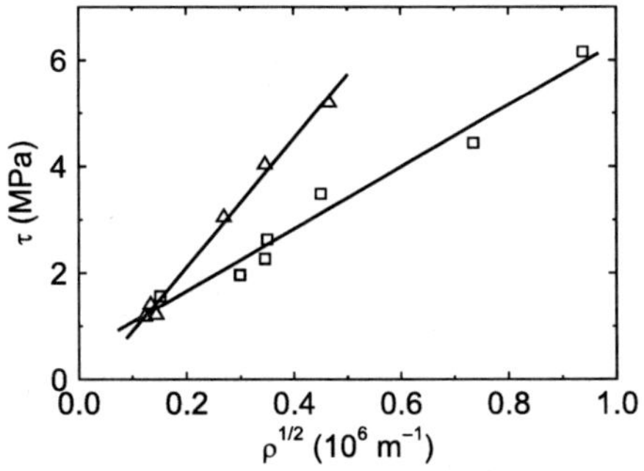

Fig. 5.15. Dependence of the flow stress of NaCl single crystals on the square root of the dislocation density. Variation of the stress by strain hardening with plastic strains between 1 and 12% in a material with 26 ppm of divalent cationic impurities (Ca^{++}) (*squares*) and by changing the impurity concentration between 1.7 and 256 ppm at a constant strain of 1% (*triangles*). Data from [302]

of high density. Such structures have been called cell structures. Very regular heterogeneous structures designated as persistent slip bands (PSBs) form during cyclic deformation (fatigue) as shown in Fig. 5.16. If the dislocations in the dense regions form dipolar structures, the long-range stresses cancel and the outer cut-off radius can be set to the distance between the dislocations in the dipoles. As this distance is considerably smaller than the average distance between the dislocations, the total strain energy is lower than that for a uniform distribution. Such low energy dislocation structures (LEDs) without long-range stress fields were the basis of work-hardening theories represented mainly by the work of Kuhlmann-Wilsdorf (e.g., [86, 303]).

On the other hand, dislocations can also form structures with an energy higher than that for the uniform distribution in the Taylor hardening model. This happens when dislocations of the same sign, for example, those generated by a localized Frank-Read source, pile up against some obstacle to slip. In this case, the stress fields of the individual dislocations superimpose to form far-reaching stress fields. The obstacles to glide may be sessile products of certain dislocation reactions, which are not discussed here. In particular, they are grain and phase boundaries. A pile-up of $N = 5$ edge dislocations with positions $x_1, x_2, x_3 ...$ against a boundary is shown in Fig. 5.17. The repulsive interaction energy between the dislocations solely depends on the relative spacings,

$$W_i = W(x_2 - x_1, x_3 - x_1, ..., x_\mathrm{N} - x_1) \,.$$

The head dislocation is in elastic equilibrium with the repulsive force of the obstacle and with the interaction forces of all the other dislocations. In equilibrium, the force on all other dislocations is zero

$$\tau b = \frac{\partial W}{\partial x_2} = \frac{\partial W}{\partial x_3} = \cdots = \frac{\partial W}{\partial x_\mathrm{N}} .$$

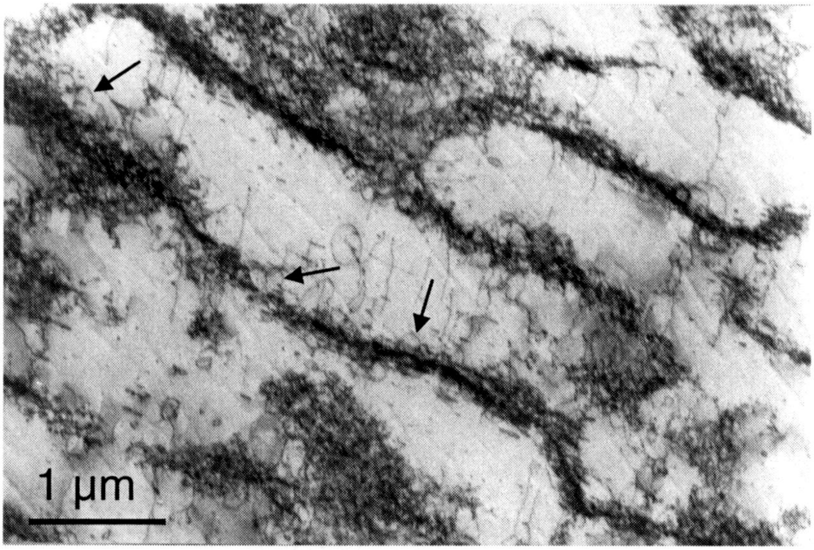

Fig. 5.16. Persistent slip band (PSB) structure formed during cyclic deformation of an Ni single crystal unloaded after the compression cycle followed by in situ tensile deformation in an HVEM. *Arrows*: dislocations with strong curvature near dense regions. Micrograph from the work in [304]

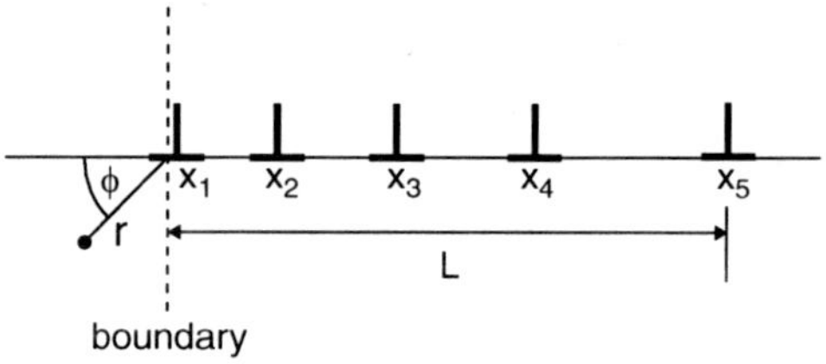

Fig. 5.17. Pile-up of five edge dislocations against a boundary

The equilibrium positions can be found by numerically solving the set of algebraic equations of the interaction forces. The force exerted on the head dislocation 1 by the dislocations 2 to N is given by

$$-\frac{\partial W}{\partial x_1} = \frac{\partial W}{\partial x_2} + \frac{\partial W}{\partial x_3} + \cdots + \frac{\partial W}{\partial x_{\mathrm{N}}} = (N-1)\,\tau b\ .$$

As the external stress acts on the head dislocation, too, the total force on it per unit length is

$$f_{\mathrm{head}} = \tau N b. \tag{5.12}$$

Thus, the force acting on the head dislocation is the same as that on a single dislocation of an N-fold Burgers vector. This rule holds for pile-ups of all types of dislocations in homogeneous stress fields. The elastic far-field of the pile-up converges for distances greater than the length of the pile-up against that of the single dislocation of N-fold Burgers vector. The stress at a point A with coordinates r, ϕ in front of the head dislocation and close to it (as shown in Fig. 5.17) can be written as

$$\tau_{\text{pileup}}(r, \phi) = f(\phi) \sqrt{\frac{L}{r}}\, \tau, \tag{5.13}$$

where $f(\phi)$ is a function of ϕ, and L is the length of the pile-up. Remarkable is the weak square root decrease of the stress with increasing radius r. This near-field resembles that of a crack.

The establishment of long-range stress fields by dislocation arrangements like pile-ups was the basis of work-hardening theories put forward by Seeger and coworkers (e.g., [87]). These theories are controversial to those of the formation of low-energy dislocation structures by Kuhlmann-Wilsdorf, mentioned earlier. In any case, pile-ups form in materials with grain or phase boundaries when dislocations emitted from localized sources queue in front of the grain or phase boundaries. The strong stress concentrations ahead of the pile-ups may then initiate slip in the neighboring grains or lead to the formation of cracks. Pile-ups of a few dislocations are regularly observed in connection with localized Frank–Read sources as the video sequences Videos 8.8 or 9.16 illustrate. The fresh dislocations pile up in front of the source. Their back-stress blocks the source until the head dislocation breaks through its obstacle, a boundary, or simply a dense region of dislocations.

A more realistic approach to the heterogeneous dislocation structures formed by deformation and their internal stresses is found in the composite model established by Mughrabi (e.g., [305]). Similar ideas were also developed by Holste and coworkers [306]. In the composite model, a deformed crystal with cell or wall structures is treated as a material composite consisting of regions of high and low dislocation densities as outlined in Fig. 5.18. According to the different dislocation densities, the local flow stresses are different, too. They are denoted τ_{h} in the hard walls of high dislocation density, and τ_{s} in the soft regions. Because of the necessary compatibility, the hard and soft regions have to be sheared in parallel so that the total shear strains, that is, elastic plus plastic ones, are equal. Thus, under low applied stress τ, both phases deform elastically. When the stress reaches τ_{s}, the soft regions start to deform plastically but the hard regions still resume elastic deformation. Only when the total shear strain reaches $\gamma_{\text{t}} = \tau_{\text{h}}/\mu$, where μ is the shear modulus, both phases deform plastically. The flow stress of the composite is then given by

$$\tau = x_{\text{h}}\tau_{\text{h}} + x_{\text{s}}\tau_{\text{s}},$$

where x_{h} and x_{s} are the area fractions of the hard and soft regions with $x_{\text{h}} + x_{\text{s}} = 1$. It can easily be shown [305] that $\tau_{\text{h}} > \tau$ and $\tau_{\text{s}} < \tau$ and that

$$x_{\text{h}}\left(\tau_{\text{h}} - \tau\right) + x_{\text{s}}\left(\tau_{\text{s}} - \tau\right) = 0.$$

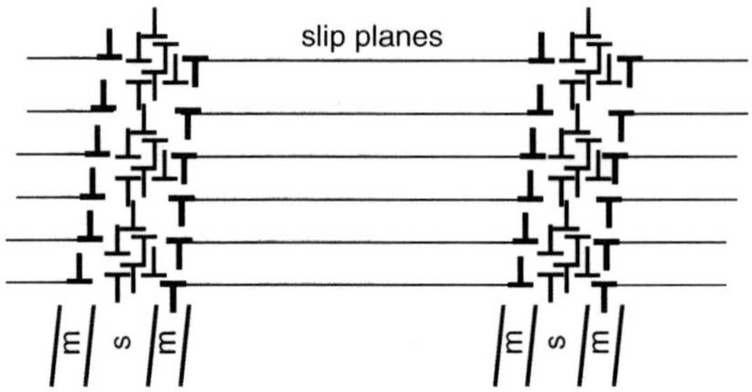

Fig. 5.18. Composite model of a sheared structure of hard and soft regions after Mughrabi [305]. Dislocations in the soft regions are not shown. *s* stored dislocations, *m* misfit dislocations

The local differences in the acting stress can be understood as the consequence of long-range internal stresses. With no external stress acting, the hard dislocation walls consist of statistically stored dislocations, mainly in low-energy dipole configurations. These are the dislocations in regions s with thin symbols in Fig. 5.18. When the deformation starts in the soft cell interiors, the dislocations are blocked at the hard cell walls and form the misfit dislocations in regions m marked by bold symbols. These geometrically necessary dislocations (GNDs) accommodate the elastic mismatch between the hard and soft regions. They are the sources of the long-range internal stresses necessary for the simultaneous compatible deformation of the whole crystal, as proved experimentally in [307]. If the stress is high enough for the deformation of the hard walls, the dislocation flux becomes equal in the walls and the soft cell interiors [308]. Dislocations in cell walls formed during multiple slip do not represent GNDs. These walls may be called incidental dislocation boundaries, whereas the kink walls and dislocation sheets forming during deformation in work-hardening stage II consist of the GNDs.

In the channels between persistent slip bands, the local stresses were measured from the curvature of dislocations, which were pinned under load by neutron irradiation [309]. As described in Sect. 3.2.7, there is the problem of the adequate values of the line tension. In [309], the line tension was scaled so that the average of the local stresses equals the macroscopic external stress. In this frame, the study shows how the local stress varies across the channels in the persistent slip band structure. The stress at the center of the channels is approximately half the external stress and near the walls it is about three times of it. Strongly curved segments near the walls are also marked by arrows in Fig. 5.16. The variation of the local stresses in a PSB structure was also estimated from dislocation curvatures during in situ straining experiments on Ni single crystals [304]. However, the line tension data used there have to be corrected (see the discussion in Sect. 5.2.3).

The composite model certainly represents a bridge between the low-energy dislocation structure and the pile-up theories of work-hardening. In this model,

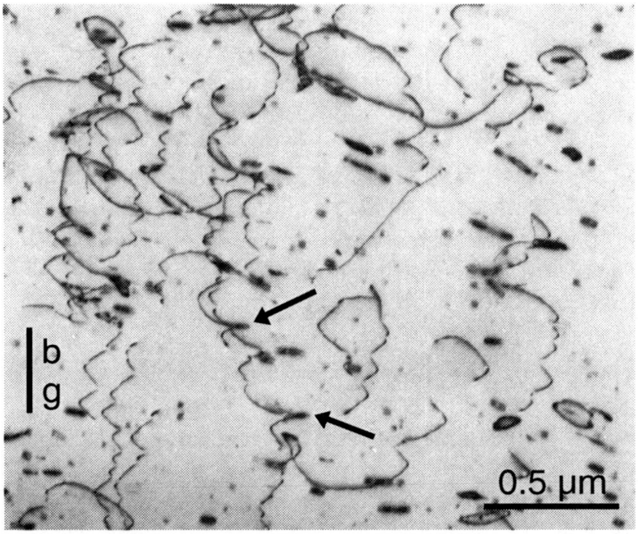

Fig. 5.19. Dislocations and dislocation debris in an NiAl single crystal during in situ deformation in an HVEM. Image normal [110], $\boldsymbol{g} = (\bar{1}10)$, $\boldsymbol{b}$ projection of [010] Burgers vector. From the work in [287]

long-range internal stresses result from the inhomogeneity of the dislocation density, without the necessity of evoking the existence of pile-ups with very high stress concentrations.

The hardening models discussed so far have all been concerned with long-range internal stresses. An exception is the production of so-called dislocation "debris" during deformation. The debris consists of small dislocation dipoles produced by double-cross slip events of low cross slip height, which do not meet the criteria for dislocation multiplication, as described above in Sects. 5.1.1 and 5.1.2 (Figs. 5.5 and 5.8). The idea of dislocation dipoles acting as hardening agents was introduced by Gilman [310]. Figure 5.19 exhibits a dislocation structure with many short dipoles during in situ deformation of the intermetallic alloy NiAl. Arrows mark two dislocations that are just trailing dislocation dipoles.

Dislocation debris occurs only in materials that generate the new dislocations by the double-cross slip mechanism. Within the single specimens, the density of the debris is frequently not uniform. A certain part of the flow stress is necessary to produce the debris. It can be estimated from the Frank–Read criterion for fully bowing dislocation segments (3.47), with L being the distance between jogs trailing dipoles. It may be concluded from Fig. 5.19 that the distances between other pinning agents along the dislocations (the cusps in the dislocations) are much shorter than those between the trailing jogs, so that the flow stress contribution from generating the debris is mostly small. Existing debris can contribute to the flow stress by elastic interactions with the dislocations passing by on parallel planes and, for multiple slip, by cutting the debris. As the stress field of finite dipoles decreases rapidly with increasing distance, it is suggested that the debris acts similarly to precipitates. According to [311], the force between a dislocation and a nonparallel dipole is small as long as the dislocation is not located between the dipole arms. The passing of terminated dipoles by dislocations moving in an oscillating stress field is simulated in [312]. The dipoles may be annihilated or moved resulting in the formation of dipole clusters. In general, the mechanism of debris hardening is not well understood. The contribution to the flow stress is probably mostly small.

5.2.2 Thermal and Athermal Components of the Flow Stress

As described in the foregoing section, the grown-in dislocations and the dislocations stored in different structures during plastic deformation cause a field of spatially varying long-range internal stresses. These stresses promote the dislocation motion in some regions but impede it in others. Thus, these stresses represent also obstacles to dislocation glide. According to Albenga's law, the spatial average of the internal stresses is zero. It has been discussed in Sect. 4.1 that at a finite temperature the dislocation motion can be supported by thermal activation. However, thermal activation is confined to processes where the energy necessary to surmount the obstacle at the actual stress is less than about $40\,kT$, corresponding to about $1\,\mathrm{eV}$ at room temperature, and where the volume involved contains less than a few hundred atoms. To move a dislocation segment of length Δy in the long-range stress field over a distance Δx against an internal stress τ_i requires an energy of $\Delta x \Delta y\, \tau_\mathrm{i} b$. With $\Delta x = \Delta y = 0.5\,\mu\mathrm{m}$, $\tau_\mathrm{i} = 10^{-3}\,\mu$, $b = 3 \times 10^{-10}\,\mathrm{m}$, and $\mu b^3 = 5\,\mathrm{eV}$, the energy becomes several thousand electron volt in contrast to $40\,kT \approx 1\,\mathrm{eV}$ at room temperature. Thus, this energy is far too high to be overcome by thermal activation. Likewise, also the number of involved atoms is far too high. As a consequence, long-range internal stress fields cannot be overcome by the action of thermal activation. It is therefore useful to distinguish between components of the flow stress that can be surmounted by the aid of thermal activation and those which cannot. The first ones, the thermal components, result from most mechanisms with short-range stress fields discussed in Chap. 4. Their temperature dependence is of Arrhenius type and depends on the particular interaction potential. The second ones, the athermal components, originate from the long-range internal stress fields and some other processes requiring high energies and being not restricted to small volumes like the Orowan process. In this case, the temperature dependence is weak. It equals that of the shear modulus as the interaction stresses are proportional to it (or perhaps some other elastic constant). Accordingly, the thermal contributions to the flow stress are also denoted temperature-dependent components and the athermal components temperature-independent ones. Figure 5.20 presents a schematic plot of the superposition of long-range internal stresses (the sine-like

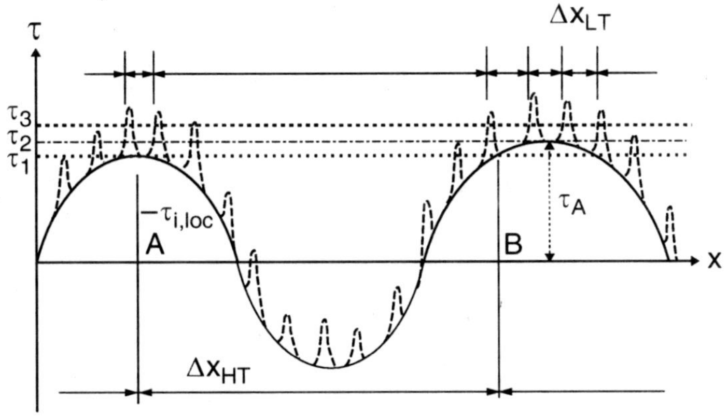

Fig. 5.20. Schematic drawing of the superposition of long-range internal stresses and short-range interaction profiles

function) and the short-range fields of localized obstacles. These stresses are plotted with negative sign. The internal stresses may be characterized by their amplitude τ_A and wavelength L. Three levels of the applied stress are characterized by the horizontal lines τ_1, τ_2, and τ_3. The local stress τ_{loc} acting on the short-range obstacles is given by the sum of the applied stress τ and the local internal stress $\tau_{i,loc}$

$$\tau_{loc} = \tau + \tau_{i,loc} \, . \tag{5.14}$$

At first, the situation may be considered without the presence of the short-range obstacles. At zero external stress, the dislocation may be positioned at the beginning of the $\tau_{i,loc}$ vs. x curve. With increasing stress, the dislocation moves forward along this curve. When the maximum τ_1 of the negative (impeding) internal stress is reached at A, the dislocation can jump at a high velocity to position B near the next maximum. As this maximum is higher than that at A, the dislocation cannot surmount it. The only way to resume the dislocation motion is to increase the stress again to the level of the maximum τ_2. If the stress is high enough, the dislocation can pass all internal stress maxima. This is a pure athermal motion of the dislocation. The situation is similar for high temperatures, even if short-range obstacles are present. At a sufficiently high temperature, the waiting times at the obstacles will be very short so that these obstacles do not impede the dislocation motion. In this athermal situation, the jump distances of the dislocation Δx_{HT} will be given by the wavelength of the internal stress fields.

At low temperatures, the waiting times at the short-range obstacles cannot be neglected. At a stress level τ_3, the dislocation coming from the left will stop at the first obstacle that exceeds the applied stress, marked by a thin vertical line. After thermal activation, it moves to the next obstacle. When this is overcome, the dislocation reaches a region of very high local stress, where it flies to the next maximum of the impeding internal stress. There, the obstacles again have to be surmounted individually by thermal activation. The jump distances Δx_{LT} are now the distances between the short-range obstacles, except in regions where the dislocation does not feel the short-range obstacles. These regions become unimportant if the amplitude of the internal stresses is small.

In summary, the kinematic behavior of the moving dislocations can be characterized as follows. At intermediate temperatures, the dislocation motion is governed by thermal activation and may be jerky on the small scale of localized obstacles. If the Peierls stress is rate-controlling, the dislocation motion may even be very smooth. The transition between the latter two mechanisms was described in Sect. 4.6. At high temperatures, the localized obstacles are overcome very quickly so that they do not control the dislocation motion. The latter is governed by the long-range internal stress fields. It becomes athermal and is jerky on the large scale of the wavelength of the stress fields. The transition between thermal and athermal behavior is not sharp. In more general terms, the character of the dislocation movement is athermal if a great part of the energy necessary to overcome the glide obstacles is supplied by the external stress. In view of the force–distance curve of Fig. 4.1, a process is mainly athermal if the work term ΔW comprises most of the total activation energy ΔF, or if the Gibbs free energy of activation ΔG is small. This is expressed also in Fig. 4.18, where the area swept after a single thermal activation event increases strongly when the applied stress approaches the athermal flow stress of the obstacle array. In this sense, also dislocation motion at low temperatures becomes athermal, because at the high stress level many obstacles are overcome spontaneously. All these considerations are valid for temperatures where diffusion can be neglected. At very high temperatures the dislocation motion will become smooth again owing to the diffusion processes described in Sect. 4.11.

As discussed in detail in Chap. 4, the actual dislocation velocity is a function of the local shear stress. The question arises as to the influence of the internal stress fields on the average velocity of a dislocation moving under the action of the macroscopic external stress τ and supported by thermal activation. A very simple argumentation goes back to Seeger [313]. The dislocation spends most of its time in the regions of low local forward stress given by the difference between the applied stress and the amplitude of the internal stress fields, $\tau - \tau_{\mathrm{A}}$. Consequently, its average velocity will mainly be determined by these jumps. In this model, the spatially varying local stresses and the internal stresses are replaced by a constant so-called effective stress τ^* and a constant internal back-stress τ_{i}. In contrast to (5.14) where the applied stress and the varying internal stress add up to the local stress, now the effective stress is the difference between the applied stress and the internal stress

$$\tau^* = \tau - \tau_{\mathrm{i}}. \tag{5.15}$$

This procedure had been adopted in most sections before. It is of highly qualitative nature. It has to be expected that τ_{i} is not equal to the amplitude

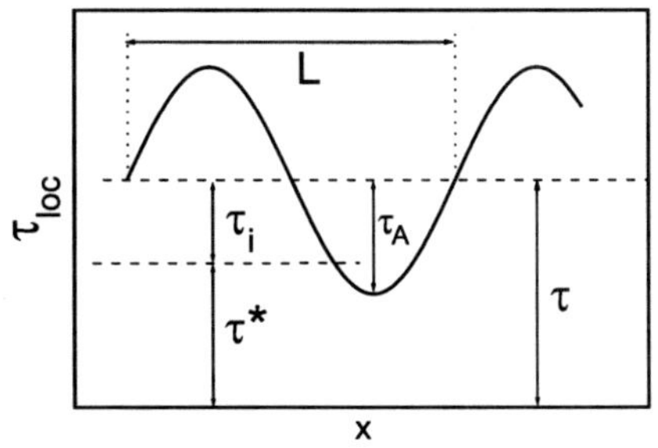

Fig. 5.21. Model of a spatially periodic internal stress field according to [315]

of the internal stresses τ_A. A more sophisticated approach to the effect of internal stresses on the movement of dislocations is the calculation of the average velocity of a dislocation moving in a spatially periodic internal stress field, as suggested by Chen et al. [314] and Li [315]. The latter is outlined here as it has the advantage of an analytical solution. The locally acting stress τ_{loc} is described by a sine function of the position x of the dislocation as illustrated in Fig. 5.21, that is,

$$\tau_{loc} = \tau - \tau_A \sin\left(\frac{2\pi x}{L}\right) .$$

Here, τ is the applied stress, τ_A the amplitude, and L the wavelength of the internal stress field. It was shown in [316] that the internal stress fields of periodic square arrays of screw dislocations of both the same sign and of opposite sign can well be approximated by the sinusoidal function of the position of the moving dislocation. In the calculations by Li, the instantaneous velocity of the dislocation is given by the power law (4.10) or (4.103), in abbreviated form

$$v_d = B\,\tau_{loc}^m .$$

The average velocity $\bar{v}_d$ is calculated by integrating $1/v_d$ over one period L and dividing the result by L. The analytical solution contains the Legendre function. Figure 5.22 presents the dependence of the average velocity $\bar{v}_d$ normalized by the velocity $v_d(\tau, \tau_A = 0)$ at the same stress but without internal stresses on the amplitude of the internal stress normalized by the applied stress (τ_A/τ). The average velocity is always lower than the velocity without the internal stress field. The deviations increase with increasing stress exponent m.

The results can be compared with the simple model of a constant back stress $\tau_i = \tau - \tau^*$ discussed above by setting the average velocity equal to the velocity at $\tau^* = \tau - \tau_i$:

$$\bar{v}_d = B\,{\tau^*}^m = B\,(\tau - \tau_i)^m .$$

τ_i may be called an apparent internal stress. It follows that

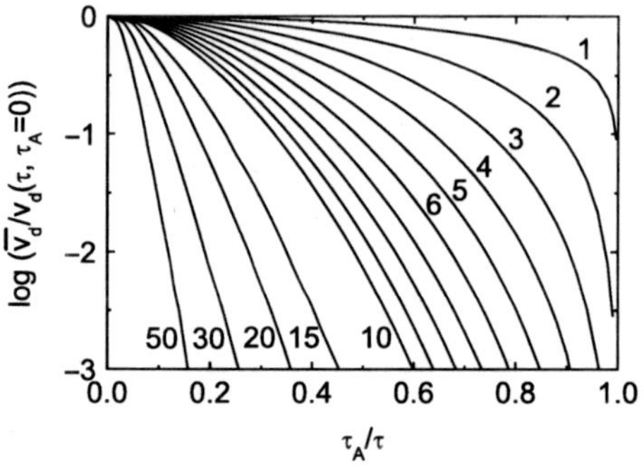

Fig. 5.22. Dependence of the normalized average dislocation velocity in a periodic internal stress field on the amplitude of the internal stresses. Parameter: stress exponent m. Data from [315]

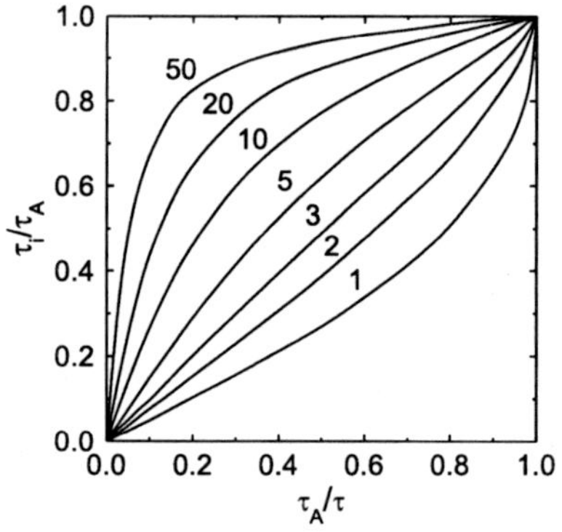

Fig. 5.23. Dependence of the apparent internal stress on the amplitude of the internal stresses. Parameter: stress exponent m. Data from [315]

$$\tau_i = \tau \left[1 - \left(\frac{\bar{v}_d}{v_d(\tau, \tau_A = 0)} \right)^{1/m} \right] .$$

This relation is plotted in normalized form in Fig. 5.23. The apparent internal stress is lower than the amplitude of the internal stress fields τ_A but approaches it for $\tau_A \to \tau$. The deviations are significant only if both τ_A and m are small.

In general, the amplitudes of the internal stress fields are not known, neither are they constant as assumed here. For studies of the dislocation dynamics by measuring the activation parameters during plastic deformation, the problem of the internal stresses is important as the formalism of the thermally activated dislocation motion is based on the locally acting effective stress, but not on the applied stress. Using the differential techniques like strain rate cycling or stress relaxation tests (4.9) and (4.11), one usually assumes that in addition to the pre-exponential factor (e.g., $\dot{\gamma}_o$ in (4.8)) also the internal stress

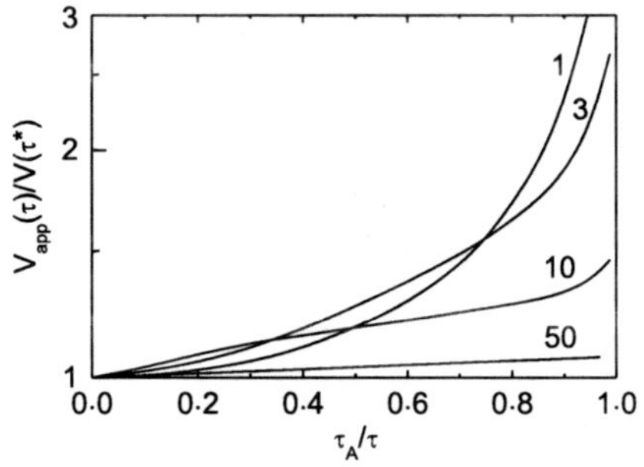

Fig. 5.24. Dependence of the normalized apparent activation volume on the amplitude of the internal stresses. Parameter: stress exponent m. Data from [315]

remains constant so that $\partial\tau^* = \partial\tau$. However, that what remains constant during an instantaneous change of the deformation conditions is not the apparent internal stress τ_i but the amplitude of the internal stress fields τ_A. Accordingly, the measured or apparent activation parameters like the activation volume and the activation enthalpy have their well defined thermodynamic meanings but they do not represent the quantities occurring in the respective models. Based on the calculation of the average dislocation velocity plotted in Fig. 5.22, the plot in Fig. 5.24 shows the ratio between the apparent activation volume $V_{app}(\tau)$ measured at the stress τ and the activation volume $V(\tau^*)$ defined by the microscopic models measured at the stress τ^*. The apparent activation volume is always larger than the true one. The deviations increase with increasing amplitude of the internal stress fields. However, they are mostly less than a factor of two. They are small for high stress exponents m and for low temperatures where τ_A/τ is small. The relation between the true and the apparent activation volumes was treated in a more general way in [317].

Similarly, also measurements of the stress exponent $m = \partial \ln v_d / \partial \ln \tau^*$ are influenced by the presence of internal stresses. The calculations show that the apparent stress exponent $m' = \partial \ln v_d / \partial \ln \tau$ is always greater than m. For high values of m, the effect of the varying internal stress of amplitude τ_A is almost equal to that of the presence of a constant back stress $\tau_i = \tau_A$. It follows then for the apparent stress exponent

$$m' = \frac{m}{1 - \tau_A/\tau}. \tag{5.16}$$

In conclusion, the influence of the presence of spatially varying internal stress fields on the experimental determination of the dynamic dislocation properties cannot exactly be described by a constant internal back stress τ_i. However, as the properties of the internal stress fields are not well known, the assumption of a locally constant back stress is sufficient for rough estimates from experimental data. Nevertheless, it should be kept in mind that the

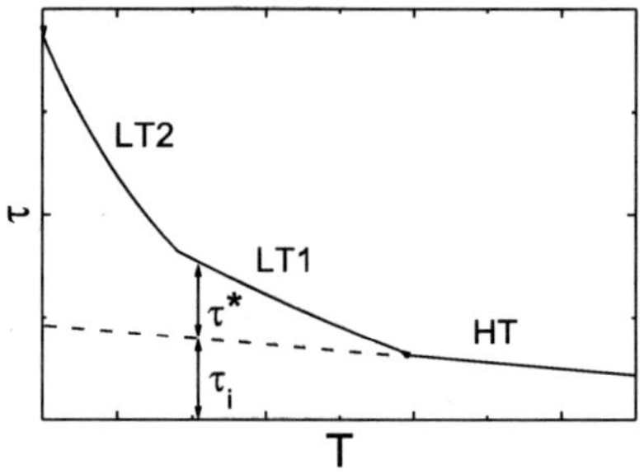

Fig. 5.25. Schematic plot of the temperature dependence of the flow stress in the ranges where diffusion is not important

activation volume and the stress exponent related to the dislocation mobility models are smaller than the apparent values obtained from experiments.

In a schematic way and if diffusion processes are neglected, the temperature dependence of the flow stress may look like that in Fig. 5.25. At high temperatures (HT), the deformation is athermal. The flow stress depends only weakly on the temperature owing to the temperature dependence of the shear modulus. Changing to lower temperatures may result in one or more thermal ranges with stronger temperature dependencies, for example, LT1 with localized obstacles and LT2 with the Peierls mechanism. It is very common to extrapolate the athermal high-temperature range to low temperatures to obtain the internal back stress τ_i also for lower temperatures, as indicated in the figure. However, this treatment is based on the supposition that the internal structure remains constant at all temperatures. This may approximately be true if a specimen is deformed at a high temperature, then cooled down to a low temperature and deformed again. In the general case where individual specimens are deformed at different temperatures, the dislocation microstructure, in general, and the dislocation density, in particular, are different at each temperature. In the first stage of deformation, the dislocation density increases very rapidly. As the multiplication rate depends on the effective stress (5.2), and this in turn on the temperature, the dislocation densities at the yield point are expected to drastically increase with decreasing temperature. This is shown in Fig. 5.26 using the data of NaCl single crystals deformed at room and liquid nitrogen temperatures. The dislocation densities are considerably higher in the crystals deformed at liquid nitrogen temperature. Thus, the assumption that the athermal part of the flow stress does not, or only weakly, depend on temperature is generally not fulfilled. A better approximation is certainly the validity of the Cottrell–Stokes law [318], that is, the proportionality between the thermal and athermal flow stress parts. This law was originally formulated for the stress variation by work-hardening but may apply also to other cases.

Most mechanisms described in Chap. 4 controlling the dislocation mobility and causing a thermal contribution to the flow stress do not depend on the

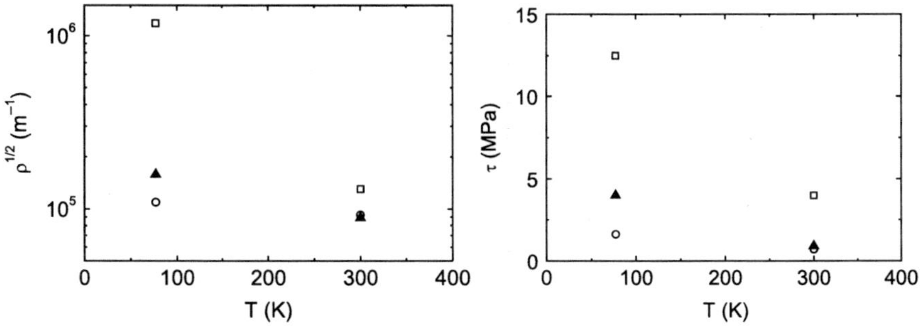

Fig. 5.26. Dependence of the square root of the dislocation density and the (shear) flow stress on the temperature in deformed NaCl single crystals of different divalent impurity concentrations: 1.7 ppm residual impurities (*open circles*), 32 ppm Ca^{++} (*full triangles*), 136 ppm Ca^{++} (*open squares*). Data from [177]

plastic strain. However, in many materials also the thermal part of the flow stress increases during work-hardening, for example, being manifest by an increasing strain rate sensitivity with increasing strain. There are only few mechanisms explaining such a behavior. One of them is dislocation cutting where part of the energy may be supplied by thermal activation. Frequently, the obstacle distance obtained from the strain rate sensitivity is in the range of the distance between forest dislocations and scales with the reciprocal stress (e.g., [319]). The other possible mechanism is the interaction between moving dislocations and debris. No detailed data as to this possibility are available yet.

In the literature, there are several attempts to link the dynamics of the motion of individual dislocations with the continuum theory of plastic deformation by the motion of many dislocations as it was founded by Kröner [320]. Recently, Hartley [321] extended these theories by introducing a dislocation density vector for a particular slip system, which describes the net length of the dislocation lines with the same Burgers vector in the respective volume element. The orientation of the vector is given by the ratio of the total length of screw and edge components. Besides, a dislocation mobility tensor relates the velocity of a dislocation configuration in the volume element to the net Peach-Koehler force on the configuration. As an example, the cutting of a forest dislocation is treated yielding a plastic potential to establish a flow law.

5.2.3 Experimental Determination of the Stress Components

Several special sequences of changing the macroscopic deformation parameters had been designed to determine the effective stress. They will not be discussed here in detail because of their restrictive assumptions involved. Looking at Fig. 5.20, one may conclude that dislocation motion is possible only as long as the applied stress is higher than the amplitude of the internal stresses, $\tau > \tau_A$, or, alternatively the applied stress is higher than the internal back stress, $\tau > \tau_i$. Thus, in a stress relaxation experiment, the relaxation rate should tend

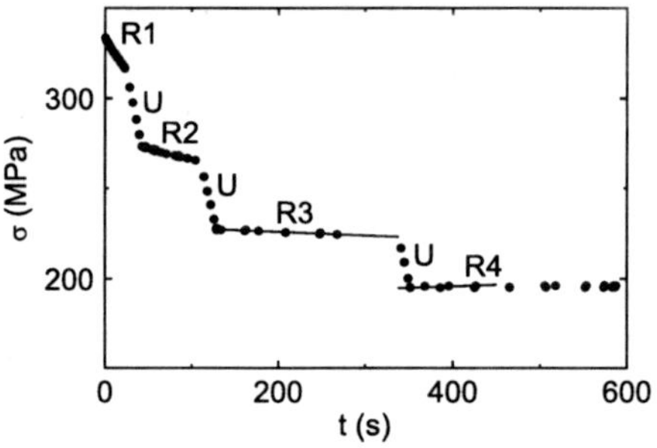

Fig. 5.27. Stress dip test on a cubic zirconia single crystal deformed at 1,180° C, consisting of short stress relaxations R_n and partial unloadings U. Data from the work in [324]

to zero when the stress reaches the athermal level. These long-time relaxation tests require a very stable testing machine to record strain rates of the order of $10^{-8}\,s^{-1}$. Because of (2.7), the relaxation time is short if the stiffness S of the testing machine (plus specimen) is high. A high machine stiffness is reached by electronic strain control. Nevertheless, even if the relaxation rate is low, there may still be an appreciable effective stress. The time of the experiment can be reduced by so-called stress dip tests [322]. A recent application is described in [323]. In these tests, the machine is stopped for a relaxation test. After sufficiently long recording to observe the actual relaxation rate, the stress is reduced by partial unloading followed by another period of relaxation, and so on. With decreasing load, the relaxation rate decreases but at a stress below the athermal flow stress, the relaxation rate changes its sign as the dislocations rearrange and move backwards in the internal stress field. In this way, it is possible to approach the athermal stress level from both sides. An example is shown in Fig. 5.27. Of course, one prerequisite to these methods is that recovery does not take place during the time of the experiment.

In heterogeneous dislocation structures, the internal stresses can be studied by X-ray diffraction. Changes in the lattice parameter as a consequence of internal stresses cause an asymmetrical broadening of the diffraction profiles because of the different volume fractions of the hard and soft regions with their different internal stresses. This method has been applied to cell structures and subgrain structures [325, 326].

One way of estimating the athermal stress component also in homogeneous dislocation structures consists in the determination of the dislocation density ϱ and the application of (5.11) with a suitable interaction constant α. The dislocation density is defined as the dislocation length per unit volume. Frequently, it is measured as the number of points of dislocations emerging through the unit area of an external surface or a cross-sectional plane of a deformed specimen. The relation between both definitions depends on the slip geometry. In most cases, setting the dislocation density equal to twice the density of emergence points is a good approximation [327]. The points of dislocations emerging through surfaces can be marked by etching a cleavage or polished surface as mentioned in Sect. 2.2. The etch pits can be observed by optical microscopy up to a density of about $10^{11}\,\mathrm{m}^{-2}$. Replicas of smaller etch pits produced by the heavy metal carbon shadowing technique can be observed in the transmission electron microscope. This method allows one to count representative values of intermediate dislocation densities up to $10^{13}\,\mathrm{m}^{-2}$.

Higher densities of dislocations are determined from diffraction contrast TEM images of thin films of the deformed samples. From the total projected length Λ imaged within the area A, the dislocation density is calculated by $\varrho = 4\Lambda/(\pi A t)$ [328]. Here, t is the foil thickness, which can be determined from the emergence points through both surfaces of a dislocation lying on a known plane, or by other methods of diffraction contrast TEM. A simpler method consists in superimposing a rectangular net of two sets of parallel straight lines of nonconstant spacing onto the micrograph. The dislocation density is determined by counting the numbers of intersections N_1 and N_2 of the dislocations along the two respective sets of grid lines. With the total lengths L_1 and L_2 of the grid lines, the dislocation density becomes $\varrho = (N_1/L_1 + N_2/L_2)\,/\,t$ [329]. If randomly oriented lines are used, the dislocation density is estimated by $\varrho = 2N\,/\,(Lt)$ [330]. TEM extends the range of dislocation density measurements up to about $10^{16}\,\mathrm{m}^{-2}$. It has to be considered, however, that part of the dislocations may be invisible because of contrast extinction at the particular imaging vector. Depending on the latter, in the f.c.c. lattice, the fraction of the extinguished dislocations may amount up to 50%.

High dislocation densities can also be determined from the broadening of X-ray diffraction profiles. A high sensitivity is necessary, which can be achieved in the so-called self-focusing arrangement where the contribution of the wavelength dispersion to the instrumental line-broadening disappears [331]. The full width at half-maximum of the X-ray lines can be represented by the modified Williamson–Hall plot [332] according to

$$\Delta K \approx 0.9/D + (\pi/2)^{1/2}\, Mb\,\varrho^{1/2}\, KC^{1/2} + \text{higher terms in } K^2C,$$

where θ is the diffraction angle, λ the wavelength of the X-rays, $K = 2\,\sin\theta/\lambda$, and $\Delta K = 2\,\cos\theta(\Delta\theta)/\lambda$. D is the particle or subgrain size, b is the magnitude of the Burgers vector, and ϱ is the dislocation density. Furthermore, M is a constant depending on the outer cut-off radius of the dislocations, and C is the dislocation contrast factor, which depends on the relative orientations between the Burgers and line vectors of the dislocations and the diffraction vector as well as on the elastic constants. For cubic crystals, C values have been calculated in [333]. The contributions to the line broadening from the particle size and from the strains due to dislocations can be separated by plotting ΔK vs. $KC^{1/2}$. The method has been applied, for example, to determine the gradients in the dislocation density between the bulk and surface regions of deformed crystals.

The most sensitive and spatially resolving technique is the measurement of the effective stress by using the dislocations themselves as a probe, that is, by measuring their local curvature, as discussed in Sect. 3.2.7. With (3.14), (3.38), and (3.42), the relation between the effective stress τ^* and the local radius of curvature r can be written as

$$\tau^* = \frac{\Gamma_0(\beta)\,(\ln(l/r_0) + C)}{r(\beta)\,b}, \tag{5.17}$$

where $\Gamma_0(\beta)$ is the elastic part of the line tension. It can be calculated by anisotropic elasticity theory with the formulae for straight dislocations. For isotropic solids, it is $\Gamma_0(\beta) = \mu b^2 \left(1 + \nu - 3\nu \sin^2 \beta\right) / (4\pi(1-\nu))$.

A suitable method for determining the curvature of bowed dislocation segments consists in comparing the images of the dislocation segments with calculated dislocation loops of different size. The effective stress is then

$$\begin{aligned} \tau^* &= \frac{E_{0\mathrm{e}}\,(\ln(l/r_0) + C)}{x_0\,b} \\ &= \frac{E_{0\mathrm{s}}\,(\ln(l/r_0) + C)}{y_0\,b}, \end{aligned} \tag{5.18}$$

with $E_{0\mathrm{e}}$ and $E_{0\mathrm{s}}$ being the elastic (prelogarithmic) parts of the line energy of edge and screw dislocations, and x_0 the half-axis of the loop in the direction of the Burgers vector and y_0 perpendicular to it. x_0 corresponds to e in the case of isotropic elasticity.

For most materials, the elastic part of the line tension or energy is known quite accurately. The problem is the logarithmic factor with the segment length l, the inner cut-off radius r_0, and the numerical constant C. Both l and C depend on the particular dislocation configuration because of the self-interaction between the curved segment under consideration and its adjoining segments. In the line tension approach, this self-interaction can be considered only by suitably choosing l and C. Reasonably well defined is the configuration of a dislocation pinned by localized obstacles. As discussed in Sect. 3.2.7, l is then the segment length of the bow-out. For half and full loops, l is the diameter of the loops. Not well defined are long dislocations bowing out between cell walls or other sources of internal stress. Unfortunately, the configuration influences also the ratio between the radii of curvature of edge and screw segments. To each configuration, a different constant C has to be applied. In general, the straighter the configuration is, the smaller is C (more negative).

To avoid this problem, some authors try to determine the line tension experimentally. According to Albenga's law, the spatial average of the internal stresses has to vanish so that the average of the local stresses equals the applied stress, that is,

$$\langle \tau_{\mathrm{loc}}(x,y) \rangle = \tau = \frac{\Gamma}{b} \left\langle \frac{1}{r_{\mathrm{loc}}} \right\rangle .$$

In [309], this procedure was applied separately to near-edge and near-screw dislocations to find the line tensions Γ of edges and screws. This method

Table 5.2. Comparison of the macroscopic shear flow stress of two MgO single crystals with the effective stresses determined from dislocation curvatures measured by a photometric method. Data from [204]

Macroscopic flow stress	From e, l, and $C = -1.61$	From M
39.0	125	50.6
42.6	111	56.4

implies the problem that the moving dislocations will stay in the regions of low effective stress so that their average will not represent the true spatial average of the local stresses. If it did, there would not be any sense in measuring the effective stress as its average always had to be equal to the applied stress.

A detailed study of the radii of curvature of dislocation segments bowing out between localized obstacles was performed based on the results of in situ straining experiments in an HVEM on MgO single crystals [104], as described in Sect. 3.2.7. From the local radii of curvature measured by a photometric analysis of the dislocation contrast profiles, the major half-axes of corresponding ellipses were calculated. The effective stresses are then given by (3.45) either from sets of e and l using the theoretical constant $C = -1.61$, or from the slope M of the e vs. $\ln l$ diagrams, which corresponds to the experimental constant $C = -5.19$. The parameters $r_0 = 0.4\,b$ [114] and $E_{0\mathrm{e}} = 1.03 \times 10^{-9}\,\mathrm{N}$ were used. Some data are collected in Table 5.2 together with the macroscopic shear flow stresses. Because of radiation hardening and a possible size effect, the flow stresses in the in situ specimens may be about 30% higher than the macroscopic ones [334]. The table shows that the data from the curvature measurements using the theoretical constant $C = -1.61$ do not agree with the macroscopic flow stresses, whereas those with the experimental value of the logarithmic factor of the line tension do. The data in the first line of Table 5.2 refer to an experiment where the specimen was unloaded and reloaded and, in addition, the curvature of many segments in the same specimen area was measured by fitting ellipses to the bowed-out segments. The load measured on the in situ stage was scaled by the macroscopic flow stress. Of these additional measurements, Fig. 5.28 presents the relation between the average effective stress τ^*, calculated with the experimental constant $C = -5.19$, and the shear stress τ, followed from the specimen load without considering radiation hardening. The effective stresses are smaller than the macroscopic ones, as it is expected, as the dislocations stay mostly in regions of low effective stress. Within the accuracy of this experiment, both stresses are proportional to each other. The differences between the data in the table and those in the figure are certainly not due to the different methods of determining the curvature but rather to the choice of the segments to be evaluated. In the table, well bowed-out segments were chosen for the photometric evaluation, whereas most of the segments were evaluated by fitting ellipses to the segments. This underlines the

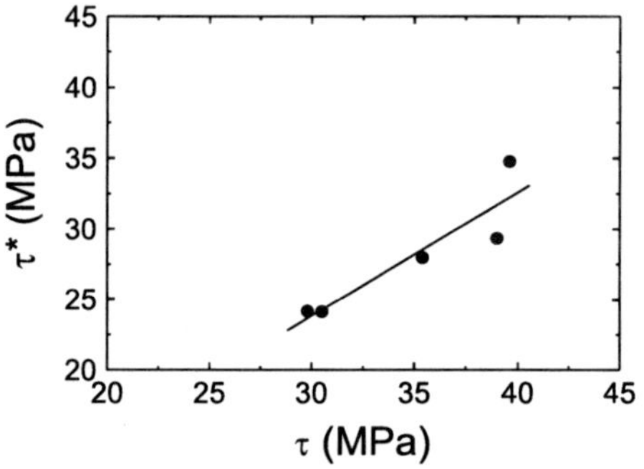

Fig. 5.28. Dependence of the effective stress on the applied stress in an MgO single crystal taken during a loading series of an in situ straining experiment in an HVEM. Dislocation curvatures measured by fitting ellipses to the bowed-out segments. The applied stress is not corrected for radiation hardening. Data from [204]

problem of a proper choice of curved segments for determining the effective stress. The latter shows a relatively wide symmetric frequency distribution with a standard deviation of $\Delta\tau^* \approx 30\,\text{MPa}$. This scattering reflects only partly the varying effective stress but also the influence of the self-stress of the neighboring segments on the curvature of the segment considered.

Summarizing, it may be stated that it is very difficult to measure the internal stresses. A rough estimate can be obtained by counting the dislocation density and applying the Taylor equation (5.11). The importance of the internal stresses is well recognized in studies of work-hardening. However, in phenomena controlled by the mobility of dislocations like the low-temperature deformation (e.g., [335]) or the flow stress anomaly in intermetallic alloys (Sect. 9), the influence of the internal stresses on the total flow stress is widely neglected.

5.2.4 Steady State Deformation

A combination of the dislocation kinetics controlling the evolution of the dislocation structure and accordingly also the internal stress with the dislocation dynamics, that is, the relation between the dislocation velocity and the effective stress, may serve as the basis for a system of constitutive equations governing the plastic deformation. As introduced in this book, the set of equations may have the following form where, of course, alterations will be necessary to describe the deformation in particular materials or in particular ranges of temperature and strain rate. In Sect. 10.5.7, the equations will be applied to the high-temperature deformation of i–Al–Pd–Mn single quasicrystals by the climb motion of dislocations. They are therefore formulated in terms of normal stresses σ and strains ε.

These equations are the following:
The kinetic equation of the dislocation density (5.8)

$$\frac{\mathrm{d}\varrho}{\mathrm{d}t} = w\sigma^* \varrho v_\mathrm{d} - q\,\varrho^2 ,$$

where, at high temperatures, the annihilation coefficient q will be a function of temperature (5.9)

$$q = q_0 \exp\left(-\Delta G_\mathrm{ann}/kT\right),$$

and the dynamic equation, for example, in the form of the (slightly simplified) power law (4.103)

$$v_\mathrm{d} = v_\mathrm{d0} \exp\left(-\frac{\Delta G_0}{kT}\right) \sigma^{*m} .$$

These equations have to be supplemented by the relation between the applied stress and the athermal and thermal flow stress components (5.15)

$$\sigma = \sigma_\mathrm{i} + \sigma^* ,$$

the Taylor equation for the internal stress (5.11)

$$\sigma_\mathrm{i} = \alpha \frac{\mu b}{2\pi} \sqrt{\varrho},$$

the machine equation (2.5)

$$\dot{\varepsilon}_\mathrm{t} = \dot{\varepsilon}_\mathrm{el} + \dot{\varepsilon} = \dot{\sigma}/S + \dot{\varepsilon},$$

and the Orowan equation between the plastic strain rate and dislocation velocity and density (3.5)

$$\dot{\varepsilon} = b\varrho v_\mathrm{d} .$$

This set of equations considers all dislocations to be mobile, thus being restricted to small plastic strains where immobile dislocation structures do not yet form.

The above equations describe the evolution of the dislocation density at the beginning of a deformation test. For a constant total strain rate $\dot{\varepsilon}_\mathrm{t}$ and a small initial dislocation density ϱ_0, the stress increases almost linearly with time, $\sigma \approx S\,\dot{\varepsilon}_\mathrm{t}\,t$, where the internal stress σ_i is small almost up to the yield point. After the yield point is passed, the deformation achieves a steady state where hardening is compensated by recovery.

For applying the model to the temperature and strain rate dependence of the steady state deformation of i–Al–Pd–Mn single quasicrystals (Sect. 10.5.7), the steady state solution is found in analytical form [336]

$$\varrho_\mathrm{ss} = \frac{\varrho_\mathrm{c}}{D^{m/(2m+1)}}$$

and

$$\sigma_{\text{ss}} = \sigma_{\text{c}} \left(D^{1/(2m+1)} + \frac{a}{D^{m/(4m+2)}} \right) . \tag{5.19}$$

Here,

$$\varrho_{\text{c}} = \left(\dot{\varepsilon} / \left[v_{\text{d}0} \exp\left(-\Delta G_0/kT\right) \right] \right)^{2/(m+2)} \left(w/S \right)^{m/(m+2)}$$

and

$$\sigma_{\text{c}} = \left(S\dot{\varepsilon} / \left[w\, v_{\text{d}0} \exp\left(-\Delta G_0/kT\right) \right] \right)^{1/(m+2)}$$

are scaling factors of the dislocation density and the flow stress, and D and a are dimensionless parameters characterizing the effects of recovery $D = (q/b)(w/\dot{\varepsilon})^{(m-1)/(m+2)} S^{-(2m+1)/(m+2)} \left[v_{\text{d}0} \exp\left(-\Delta G_0/kT\right) \right]^{-3/(m+2)}$ and hardening $a = \alpha\mu(bw/S)^{1/2}$.

Figure 5.29 compares the calculations with the experimental data. The parameters of the curves are the hardening factor $\alpha = 2\pi$, the stress exponent $m = 4$, the activation energy of dislocation motion $\Delta G_0 = 3\,\text{eV}$, and the activation energy of dislocation annihilation $\Delta G_{\text{ann}} = 4\,\text{eV}$. The activation parameters fit the data obtained by stress relaxation and temperature change experiments. The remaining parameters $v_{\text{d}0}$ and q_0 have been chosen to make the dislocation density consistent with the values measured at the lower yield point. The figure demonstrates that the model quite well represents the temperature and strain rate dependences of the steady state flow stresses as well as the temperature dependence of the dislocation density. A closer inspection of the formulae reveals that the apparent activation parameters are not equal to the model parameters. If the first term in (5.19), i.e., σ^*, dominates the deformation, then the apparent activation energy is

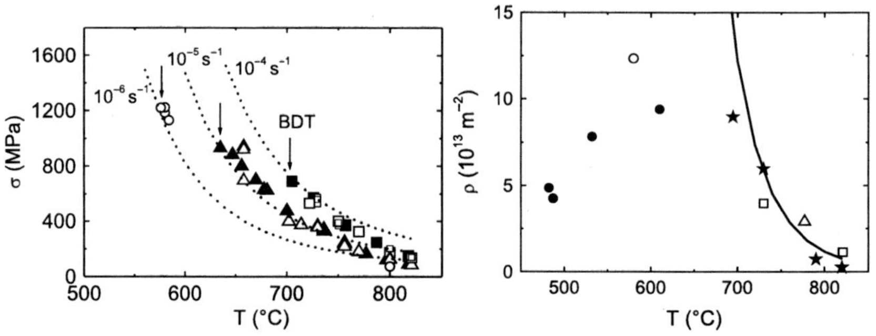

Fig. 5.29. Comparison of the steady state flow stresses and the dislocation densities of i–Al–Pd–Mn single quasicrystals with the results of the set of constitutive equations. (**a**) Temperature and strain rate dependence of the steady state flow stresses. Compression axes: twofold (*open symbols*), fivefold (*full symbols*). (**b**) Temperature dependence of the dislocation density. Data from steady state deformation (*open symbols*), from work-hardening range (*full symbols*), dislocation density data from [338] (*asterisks*). Strain rates $\dot{\varepsilon} = 10^{-6}\,\text{s}^{-1}$ (*circles*), $10^{-5}\,\text{s}^{-1}$ (*triangles*), $10^{-4}\,\text{s}^{-1}$ (*squares*). *Curves*: predictions of the model, *arrows* BDT: brittle-ductile transition. Data from [336] and [337]

$\Delta G_{\text{ex}} = 2\Delta G_0 - \Delta G_{\text{ann}}$. If the second term, that is, σ_{i}, is dominant, then $\Delta G_{\text{ex}} = (\Delta G_0 + m\Delta G_{\text{ann}})/(m+1)$. Thus, the model can explain a change in the experimental activation energy with the increasing influence of recovery at higher temperatures. Similarly, also the apparent stress exponent is not equal to m of the dislocation mobility.

Steady state deformation conditions can be attained in experiments at a constant strain rate as well as under constant load. The latter experiments are called creep tests where the steady state range corresponds to a minimum in the creep (strain) rate. A prerequisite to a steady state range is that the dislocation density recovers, which balances the generation of new dislocations during straining. Thus, steady state deformation is mostly restricted to higher temperatures where diffusion allows climb and dislocation annihilation. The steady state strain rate in single crystals can frequently be described by the phenomenological equation

$$\begin{aligned} \dot{\varepsilon} &= A\frac{\mu b}{kT} D \left(\frac{\sigma}{\mu}\right)^n \\ &= A\frac{\mu b}{kT} D_0 \exp\left(-\frac{\Delta G}{kT}\right) \left(\frac{\sigma}{\mu}\right)^n . \end{aligned} \tag{5.20}$$

A is a dimensionless factor, D is the diffusion coefficient as in (4.86) with the pre-exponential factor D_0 and the activation energy ΔG. The other quantities have their usual meaning. There may be very different processes controlling steady state creep, dislocation motion controlled by glide or climb, climb of dislocations in a three-dimensional network, and many others. An early review is given in [339]. The paper [340] lists a great number of references on the mechanisms and experimental data on ceramics. A review on the behavior of metals is given in [341]. The diffusion coefficient may be that of self-diffusion or pipe diffusion along dislocations. The stress exponents n are between about 2 and 6 with preference values around either 5 or 3. $n \approx 5$ mostly occurs in pure metals, and $n \approx 3$ in solid solution alloys. As stated earlier, dislocation recovery and annihilation play an important role. In [342], the change from power law creep according to (5.20) to a range, with the stress exponent increasing and the activation energy decreasing with increasing stress, is explained by a change from the annihilation of dipole-like configurations by climb to that by glide-controlled dislocation reactions. In polycrystals, creep is frequently controlled by grain boundary processes.

Plastic deformation at high temperatures is certainly one of the most complex processes of dislocation motion. Both its modes take place, glide, and climb. They do not only carry the plastic deformation but they also realize the recovery of the dislocation density. Frequently, the dislocations form a relatively regular network where the motion of links in the network is governed by the link length like the moving branch of a Frank–Read source, for example, in the model of [343], explaining a transition at decreasing temperature from climb-recovery controlled deformation to viscous glide governed by

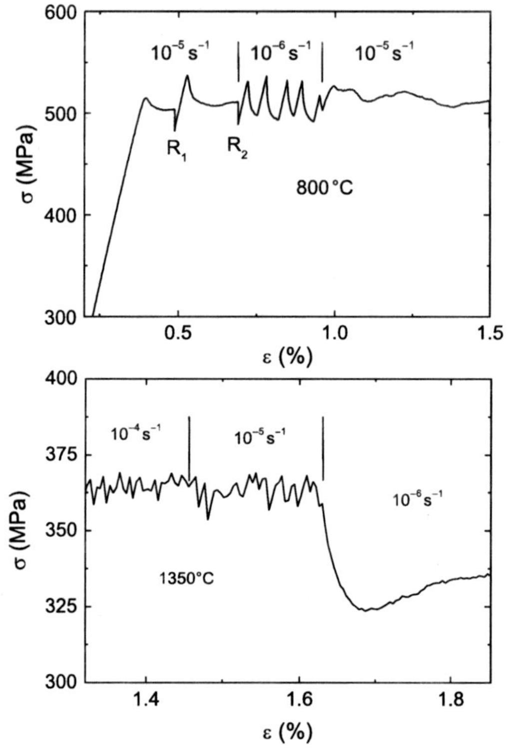

Fig. 5.30. Sections of stress–strain curves of ZrO_2-15 mol% Y_2O_3 single crystals containing changes of the strain rate from 10^{-5} to $10^{-6}\,s^{-1}$ resulting in changes of the stability state. (**a**) 800°C. R_1 and R_2 are stress relaxation tests. (**b**) 1,350°C. Data from [344]

moving point defect atmospheres. The problem is that many processes involve lattice diffusion so that the experimental identification of the mechanisms is quite difficult.

5.3 Plastic Instabilities

Up to this point, plastic deformation has been treated as a stable process, that is, the stress–strain curve has been considered smooth. This is very frequently the case but there are also exceptions in certain temperature and strain rate ranges where the stress–strain curve shows regular or irregular drops and re-increases, which is called jerky flow or serrated yielding. Figure 5.30 illustrates the influence of the strain rate on the stability of deformation in ZrO_2 single crystals. At two temperatures, the strain rate is changed from 10^{-5} to $10^{-6}\,s^{-1}$. At the lower temperature of 800°C (Fig. 5.30a), the deformation changes from stable to unstable, but at 1,350°C (Fig. 5.30b) it changes in the opposite direction. Similarly, the instability ranges and the shape of the serrations depend on the temperature. The latter is demonstrated in the stress-time records of Fig. 5.31. Figure 5.31a taken at 860°C is characteristic also of higher temperatures. First, the stress increases linearly, with the slope equalling that of the elastic line as measured during unloading, for instance. The elastic slope is indicated by the straight line. Thus, the loading takes place as purely elastic deformation. When a certain stress level is reached, plastic deformation sets in at a high rate with an abruptly decreasing load. At lower temperatures, plastic deformation does not start at a very high rate so that the unloading parts of the load-time curve get curved similarly to those of stress relaxations, as shown in Fig. 5.31b for 800°C. At an even lower temperature in Fig. 5.31c, also the loading parts become rounded at their tips, which indicates plastic deformation even during loading. In general, instability ranges can be described in a plot of strain rate vs. temperature where instabilities occur in limited regions.

Plastic instabilities are systematically classified in [345], according to which they may occur by strain softening, by a negative strain rate sensitivity (strain rate softening instabilities), or by localized heating (thermomechanical instabilities). Strain softening may occur, for example, when the moving dislocations destroy ordered states in alloys. Strain rate softening instabilities are frequently connected with the Portevin–LeChatelier (PLC) effect [346, 347] due to dynamic strain ageing described in Sect. 4.11. Thermomechanical instabilities result from a reduction of the flow stress owing to local heating in slip bands at low temperatures.

Plastic instabilities are discussed theoretically at different degrees of sophistication. Early linear stability analyses of a constitutive model of plastic flow by Kubin and Estrin [345, 348–350] have been extended by including transient effects, as reviewed by Zaiser and Hähner [351], as well as by the evolution of the mobile and forest dislocation densities [352]. According to [351], a change in the flow stress can be written as

$$\mathrm{d}\sigma = \Theta \mathrm{d}\varepsilon + r \,\mathrm{d} \ln \dot{\varepsilon} + \mathrm{d}\sigma_\phi , \tag{5.21}$$

where Θ is the work hardening coefficient and $\mathrm{d}\sigma_\phi$ is a change in the flow stress due to the relaxation of an internal parameter ϕ which, after a change in the deformation conditions, attains a new quasi-steady state value after a characteristic time t_ϕ. Including $\mathrm{d}\sigma_\phi$ takes into account the transient effects observed experimentally after changing the deformation conditions as described in Sect. 2.1, for example, the difference between the instantaneous and steady state stress increments $\Delta\sigma_{\mathrm{in}}$ and $\Delta\sigma_{\mathrm{ss}}$ in strain rate cycling tests, or the difference between an original and a repeated stress relaxation. With a linear approximation of the relaxation process, a linear stability analysis yields the condition for unstable deformation [351]

$$\frac{1}{t_\phi} \frac{r_{\mathrm{in}} - r_{\mathrm{ss}}}{r_{\mathrm{in}}} > \frac{\dot{\varepsilon}\,(\Theta - \sigma)}{r_{\mathrm{in}}} + \frac{1}{t_\phi} . \tag{5.22}$$

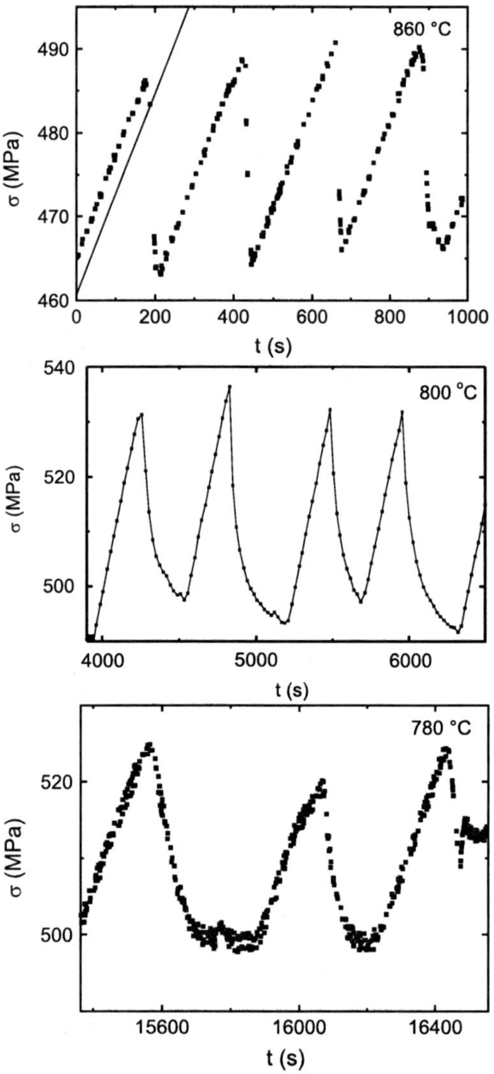

Fig. 5.31. Sections of stress-time records of ZrO_2-15 mol% Y_2O_3 single crystals in the instability range at $10^{-6}\,s^{-1}$. (**a**) Abrupt transitions between elastic and plastic deformation and vice versa at 860°C. At lower temperatures, temperature-dependent plastic deformation develops. (**b**) 800°C. (**c**) 780°C. Data from [344]

$r_{\text{in}} = \Delta\sigma_{\text{in}}/\Delta \ln \dot{\varepsilon}$ and $r_{\text{ss}} = \Delta\sigma_{\text{ss}}/\Delta \ln \dot{\varepsilon}$ are the instantaneous and steady state strain rate sensitivities. This more general stability criterion includes the criteria of the simpler theories as approximations under simplifying conditions. Assuming a short relaxation time t_ϕ so that

$$\frac{\dot{\varepsilon}\,(\Theta - \sigma)}{r_{\rm in}} \ll \frac{1}{t_\phi},$$

instability occurs for $r_{\rm ss} < 0$ at $\Theta - \sigma > 0$. This is equivalent to the adiabatic solution of the constitutive equation (5.21), that is, by canceling the term $d\sigma_\phi$ and setting $r_{\rm ss} = r$. These kinds of instabilities are called strain rate softening instabilities, which may occur if $r_{\rm ss}$ is negative, but not $r_{\rm in}$, which is always positive. The inequality (5.22) can also be written as $-r_{\rm ss} > t_\phi\dot{\varepsilon}(\Theta - \sigma)$, where $t_\phi\dot{\varepsilon} = \Delta\varepsilon_{\rm ss}$. It can be fulfilled not only by sufficiently negative values of $r_{\rm ss}$, but also by sufficiently negative values of $\Theta - \sigma$. The latter kinds of instabilities are called strain softening instabilities.

The different physical processes causing strain rate softening instabilities can be discussed by assigning different physical variables to the structural parameter ϕ. One possibility is to identify ϕ with the local specimen temperature T, as an increase in T yields an increase in the strain rate because of the thermally activated character of the plastic deformation. Thus, instabilities may occur by a feedback between the heat locally released by the plastic deformation and the resulting increase in the strain rate. These instabilities are called thermomechanical instabilities, being a particular case of the strain rate softening instabilities.

In many cases, the plastic instabilities are of the strain rate softening type. As discussed above, the simple theories then require that $r_{\rm ss} < 0$ for $\Theta - \sigma > 0$. Frequently, however, $r_{\rm ss}$ is small but not negative. The following shows that this criterion is relaxed in the more sophisticated theory in [351]. In alloys, serrated yielding owing to strain rate softening is frequently connected with diffusion processes of alloying elements in the stress fields of the dislocations. This is called dynamic strain ageing as it reduces the dislocation mobility and gives rise to the Portevin–LeChatelier effect [346, 347]. The formation and dragging of a dynamic cloud of point defects around a moving dislocation was treated in Sect. 4.11.

The models of the Portevin–LeChatelier effect reviewed in [351] use a different approach. The dislocation motion is assumed to be jerky on a mesoscopic scale owing to obstacles to glide, which are overcome by stress-assisted thermal activation. During the waiting time at the obstacles, the dislocations are gradually aged by the formation of the solute clouds, leading to an increase in the activation energy of dislocation motion, which depends on the waiting time $t_{\rm w}$ at the obstacles [353–355]. According to [356], the Gibbs free energy of activation can be expressed as

$$\Delta G = \Delta F - V\tau^* + \Delta g\,\left(1 - \exp\left[-(\eta t_{\rm w})^n\right]\right). \tag{5.23}$$

Like in (4.101), ΔF, V, and τ^* are the stress depending free energy of activation, the stress dependent activation volume, and the effective stress. Δg is the increase in the activation energy owing to ageing, $t_{\rm w}$ the time constant of ageing, and n an ageing exponent. ΔF and V are supposed to describe the temporary pinning of the dislocations, which is caused by processes different from the diffusing solutes, for example, by forest dislocations [350] or precipitates. After the dislocation has overcome the pinned and aged configuration, the activation energy takes the lower value of $\Delta G = \Delta F - V\tau^*$ so that the dislocation may move a greater distance until it is pinned again. The plastic strain rate is then given by

$$\begin{aligned}\dot{\varepsilon} &= \nu_0 \Omega \exp\left[-\frac{\Delta G}{kT}\right] \\ &= \nu_0 \Omega \exp\left[-\frac{\Delta F - V\tau^* + \Delta g\left(1 - \exp\left[-(\eta t_{\mathrm{w}})^n\right]\right)}{kT}\right] . \end{aligned} \tag{5.24}$$

ν_0 is a vibrational frequency, and Ω an elementary strain, that is, the strain occurring if all dislocations overcome an obstacle. Here, the waiting time for ageing is set equal to the average waiting time for thermal activation, $t_{\mathrm{w}} = \Omega/\dot{\varepsilon}$. Equation (5.24) can be solved for $\dot{\varepsilon}$, if the stress dependencies of ΔF and V are known. A resulting schematical plot of τ^* vs. $\ln \dot{\varepsilon}$ or $\ln \dot{\gamma}$ was presented in Fig. 4.39 illustrating the three ranges discussed already. In range A at low strain rates, τ^* increases owing to an increasing efficiency of ageing. In range B, the dislocations move too fast for effective ageing so that τ^* decreases. Finally, τ^* increases again in range C due to the increasing resistance of the obstacles, which temporarily pin the dislocations. The slope of the curve equals the strain rate sensitivity r, which is positive in range A, negative in range B, and positive again in range C as discussed in Sect. 4.11. Plastic deformation cannot be stable in range B. Thus, if the macroscopic strain rate corresponds to range B, the specimen may deform at a lower rate in range A, leading to an increase in τ^*. When it reaches its maximum value at the transition to range B, the deformation can continue under the same stress at a high rate in range C. This leads to a decrease in τ^* until the transition to range B is reached where the deformation switches back to range A as indicated by the thick broken arrows in Fig. 4.39. The cycle described corresponds to both an increase and a drop in stress during serrated yielding. The figure predicts also the dependence of the strain rate sensitivity r on the strain rate or the stress in the stable ranges. In the normal range C of thermally activated overcoming of obstacles, the curve is bent upwards, which corresponds to an increase in r or a decrease of the activation volume $V = lb\Delta d$ with the stress increasing owing to a decrease of the activation distance Δd. In Friedel statistics (Sect. 4.5.1), also the obstacle distance l decreases with increasing stress. In the diffusion-controlled range A, however, the curve in Fig. 4.39 must bend downwards to form the maximum at the transition to range B, that is, r decreases with increasing stress. The dependence of r on the stress is reflected in the shapes of stress relaxation curves as it will be discussed, for example, in connection with the flow stress anomaly in intermetallics (Sect. 9). Besides, the time dependence of the activation energy as expressed in (5.23) leads to transient effects under nonconstant deformation conditions.

As described so far, all dislocations may move at the same velocity. The theory reviewed by Zaiser and Hähner [351] includes two important extensions.

- The assumption of equal waiting times t_w for all dislocations is replaced by a distribution function of the waiting times.
- Dislocations moving on parallel slip planes are considered coupled by long-range elastic interactions. This leads to a synchronization of the waiting times, that is, to a collective mode of dislocation motion.

Plastic instabilities require both the occurrence of dynamic strain ageing and the coupling of dislocations. Thus, the Portevin–LeChatelier effect consists in a localization of slip in time and space. The most important result of the extended theory is that the strain rate sensitivity r need not be negative as in the earlier theories. It is sufficient if r is close to zero. The low strain rate sensitivity is necessary for the synchronization of the waiting times.

In the extension of the earlier Kubin–Estrin model in [352], equations similar to those described above are coupled with equations describing separately the evolution of the densities of mobile and forest dislocations. While dislocation pinning and strain ageing take place on a fast time scale, the evolution of the dislocation structure occurs on a slow time scale. The model reproduces the negative strain rate sensitivity of the threshold stress for the instability.

While many materials under most conditions deform in a stable way, serrated flow may occur under certain conditions of temperature and strain rate. The unstable flow results from a combination of diffusion-controlled ageing and pinning and thermally activated surmounting of stronger obstacles. In this respect, the literature mainly considers the cutting of forest dislocations. However, jerky flow may occur also in single slip so that other obstacles like precipitates may act as pinning centers. Characteristic of dynamic strain ageing is the low strain rate sensitivity. Plastic instabilities are a collective dislocation process where many dislocations act simultaneously. This requires a coupling in time and space. Accordingly, slip is mostly concentrated in narrow slip bands. A prerequisite to the spatial localization is the generation of many dislocations in a restricted volume. This is possible by the localized Frank–Read sources or by multiplication at a high stress where the new dislocations appear near the multiplying ones. Experimental details of plastic instabilities will be described in Sect. 7.3.

Part II

Dislocation Motion in Particular Materials

Part II of the book presents results on a number of selected materials mainly chosen from experiments performed in the group of the author, whenever suitable data, micrographs, and video recordings were available. Thus, these chapters cannot be considered review articles of the respective classes of materials. Nevertheless, the selection covers most of the processes controlling dislocation dynamics in the range of thermal activation, which were described in Part I. When necessary, the text is supplemented by other data to complete the view on the particular class of materials. In the elemental semiconductors, the double-kink or Peierls mechanism controls the dislocation mobility over a wide range of temperatures (Sect. 6). Ceramic single crystals (Sect. 7) are partly brittle at low temperatures because of a high Peierls stress. At high temperatures, they may show an athermal behavior and several other mechanisms governing the dislocation velocity. Pure metals are plastic down to very low temperatures, but alloying with other elements causes solution and precipitation hardening (Sect. 8). Many intermetallic alloys show a flow stress anomaly, i.e. an increase in the flow stress with increasing temperature. The anomaly may be caused by particular configurations of the dislocation cores but also by local diffusion phenomena (Sect. 9). Finally, quasicrystals are a special case of intermetallic materials with long-range order but without lattice periodicity. Here, the structure is described in a higher-dimensional hyperspace, which determines also the properties of dislocations. However, diffusion processes seem to control the dislocation mobility (Sect. 10). The following chapters will present the crystal structures of all described materials, their dominating slip systems, their microscopic and macroscopic deformation properties, and finally the processes controlling the dislocation dynamics. If available, the dislocation motion is illustrated by video sequences accompanying this book. These video clips are commented in the text. Within the video sequences, special events are marked by letters or other labels.

It is hoped that the reader who is not very familiar with dislocation properties will obtain an overview of the processes governing dislocation dynamics in the different materials and that the video sequences and their present interpretation will stimulate researchers in the field to reconsider some of the established opinions.

Comments on the Video Sequences Available in the Internet

The video sequences can be downloaded via http://extras.springer.com/2010/978-3-642-03176-2 as a zip archive. Because of the large size of the file, DSL is essential. Unpacking the zip file yields several folders containing the video files organized on an internet browser basis. It is recommended to employ Firefox as the internet browser and the Windows Media Player as the default player. The videos will also run on a more recent Quicktime player. Viewing the videos can be started by double-clicking the file START.html.

In all video recordings, the tensile direction is inclined with respect to the vertical direction. The angle amounts to 36° in clockwise sense at a magnification where the frame width corresponds to 6 μm. The angle is slightly smaller for higher magnifications and larger for lower ones. This relation between the images and the tensile direction holds also for many micrographs from in situ experiments but not for all. Sometimes the tensile direction is indicated. As a measure of the magnification, the frame width is included in the titles of the video clips.

In the video recordings, the images of quickly moving dislocations are blurred due to the afterglow of the luminescence screen, which the videos were taken from. In many video clips, the place of observation was intentionally shifted from time to time to follow special events or to move to more active specimen regions. Besides, it was frequently tried to improve the imaging conditions by slightly tilting the specimen during the video recording. This is frequently connected with strong changes in the contrast of the dislocations and mostly also with a shift of the image position. These manipulations were done trying to improve the image quality.

If the Burgers vectors of the moving dislocations have a component out of the specimen surfaces, the dislocations trail a slip step, which requires additional surface energy causing a drag force on the dislocations. Therefore, the dislocations may bow out near the surfaces. The same happens if the specimens are covered with a contamination layer, which is mostly the case. Then, the moving dislocations have to deform the surface layer. This process may be treated as the creation of dislocations parallel to the surface between the specimen and the surface layer marking the traces of the moving dislocations. The elastic strain fields of the surface dislocations are imaged in the diffraction contrast as so-called slip trails or slip traces. They mark the intersection lines between the slip planes and the surfaces and can be used to identify slip, cross slip, and climb as well as to index the slip planes.

For viewing the video clips, it is very useful to employ a video player with a shuttle control, where it is possible to shift a certain sequence forward and backward to analyze the process in detail.

6

Semiconductors

Semiconductor crystals show a strong directional covalent bonding and are therefore prime examples of the control of the dislocation dynamics by the double-kink or Peierls mechanism. This is connected with a strong temperature dependence of the dislocation mobility and the macroscopic flow stress. At low temperatures, the semiconductor crystals are brittle, like most of the other materials with a high Peierls stress. After the first review by Alexander and Haasen [279], plastic deformation and dislocation properties have been summarized in several articles, for example, [357–359]. The knowledge of the dislocation and plastic properties of semiconductor crystals is important for the production of electronic devices with their manifold thermal treatments. As dislocations are mostly detrimental to the operation of devices, respective production processes have to minimize the number of dislocations in them. This requires information about dislocation mobility and annealing properties.

6.1 Crystal Structure and Slip Geometry

The elemental semiconductors like silicon and germanium, which are the main topic of this chapter, crystallize in the cubic diamond structure and many compound semiconductors in the sphalerite structure. The diamond structure is built of $\{111\}$ double layers (a′b, b′c, c′a, etc.) of tetrahedrally coordinated atoms stacked in an ...ABCABC... sequence like the f.c.c. metals, as shown in Fig. 6.1 in a projection along $[\bar{1}10]$. In the sphalerite structure, the a, b, c planes are occupied by one sort of atoms, while the a′, b′, c′ planes are occupied by the other one. The $\{111\}$ planes are also the slip planes with $1/2\langle 110\rangle$ Burgers vectors. Thus, Fig. 6.1 is the projection in the direction of one of the three Burgers vectors on the (111) plane. In total, there are 12 slip systems. As outlined in Sect. 4.2.1 and described by (4.19), the Peierls stress should be small for small Burgers vectors and large interplanar distances. Dislocations should therefore glide between the widely spaced atom planes aa′, bb′, or cc′. These interatomic planes are called the *shuffle set*. In contrast to

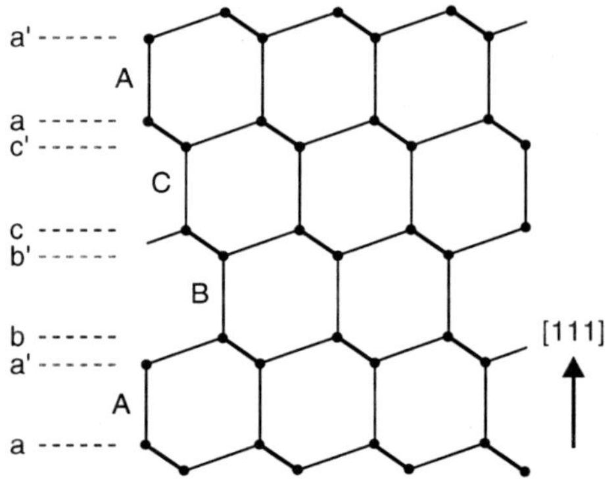

Fig. 6.1. $[\bar{1}10]$ projection of the diamond structure consisting of an ...ABCABC... stacking of (111) double layers. *Thin lines* mark single bonds. *Thick lines* represent two bonds starting from an atom, one directed forward and the other backward so that each atom has four bonds in a tetrahedral configuration

that, dislocations moving on the narrow interatomic planes belong to the *glide set*. The distinction between the glide and the shuffle sets was made in [12]. The interplanar distance of the shuffle set is three times that of the glide set. Weak beam and high resolution TEM indicate that the dislocations dissociate into Shockley partial dislocations bounding an intrinsic stacking fault on the plane of the glide set. The reaction is the same as that in f.c.c. crystals (3.50) outlined in Sect. 3.3.2 [360–364]. In situ TEM studies of dislocations in Si generated at the edge of the perforation of the specimen show the successive creation of two partial dislocations [365]. Thus, the dislocations are created and move in the dissociated configuration on the narrow interatomic planes of the glide set. The presence of a dislocation in a semiconductor crystal is connected with broken (dangling) bonds. However, neighboring dangling bonds saturate to reconstructed covalent bonds. These reconstructions have to be broken if the dislocation moves.

Because of the action of the Peierls mechanism, the dislocations tend to orient along the Peierls valleys. These are the most closely packed $\langle 110 \rangle$ directions. Thus, full dislocation loops take the shape of hexagons as illustrated in Fig. 6.2. The loop consists of two screw segments and four segments of 60° character. The screw segments are split into two partial dislocations of 30° character, while the 60° segments dissociate into a 30° and an edge partial. According to (3.51), the dissociation width is larger for the 60° segments, that is, 5.8 nm compared to 3.6 nm for screws in Si, and 3.9 or 2.4 nm, respectively, in Ge [366]. It is assumed that the 30° and 90° partials have different mobilities. In an expanding dislocation loop, two pairs of 60° segments have a different sequence of the leading and trailing partials. The consequences are a slightly different mobility of the respective pairs of segments and a different dissociation width [367].

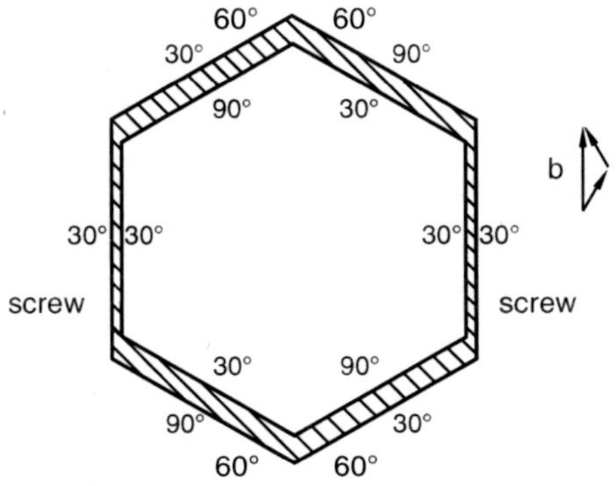

Fig. 6.2. Hexagonal dislocation loop consisting of two Shockley partials comprising an intrinsic stacking fault

6.2 Microscopic Observations

As discussed above and shown in Fig. 4.6, dislocations at rest in silicon assume a relatively straight course along the ⟨110⟩ directions, in accordance with the expectation from the Peierls mechanism. In the corners between the straight segments, the dislocations may be curved. During their motion, the morphology of the dislocations may deviate from the straight shape as illustrated in the following video clips taken from HVEM in situ straining experiments on polycrystalline silicon described in [163].

Video 6.1. *Dislocation motion in polycrystalline silicon at about* 500 *and* 550°C*:* This clip consists of two sequences.

1. 500°C: At this temperature, which is relatively low for silicon, the dislocations move in an essentially straight configuration oriented along the Peierls valleys in a steady viscous way. Several dislocations have jogs (angles in the dislocation lines) that impede the dislocation motion and move together with the dislocations.
2. 550°C: This is an intermediate temperature, where several moving dislocations are still straight but others are curved. Many moving dislocations are impeded at their emergence points through the surface and bow out there. On the right of the blue label A, dislocations move with jogs, which are dragged behind the main dislocation line. On the right of label B, a jog in a dislocation near the upper edge of the image trails a dipole.

Video 6.2. *Dislocation motion and generation in polycrystalline silicon at about* 650°C*:* At this relatively high temperature, the dislocations move still viscously. In motion, they are no longer oriented along the Peierls valleys. On the left of the blue label A, a dislocation moves upwards before it reacts with a fixed straight dislocation in 45° orientation in the image, where parts of both dislocations are annihilated. It is supposed that both dislocations are arranged on different glide planes but have the same 1/2⟨110⟩ Burgers vector oriented along the intersection line between both

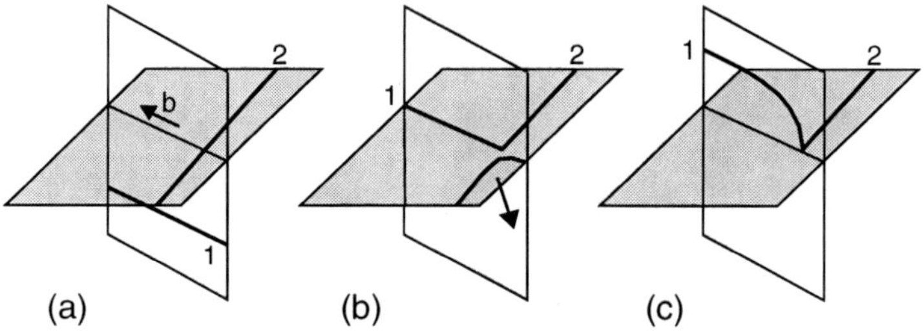

Fig. 6.3. Formation of a dislocation source by the reaction of two dislocations of the same Burgers vector

{111} slip planes, as outlined in Fig. 6.3a. When the moving dislocation 1 meets the fixed one 2 at the intersection line, two angular dislocations may form as sketched in Fig. 6.3b. One of them marked by an arrow may move out of the specimen. If the remaining part of the dislocation 1 resumes moving, it is pinned by the rest of the fixed dislocation 2 at the intersection point. It will bow out as outlined in Fig. 6.3c and revolve around the pinning point to emit a number of new dislocations.

When the source dislocation stops moving after label B, the curvature disappears and the dislocation at rest orients again along a Peierls valley.

A series of in situ experiments have been performed on pure germanium and Ge–Si solid solutions [368]. On the one hand was Ge alloyed with 4% Si, and on the other was Si with 5% Ge. The crystals were predeformed to reduce their brittleness. All specimens had (111) foil surfaces. They were deformed along ⟨123⟩. Qualitatively, the alloying changes the dislocation behavior only little so that these experiments give some additional information on the general dislocation behavior in the elemental semiconductors. However, as will be described later, the mobility of the dislocations is reduced with respect to the pure elements so that the deformation conditions change. The following two clips present pure Ge. In accordance with the lower melting point of Ge with respect to Si, the deformation temperatures are also lower than those of Si.

Video 6.3. *Several slip bands in* Ge *at about* 430°C*:* The quality of this clip is poor because of strong adjustments of the imaging conditions during recording. Nevertheless, the clip shows at this low temperature that only part of the dislocations move in the straight oriented configuration.

Video 6.4. *Motion of dislocations in a dislocation network in* Ge *at about* 610°C*:* In pure Ge at an intermediate temperature, dislocations in a relatively dense network move in curved configurations. This underlines the observation that dislocations mostly do not move in the straight shape predicted by the Peierls model.

The next video shows a sequence of the Ge-4 at% Si alloy. As it will be described in Sect. 6.5, the alloying increases the flow stress so that the adequate deformation temperatures are higher than those of the pure elements.

Video 6.5. *Dislocation band in a network in* Ge-4 at% Si *at about* 655°C*:* In this sequence, moving dislocations are slightly pinned at the emergence points through the surface and bow out in-between. It was shown in the Video 6.2 that dislocations in semiconductor crystals are generated by localized sources. Here, dislocations emitted from a source on the right of the imaged area move in a collective way in a narrow slip band. The collective behavior mainly results from the elastic interactions between the piled dislocations in the band.

In the Si-5%Ge alloy, the deformation temperatures are higher than in Ge, as stated earlier. The slip geometry and the shape of dislocations in the partly unloaded state are demonstrated in Fig. 6.4 taken at different temperatures. The slip geometry is indicated in Fig. 6.4b. There are two traces of slip planes, trace 1 along $[01\bar{1}]$ and trace 2 along $[\bar{1}10]$. Burgers vectors of these directions have the highest orientation factors of 0.47 and 0.29. The two shorter edges of the triangles mark the projections of the other two $\langle 110 \rangle$ directions on the respective slip planes. Thus, dislocations along these directions like the dislocations a and a′ in Fig. 6.4b should be 60° dislocations. These are the dislocations mainly moving in the following video sequences. The straight dislocation segments labeled a and b in Fig. 6.4a belong to the slip systems 1 and 2 and are of screw character. They are less mobile than 60° and mixed segments, so that they are dragged by them.

The following videos are from the same in situ straining experiment.

Video 6.6. *Dislocation bands in* Si-5 at% Ge *at about* 745°C*:* This clip consists of three sequences of neighbored specimen areas taken under very similar conditions.

1. Dislocations move in two slip bands with traces 1 and 2 at an intermediate temperature at a relatively high speed. The images of the moving dislocations are therefore blurred by the afterglow of the luminescent screen of the recording system. According to the high stress corresponding to the high speed, many dislocations are quite straight, especially when they are at rest. The dislocation at A is a forest dislocation for the dislocations of slip system 2. The dislocations of the latter system are pinned temporarily and wind around the forest dislocation, showing that the forest dislocations are strong obstacles.
2. This sequence continues the foregoing one. Above B is an obstacle, probably a forest dislocation arranged approximately end-on. Dislocations interacting with the obstacle acquire a jog with the adjacent segments strongly bowing out.
3. In the lower part of the last sequence, a dislocation moving from left to right is impeded by other dislocations. It forms a localized dislocation source emitting some new dislocations.

Video 6.7. *Dislocation band in* Si-5 at% Ge *at about* 835°C*:* The clip shows the same specimen as before at a high temperature. The dislocations in the slip band are pinned by many jogs, which are most probably formed by cross slip as described in Sects. 5.1.1 and 5.1.2 (Fig. 5.5). During the motion, the jogs glide along the dislocations in the direction of the Burgers vector, marked by the blue line. The segments between the jogs bow out, indicating that jogs performing a combined conservative motion in forward direction and sidewise along the dislocations represent obstacles to the dislocation motion. In some places, debris (small dislocation dipole loops) is left, for example, at A.

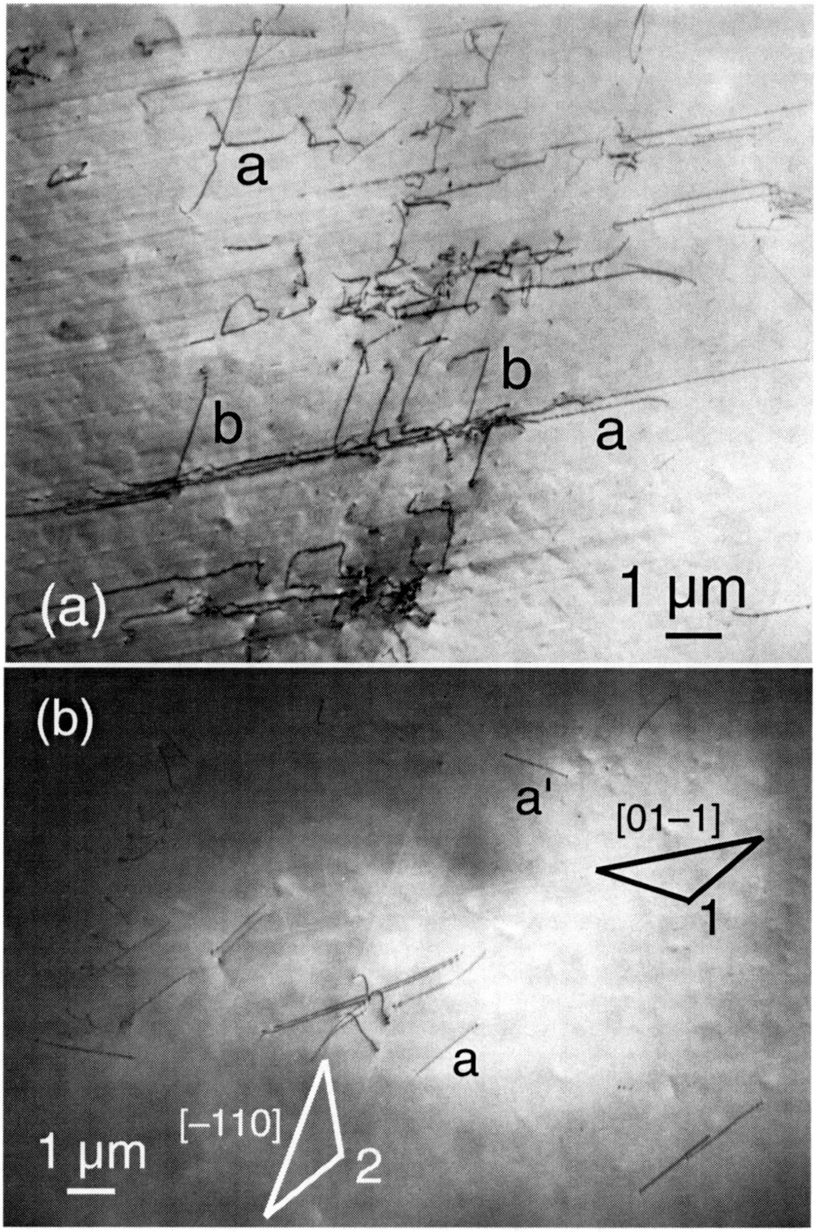

Fig. 6.4. Dislocation structures in Si-5 at% Ge taken during an in situ straining experiment in the partly unloaded state. (**a**) 610°C, (**b**) 745°C. From the work in [368]

In summarizing, it may be stated that dislocations arrange along the Peierls valleys at rest and during the motion at low temperatures, that is, at high stresses. 60° dislocations are more mobile than screws. Dislocations moving at higher temperatures assume a curved shape. They have many constrictions, which are mostly connected with jogs [361, 369]. The latter move with the dislocations and impede their motion. Between these obstacles, the dislocations bow out. This behavior is in contrast to the general assumption that dislocations moving under the control of the double-kink mechanism are very straight and oriented along the Peierls valleys. In elemental semiconductors, dislocations always move in the dissociated state [370] in a smooth viscous way.

6.3 Dislocation Dynamics

The dependence of the dislocation velocity v_d on the stress and temperature in semiconductors is commonly described by the power law (4.103), which is an approximation of the Arrhenius equation with correct behavior at low stresses. Data of the stress dependence of high-purity silicon by Imai and Sumino [262] were presented above in Fig. 4.42. These data were obtained by observing the dislocation motion on a mesoscopic scale by X-ray topography (Sect. 2.5.1). As observed in the above videos, 60° dislocations are generally slightly more mobile than screw dislocations. The stress exponent m is close to unity for both. Other authors find higher stress exponents between 1 and 2 by using the stress pulse-double etching technique (Sect. 2.2) (e.g. [371, 372]). It is argued in [358] that the higher exponents are due to the pinning of the dislocations before their motion, thus being an artefact of the etching technique. However, stress exponents near 1.5 were also observed in germanium and in compound semiconductors (see Table 3 of [359]).

The activation energy ΔG_0 is obtained from Arrhenius plots of the logarithm of the dislocation velocity at constant stress vs. the reciprocal absolute temperature. For high-purity silicon, ΔG_0 amounts to 2.2 eV for 60° dislocations and to 2.35 eV for screws [358].

Special stress pulse-etching experiments were performed by Farber et al. [373] on 60° dislocations in Si to separate the processes of double-kink formation and kink migration. They applied sequences of load pulses of the duration t_l separated by load pauses of duration t_p. The total time of loading $\sum t_\mathrm{l}$ was always equal to the time of static loading necessary to attain a traveling distance of about 30 µm. The duration t_l of the load pulses was chosen in the range of the traveling times of the dislocations from one Peierls valley to the next, $t_\mathrm{l} \approx h/v_\mathrm{d}$, where h is the distance between the Peierls valleys and v_d is the dislocation velocity. For a series of experiments at 600°C and a stress of 7 MPa with $t_\mathrm{l} = t_\mathrm{p}$, the maximum of the distribution of the traveling distances shifts with t_l decreasing from the 30 µm of static loading to lower values. With t_l further decreasing, undisplaced dislocations appear with increasing frequency until at $t_l < 10\,\mathrm{ms}$ there are no more dislocations moving. In another series of experiments, the duration of the pauses was changed at constant $t_l = 94\,\mathrm{ms}$. With increasing t_p/t_l, the average traveling distances decrease toward zero for t_p/t_l being about 3–5.

These observations are interpreted in the following way. The total time t_t necessary for the dislocation to pass over to a neighbored Peierls valley is the sum of the time t_f to form a kink pair of the critical size, and the time t_m for the two kinks to travel until they are annihilated by kinks of opposite sign of other kink pairs, $t_t = t_f + t_m$. Different cases may appear.

1. If the pulse duration is smaller than the formation time of kink pairs ($t_l < t_f$), double-kinks having already formed do not have time to expand. They will shrink and annihilate again during $t_l + t_p$, not resulting in dislocation motion.
2. If the loading time is between the time of a kink pair to generate and the time necessary to complete the transition to a new Peierls valley ($t_f < t_l < t_t$), stable kink pairs may form and start to expand but, during the pause, they become unstable and start to shrink. If the pause is long enough ($t_p > t_l$), they recombine also not causing dislocation movement.
3. If $t_f < t_l > t_t$, stable double-kinks form and the pulse separation is not long enough for all double-kinks to recombine so that the remaining ones can further travel during the next loading pulse. This is the range of strongly increasing traveling distances.
4. If finally $t_l > t_t$, most kink pairs have enough time to travel and to annihilate with kinks of adjacent kink pairs. This is the region where the traveling distance approaches the value of static loading.

The behavior described is possible only if the migration energy of the kinks is not small. If $\Delta F_{mk} < kT$, the kink velocity is very high and controlled by viscous damping (see Sect. 4.2.2). Then, dislocation motion in case 3 above would not be possible as all kink pairs would immediately annihilate in the pause phases. Thus, in silicon, kink migration is also a thermally activated process and the dislocation velocity of long dislocation segments is controlled by (4.46), with the activation energy being the sum of the kink formation and migration energies,

$$\Delta G_0 \approx \Delta F_{fk} + \Delta F_{mk}.$$

A quantitative estimate yields $\Delta F_{mk} \approx 1.6\,\mathrm{eV}$, which implies $\Delta F_{fk} \approx 0.6\,\mathrm{eV}$ with the value of $\Delta G_0 = 2.2\,\mathrm{eV}$ quoted above [358]. The very high value of ΔF_{mk} is problematic as the secondary Peierls potential cannot be higher than the primary one. The authors of [373] therefore assume that extrinsic defects control the mobility of kinks.

Adding foreign atoms to pure elemental semiconductors changes the dislocation mobility slightly, as reviewed in [358]. Electrically inactive light elements in Si like C, O, or N reduce the dislocation mobility at low stresses leading to a higher stress exponent m. Below a critical stress, the motion is

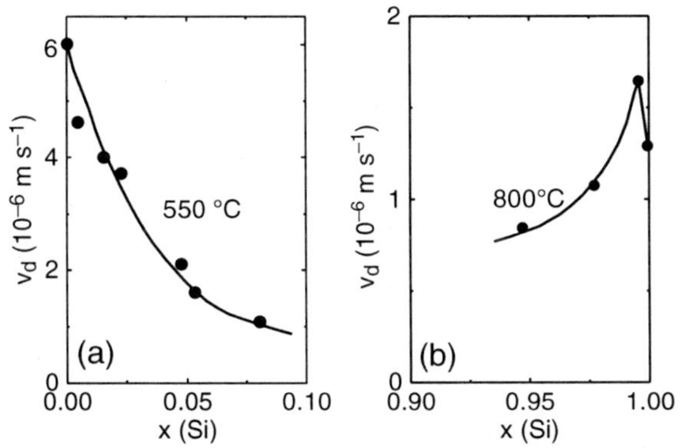

Fig. 6.5. Dependence of the dislocation velocities in Ge–Si alloys on the Si content. (**a**) Ge-rich alloys at 550°C. (**b**) Si-rich alloys at 800°C. Stress $\tau = 20\,\text{MPa}$. Data from [374]

totally blocked. Electrically active donor impurities like P, As, and Sb increase the dislocation velocity in the high-stress region without changing m, but decrease it at low stresses yielding an increase in m. The effect depends only on the concentration, but not on the particular element. Acceptor impurities like B influence the dislocation dynamics very little. Many models not treated here have been developed to explain these observations.

In the Ge–Si alloys of elements of equal valency, the dislocation mobility is reduced with respect to the pure constituents [374], as demonstrated by the plots in Fig. 6.5, showing the dislocation velocities at fixed stress in dependence on the concentration x of Si atoms. The interpretation will be given below (Sect. 6.5) in connection with the macroscopic deformation data.

The in situ straining experiments on polycrystalline silicon [163] allowed the quantitative determination of the activation parameters of the dislocation mobility by observing individual moving dislocations and by measuring the dependence of the average dislocation velocity as a function of the applied load at a constant temperature, and on the temperature. The low stress exponent of semiconductor materials is of advantage for carrying out such quantitative in situ straining experiments in a TEM, as changes in the dislocation velocity are achieved by comparatively strong changes in the specimen load. At each temperature, the load exerted on the specimen was adjusted to achieve dislocation velocities in a range suitable for video recording. The average effective stress was measured by matching the shape of temporarily pinned dislocation segments to loops calculated by the line tension model, as described in Sect. 3.2.7 and in [101] for MgO single crystals. The loops calculated for the most prominent slip system $(111)\,(a/2)[10\bar{1}]$ were plotted in the projection of the actual video recordings onto a $(\bar{1}12)$ image plane. Dislocation segments were selected, which fitted the ellipticity of the calculated loops. The relation between the load and the effective stress is shown in Fig. 6.6.

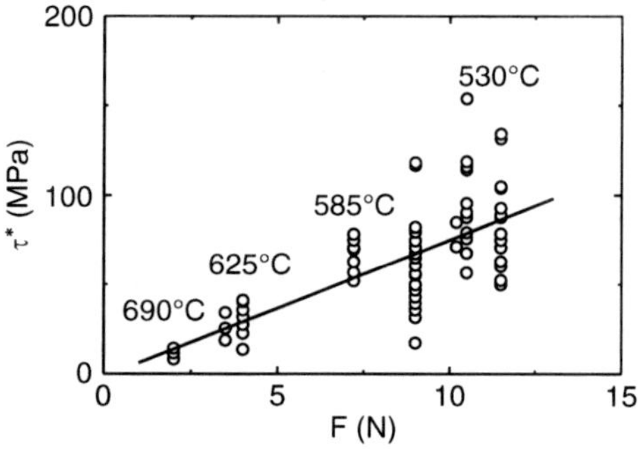

Fig. 6.6. Dependence of the effective stress τ^* in polycrystalline Si on the load applied to the in situ tensile specimen. Data from [163]

The stress exponent was determined to be $m = 1.6$ by applying five different loads at a constant temperature of 530°C and by plotting the logarithm of the dislocation velocities v_d vs. the logarithm of the effective stress τ^*. Similar to the X-ray topography experiments in [262], the velocities were measured on continuously moving dislocations so that the difference between the present value and the exponent of unity found in [262] is not due to pinning as in the etching studies. It may result from a lower purity of the present solar grade crystals, the pinning of the dislocations by jogs, or the effects of radiation in the HVEM, as will be described later. To obtain an Arrhenius plot of the logarithm of the dislocation velocity vs. the reciprocal temperature, dislocation velocities measured at different temperatures and under different loads were normalized to a constant effective stress of $\tau^* = 30\,\text{MPa}$ using the above stress exponent and (4.103). The resulting Arrhenius plot is presented in Fig. 6.7. From the slope of this plot follows an activation energy of $\Delta G_0 = (1.6 \pm 0.3)\,\text{eV}$. This energy is essentially lower than that determined by other techniques. The dislocation velocities plotted by large open squares in Fig. 4.42 are approximately equal to those measured by Imai and Sumino [262] at high temperatures. At low temperatures, however, the dislocation velocities in the in situ experiments are more than one order of magnitude higher than in the X-ray topography experiments.

An earlier in situ study of the dislocation mobility in silicon by Louchet [375] proved the dependence of the dislocation velocity on the length of the gliding dislocations for short dislocation segments ($<0.4\,\mu\text{m}$), as it is predicted by the theory of the Peierls mechanism (4.47). More detailed data on this effect shown in Fig. 6.8 were given in [376] for germanium and in [377] for the compound semiconductor ZnS. The early in situ experiments on semiconductors are reviewed in [378]. These papers and theoretical ab-initio calculations [379] confirm the high kink migration energy in elemental semiconductors mentioned earlier.

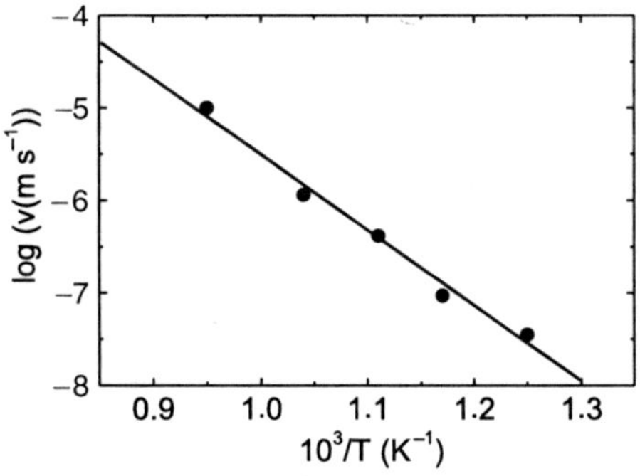

Fig. 6.7. Arrhenius plot of the dislocation velocities normalized to a stress of 30 MPa. Data from [163]

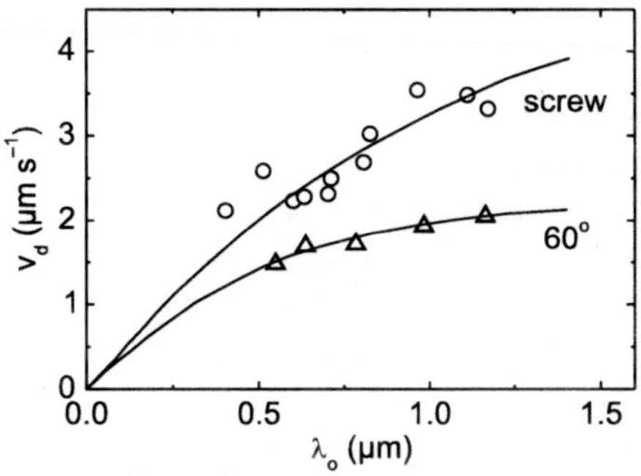

Fig. 6.8. Dependence of the velocity of screw and 60° dislocations in Ge single crystals on the length of straight segments taken during in situ straining experiments in a TEM at 430°C and a stress of 40 MPa. Data from [376]

6.4 Recombination-enhanced Dislocation Mobility

The differences between the activation parameters determined in the in situ experiments and the conventional ones may have several reasons. The Peierls mechanism alone acts only at relatively low temperatures. At high temperatures, the dislocations are bowed out between high jogs. This transition between two mechanisms is similar to that between the Peierls mechanism and the localized obstacles described in Sect. 4.6. In the transition region, the strain rate sensitivity decreases more strongly than the stress, which, according to (4.11), is connected with an increase in the stress exponent m. The nonconservative motion of the jogs results in the production of intrinsic point defects. Vacancy defects have been proved by positron annihilation studies on Si and Ge crystals deformed under low-temperature high-stress conditions (e.g. [380]). Subsequent annealing causes the vacancies to agglomerate to clusters of about ten vacancies.

The main reason of the high dislocation velocities at low temperatures during the in situ experiments in the HVEM and the corresponding low value of the activation energy (cathodo-plastic effect) is attributed to the recombination enhancement due to excess carriers [381]. The high-energy electron beam in the HVEM excites a large number of electron–hole pairs, which recombine in a nonradiative process in dislocation-related places. In principle, the released energy can enhance both kink formation or migration. Respective theories have been developed in [382, 383] and reviewed in [384]. Kink formation is more probable since a single process of enhancement is necessary for the creation of a double-kink, where the kinks afterwards spread over a longer distance, while many enhancement events are necessary to support each migration step over only one lattice distance. Recombination-enhanced dislocation motion can also be stimulated by other kinds of radiation, for example, laser irradiation.

Special experiments to study the influence of the electron beam on the dislocation mobility were performed during the in situ straining experiments in an HVEM on Ge and Si-5 at% Ge single crystals discussed above (Sect. 6.2) [368]. At 566°C, the average dislocation velocity in Ge increased by a factor of about three, while the imaging electron beam current was increased by a factor of five. Measurements of the dislocation velocities at different temperatures but at a constant electron beam current yielded again drastically lower activation energies than those experiments in which the stress pulse-double etching method was applied without radiation [385].

6.5 Macroscopic Deformation Properties

The most prominent feature of the stress–strain curves of semiconductor single crystals oriented for single slip is the strong yield drop between the upper (UYP) and lower yield points (LYP), as outlined in Fig. 6.9. After the lower yield point, there follows a region of low work-hardening. The upper and lower yield stresses show a strong dependence on the temperature, but a weak one on the strain rate. The yield drop from the upper to the lower yield point is a typical effect of both rapid dislocation multiplication and the low stress exponent. At the beginning of deformation, the dislocation density is low, leading to high dislocation velocities corresponding to a high flow stress. With increasing dislocation density, the dislocation velocity drops down and accordingly the flow stress, too. These features and the temperature dependence of the flow stress of semiconductor crystals can well be modeled by a variant of the set of constitutive equations described in Sects. 5.1.1 and 5.2 [278, 279]. Figure 6.10 presents the temperature dependence of the lower yield stress for silicon and germanium in a plot similar to an Arrhenius plot.

The temperature and strain rate dependence of the upper and lower yield stresses can be described by a phenomenological relation [279, 359]

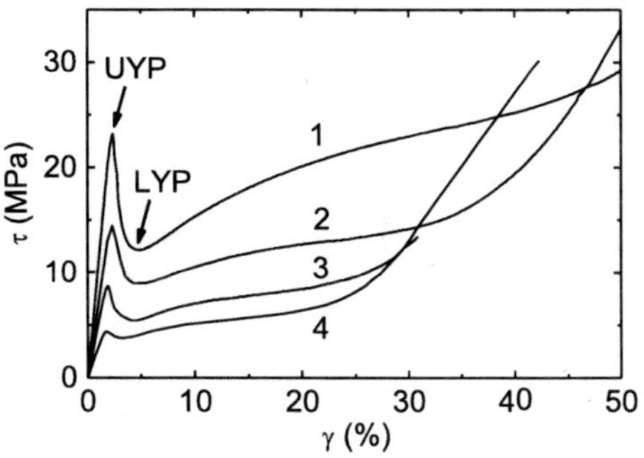

Fig. 6.9. Shear stress vs. shear strain curves for high-purity silicon single crystals grown by the floating zone method and deformed along a ⟨123⟩ tensile axis. Strain rate $1.2 \times 10^{-4}\,\mathrm{s}^{-1}$. 800°C (1), 850°C (2), 900°C (3), 950°C (4). Data from [386]

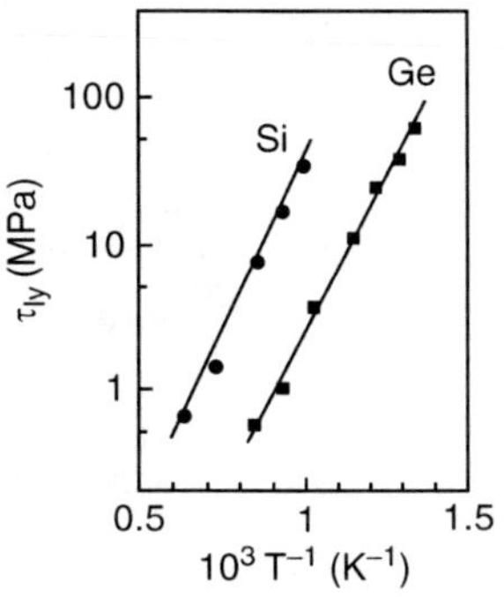

Fig. 6.10. Temperature dependence of the lower yield stress of high-purity silicon and germanium. Strain rate $2 \times 10^{-4}\,\mathrm{s}^{-1}$. Data from [359]

$$\tau_{\mathrm{uy}} \text{ or } \tau_{\mathrm{ly}} = \mathrm{A}\,\dot{\varepsilon}^{1/n} \exp\left(\frac{Q}{kT}\right) \tag{6.1}$$

where A is a constant and Q is an energy.

The latter equation is not an Arrhenius relation as it does not describe the dependence of a rate on stress and temperature. It can be derived from the power law (4.103), which is an approximation to the Arrhenius relation and the Orowan equation (3.5) between the strain rate and the dislocation velocity. However, the mobile dislocation density enters the prelogarithmic factor of the Arrhenius equation of the strain rate. As the mobile dislocation density at the yield points strongly depends on the deformation temperature, the prelogarithmic factor is not constant. According to [359], the correlation between the microscopic activation parameters m and ΔG_0 and the macroscopic phenomenological quantities n and Q is given by

$$n = m + 2,\ Q = \Delta G_0\,/(m + 2).$$

The reduction of the dislocation mobility by alloying the elemental semiconductors with additions of equal valency, as it was discussed for the Ge–Si alloys, also yields an increase in the macroscopic flow stress [374]. The yield stress maximum probably occurs at the concentration $x = 0.5$. With respect to pure Ge, the hardening is relatively weak at low temperatures but very strong at higher ones, resulting in an almost athermal behavior at $x = 0.4$. For lower concentrations, the hardening can be explained by the formation of Cottrell atmospheres (Sect. 4.11). In the concentrated alloys, ordering phenomena may also play some role.

6.6 Summary

In the elemental semiconductors, and in most compound semiconductors, too, the dislocation dynamics is controlled by the double-kink mechanism over the whole range of stress and temperature that is experimentally accessible. Then, the character of the dislocation motion is viscous and the stress and temperature dependence of the dislocation velocity can be described by the power law (4.103), with a stress exponent m between 1 and 2. The in situ experiments reveal that the dislocations are straight as expected from the Peierls mechanism only at rest and at high stresses or low temperatures. Mostly, the dislocations contain jogs, from cross slip of some segments as well as from cutting forest dislocations. The latter may be strong obstacles to glide. In mixed dislocations, the jogs glide along the dislocations in the direction of the Burgers vector. Nevertheless, this process impedes the dislocation motion owing to the high lattice resistance. Therefore, the dislocations bow out between the jogs, thus contributing to the flow stress. In addition, they may form dislocation debris. Further obstacles to the dislocation motion are impurities and their agglomerates. All these processes lead to values of m higher than unity, which is expected from the pure Peierls mechanism.

During plastic deformation, most dislocations are created in localized sources, resulting in the concentration of slip in narrow slip bands. The temperature and strain rate dependencies of the macroscopic yield stresses (upper and lower yield stress) can also be described by a power law. The stress–strain curves can be modeled by the set of constitutive equations introduced in Sect. 5.2. The strong yield drop effect is due to the low stress exponents as the strong initial increase of the dislocation density results in a similarly strong decrease of the glide resistance. At low temperatures where the glide resistance is high, semiconductor crystals are brittle.

7

Ceramic Single Crystals

Ceramic single crystals span a very wide range of materials from ionic crystals like the alkali halides with ionic bonding over oxides with mixed bonding to the carbides or nitrides with strong covalent bonding. Correspondingly, the Peierls stress and the double-kink formation energies are very different. On the one hand, these quantities are low in the alkali halides so that the Peierls mechanism controls the dislocation mobility and the flow stress only at low temperatures. On the other hand, the ceramic crystals with strong covalent bonding become plastic only at high temperatures. Typical examples of ionic crystals and oxides are presented in this chapter. Heuer and Mitchell recently presented a review of the dislocation and deformation properties [387]. Structural applications of ceramic materials are mostly in the form of polycrystals. At low temperatures, their strength is limited by the brittleness, which is controlled by flaws in the polycrystalline structure acting as crack nuclei. At high temperatures, the materials creep. The creep rates are governed partly by the properties of the grain boundaries and partly by the plastic properties of the crystal grains, the latter of which are discussed in this chapter. A review of creep parameters and models of polycrystalline ceramics is given in [340].

7.1 Alkali Halides

The dislocation processes during plastic deformation of alkali halide single crystals were extensively studied in the 1960s and 1970s. An early review of the plastic deformation is given by Sprackling [388]. These materials are considered model substances for defect properties of other solids because of several reasons. Atomistic calculations of the defect properties are relatively easy and reliable, as, for example, the theoretical estimation of the Peierls stress in [132]. The interatomic potentials to be considered here are the Coulomb attraction, Born–Mayer repulsion, and Van-der-Waals potentials. High purity alkali halide crystals can be grown so that the intrinsic deformation parameters can be studied. They can easily be cleaved, in some cases yielding almost atomically smooth surfaces (e.g., {100} faces on NaCl crystals), which facilitates the specimen preparation and enables the surface decoration of monatomic slip steps (Figs. 4.9 and 4.12). Besides, chemical etching is well suited to image the points where dislocations emerge through the surfaces. From combinations of measurements of diffusion, electrical conductivity, and dielectric and mechanical relaxation, the spectrum of point defects is comparatively well known, that is, from the intrinsic defects like vacancies in both sub-lattices (and in some materials interstitials) to associates of aliovalent impurities or additions with vacancies and their agglomerates. Thus, several interactions between dislocations and point obstacles can well be investigated. The famous studies of Johnston and Gilman [19, 276] on the dislocation mobility and kinetics in LiF single crystals certainly benefited from these advantages. A drawback is the strong sensitivity to radiation damage, so that TEM observations require cryo-microscopy.

7.1.1 Crystal Structure and Slip Geometry

Several alkali halides and oxides crystallize in the sodium chloride structure. As shown in Fig. 7.1, it consists of two f.c.c. lattices of the two ions shifted by $a/2$ with respect to each other. The Burgers vector is $a/2\langle 110\rangle$. Thus, the extra half-plane of an edge dislocation consists of two inserted half-planes to preserve electrical neutrality of the crystal. Full jogs in the dislocation line have the height of b. They are electrically neutral. Half-jogs carry an electrical charge of $\pm e/2$. Correspondingly, dislocations with unequal numbers of half-jogs of different sign are electrically charged. The dominating slip planes are of type {110} so that there exist six independent slip systems. Always pairs of two systems are perpendicular to each other. For one primary slip plane, another one is perpendicular and the four others are oblique. Accordingly, the secondary slip systems are called orthogonal or oblique slip systems. At a $\langle 100\rangle$ deformation axis, two slip systems have zero orientation factors and four have $m_s = 0.5$, while along $\langle 111\rangle$ all $\{110\}\langle 1\bar{1}0\rangle$ slip systems have zero orientation factors. Then, also $\{001\}\langle 110\rangle$ systems are activated.

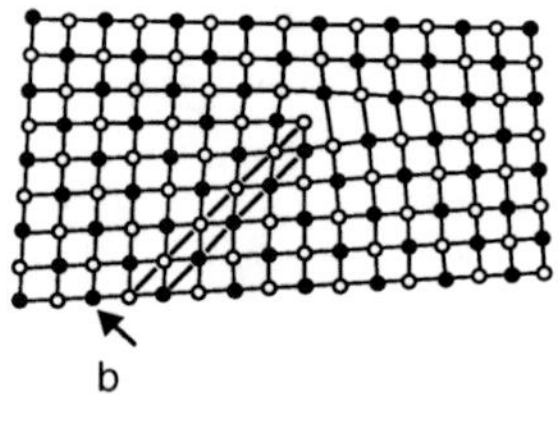

Fig. 7.1. Sodium chloride structure with an edge dislocation. The two ion species are marked by *open* and *full circles*

7.1.2 Dislocation Dynamics

In the following, the dislocation dynamics and plastic deformation of sodium chloride single crystals will be described, as a typical example also of other alkali halides. Dislocation velocities were measured as a function of the applied stress by Gutmanas et al. [389, 390] using the stress pulse-double etching technique (Sect. 2.2). In a double-logarithmic plot of v_d vs. τ, the curves look very similar to those of LiF in Fig. 4.41. At low velocities between about 10^{-9} and $1\,\mathrm{m\,s^{-1}}$, the plots follow straight lines indicating power law dependencies (4.103) with high values of the stress exponents m. Above about $10\,\mathrm{m\,s^{-1}}$, the curves bend down, which corresponds to lower stress exponents. Again, edge dislocations have a slightly higher mobility than screws. Characteristic for alkali halide crystals is a strong reduction of the dislocation mobility by the addition of aliovalent cationic impurities. The addition of the divalent impurities Sr^{++} and Ca^{++} to NaCl has been well studied. In the low-velocity region, the stress exponents range between 30 and 10, and in the high-velocity one they range between 2.6 and 1 for crystals with 10^{-2} mole% Sr^{++} and for nominally pure NaCl, respectively. As in most materials, the dislocation velocity in the high-velocity range is controlled by viscous damping. It will be shown below that at low and intermediate velocities and at temperatures not too low, the dislocation mobility is governed by the interaction between dislocations and defects containing the divalent impurities.

7.1.3 Macroscopic Deformation Properties

Ionic crystals and sodium chloride, in particular, demonstrate well in which way the different processes control the plastic deformation in the various temperature ranges. A summary was given by Haasen [391]. Several data on the deformation of NaCl single crystals have been discussed in Part I, Sects. 4.8, 5.2.1, and 5.2.2. Figure 7.2 presents the temperature dependence of the yield stress up to about room temperature for the $\{110\}$ and $\{100\}$ slip planes. The figure indicates a strong low-temperature increase of the yield stress, for the $\{110\}$ slip plane below about 50 K and for the $\{100\}$ plane already below room temperature. In addition, the yield stress is remarkably enhanced by the addition of divalent cationic impurities. As described earlier, the impurities directly influence the dislocation mobility.

The low-temperature increase of the yield stress and its difference between the slip planes are ascribed to the action of the Peierls mechanism. In ionic crystals, the elastic theory of the Peierls model as described in Sect. 4.2 fails, particularly in explaining the differences between the various ionic crystals. The experimental Peierls stresses reduced by the square of the ion valency have an inverse dependence on the sum of the polarizabilities of the matrix ions [393]. Realistic values of the Peierls stress for the $\{110\}$ slip planes were calculated by Granzer et al. [132] using an atomistic model. They obtained

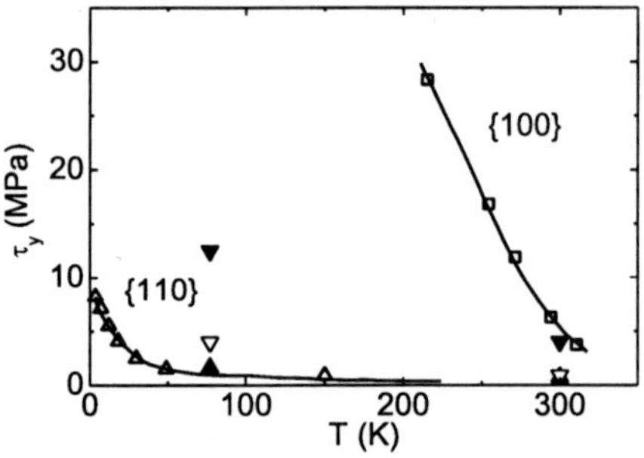

Fig. 7.2. Dependence of the shear yield stress of NaCl single crystals on the temperature. Data from [392] for nominally pure crystals, {110} slip plane (*small open upward triangles*), {100} slip plane (*open squares*). Data from [177] already shown in Fig. 5.26 for the {110} slip plane, 1.7 ppm residual impurities (*large full upward triangles*), 32 ppm Ca^{++} (*large open downward triangles*), 136 ppm Ca^{++} (*large full downward triangles*)

values between 14 and 22 MPa in the correct order of magnitude of the low-temperature increase of the yield stress in Fig. 7.2. For the {100} plane, the summation over the ion rows is difficult. According to Gilman [394], the high lattice friction stress on this plane results from the close neighborhood of always two ions of equal charge after a shift of the dislocations by $b/2$.

The strong dependence of the dislocation mobility and the yield stress on the concentration of divalent cation impurities at intermediate temperatures is due to the elastic interaction between the moving dislocations and elastic dipoles with tetragonal stress fields formed by associates of the impurities and their charge-compensating cation vacancies. The degree of association is controlled by the interaction energy between the impurities and vacancies through the law of mass action. In the respective temperature interval, practically all impurities are associated with vacancies. At high impurity concentrations, the associates may agglomerate to higher complexes. At lower temperatures, the associates act as fixed localized obstacles to the dislocations. The interaction is of Fleischer type (Sect. 3.2.6). At higher temperatures, the dipoles may reorient in the stress field of the dislocations by diffusion jumps of the vacancies around the impurities leading to a Snoek effect interaction (Sect. 4.11). These theoretical concepts were applied to alkali halide crystals by Frank [395].

In the range of operation of these mechanisms, the macroscopic deformation of NaCl crystals containing Ca^{++} impurities has extensively been studied by Appel [396–398]. Nominally pure and doped crystals were cooled by a thermoelectric cooler, which allowed quick changes of the temperature. The activation volume V and (Gibbs free) energy ΔG were determined by strain rate and temperature cycling tests as outlined in Sect. 4.1. In addition, the effective stress τ^* was measured by long-time stress relaxation tests, supplemented by etch pit countings of the dislocation density ϱ. As the total (shear) flow stress τ as well as the effective stress τ^* depend on the plastic strain, the flow stress was assumed to consist of the following parts

$$\tau = \tau_\mathrm{i} + \tau^*,$$

with

$$\tau^* = \tau_\mathrm{p}^* + \tau_\mathrm{s}^* , \tag{7.1}$$

where τ_p^* is the chemical hardening due to aliovalent cationic impurities and τ_s^* is a work-hardening contribution to τ^*. While τ_p^* is assumed to depend only on the state of the impurities and the thermodynamic variables, τ_i and τ_s^* in addition depend on the strain and on τ_p^*. In Sect. 4.5.1, the superposition rules of two types of thermal obstacles were discussed. For obstacles of not very different strengths and concentrations, a quadratic superposition law holds (4.59), while for many weak and few strong obstacles, the stress parts superimpose linearly (4.63). It will be shown that the strain-depending part of the effective stress consists in the intersection of dislocations created during the deformation. This interaction is much stronger than the Fleischer-type interaction with elastic dipoles so that the linear rule (7.1) can be applied. The different stress parts were varied by changing the temperature T, the strain rate $\dot{\varepsilon}$, the strain ε, and the divalent impurity concentration c^{++}. The variation in the thermodynamic variables T and $\dot{\varepsilon}$ should be equivalent. At low plastic strains, yield drop effects do not occur after changing the deformation conditions below 250 K, indicating that the microstructure remains constant during the stress changes.

Figure 7.3a presents the dependence of the flow stress τ and the stress τ_R relaxing during long-time stress relaxation tests on the temperature for a fixed concentration $c^{++} = 92$ ppm of divalent cationic impurities. τ_R is supposed to be equal to τ^*. The figure shows that the athermal stress component $\tau_\mathrm{i} = \tau - \tau^*$ is the main part of the total flow stress, amounting to more than 70%. However, this stress part is not constant but depends, for example, on the concentration c^{++} of divalent additions, as outlined in Fig. 7.3b. It

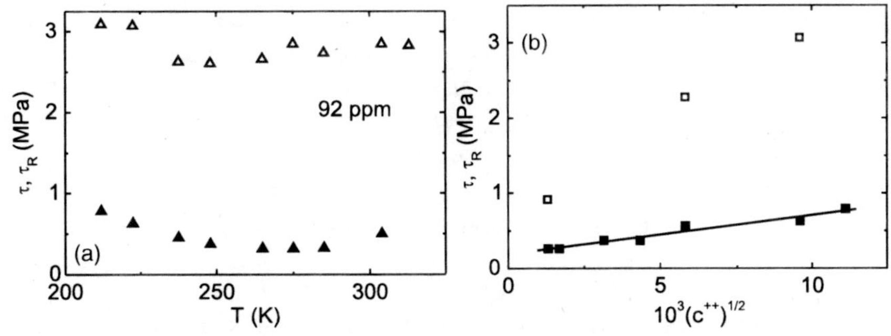

Fig. 7.3. Dependence of the flow stress τ (*open symbols*) and the relaxation stress τ_R (*full symbols*) on the temperature at a concentration of Ca^{++} of 92 ppm (**a**), and on the square root of the Ca^{++} concentration at a fixed temperature of 223 K (**b**). Strain rate $4.2 \times 10^{-4}\,\mathrm{s}^{-1}$. Data from [396]

was discussed in Sects. 5.2.1 and 5.2.2 that the athermal stress originates from long-range interactions between dislocations. The corresponding linear dependence of the flow stress on the square root of the dislocation density (5.11) was first proved in pure NaCl crystals, where τ^* is small, in [399]. Later on, the different influences of the densities of primary and secondary dislocations were demonstrated by cryo-TEM in [400]. The total dislocation density follows the Taylor law (5.11) with the numerical constant $\alpha = 0.7\pi$. For the present materials, the square root dependence was shown above in Fig. 5.15. As discussed there, this relation is not universal, and τ_i depends on τ^* and correspondingly also on the temperature and the strain rate. The effective stress follows quite well a square root dependence on the Ca^{++} concentration (Fig. 7.3b).

In Fig. 7.3a, both τ and τ^* exhibit two temperature ranges. First, they decrease with increasing temperature in the normal way, and above 250 K they remain constant or even increase. These are the ranges of the short-range Fleischer type interaction with the Ca^{++} vacancy dipoles at low temperatures, and the Snoek effect interaction at higher ones.

At first, the contribution τ_p^* of the Fleischer type interaction to the effective stress will be discussed. It is assumed that τ_s^* is small at low plastic strains so that the stress dependence of the activation parameters can be interpreted by using the Fleischer approximation (3.35). Neglecting Friedel statistics, for the dependence of the activation volume on the effective stress follows

$$V = \Delta d\, l\, b = b^2\, l \left[\left(\frac{\tau_0}{\tau^*} \right)^{1/2} - 1 \right],$$

and for the Gibbs free energy

$$\Delta G = \Delta G_\circ \left[1 - \left(\frac{\tau^*}{\tau_0} \right)^{1/2} \right]^2 .$$

Here, $\tau_0 = F_0/(l\, b)$ is the stress for athermally surmounting the obstacle array, and ΔG_0 is the activation energy at zero stress. The latter equation is identical with (4.60). The plot of V vs. $\tau^{*-1/2}$ is presented in Fig. 7.4 for crystals with a Ca^{++} concentration of 92 ppm. τ^* is varied by the temperature, the strain rate, and the plastic strain. Data resulting from changes in the temperature and the strain rate form a single curve, as expected for a single thermally activated mechanism. Below 250 K, it is a straight line (full squares) with the slope $l\, b^2\, \tau_0^{1/2}$ and the intercept $l\, b^2$. Similar plots were obtained also for the other Ca^{++} concentrations. The values of the obstacle distance l from the intercept are compared with those from the concentration c^{++}, $l_\mathrm{sq} = a\, / \sqrt{(4/3) c^{++}}$, in analogy with (4.48). a is the lattice constant. The numerical factor considers the geometry of dislocations passing elastic dipoles of different orientations on nearest planes. It turns out that $l \approx l_\mathrm{sq}$ for low c^{++}, and $l > l_\mathrm{sq}$ for higher ones. The disagreement for higher concentrations does not indicate the beginning of agglomeration of the individual elastic dipoles but results

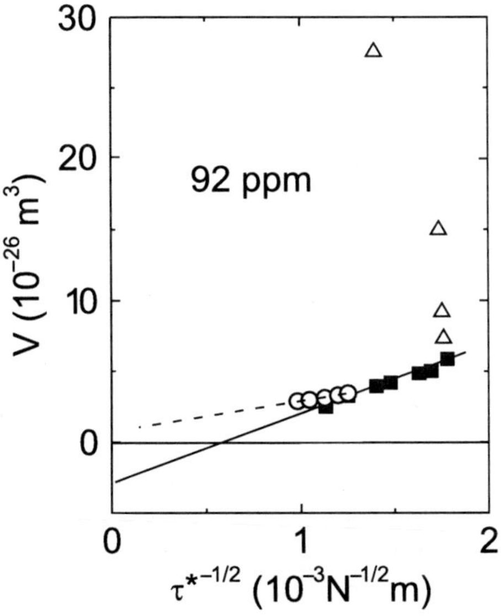

Fig. 7.4. Dependence of the activation volume on the reciprocal square root of the effective stress in NaCl single crystals of 92 ppm Ca^{++}. Effective stress varied by applying different temperatures and strain rates at a constant low plastic strain, temperatures between 212 and 248 K (*full squares* and *solid extrapolation line*) and between 265 and 304 K (*open triangles*). Effective stress varied by increasing plastic strain at 223 K and $4.2 \times 10^{-4}\,s^{-1}$ (*open circles* and *dashed extrapolation line*). Data from [396]

from changes in the pre-exponential factor in the Arrhenius equation, which are reflected in changes in the Gibbs free energy of activation [302]. Besides, agglomeration contradicts the linear plot of τ^* vs. $c^{++1/2}$ in Fig. 7.3b. Slope and intercept together of the plot in Fig. 7.4 yield ΔG_0. The consistency of the results was checked by plots of V vs. T following from the Fleischer potential, and by comparing the activation energies calculated from V vs. $\tau^{*-1/2}$ by the formalism above, with the values from temperature and strain rate cycling tests and applying (4.13) and (4.15). The agreement is satisfactory. The Gibbs free energies of activation for small strains increase from 0.22 to 0.34 eV with an increasing concentration of Ca^{++} [302].

At temperatures above 250°C, the activation volumes (open triangles in Fig. 7.4) strongly deviate from those of the low-temperature Fleischer type range. They show an inverse dependence on the effective stress, that is, a decreasing strain rate sensitivity with increasing stress, which contradicts the usual thermally activated mechanisms. The inverse dependence points at diffusion processes involved in the dislocation motion and will be discussed in more detail in connection with the occurrence of yield stress anomalies in intermetallic alloys in Chap. 9. The effect is connected with strain ageing and is ascribed to the induced Snoek effect (Sect. 4.11) [395,398]. After stress relaxation tests (stopping the deformation machine) for different durations, sharp

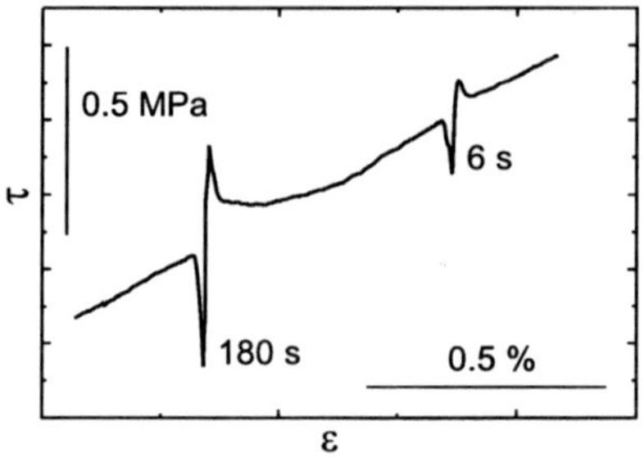

Fig. 7.5. Section of a stress–strain curve showing yield point effects due to strain ageing of a crystal with 136 ppm Ca^{++} at 266 K and a strain rate of $10^{-4}\,s^{-1}$. Data from [398]

yield points arise during reloading the samples as illustrated in Fig. 7.5. The stress increments of the yield points depend on the concentration c^{++}, the temperature, and the ageing time [398]. Arrhenius plots of the time constant of the saturation process and the times to saturation yield an activation energy of about 0.55 eV. This is slightly lower than the energy of 0.65–0.7 eV for reorientation of the elastic dipoles, measured by dielectric relaxation. It disagrees with the Cottrell effect, which requires diffusion of the cation impurities and which has a much higher activation energy.

Figure 7.4 contains also some data points resulting from a variation of the effective stress by an increasing plastic strain under otherwise constant conditions (open circles and dashed extrapolation line) [397]. These data points suggest a strain-induced contribution to the thermal part of the flow stress, which is connected with an increase in the strain rate sensitivity. The data points deviate from the straight line belonging to the interaction with tetragonal defects discussed earlier. It was proved that the Gibbs free energy of activation is independent of the plastic strain. According to (4.8), the Gibbs free energy determines the pre-exponential factor of the Arrhenius equation. Thus, the strain dependence does not result from a change in the pre-exponential factor for overcoming the tetragonal defects. New short-range obstacles should be created as the concentration and state of agglomeration of the tetragonal defects is not expected to change during plastic deformation.

As pointed out in [206,401], a separation of the thermal stress parts is possible if these can be varied independently. This is possible first by measuring the activation parameters of crystals with different concentrations of Ca^{++} at the same small strain, and second, by studying the strain dependence on each individual crystal. Applying the Arrhenius equation (4.8) and (4.3), as well as the linear superposition rule (7.1), there follows

$$\tau^* = \frac{1}{V_\mathrm{p}} \left(\Delta F_\mathrm{p} + kT \ln \frac{\dot\varepsilon}{\dot\varepsilon_0} \right) + \frac{1}{V_\mathrm{s}} \left(\Delta F_\mathrm{s} + kT \ln \frac{\dot\varepsilon}{\dot\varepsilon_0} \right) . \tag{7.2}$$

For $\dot\varepsilon$ and T being constant, the factors $\Delta F + kT \ln(\dot\varepsilon/\dot\varepsilon_0)$ and the activation distances Δd are constant. Figure 7.6 is the respective plot of τ^* vs.

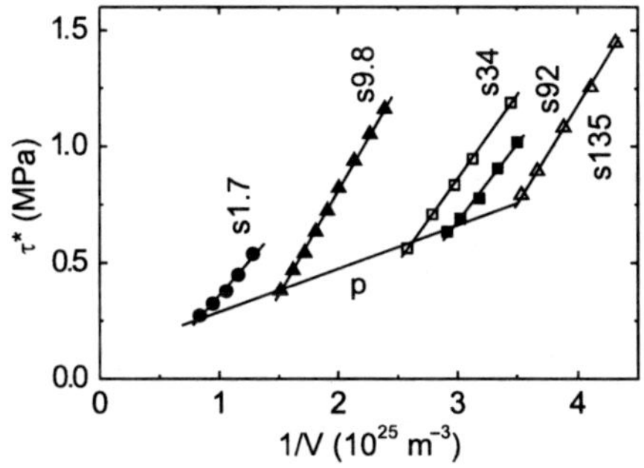

Fig. 7.6. Dependence of the effective stress on the reciprocal activation volume in NaCl single crystals at 223 K. The effective stress is independently varied by using crystals of different concentrations of Ca^{++} (extrapolation line p) and by applying different plastic strains (extrapolation lines s with indication of the Ca^{++} concentrations in ppm). Data from [397]

the reciprocal activation volume $1/V$ with the extrapolation lines p of measurements at small strains with different Ca^{++} concentrations, and s for the variation of τ^* by work-hardening. The slope of curve p yields the first term in parentheses in (7.2), that is, the work term $V_p\tau_p^* = \Delta F_p + kT \ln(\dot{\varepsilon}/\dot{\varepsilon}_0)$, whereas the slopes of the curves s equal the second term in parentheses $V_s\tau_s^*$. From the intercepts, the ratios τ_p^*/τ_s^* and V_p/V_s can be calculated. The result is that for small strains, the impurity part dominates the effective stress (as assumed above) except for small impurity concentrations. The Helmholtz free energy of activation of the strain-dependent process is obtained from the strain-independent Gibbs free energy quoted above and the work term, $\Delta F_p = \Delta G + V_s\tau_s^* \approx 0.6 \ldots 0.8\,\mathrm{eV}$.

The parameters determined fit the intersection of forest dislocations as the strain-induced hardening process. In agreement with this suggestion, the square root of the dislocation density during hardening shows a linear dependence on the reciprocal activation volume, leading to an activation distance of $\Delta d_p \approx 0.6 \ldots 0.7\,b$. This distance and the activation energy correspond well with the process of jog formation, the activation energy of which approximately equals the core energy $\mu b^3/10$ (3.16). This energy characterizes only the thermal part of the interaction between the cutting dislocations. A great part of this interaction is of long-range character, in agreement with the large athermal flow stress component τ_i. This is supported by the observation that the latent hardening by forest dislocations discussed in Sect. 4.8 is of athermal nature.

7.1.4 Summary

In the literature, plastic deformation of ionic crystals is frequently being discussed from two different points of view. One is considering the mechanisms controlling the dislocation mobility, that is, the thermal part of the flow stress, while the other one is regarding the relation between the flow stress and the dislocation density, that is, the athermal stress part. The discussion above has shown that both aspects have to be considered simultaneously. Except at very low temperatures, the athermal stress τ_i dominates the flow stress. However, it is not a constant as supposed in many papers but depends on the effective stress τ^*, and correspondingly on the chemical hardening state and on the deformation parameters.

The dislocation mobility and thus the thermal part of the flow stress are controlled by the Peierls mechanism at very low temperatures and by the tetragonal defects (elastic dipoles) formed by aliovalent cationic impurities or additions and their charge-compensating vacancies. Up to about 250 K, these complexes have a short-range Fleischer type interaction with the gliding dislocations with activation energies between 0.25 and 0.45 eV. Above this temperature, the vacancies may rotate around the impurities in the stress field of the dislocations resulting in the induced Snoek effect. It is interesting to compare the stress exponents m determined from the dislocation mobility by means of the selective etching method and from the macroscopic deformation data. From the latter, m can be calculated by $m = \sigma^*/r = \tau^*/(m_s r)$ (4.11). Usually, τ^* is not available so that the experimental stress exponent $m' = \sigma/r = \tau/(m_s r)$ (4.12) is used. Table 7.1 compares the discussed m values. For the pure crystals, the data of m agree very well, irrespective of the temperature. For the doped crystals, the agreement holds only for room temperature, which is in the range of the Snoek effect interaction. The high m value results from the low strain rate sensitivity characteristic of the diffusion-controlled dislocation mobility. The low-temperature value from the macroscopic experiments is much lower, corresponding to the Fleischer type range. The experimental m' values calculated from the total stress τ are mostly too high. They do not represent the processes governing the dislocation mobility. One should therefore take care in interpreting m' data in connection with the dislocation mobility.

The change of the activation parameters with increasing plastic strain can be attributed to an increasing frequency of dislocation intersections. While

Table 7.1. Stress exponent m from selective etching [390] and from macroscopic deformation tests [396] calculated from the strain rate sensitivity r and the effective stress τ^* and m' from the total stress τ (see text)

c^{++} (ppm)	T (K)	Etching	m from τ^*	m' from τ
Pure	RT	10		
	248		8	40
	223		10	35
92 (Ca^{++})	RT		33	187
	223		7	36
100 (Sr^{++})	RT	30		

the main part of the energy necessary for dislocation cutting originates from the long-range dislocation interaction, which contributes to the athermal part of the hardening, the top of the interaction potential is overcome by the aid of thermal activation. The activation energy is in the order of magnitude of 0.7 eV. Thus, the assumption made for applying the linear superposition rule between the two thermal flow stress parts is justified. The obstacle spectrum consists of many weak obstacles and few strong ones.

7.2 Magnesium Oxide

Oxide crystals cover materials with very different dislocation behavior and macroscopic deformation properties. Two examples are presented here, magnesium oxide (MgO), which is plastic down to helium temperatures, and zirconia–yttria (ZrO_2–Y_2O_3) alloys, which are brittle up to about 800°C under normal deformation conditions. MgO has the same crystal structure and slip geometry as NaCl. Thus, for the specimens with $\langle 100 \rangle$ loading axes, the orientation factor is $m_s = 0.5$.

7.2.1 Microscopic Observations

After some early studies of the plastic deformation of MgO single crystals (e.g., [402–404]), joint work was carried out in the Institute of Solid State Physics (Acad. Sci. of U.S.S.R.), Chernogolovka, and the former Institute of Solid State Physics and Electron Microscopy (Acad. Sci. of G.D.R.), Halle (Saale), covering TEM and in situ straining experiments, measurements of the dislocation velocity, and macroscopic deformation experiments. All specimens were obtained from the same large single crystal block. It contained about 300 ppm aliovalent impurities, mainly Fe^{3+}. In as-grown and annealed crystals, these impurities form small precipitates. The dominating microstructural observation is the interaction of the moving dislocations, with these precipitates acting as localized obstacles. Examples and micrographs from the in situ straining experiments were presented in several sections of Part I, for example, in Fig. 4.13. Figures like this and the video recordings discussed below allowed detailed analyses of the statistical and kinematic properties of the dislocation motion in the field of localized obstacles. A problem in interpreting the results of the in situ experiments was the occurrence of radiation hardening during electron irradiation of the specimens inside the HVEM [334]. The irradiation increases the flow stress of the in situ specimens by about 30% without qualitatively affecting the dislocation motion in the observed specimen area. It is therefore assumed that the hardening does not seriously affect the semi-quantitative conclusions drawn from these experiments.

In all experiments described below, the specimens had a [001] foil normal and a [010] tension or compression direction. Orthogonal slip bands extending in [100] direction have $1/2\langle 011 \rangle$ Burgers vectors inclined with respect to the foil plane, and oblique slip bands extending in [110] or [1$\bar{1}$0] directions have 1/2⟨110⟩ Burgers vectors in the foil plane. Dislocations moving during in situ experiments in oblique slip bands are presented in Fig. 7.7. The specimens are tilted around an axis parallel to the extension of the bands to look onto the slip planes oriented edge-on in the specimens. Figure 7.7a shows an oblique slip band with edge dislocation segments E trailing long screw segments S, thus forming very elongated dislocation loops. This indicates a higher mobility of edge dislocations as observed in most materials, for example, in LiF in Fig. 4.41. Screw dislocations can well be observed in orthogonal slip bands as in Fig. 7.8. In these bands, the traveling distance of the edge dislocations is limited by the foil thickness so that they escape through the surface. The screw dislocations are strongly pinned by the localized obstacles and bow out between them as illustrated before in Fig. 4.13. The deep cusps in the dislocation line as, for example, J in Fig. 4.13, originate from jogs formed by cross slip trailing dipoles (Sect. 5.1.2). Cusps are also observed in moving edge dislocations, c.f. Fig. 7.7b. They may also be due to localized obstacles or to the drag caused by jogs moving conservatively (Sect. 4.8, Fig. 4.31). In the orthogonal slip bands as in Fig. 7.8, edge segments do sometimes not show the curly shape of locally pinned dislocations but exhibit quite straight segments connected by superkinks. This points at the additional action of the Peierls mechanism. The double-exposure of the dislocation line marked by an arrow indicates that the superkinks do not move smoothly as predicted by the Peierls mechanism. They rather move in jumps. Thus, the kinks are stabilized by localized obstacles. After their overcoming, the kinks move quickly into a new pinned position. Within the slip bands, the dislocation densities typically amount to $1.5 \times 10^{12}\,\mathrm{m}^{-2}$. With $\alpha_\mathrm{P} = \pi$ in (5.11) and a shear modulus of $\mu =$ 124 GPa, this yields an athermal component of the flow stress of $\tau_\mathrm{i} \approx 23$ MPa.

Many of the kinematic features of the jerky motion of dislocations in MgO are well illustrated by the following two video sequences. The first one also illuminates all the processes involved in the generation of dislocations by the double-cross slip mechanism (Sects. 5.1.1 and 5.1.2 and the figures therein).

Video 7.1. *Dislocation motion and generation in* MgO *single crystals at room temperature:* This video clip is of low quality. As the original record is lost, the present clip is a copy of a 16 mm movie taken from the original recording. Nevertheless, this sequence shows many processes in a clear way. The projection of the Burgers vector is indicated by the red line b. Owing to the pinning by localized obstacles, the dislocations move in a jerky way, usually by jumps of segments of the order of magnitude of the obstacle spacings. Many processes visible are related to the motion of jogs, the trailing of dislocation dipoles, and the double-cross slip multiplication mechanism. At C, jogs in a screw dislocation move conservatively along the dislocations. A multiplication event starts at M by forming a loop at a jog. Later, the loop increases in size and forms two new dislocations. At L, another dislocation loop develops at a jog. Much later, it is growing to a large loop before it further develops into two new dislocations at MM. Jogs JE in edge dislocations move conservatively with the dislocation. Nevertheless, they form cusps indicating an increased lattice friction, owing to the Peierls mechanism acting simultaneously. A small loop collapses at SL. Many of these processes occur simultaneously in all regions of the image.

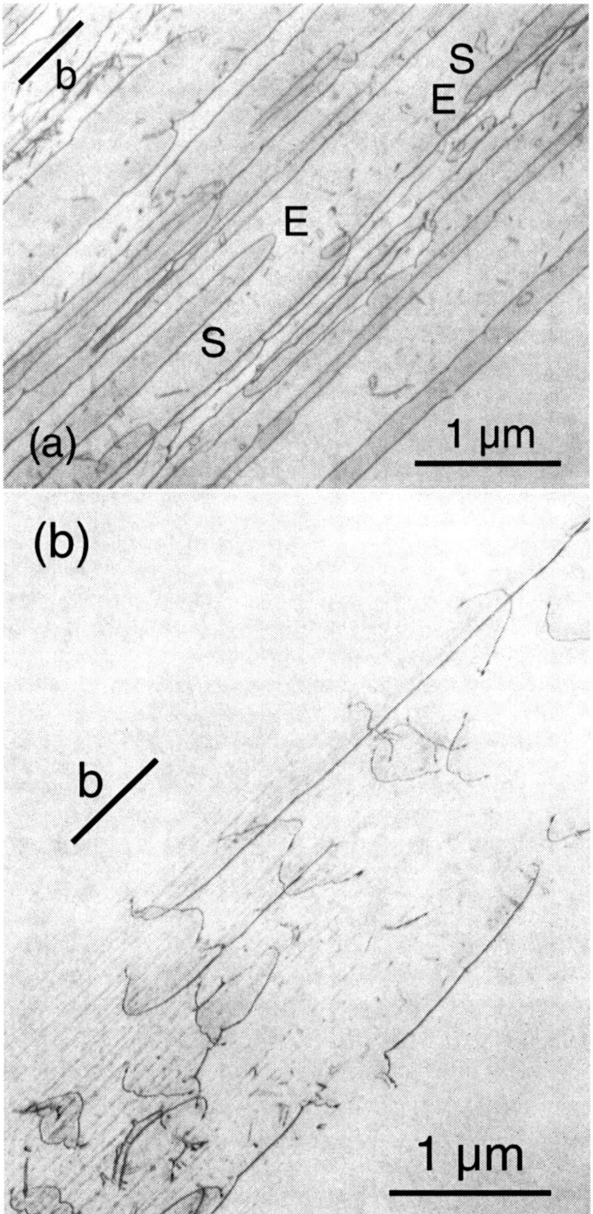

Fig. 7.7. Oblique slip bands in MgO single crystals during in situ deformation in the HVEM at room temperature. Trailing of long screw dislocations by moving edge dislocations (**a**). Moving edge dislocations (**b**). The specimens are tilted by about 15° away from the [001] foil normal in [1$\bar{1}$0] direction. Tensile direction vertical, ***b*** Burgers vector. From the work in [281]

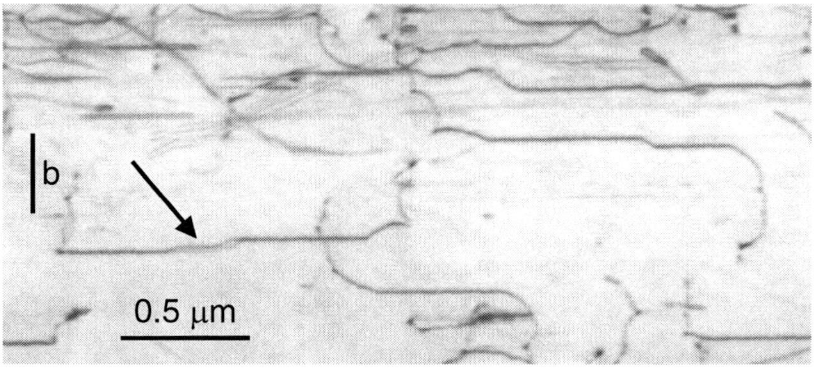

Fig. 7.8. Slip band in an MgO single crystal during in situ deformation in the HVEM at room temperature. Tensile direction vertical, ***b*** projection of Burgers vector. From the work in [281]

Video 7.2. *Motion of a dislocation in* MgO *at room temperature:* At high magnification, details can be observed of the motion of a dislocation in an array of local obstacles in MgO. The screw dislocation is pinned locally. Between the stable positions, it moves at different velocities. Sometimes, the dislocation jumps to the next stable position during the time interval between successive frames. In Fig. 7.9a, taken from this video sequence, the dislocation is pinned in a stable position. After thermal activation in the next frame (in b), the dislocation is imaged in several positions in a blurred way. It reaches the next stable position in (c). The same repeats in the next two frames (d, e). By shuttling from frame to frame, the transition between stable and very dynamic states can be observed several times within this sequence. Frequently, however, the dislocation moves also in a smooth way, again indicating the action of the Peierls mechanism. A jog marked by a deeper cusp in the dislocation line at C moves conservatively along the dislocation.

Apparently, the motion of dislocations in MgO at room temperature is controlled by both the localized obstacles and the Peierls mechanism. As discussed in Sect. 4.6, the relative significance of both processes can be judged by comparing the waiting times t_w at the obstacles with the traveling times t_t between them. In the Video 7.2 showing a dislocation just in motion, both are of the same order of magnitude. In general, however, as in Video 7.1, most dislocations do not move temporarily, pointing at long waiting times at the obstacles. Observing the specimen area over a long time illustrates that practically all dislocations are mobile. Thus, it may be justified to evaluate the dislocation motion mainly in terms of localized obstacles but considering also the Peierls mechanism.

7.2.2 Statistics of Overcoming Localized Obstacles

Micrographs from the in situ straining experiments on MgO crystals offered the unique opportunity to determine the statistical properties of the interaction

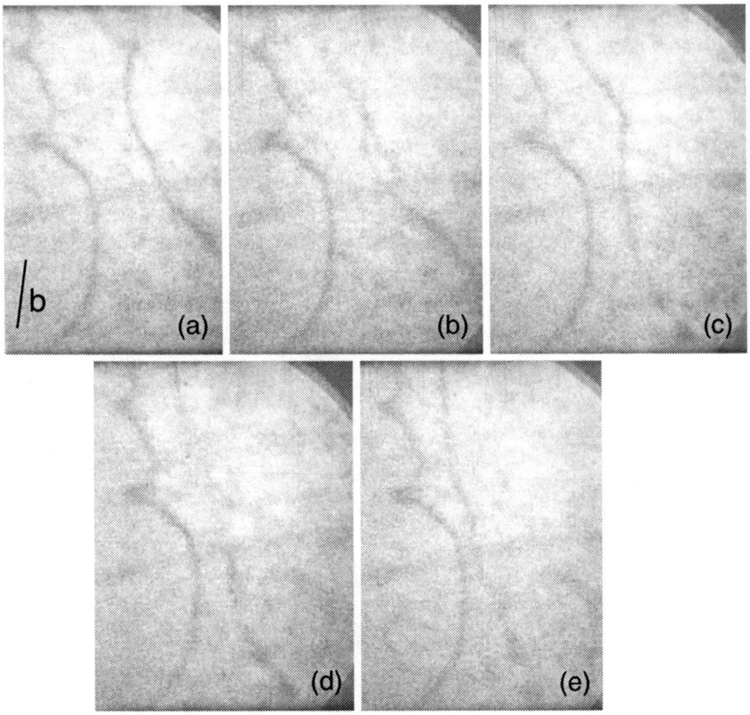

Fig. 7.9. Successive frames of the motion of a dislocation in MgO at room temperature at high magnification. From the work in [281]

between dislocations and localized obstacles, that is, the frequency distributions of the forces acting upon the obstacles, the obstacle distances (segment lengths), and the local effective stresses. Positive copies of the micrographs were projected onto the screen of a digitizer. The bowed dislocation segments were visually compared with ellipses of different sizes calculated by isotropic line tension theory according to the anisotropic energy factors of edge and screw dislocations in MgO. The measuring procedure [101] consisted in digitizing the position of a cusp in the dislocation line representing an obstacle, fitting an ellipse of suitable size to the adjoining dislocation segment and repeating this procedure along the dislocation line. The obstacle distances are the distances between the cusps, and the forces on the obstacles were calculated from the vector sum of the line tensions of the segments neighboring the obstacles (3.39), $\Gamma = E_e\,(1 + \nu - 3\nu \sin^2 \beta)$, where β is the orientation angle of the dislocation arcs touching the obstacle. In the respective line tension approximation, the local effective stress is given by (5.18)

$$\tau^* = \frac{E_e}{b}\,S = \frac{E_{0e}}{b}\,(\ln{(l/r_0)} + C)\;S. \tag{7.3}$$

$S = x_0^{-1} = e^{-1}$ is the reciprocal major half-axis of the fitting ellipses (Sect. 3.2.7).

The experimental distributions of the forces and obstacle distances were first published in [102, 405] and compared with the results of computer simulations of thermally activated dislocation motion in the field of Fleischer type obstacles in [406] as mentioned above in Sect. 4.5. In the simulations, the bowed dislocation segments were treated in the approximation of constant line tension, that is, of constant radius of curvature r. It is discussed in [406] that for low forces $f_0 < 0.7$ and low applied stresses $\tau'/\tau_0' < 0.3 \ldots 0.4$ (i.e., thermal activation at all obstacles), the frequency distributions can be normalized also in the thermally activated case according to the scaling laws

$$g(f)_{R,l_{sq}} = A\,G(Af) = A\,G(u)$$
$$h(l)_{R,l_{sq}} = B\,H(Bl) = B\,H(v),$$

with

$$u = Af \text{ and } v = Bl$$

and the scaling factors

$$A = (2/(3\tau'^2))^{1/3} \text{ and } B = l_{sq}^{-1}(2\tau'/3)^{1/3}. \tag{7.4}$$

Accordingly, the averages of the physical and normalized quantities are related by

$$\int_0^\infty uG(u)\,\mathrm{d}u = \langle u \rangle = A\langle f \rangle$$
$$\int_0^\infty vH(v)\,\mathrm{d}v = \langle v \rangle = B\langle l \rangle \tag{7.5}$$

To normalize the experimental histograms $g(f)$ and $h(l)$, the scaling factors may be derived from the experimental and theoretical averages. Special computer runs were performed to satisfy the experimental conditions as best as the available computer program allowed. This included the choice of the temperature and the dislocation length, the latter of which influences the shape of the distributions. The average values slightly decrease with increasing reciprocal temperature and dislocation length but they do not depend on the stress for $\tau'/\tau_0' < 0.35$ and $f_0 < 0.7$. In the present case, $\langle u \rangle = \langle v \rangle = 0.71$. Figure 7.10 presents the result of the comparison between the experimental and theoretical distributions.

For the evaluations, a set of micrographs was chosen showing the same specimen area, with the applied stress increasing from 6.06 to 8.06 (in relative units). More than 200 segments were measured for each micrograph. The theoretical histograms correspond to "short" dislocations, that is, those containing 20 segments. This length agrees with the length of dislocations in the foil specimens. The histograms for short dislocations resemble the experimental distributions more closely than those for long dislocations, calculated

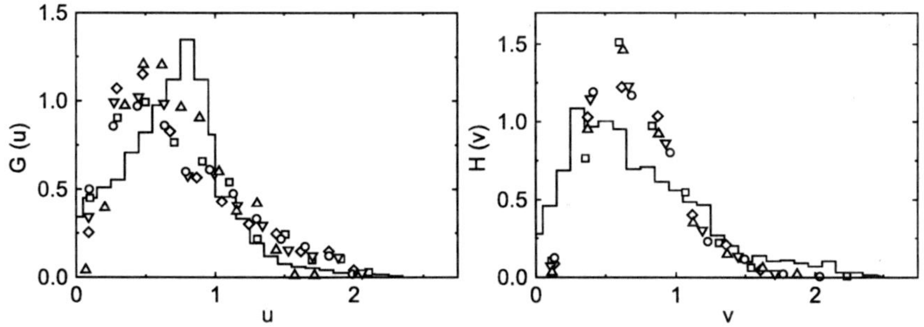

Fig. 7.10. Frequency distributions of the normalized distributions of forces acting on the obstacles (**a**) and obstacle distances (**b**) in an MgO single crystal. Histograms: result of computer simulation; data points: results from in situ experiments. Stresses applied to the tensile specimen in relative units: *squares* 6.06, *diamonds* 6.22, *downward triangles* 7.21, *circles* 7.95, *upward triangles* 8.06. Data from [406]

before also by other authors [194, 198]. According to [407], the maximum is always at about $u = 0.8$ for intermediate temperatures.

The theoretical force distribution follows the schematic shape shown in Fig. 4.17, except that at high forces the tail is longer. The experimental and theoretical force distributions resemble each other, but show also characteristic differences. In particular, the maximum of the experimental distribution is shifted to lower forces, which may have several reasons. While some parameters of the simulation agree with those of the experiment as, for example, the temperature and the relative stress $\tau'/\tau_0' \approx 0.28$, the strength of the obstacles is quite different, $f_0 = 0.26$ in the simulation and $f_0 \approx 0.8$ in the experiment, where for f_0 approximately the upper bound of the force distribution is used, which certainly is an overestimate. Besides, the resolution power of the electron micrographs is limited, resulting in a decrease of the frequencies of small obstacle distances and forces. Nevertheless, it is not very probable that these effects caused the deviations between the histograms from the simulations and the experimental results. A main reason for this difference will be the assumption of a single type of obstacles for the calculations and the presence of a spectrum of obstacle strengths and sizes in the experiments, which leads to a shift of the maximum to lower forces and to a longer tail after the maximum. The obstacle spectrum will be discussed in Sect. 7.2.5.

The segment length distribution has a typical asymmetric shape. This can be used to discuss the mechanisms causing the length distributions in other materials. According to the theory, the average obstacle distances depend on the acting stress according to the Friedel relation (4.53) or (4.54), respectively. With the normalizing relations for the case of thermal activation (7.4) and (7.5), and r equal to $1/S$, it reads

$$\langle l \rangle = \langle v \rangle \, l_{sq} (2\tau'/3)^{-1/3} = 0.812 \, l_{sq}^{2/3} \, (S/2)^{-1/3}, \qquad (7.6)$$

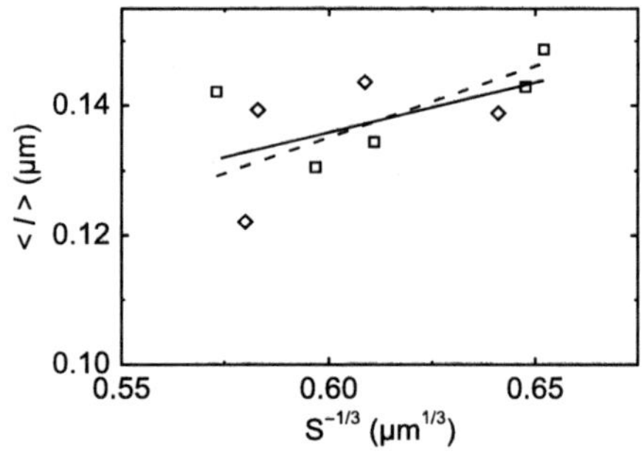

Fig. 7.11. Dependence of the average obstacle distance on the dislocation curvature according to the Friedel relation for two series of load changes during in situ straining of MgO. Data from [204]

Table 7.2. Average obstacle data for an MgO single crystal from an in situ straining experiment at a relative stress of 7.95

$\langle F \rangle$	$\langle l \rangle$ (µm)	$\langle 1/S \rangle$ (µm)	l_{sq} (µm)	l_{sq}/b	r (µm)	τ'
0.273	0.130	0.213	0.093	320	0.239	0.22

Measured from micrograph: $\langle F \rangle$ force, $\langle l \rangle$ segment length, $\langle 1/S \rangle$ dislocation curvature. Calculated from distributions by (7.4) and (7.5): l_{sq} square lattice distance, r radius of curvature of segments, τ' normalized effective stress. Data from [406]

which only slightly differs from (4.53). Figure 7.11 presents the respective plot for two independent series of load changes during in situ straining. The full line is a linear regression including an intercept. The correlation coefficient of 0.57 yields a confidence probability of the correlation of 95%. The dashed line meets the origin and, apart from a slight difference in the numerical constant, it represents the Friedel relation. According to the author's knowledge, there is no other microscopic evidence of the Friedel relation.

Table 7.2 presents average values of the parameters of the distributions at the relative stress of 7.95. Equations (7.4) and (7.5) and $\tau' = l_{sq}/(2r)$ were applied to calculate these values. The data at different stresses proved the constancy of the square lattice distance $l_{sq} \approx 320\,b$ except at the highest stress where the dislocations had moved considerably. The small difference between $\langle l \rangle$ and l_{sq} results from the high strength of the obstacles and the related strong bowing of the dislocation segments. l_{sq} defines an effective concentration of the obstacles on the slip plane via $b^2/(\sqrt{2}l_{sq}^2)$. With the obstacles having a width w, each of them acts on w/b slip planes so that the number of obstacles per lattice site becomes $c = b^3/(\sqrt{2}wl_{sq}^2)$. As it will be discussed in Sect. 7.2.5, the obstacles are supposed to be spinel particles of octahedral shape. Then, the edge length w can be expressed by l_{sq}, and the total concentration of trivalent impurities c_t via $w = 2^{1/4}l_{sq}\sqrt{3c_t}$. With $c_t = 300$ ppm, w amounts to $12b$. With this small size and high strength of the obstacles, the necessary condition for applying Friedel statistics (4.52) is fulfilled, $\xi_0 = (x_0/l_{sq})f_0^{-1/2} = 0.04 \ll 1$.

As it is expected, the values of S representing the effective stress τ^* increase with increasing applied stress as discussed in Sect. 5.2.3. The relation between τ^* and τ was demonstrated in Fig. 5.28. The dependence of the dislocation curvature on the segment length being a consequence of the dislocation self-interaction was plotted in Fig. 3.21. In contrast to the computer simulations with a constant radius of curvature of the dislocation segments, the values of S show a relatively wide symmetric distribution with a standard deviation ΔS of the individual values slightly increasing with increasing S, that is, $\Delta S \approx (0.7 + 0.2\,S)\,\mu\text{m}^{-1}$. This spread of dislocation curvatures is only partly due to the spatially varying internal stress but results also from the dislocation self-interaction. Measurements of the dislocation curvature as a probe of the local effective stress to study the stress distribution around a crack in MgO under load [408] will be described in Sect. 7.2.6.

7.2.3 Kinematics of Overcoming Localized Obstacles

The video recordings of the dislocation motion also yield information on the kinematic dislocation behavior. The theory of thermal activation as treated in Sect. 4.1 describes the waiting time of the dislocation at an obstacle in dependence on the acting force on the condition that the segments adjoining the obstacle do not change their positions. When a certain obstacle is surmounted, however, the dislocation position changes at both neighboring obstacles so that the time t_l the dislocation is in contact with an obstacle is usually not identical with its waiting time. t_l is called the lifetime of the obstacle at the dislocation. The distribution of the lifetimes, which is related to the obstacle spectrum, defines the jerkiness of the dislocation motion as well as the average dislocation velocity. Although this distribution can easily be obtained from computer simulations, experimental data are very scarce. Because of the different lifetimes at the individual obstacles, the dislocations do not advance simultaneously over their whole length. Frequency distributions of the local slip distances in dependence on the observation time intervals t_o contain information on the spatial arrangement of the obstacles.

For kinematic evaluations, photographs were taken from the video screen in a fixed time sequence. To evaluate numerical data, a copy of one state was superimposed onto another one of the state after a certain observation time t_o. An example is given in Fig. 7.12. Pairs of points on dislocations in the initial and final positions were measured by a coordinate digitizer dividing the swept area into tetragons. On resting dislocation segments, the same points were digitized twice. The following quantities were calculated: the segment length λ_i belonging to a single tetragon i as the average of the lengths in both positions, and the total dislocation length λ_0; the area ΔA_i of the single tetragons, and the total area A swept during t_o; the local slip distances $p_\text{i} = \Delta A_\text{i}/\lambda_\text{i}$, and their average during t_o; and finally, the length of dislocations λ_m moving during t_o. Segments resting during t_o are characterized by $p_\text{i} = 0$. The evaluations comprised a specimen region of Video 7.1 in which mainly screw dislocations

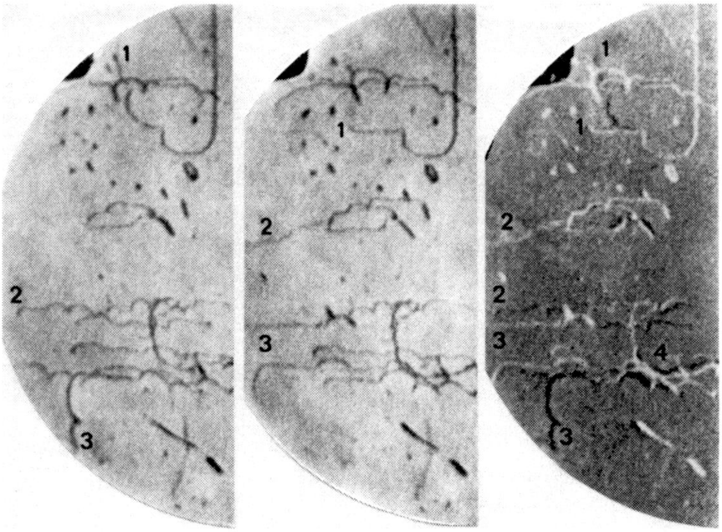

Fig. 7.12. Two stages of dislocation motion in MgO taken from the Video 7.1. *Left*: initial state. *Middle*: state after t_o. *Right*: superposition of the initial and subsequent states, the latter as a negative. Moving dislocations appear as *dark* and *bright lines*, resting ones are imaged with reduced contrast. From [409]. Copyright (1988) Pergamon Press, Oxford

of 19 μm in length moved for 110 s as well as a slip band with dislocations of a large edge component of 33 μm in length observed for 18 s.

The total swept area proved to be proportional to the observation time, indicating a homogeneous dislocation motion in the observed specimen region of the screw band at an area velocity of $\dot{A} = 0.13\,\mu\mathrm{m}^2\,\mathrm{s}^{-1}$ and an average dislocation velocity of $v_\mathrm{d} = \dot{A}/\lambda_0 = 7\times10^{-9}\,\mathrm{m\,s}^{-1}$ corresponding to a local strain rate of $6 \times 10^{-6}\,\mathrm{s}^{-1}$. The frequency distributions of the slip distances in the screw band are plotted in Fig. 7.13 for four different observation times t_o. For short t_o, the frequencies rapidly decrease with increasing slip distances. After about 10 s, a maximum appears at about $p = 0.35\,\mu\mathrm{m}$. A second maximum occurs at $p = 0.85\,\mu\mathrm{m}$ after longer times.

For a single type of localized obstacles, something like a Poisson distribution may be expected theoretically, with the maximum continuously shifting towards larger p with increasing t_o. The occurrence of maxima in fixed positions points at a preference of certain slip distances. The smallest preferred distance should be the square lattice distance l_sq, determined to be 0.093 μm in the preceding section. This distance is represented by its high frequencies for short observation times. From the shape of the force distribution in Fig. 7.10a, it was concluded that the obstacles are not of unique type. The preferred slip distance of 0.35 μm may therefore correspond to the distances between few of strong localized obstacles. The second maximum at 0.85 μm can be interpreted as the wavelength of the long-range stress fields of the dislocations $1/\sqrt{\varrho} = 1/\sqrt{1.5 \times 10^{12}\,\mathrm{m}^{-2}} \approx 8 \times 10^{-7}\,\mathrm{m}$.

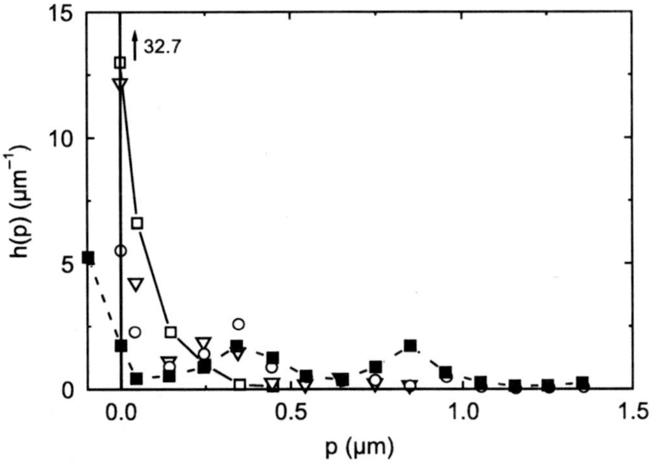

Fig. 7.13. Frequency distribution of slip distances in a screw dislocation band for different observation times. t_o = 3.9 s (*open squares*), 11.6 s (*triangles*), 30.8 s (*circles*), 100 s (*full squares*). Data point at negative p: dislocation segments which had moved out of the frame. Data from [409]

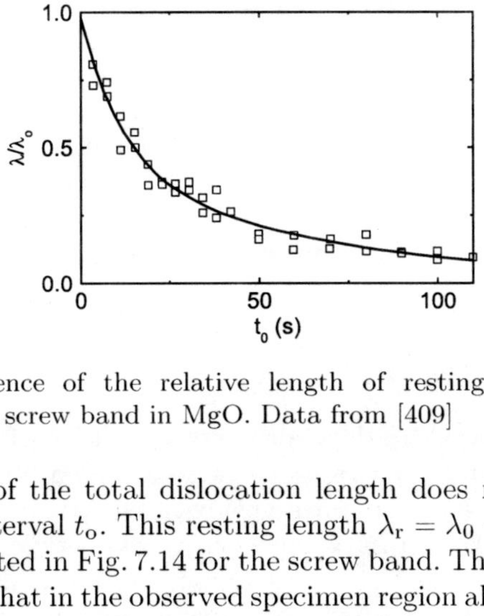

Fig. 7.14. Dependence of the relative length of resting dislocations on the observation time in a screw band in MgO. Data from [409]

A certain part of the total dislocation length does not move during an observation time interval t_o. This resting length $\lambda_r = \lambda_0 - \lambda_m$ decreases with increasing t_o as plotted in Fig. 7.14 for the screw band. The decrease down to a low value indicates that in the observed specimen region almost all dislocations are mobile. The decrease can well be described by two exponentials with different time constants

$$\lambda_r(t_o)/\lambda_0 = C_1 \exp(-B_1/t_o) + C_2 \exp(-B_2/t_o), \tag{7.7}$$

with $B_1 = 0.0909\,\mathrm{s}^{-1}$ and $B_2 = 0.0155\,\mathrm{s}^{-1}$. There ought to be a theoretical connection between the dependence of $\lambda_r(t_o)/\lambda_0$ and the frequency distribution of the lifetimes $h(t_l)$. Several proposals [204,410,411] have been discussed in [204] but a coherent theoretical formulation is still missing. According to [410], the fraction of dislocations not moving during t_o is given by the cumulated frequency of the lifetimes with $t_l \geq t_o$,

$$\lambda_r(t_o)/\lambda_0 = \int_{t_o}^{\infty} h(t_l)\mathrm{d}t_l.$$

After computer simulation experiments, the cumulated frequency is given by an exponential function decreasing with t_o for long dislocations at high temperatures. For short dislocations at low temperatures, the course can be described by the sum of two exponentials as found experimentally in (7.7). By differentiation, there follows that

$$h(t_l) = B_1C_1 \exp(-B_1t_l) + B_2C_2 \exp(-B_2t_l).$$

In all the models, the time constants in (7.7) equal the reciprocal lifetimes. The existence of two lifetimes again indicates the occurrence of a spectrum of obstacles of different strengths.

The mode of dislocation motion observed experimentally where most obstacles are overcome individually but some are passed spontaneously allows conclusions to be drawn on the relation between the acting effective stress τ^* and the athermal stress τ_0 necessary to surmount the obstacle array without thermal activation, or between the normalized quantities τ' and τ_0', respectively. It follows from Fig. 4.18 that then $\tau_0' \approx 3\,\tau'$, or with $\tau' = 0.22$ from Table 7.2 that $\tau_0' \approx 0.66$. This agrees with the estimation of $f_0 \approx 0.8$ from Sect. 7.2.2 and (4.57), which is quite a high strength.

In conclusion, in situ deformation of MgO single crystals with aliovalent impurities reveals a wealth of detailed information on the statistics and kinematics of the motion of dislocations in the field of localized obstacles obeying Friedel statistics.

7.2.4 Dislocation Dynamics

The stress and temperature dependence of the dislocation velocity in nominally pure crystals and crystals doped with 150 ppm Fe^{3+} was measured in [403] at and above room temperature using the stress pulse etching technique. In the pure crystals, the stress exponent m varied between 3 and 6.5, depending on the oxidation state. The doped crystals were annealed and quenched to solve the impurities. The stress exponent was then about 11. The edge dislocations turned out to be much more mobile than screws with stresses to reach dislocation velocities near $10^{-6}\,\mathrm{m\,s^{-1}}$ of about 10 and 15 MPa. Similar measurements were carried out on the same more impure material used also for the other data in this section [412]. Problematic were the large slip distances of edge dislocations above 100 K, which interfered with the screw dislocations so that the latter could not be identified. Therefore, the velocity of head dislocations in screw bands was determined additionally. As discussed in Sect. 5.1.1, head dislocations move faster than individual ones under the

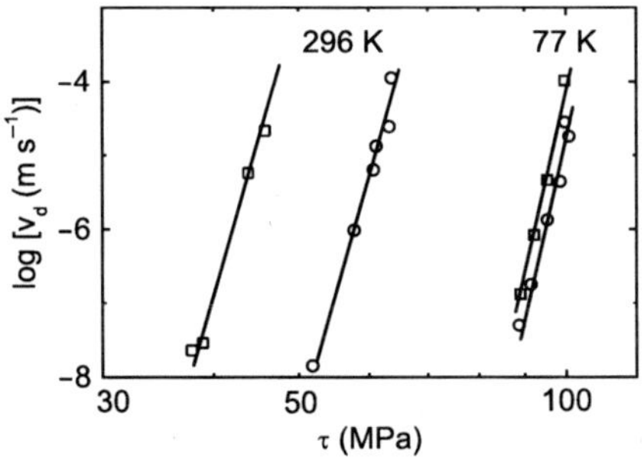

Fig. 7.15. Stress dependence of the velocity of head dislocations in slip bands in MgO. Edge dislocations (*squares*), screw dislocations (*circles*). Data from [413]

same stress. For the present material, the velocity ratio was less than 10, corresponding to only a small shift in the stress. The data at room temperature and 77 K are plotted in Fig. 7.15. The slope of the curves represents the stress exponent $m = \partial \ln v_{\mathrm{d}} / \partial \ln \tau$. For screw dislocations at 77 K, it amounts to almost 50. Similar values result for the other conditions. The figure verifies the great difference between the mobility of edge and screw dislocations, particularly at room temperature. This leads to the formation of loops with very long screw dislocations, which finally control the flow stress. Elongated loops were also observed in the in situ experiments in the oblique slip bands where the edge components do not move out of the foil (Fig. 7.7).

For comparing the results of the velocity measurements with those of macroscopic deformation tests, some parameters of both methods were plotted together. In Fig. 7.16, the temperature dependence of the stress necessary to reach a velocity of individual edge dislocations of $v_{\mathrm{d}} = 10^{-7}\,\mathrm{m\,s^{-1}}$ is plotted together with the macroscopic yield stress. This dislocation velocity corresponds to that in macroscopic experiments at a strain rate of $10^{-4}\,\mathrm{s^{-1}}$ and a dislocation density of $1.5 \times 10^{12}\,\mathrm{m^{-2}}$ quoted before. Both data sets fit well the same curve. The activation volume was calculated by $V = kT\,(\partial \ln v_{\mathrm{d}} / \partial \tau)_T$. Figure 7.17 compares the data with those of the macroscopic tests, revealing a satisfactory agreement.

7.2.5 Macroscopic Deformation Properties and Discussion

The temperature dependence of the yield stress of the MgO crystals (Fig. 7.16) shows three ranges: a steep decrease at low temperatures, a more moderate decrease between about 100 and 750 K, and a plateau above this temperature. These ranges are explained by the dominance of the Peierls mechanism at low temperatures, by the interaction with localized obstacles at intermediate temperatures, and by long-range dislocation interactions at high ones. The values of the stresses to reach a velocity of edge dislocations of $10^{-7}\,\mathrm{m\,s^{-1}}$ are slightly lower. Considering the difference between the mobilities of edge and screw dislocations suggests that the respective stresses for screw dislocations

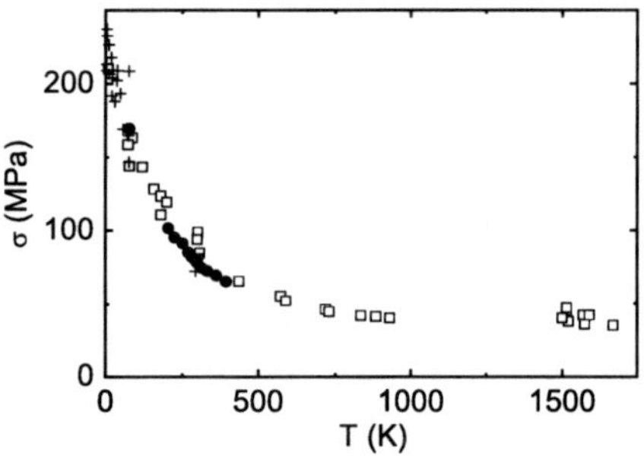

Fig. 7.16. Yield stress of MgO single crystals with data from [414] (*open squares*) and [415] (*crosses*) as well as stress to reach the velocity of edge dislocations of $10^{-7}\,\mathrm{m\,s^{-1}}$ (*full circles*), data from [413]

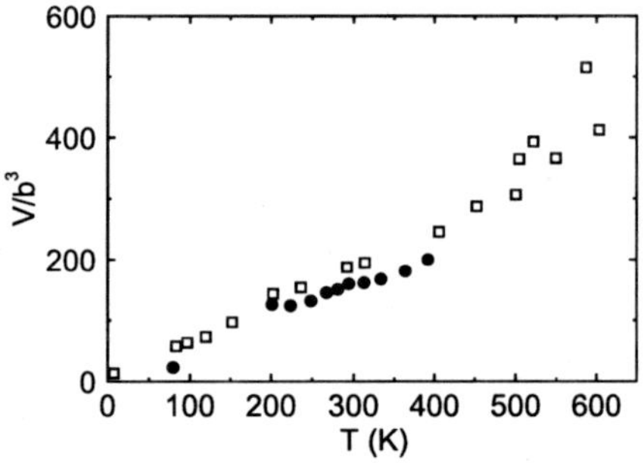

Fig. 7.17. Activation volume in MgO from macroscopic deformation tests [414] (*open squares*) and from measurements of the velocity of individual edge dislocations [413] (*solid circles*)

should be much higher. Nevertheless, screw dislocations are considered to control the deformation over a wide range of intermediate temperatures.

The strain rate sensitivity was measured by strain rate cycling and stress relaxation tests. It amounts to about $r = 0.8\,\mathrm{MPa}$ near helium temperature, but it is constant at the low value of 1.5 MPa over a very wide temperature range between liquid nitrogen temperature and 600 K. Above this temperature it decreases again. Accordingly, the activation volume is almost proportional to the temperature as plotted in Fig. 7.17. Reference [413] contains also measurements of the activation enthalpy by a combination of temperature and strain rate change tests between about 77 and 600 K and the resulting Gibbs free energy calculated by (4.15). The latter is proportional to the temperature with $\Delta G/T = 2.17 \times 10^{-3}\,\mathrm{eV\,K^{-1}}$ as quoted before. This proportionality is a necessary prerequisite to the validity of the Arrhenius equation if the deformation mechanism does not change within the temperature interval.

The low-temperature deformation data can be analyzed in terms of the Peierls mechanism. The motion of edge dislocations by the double kink

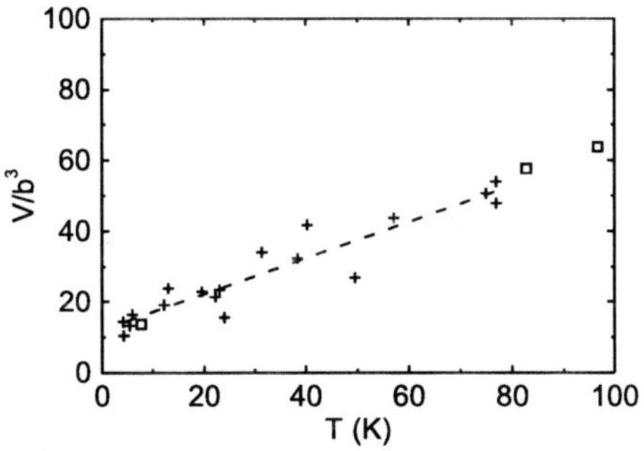

Fig. 7.18. Activation volume in MgO from macroscopic deformation tests at low temperatures. Data from [414] (*squares*) and from [415] (*crosses*)

mechanism was studied by means of atomistic models in [114, 133], yielding Peierls stresses τ_p of the order of magnitude of 90–150 MPa and activation volumes of $60\,b^3$ to $120\,b^3$ (quoted from [415]). τ_p agrees well with the yield stress extrapolated to zero K of $\sigma = 2\tau \approx 250$ MPa. Also the experimental activation volumes at low temperatures plotted in Fig. 7.18 are well within the range of the theoretical expectation.

After the continuum model in [151], the temperature dependence of the yield stress in the Peierls region can be described by

$$\tau = \tau_0 \left(1 - \left(\frac{T}{T_0}\right)^{1/2}\right) ,$$

where τ_0 is the flow stress at zero temperature and T_0 the temperature at which the Peierls mechanism vanishes. The corresponding plot is shown in Fig. 7.19, yielding $T_0 = 574$ K, which means that the double kink mechanism should act still far above room temperature. This plot does not consider the athermal stress contribution τ_i so that T_0 should be lower. The dislocation velocity measurements in nominally pure MgO in [403] are also interpreted by the Peierls mechanism. The shear stress to move screw dislocations at room temperature at a velocity of $10^{-7}\,\text{m s}^{-1}$ amounts to about 5 MPa in contrast to a much higher shear flow stress of about 45 MPa. However, the microstructural observations described in Sect. 7.2.1 also hint at this mechanism, the occurrence of straight crystallographically oriented edge dislocations and the viscous motion of screw dislocations between localized obstacles.

However, there is clear microstructural evidence that the overcoming of localized obstacles dominates the dislocation motion at room temperature (Fig. 4.13 and Sect. 7.2.1). This is in accordance with the high activation volume of the order of $200\,b^3$ or of the corresponding stress exponent of $m' \approx 60$. A model explaining many of the observed features was suggested by Reppich and Knoch [218, 416]. According to them and the literature quoted, in crystals doped with Fe in the 1,000 ppm range, particles of an octahedral shape with {111} habit planes form, having an inverse spinel structure. The particles

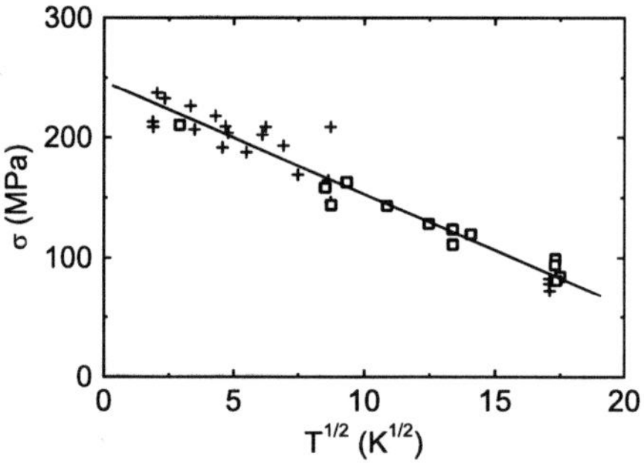

Fig. 7.19. Temperature dependence of the yield stress according to the model of the double kink mechanism in [151]. Data from [414] (*squares*) and from [415] (*crosses*)

are incorporated coherently. During the cutting of the particles, an antiphase boundary (APB) has to be formed. The energy required represents the main obstacle friction mechanism. It is suggested here that the localized obstacles observed in the TEM and during the in situ investigations consist of these octahedral particles. The mechanism was discussed already in Sect. 4.7. The shape of the cutting plane is demonstrated in Fig. 4.26 with the coordinates being chosen for the intersection by an edge dislocation. For the cutting by a screw dislocation, the particle has to be rotated by 90°. If the obstacle resistance is given only by the fault energy γ_{APB}, the obstacle force amounts to $F = \gamma_{\mathrm{APB}}\, 2z$, where z is half the width at the actual force, and the activation distance equals the difference between the equilibrium x values at the entrance and exit sides $\Delta d = x_{\mathrm{a}} - x_{\mathrm{e}}$. The outline of the particle cutting plane represents the central cut and the hatched area an "average" off-center cut. The noncentral cuts form a spectrum of varying particle widths. The situation is different for edge and screw dislocations as illustrated by the force–distance curves in Fig. 7.20. At the central cut, edge dislocations experience the larger maximum force F_{me} in accordance with the larger maximum width. The noncentral cuts generate a spectrum of different obstacle strengths because of the different widths in z direction. The average obstacle has the strength $F_{\mathrm{ae}} = F_{\mathrm{me}}/2$. All particles have the same maximum activation distance Δd_{me}. The central cut for screw dislocations has a lower maximum force F_{ms} than that of edge dislocations, but all noncentral cuts exhibit the same maximum strength. The varying quantity is now the activation distance. The "average" obstacle has the width $\Delta d_{\mathrm{as}} = \Delta d_{\mathrm{ms}}/2$. Thus, while a few particles form strong obstacles to edge dislocations, many particles cut off-center have low strengths and may be overcome spontaneously. For screw dislocations, the particles all have the same strength, though they have different activation distances and energies. These features may explain why screw dislocations are pinned by localized obstacles with small obstacle distances, while edge dislocations are pinned having larger segment lengths, resulting in a higher mobility.

It is difficult to analytically describe the trapezoid shape of the "average" obstacle interaction potential. The potentials are therefore approximated

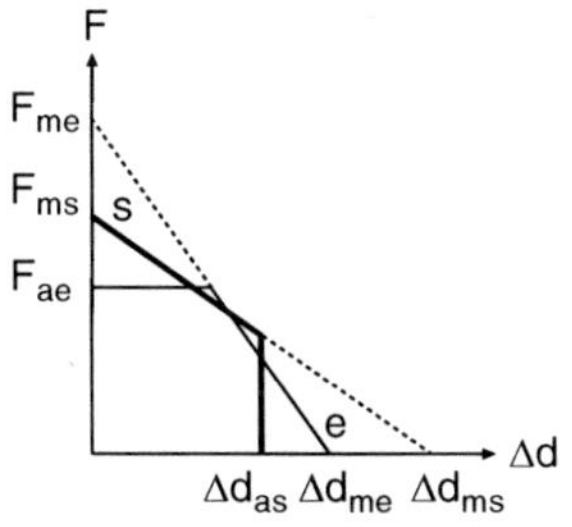

Fig. 7.20. Force–distance curves for particles of octahedral shape interacting with dislocations by the formation of antiphase boundaries inside the particles. Interaction potential of an "average" obstacle for screw dislocations (*thick solid line*), for edge dislocations (*thin solid line*). Curves for the central cut with maximum interaction parameters (*dotted curves*)

by the simple box potential with a constant activation distance Δd. This approximation is certainly not too bad for screw dislocations controlling the deformation rate. The Gibbs free energy of activation of the "average" obstacle can then be written as

$$\Delta G = \Delta G_0 \left(1 - \frac{F}{F_0}\right),$$

with $\Delta G_0 = (w/2)F_0 = (3/8\sqrt{2})\, w^2 \gamma_{\mathrm{APB}}$, $F_0 = (w/2)\gamma_{\mathrm{APB}}$ for edge dislocations, and $F_0 = (1/\sqrt{2}) w \gamma_{\mathrm{APB}}$ for screws. w is the edge length of the octahedral particles (equal to the maximum x difference of the central cutting plane for screw dislocations). Applying the Friedel relation for a random arrangement of strong, wide-spaced obstacles (4.53) and setting the numerical constants $k_1 = k_2 = 1$ yields

$$\Delta G = \Delta G_0 \left(1 - \left(\frac{\tau^*}{\tau_0^*}\right)^{2/3}\right),$$

and with (4.5) and (4.8)

$$\left(\frac{V}{b^3}\right)^{-2} = \left(\frac{3}{2}\frac{b^3}{V_0}\right)^2 \left(1 - \frac{T}{T_0}\right). \tag{7.8}$$

Using (4.56) and (4.50), the parameters are

$$V_0 = \frac{\Delta G_0}{\tau_0} = (w/2) b l_{\mathrm{sq}} f_0^{-1/2} \text{ and } T_0 = \frac{\Delta G_0}{k \ln(\dot{\gamma}_0/\dot{\gamma})}. \tag{7.9}$$

In Fig. 7.21, the activation volume is plotted according to (7.8). Here, the factor of 3/2 in (4.55) resulting from Friedel statistics is considered, which has not been regarded in the figures before. The full line is a linear regression

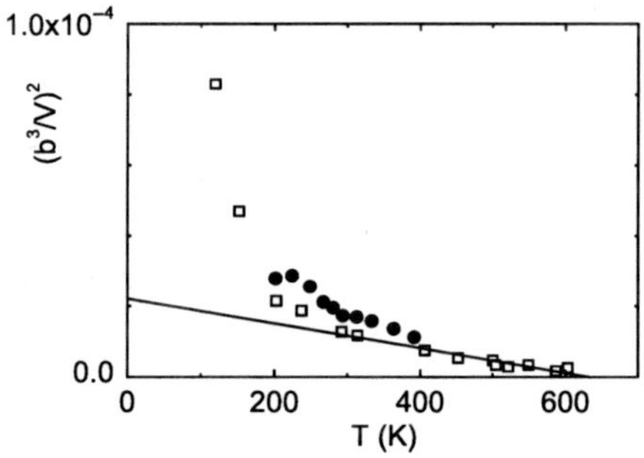

Fig. 7.21. Temperature dependence of the activation volume in MgO according to (7.8). Data from macroscopic deformation tests [414] (*open squares*) and from measurements of the velocity of individual edge dislocations [413] (*solid circles*). *Straight line*: extrapolation of the data from macroscopic measurements at and above room temperature

of the data at and above room temperature. The parameters are $V_0 \approx 320\,b^3$ and $T_0 \approx 635\,K$. In the figure, the data points for temperatures below room temperature deviate strongly from the extrapolated course towards smaller activation volumes, which should be due to the transition to the action of the Peierls mechanism. With the value of $f_0 \approx 0.8$ and $l_{\text{sq}} = 320\,b$ quoted in Sect. 7.2.2, from (7.9) it follows that w amounts to a few b. On the other hand, from the concentration of impurities, a higher value of w of about $12\,b$ was estimated. The "true" activation volume at room temperature $V = \Delta d\, lb$ from Fig. 7.17 takes a value of $300\,b^3$ if the factor of 3/2 from Friedel statistics is considered, leading to an activation distance of $\Delta d \approx 1\,b$. All estimations show that the particles are quite small but that the activation distance is smaller than the width of the particles.

The total activation energy is the sum of the energy at the given effective stress plus the work term $\Delta F = \Delta G + V\tau^* \approx \Delta G_0$ (4.3). Considering the yield stress of $\tau \approx 40$ MPa at room temperature, the estimate of the internal stress of $\tau_{\text{i}} \approx 23$ MPa in Sect. 7.2.1, and the measurements from dislocation curvatures in Fig. 5.28 corrected for radiation hardening, a reasonable value of the effective stress is $\tau^* \approx 22$ MPa. Together with the relevant data $\Delta G(300\,\text{K}) = 0.65\,\text{eV}$ and $V = 300\,b^3$, it follows that $\Delta G_0 \approx 1.7\,\text{eV}$. This characterizes the obstacles as strong ones, which are mainly surmounted by the applied (effective) stress aided by thermal activation. With an obstacle width of $w = 10\,b$, this corresponds to a very reasonable APB energy of $\gamma_{\text{APB}} \approx 125\,\text{mJ}\,\text{m}^{-2}$. ΔG_0 is in rough agreement with the value calculated from the limiting temperature T_0 together with $k \ln(\dot\gamma_0/\dot\gamma) = \Delta G/T = 2.17 \times 10^{-3}\,\text{eV}\,\text{K}^{-1}$ from the measured activation energies, yielding $\Delta G_0 \approx 1.4\,\text{eV}$.

The strength τ_0 of the obstacle array can roughly be estimated from the effective stress $\tau^* \approx 22$ MPa and the statement that $\tau'/\tau'_0 \approx 0.3$ obtained from the kinematic dislocation behavior (Sect. 7.2.3). According to this, the obstacle strength may be of the order of magnitude of $\tau_0 \approx 3\,\tau^* \approx 66\,\mathrm{MPa}$. This is a high strength suggesting that the precipitation hardening mechanism acts also at temperatures below room temperature. Thus, there is probably a very wide range of the simultaneous action of both the Peierls mechanism and localized obstacles. There is some disagreement between the obstacle strength data from the macroscopic measurements and from the in situ experiments. The data of the latter are considerably higher. This may only partly be due to the radiation hardening observed during the in situ experiments.

In [218, 416], the dislocations were observed to move in pairs owing to coupling by the APBs. In Sect. 8.1.3, this mechanism will be described for alloys containing precipitates of intermetallic phases. But in the present MgO crystals with small precipitates it does not act.

At temperatures between about 800 and 1,600 K, the yield stress remains constant at $\tau_y = 20\,\mathrm{MPa}$. It can be explained by long-range dislocation interactions. In Sect. 7.2.1, the internal stress component was estimated to amount to $\tau_i = 23\,\mathrm{MPa}$, well agreeing with the measured yield stress. Above about 1,600 K, the yield stress may drop to lower values owing to the onset of diffusion-controlled recovery. This process will be described in the section about zirconia.

7.2.6 Dislocations in the Plastic Zone of a Crack

In a perfectly elastic solid, fracture occurs when the stress at the crack tip reaches the theoretical cohesive stress [417, 418]. By analogy with the theoretical shear stress in Sect. 1.1, the latter is estimated to $\sigma_{id} \approx E/10$, with E being Young's modulus. The flow stress of most of the crystalline materials, however, is essentially lower than this value so that plastic flow is expected to occur in the region of the highest stress concentration ahead of the crack. This localized plastic flow increases the resistance of the material to unstable fracture [419, 420] and the crack propagation is essentially determined by the size, the structure, and the internal stress distribution of the plastic zone.

During one straining experiment inside an HVEM on an MgO single crystal, the stable growth of a crack on a cube face had been observed in situ [408]. In the stress field of the crack, a plastic zone developed with increasing size and dislocation density. An intermediate state is presented in Fig. 7.22. In front of the crack, the plastic zone is elongated with the dislocation density decreasing continuously with increasing distance from the crack tip.

Under the tensile load in the in situ stage along the [010] axis, the crack with its tip along [001] extended on a (010) plane. The tensile stress σ_{22} of the stress field of the crack activated the $(011)[01\bar{1}]$ and $(01\bar{1})[011]$ slip systems. The respective dislocations bowed out under the stress. By measuring the curvature of the bowed segments using (7.3) of Sect. 7.2.2, the dislocations may serve as a probe to determine the spatial distribution of the stress around the crack. As a measure of the stresses, the quantity $S\,(\ln{(l/r_0)} + C)$ can be taken. The curvatures and segment lengths l were measured on the individual

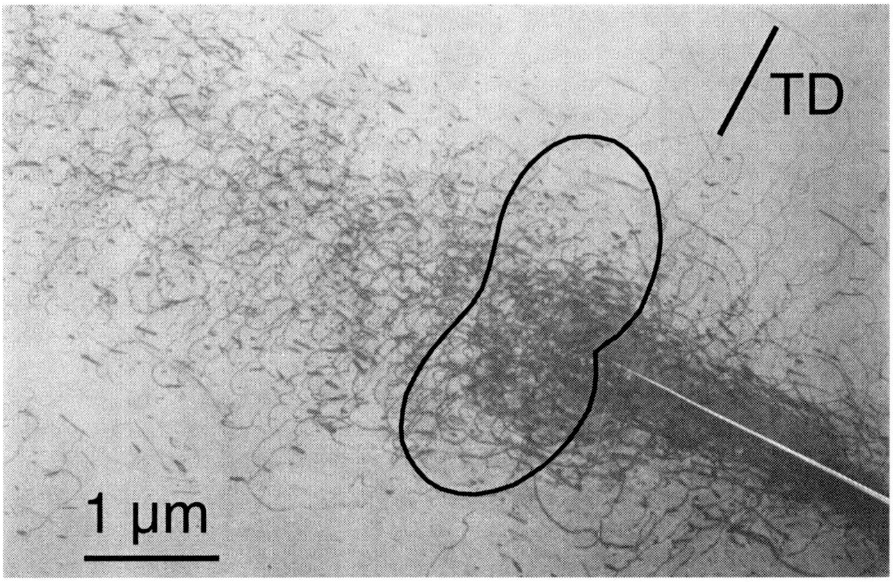

Fig. 7.22. Plastic zone on a crack in an MgO single crystal during in situ straining in an HVEM. TD tensile direction. From the work in [408]

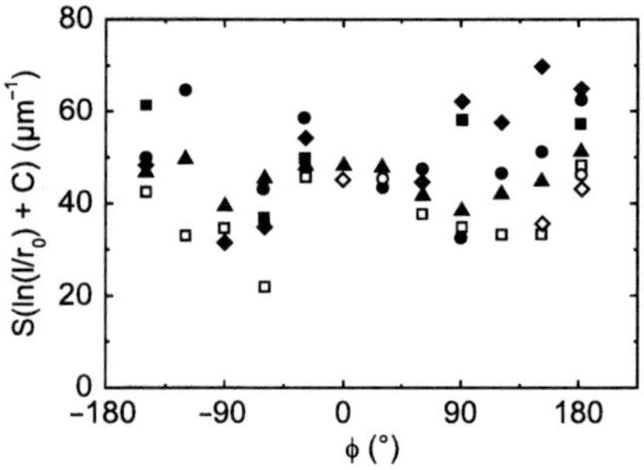

Fig. 7.23. Angle distribution of the stress parameter S $(\ln(l/r_0) + C)$ in the plastic zone of a crack in an MgO single crystal during in situ straining inside an HVEM. Parameter: radius $r = 0.5\,\mu m$ (*solid squares*), $1.5\,\mu m$ (*solid diamonds*), $2.5\,\mu m$ (*solid circles*), $3.5\,\mu m$ (*solid triangles*), $4.5\,\mu m$ (*open squares*), $5.5\,\mu m$ (*open diamonds*), $6.5\,\mu m$ (*open circles*). Data from [408]

segments. The angular distribution of the stresses is plotted in Fig. 7.23 with the distance from the crack as a parameter. The xy plane around the crack tip is sectioned into zones of polar coordinates, $1\,\mu m$ along the radius and $30°$ wide. The x axis points into the crack. The values were calculated using the theoretical constant of $C = -1.61$ in (7.3) or (5.18). The figure shows only a weak maximum at $\pm 150°$ in front of the crack and a slight decrease with increasing distance from the crack. The stress quantity far from the crack amounts to about $S\,(\ln{(l/r_0)} + C) = 60\,\mu\mathrm{m}^{-1}$. Thus, everywhere the effective stress within the plastic zone of the crack roughly equals the flow stress of the material.

This result can be compared with the predictions of the elastic theory of fracture. Fracture can take place in a brittle way when the solid undergoes only elastic deformations so that the energy necessary to extend the crack is solely given by the energy of the additional surfaces to be created. Then, the stress field of the crack can be modeled by a pile-up of dislocations with Burgers vectors perpendicular to the crack plane analogous to the shear pile-up in Sect. 5.2.1. Instead of (5.13), the internal stress field of the crack is then given by

$$\sigma_{\mathrm{i}}(r,\phi) = f(\phi)\,\sqrt{\frac{c}{r}}\,\sigma, \tag{7.10}$$

where the length of the pile-up is now called the crack length c and $f(\phi)$ is again a function of ϕ. Thus, the crack acts as a stress concentrator with the stresses decreasing only slowly with the distance as $1/\sqrt{r}$. The critical fracture stress σ_{f} necessary to extend the crack can be estimated by assuming that σ_{i} approaches the theoretical tensile strength σ_{id} of an ideal crystal at the smallest possible distance in the crystal $r = a$, where a is the lattice constant. As mentioned earlier, the latter is estimated as $\sigma_{\mathrm{id}} \approx E/10$. For the fracture stress there follows

$$\sigma_{\mathrm{f}} \approx \sqrt{\frac{E^2\,a}{100c}}.$$

The quantity $E\,a/20$ can be replaced by an estimate of the surface energy γ_{s}, so that

$$\sigma_{\mathrm{f}} \approx \sqrt{\frac{E\,\gamma_{\mathrm{s}}}{5c}}.$$

Apart from the numerical constant ($1/5$ instead of $2/\pi$), this is the Griffith relation for the fracture stress of an elastic solid [417]. The latter is no material parameter as it depends on the crack length c. The respective material parameter is called the critical stress intensity factor defined as

$$K_{\mathrm{c}} = \sigma_{\mathrm{f}}\sqrt{\pi\,c} = \sqrt{2\gamma_{\mathrm{s}}E}.$$

The quantity $K_{\mathrm{c}}^2/E = G_{\mathrm{c}} = 2\gamma_{\mathrm{s}}$ is called the crack extension force or energy release rate. If the work necessary to create the plastic zone around the crack is included in G_{c}, it can be considered a fracture parameter also for elastic/plastic fracture.

For $\sigma < \sigma_{\mathrm{f}}$ and with (7.10), the stress intensity factor $K = \sigma\sqrt{\pi\,c}$ describes the (elastic) stress field around the crack. This equation may then read

$$\sigma_{\mathrm{i}}(r,\phi) = f(\phi)\,\frac{K}{\sqrt{2\pi r}}. \tag{7.11}$$

For a crack in an isotropic medium and for the tensile stress σ_{22}, the function $f(\phi)$ becomes (e.g., [421])

$$f(\phi) = \left(\frac{5}{4} \sin \frac{\phi}{2} - \frac{1}{4} \cos \frac{\pi - 5\phi}{2} \right) . \tag{7.12}$$

To obtain the actual stress intensity of the crack of Fig. 7.22, the crack opening profile $v(r)$ was measured and compared with the theoretical curve

$$v(r) = \frac{1 - \nu}{\mu} \sigma \sqrt{2cr} = \frac{1 - \nu}{\mu} K \sqrt{\frac{2r}{\pi}},$$

yielding $K = 0.55\,\mathrm{MPa\,m^{1/2}}$. This allows one to calculate profiles of equal stress σ_{22} around the crack. The profile plotted in Fig. 7.22 is based on the flow stress of $\sigma_{22} = \tau^*/m_s = 2\tau^*$ with $\tau^* = 115\,\mathrm{MPa}$ from Table 5.2, again with the theoretical value of $C = -1.61$, which overestimates the true flow stress but may consider the irradiation hardening during the in situ experiment. In this approximation, the curve fits the width of the actual plastic zone but not its elongated shape. The latter may depend on the slip geometry. The present slip bands grow in length by glide of the dislocations generated by multiplication of existing dislocations via the double-cross slip mechanism (Sects. 5.1.1 and 5.1.2). To increase the width of the bands would require extensive cross slip.

In conclusion, one can say, inside the plastic zone of the crack the high stresses predicted by the elastic theory of fracture are relaxed down to the average flow stress of the material owing to dislocation multiplication and motion. The shape of the plastic zone depends on the ability of the dislocations to slip and cross slip.

In the literature, the question had been discussed whether in crystals of low dislocation density, cracks may grow accompanied with a dislocation-free zone or whether dislocations are nucleated in front of the crack due to the high stress concentration. Respective in situ straining experiments in a TEM were performed and interpreted by Ohr [422]. In these experiments, dislocations are generated mainly under shear loading, in contrast to the experiment described above with tensile loading.

7.2.7 Summary

MgO single crystals with small precipitates have proved an ideal material to experimentally study the statistical and kinematic aspects of the interaction between dislocations and localized obstacles that obey Friedel statistics. There are no other experimental data found in the literature on the frequency distribution of the forces acting on the obstacles, the relation between the curvature of bowed-out dislocation segments, and their length as well as the Friedel relation between the obstacle distance and the effective stress. The material shows all the three ranges of mechanisms presented schematically in Fig. 5.25, the Peierls mechanism at low temperatures, the thermally activated overcoming of obstacles at and above room temperature, and the dominance of the athermal stress component owing to long-range dislocation interactions at high temperatures.

7.3 Zirconia Single Crystals

Zirconia may serve as a constituent of materials with attractive properties for structural applications over a wide range of temperatures. The tetragonal-to-monoclinic transformation is used for designing tough materials around ambient temperature [423, 424]. At high temperatures, partially stabilized zirconia (PSZ) may have a very high yield stress due to precipitation hardening (e.g., [425, 426]). In addition, ferroelastic deformation observed in so-called t′ zirconia may contribute to a high toughness at high temperatures (e.g., [427, 428]). This chapter is concerned with the dislocation processes in cubic zirconia single crystals, the ferroelastic deformation in t′ and partially stabilized zirconia, and the subsequent plastic deformation.

7.3.1 Crystal Structure and Slip Geometry of ZrO_2–Y_2O_3 alloys

Heuer and coworkers have presented an early review of the structure and microstructure of ZrO_2–Y_2O_3 materials [429]. Pure zirconia (ZrO_2) has the CaF_2 structure at temperatures above about 2,400°C as shown in Fig. 7.24. While the zirconium ions occupy the sites in a face-centered cubic unit cell, the oxygen ions are situated in $1/4\langle 111\rangle$ and equivalent positions. Below 2,400°C, ZrO_2 undergoes a phase transformation to the tetragonal phase. In addition to a small elongation by about 2% in the direction of the c axis, rows of oxygen ions are shifted in upward and downward directions as indicated by the arrows in the figure.

These high-temperature phases can be stabilized down to low temperatures by adding aliovalent cations. According to the ZrO_2–Y_2O_3 phase diagram, the cubic phase (c-ZrO_2) is stabilized at Y_2O_3 concentrations higher than about 8–10 mol% (fully stabilized zirconia, FSZ). At lower yttria concentrations, yttria-lean precipitates of the tetragonal structure form in an yttria-rich

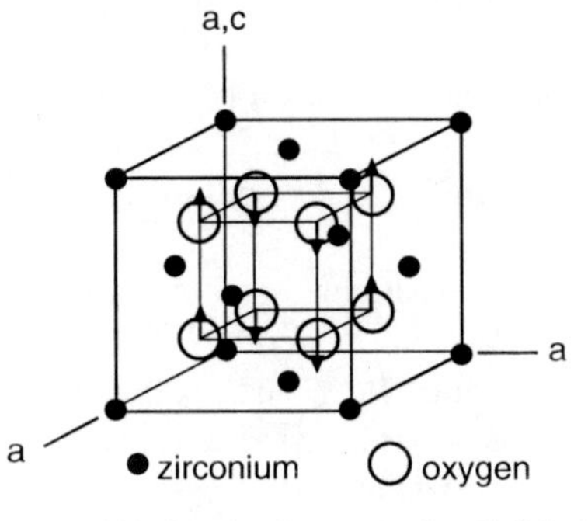

Fig. 7.24. Crystal structure of ZrO_2 single crystals. Cubic ZrO_2: oxygen ions in the drawn $1/4\langle 111\rangle$ positions. Tetragonal ZrO_2: rows of oxygen ions shifted as indicated by *arrows*

cubic matrix (partially stabilized zirconia, PSZ). The diffusion-controlled precipitation reaction proceeds very slowly owing to the sluggish cation diffusion requiring long times to attain the equilibrium concentration of the precipitated tetragonal phase. A variety of microstructures may form depending on the yttria concentration, annealing temperature, and duration [430]. The diffusive transformation cannot take place during quenching from the cubic phase field. The crystals then undergo a displacive phase transformation to a metastable tetragonal material of uniform yttria concentration, the so-called tetragonal polydomain zirconia or t′ zirconia.

As cubic zirconia may constitute the matrix of the two-phase materials, its deformation behavior forms the basis of the understanding also of the deformation of the more complex materials. The Burgers vectors in c-ZrO_2 are of type $1/2\langle 110\rangle$ with slip on the easy $\{100\}$ planes. Most experiments are carried out along a soft $\langle 211\rangle$ loading axis favoring single slip on one cube system with an orientation factor of $m_s = 0.47$. At high temperatures, $\{110\}$ and $\{111\}$ planes may be activated as secondary planes with orientation factors of 0.5 and 0.41. These planes are also activated if cube slip is suppressed by a hard $\langle 100\rangle$ deformation axis.

7.3.2 Microscopic Observations in Cubic ZrO_2

Zirconia single crystals are brittle at low temperatures. The lowest temperatures where plastic deformation of cubic zirconia with 10 mol% yttria in compression in the soft orientation was successful is 400°C under usual deformation conditions [213], and 250°C under confining hydrostatic pressure [214]. All figures and data following refer to this material if not stated otherwise. As shown in Fig. 7.25 for 500°C, slip is localized in narrow slip bands consisting mainly of screw dislocations pinned at localized obstacles. Most obstacles are supposed to be small precipitates, probably containing nitrogen [431]. The obstacle distances l are of the order of magnitude of 0.1 μm and decrease with decreasing temperature and thus with increasing effective stress, but they do not fulfil the Friedel relation (4.54), probably because of a different obstacle spectrum at different temperatures and because of the presence of many jogs marked J, trailing dislocation debris. Similar dislocation structures are found also at 700°C as presented in Fig. 7.26, where the TEM foil was cut parallel to the primary slip plane.

At higher temperatures, the localized obstacles are no longer active . The dislocation structure is quite homogeneous. The dislocations being pinned by jogs in their screw components bow out to larger arcs as demonstrated in Fig. 7.27 for 1,000°C. Stereo pairs reveal that the dislocations are arranged not solely on cube planes and that the cross slip planes probably are $\{111\}$ planes [432]. In addition, many dislocation loops are present.

In the same temperature range, also in situ straining experiments in an HVEM were successful [282]. As shown in Fig. 7.28, the dislocation structures consist of long edge dislocations and screw dislocations pinned by jogs in the

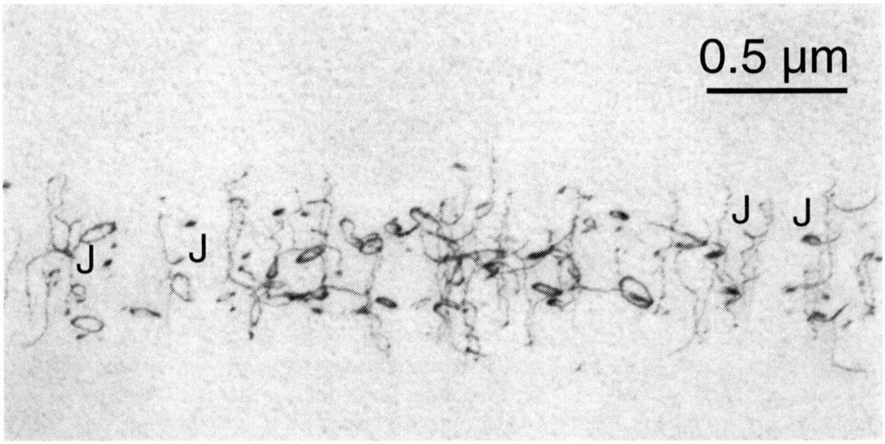

Fig. 7.25. Screw dislocations pinned at localized obstacles in c-ZrO_2 deformed along $[\bar{1}1\bar{2}]$ at 500°C. $(\bar{1}11)$ foil plane. Jogs trailing debris dipoles are marked *J*. From the work in [213]

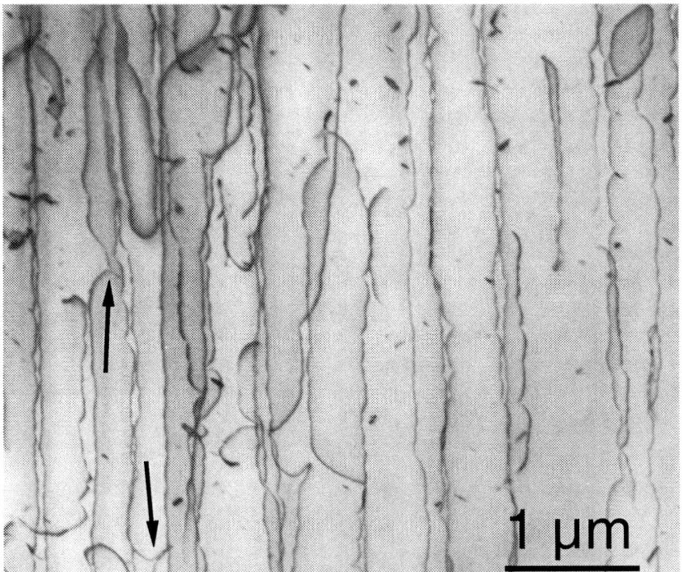

Fig. 7.26. Screw dislocations pinned at localized obstacles in c-ZrO_2 deformed along $[\bar{1}1\bar{2}]$ at 700°C. Foil plane parallel to (001) slip plane. Jogs trailing debris are marked by *arrows*. From the work in [213]

"α" like configuration typical of dislocation multiplication by the double-cross slip mechanism (Fig. 5.5). The equilibrium shape of the dislocations can be calculated by the line tension approximation in anisotropic crystals as described

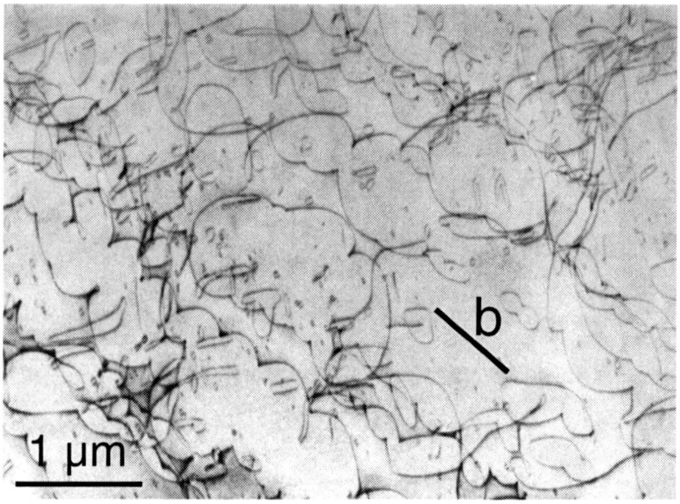

Fig. 7.27. Dislocation structure in c-ZrO_2 after macroscopic deformation along $[\bar{1}1\bar{2}]$ at 1,000°C. Specimen cooled under load. ***b*** projection of Burgers vector. From the work in [433]

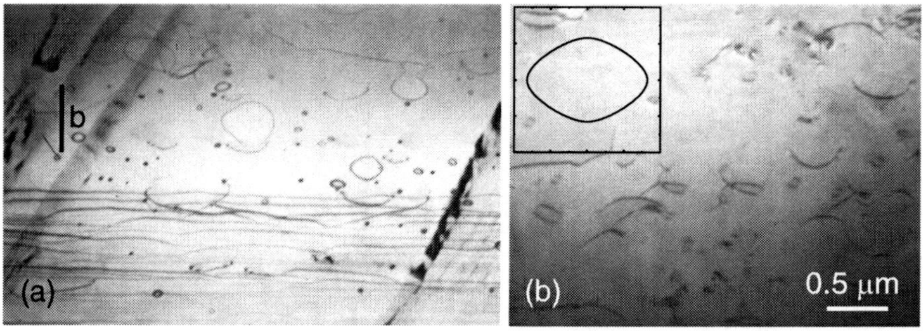

Fig. 7.28. Dislocation structure during in situ straining of c-ZrO_2 deformed along $[\bar{1}1\bar{2}]$ at 1,150°C. $(\bar{1}11)$ foil plane. Long edge dislocations (**a**) and screw dislocations with an "α"-like configuration (**b**). Inset in (**b**): Equilibrium configuration of dislocation loop on the $\{100\}$ slip plane in the projection of the micrographs. From the work in [282]

in Sect. 3.2.7. The inset in Fig. 7.28b presents the slightly angular shape of the loops on the $\{100\}$ slip plane with relatively straight segments near the cube orientations. The curved dislocations approximate the calculated shape, which is also obvious from Fig. 7.27. Fitting the size of the loops to the experimental images and using (5.18), the back stress τ_b caused by the dislocation curvature can be calculated. For the constant C, the theoretical value of -1.61 may be more suitable than the experimental value of −5.19 (see Sect. 3.2.7) because of the large size of the loops compared to small bow-outs at dislocations pinned by localized obstacles. This leads to an upper limit of $\tau_b = 75$ MPa.

The video recordings reveal the dynamic character of the dislocation motion on the easy cube plane at 1,150°C.

Video 7.3. *Dislocation motion in* c-ZrO_2 *on a* {001} *plane during deformation along* $[\bar{1}1\bar{2}]$ *at* 1,150°C*:* This video consists of four short sequences. The lines b and t at the beginning mark the directions of the projection of the $[1\bar{1}0]$ Burgers vector and the [110] trace of the (001) slip plane on the surface. The imaging is frequently obscured by strong fringes caused by the bending in the slip bands. In specimens with a $[\bar{1}1\bar{2}]$ tensile axis and a $(\bar{1}11)$ foil surface, the easy primary slip system with a (001) cube slip plane is activated. Dislocations on this system move in a very jerky way frequently over distances larger than the frame width so that the dislocations appear or disappear within two successive frames. Figure 7.29 shows the disappearance of the dislocation marked by the arrow between (a) and (b) and the sudden formation of a small loop between (c) and (d). In the second sequence, a slip band proceeds forward always followed by the bending fringe.

Thus, dislocation motion on the cube plane at 1,150°C is obviously controlled by the long-range internal stress fields. It is of athermal nature.

At high temperatures, the dislocation structures become very homogeneous, and consist of a network of dislocations on different slip systems, as demonstrated by Fig. 7.30. This points at the action of recovery-controlled deformation mechanisms.

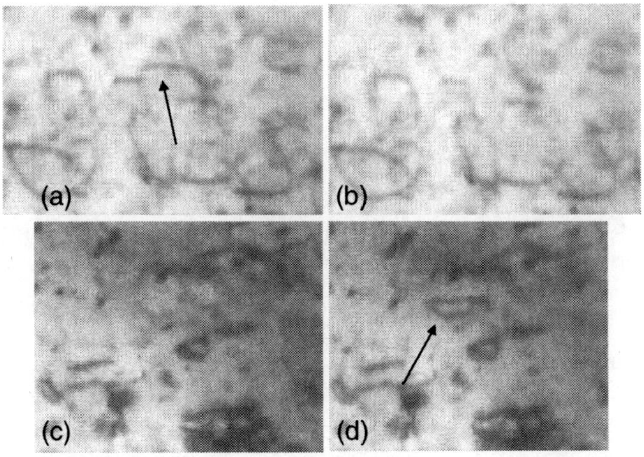

Fig. 7.29. Disappearance of a dislocation marked by an arrow within successive frames (**a**) and (**b**) of the video recording, and sudden appearance of a dislocation loop between (**c**) and (**d**)

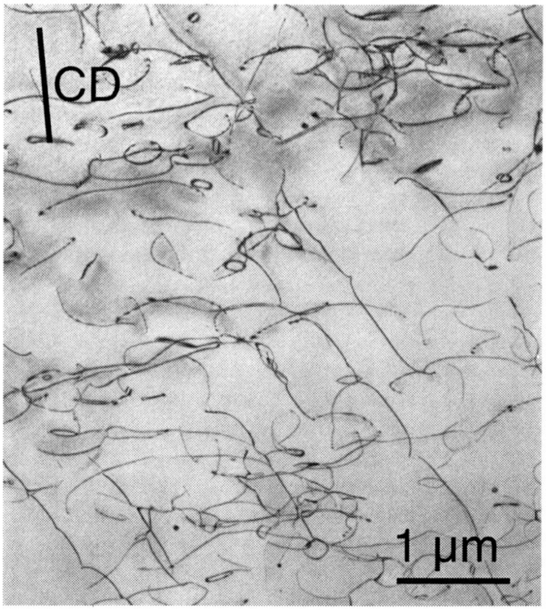

Fig. 7.30. Dislocation structure in c-ZrO_2 after macroscopic deformation along $[\bar{1}1\bar{2}]$ at 1,400°C. Specimen cooled under load. CD projection of compression direction. From the work in [434]

In the hard [100] orientation in which the slip systems with {100} planes have zero orientation factors, dislocations on {111} as well as on {110} planes are activated. Figure 7.31 presents an example with two sets of dislocations on {111} planes. In Fig. 7.31a, both sets are visible, while in Fig. 7.31b the horizontal set is extinguished. At the relatively low temperature of 800°C, the dislocations are pinned by localized obstacles and bow out between them as on {100} planes during the deformation along the $[\bar{1}1\bar{2}]$ loading axis. Some debris marked D is also produced. On the non-cube planes, these structures prevail up to higher temperatures. Above 1,200°C, three-dimensional networks form, which are characteristic of the high-temperature deformation.

Non-cube slip was also studied by in situ straining experiments in the HVEM at 1,150°C. An unusual (211) slip plane was observed in specimens deformed along the soft $[\bar{1}1\bar{2}]$ orientation but with a (110) foil surface, as demonstrated by the following video clip.

Video 7.4. *Dislocation processes in* c-ZrO_2 *on a* (211) *plane during in situ straining along* $[\bar{1}1\bar{2}]$ *at* 1, 150°C*:* The sequence presents dislocation motion in a specimen with a (110) foil surface. The Burgers vector is now $1/2[0\bar{1}1]$ and the trace t is that of the (211) slip plane. The dislocations are smoothly curved at this temperature. Frequently, they move in a jerky way over distances comparable with those between the dislocations. In contrast to the motion on the {001} easy slip plane, the dislocations also move slowly in a viscous manner. Thus, the movement is controlled by both the

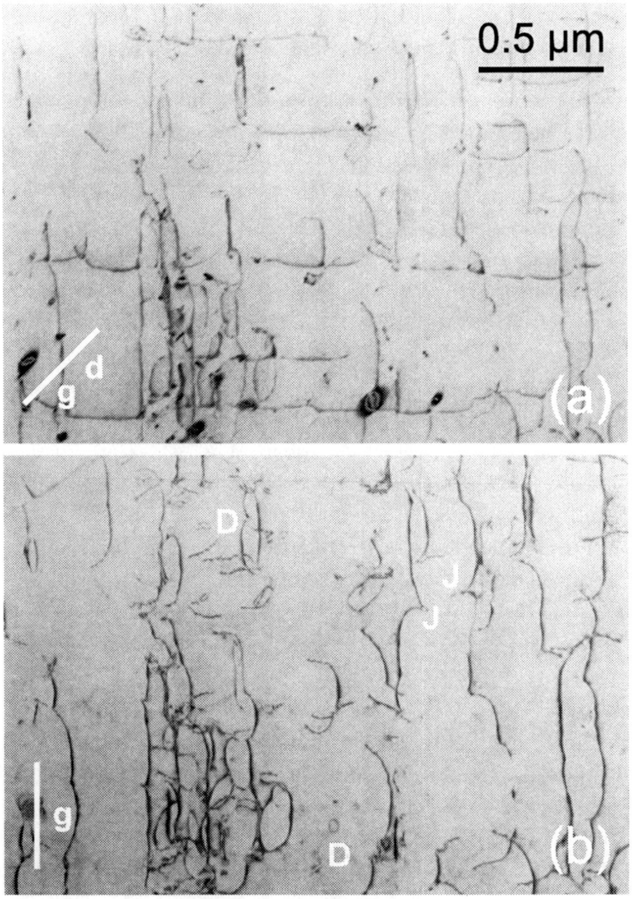

Fig. 7.31. Dislocation structure of c-ZrO_2 deformed along $d = [100]$ at 800°C up to 1.1% imaged under different $\boldsymbol{g}$ vectors at the [001] pole. $\boldsymbol{g} = (200)$ (**a**), $\boldsymbol{g} = (2\bar{2}0)$ (**b**). From the work in [435]

long-range internal stress fields and a viscous friction process. At A, two half-loops meet and partially annihilate each other. As the slip planes are not identical, jogs remain in both dislocations, which impede further motion so that the dislocations bow out. Simultaneously, a jog in a dislocation at B trails a dipole (debris) in low contrast, which later on opens to a wider dislocation loop. The dislocation at C moving to the right changes its slip plane a few times, visible by steps in the lower trace.

One in situ experiment was successful with the hard [100] tensile direction.

Video 7.5. *Dislocation motion in* c-ZrO_2 *on a* $\{10\bar{1}\}$ *plane during deformation along* [100] *at* 1,150°C*:* Loading along [100] activates dislocations with $1/2[1\bar{1}0]$ Burgers vectors to move on {110} planes in a viscous way, similar to the motion on {211} planes. At A, a dislocation multiplies and emits two new dislocations in opposite directions. A dislocation loop is formed at B. Unfortunately, strong contrast changes occur due to corrections of the imaging conditions.

To estimate the athermal stress component, dislocation densities were measured from TEM images taken at several temperatures. In specimens deformed in the soft orientation, the densities amount to about $2 \times 10^{13}\,\mathrm{m}^{-2}$ over a wide temperature range, but they rise drastically below 700°C. In the hard orientation, the densities equal about $10^{13}\,\mathrm{m}^{-2}$. They strongly decrease above 1,200°C. The calculation of the internal stress will be described later.

In conclusion, one may say that dislocation motion is impeded by small precipitates acting as localized obstacles at and above the lowest temperatures where TEM preparation was successful, that is, 500°C for slip on the soft cube plane and 700°C for slip on the {111} and {110} planes. At and above 1,000°C, dislocation segments on the cube planes are long, smoothly curved, and pinned by jogs. They move in a very jerky athermal way. At 1,150°C, dislocations on non-cube planes are also smoothly bent, but they move in a continuous viscous way. At high temperatures, that means, at 1,400°C, the dislocations form three-dimensional networks in all orientations, which suggests that the deformation is controlled by recovery processes.

7.3.3 Dislocation Dynamics in Cubic ZrO_2

The dependence of dislocation velocities on the stress was measured between 1,100°C and 1,450°C by the stress pulse etching technique on dislocations introduced by microhardness indents [21]. While edge dislocations are slightly more mobile at lower temperatures, screw dislocations are faster at higher ones. This is in accordance with the fact that screw dislocations predominate in specimens deformed at low temperatures, whereas edge dislocations prevail at high ones. The stress exponent m' can be obtained from double-logarithmic plots of the dislocation velocity vs. stress, as shown in Fig. 7.32. At 1,435°C,

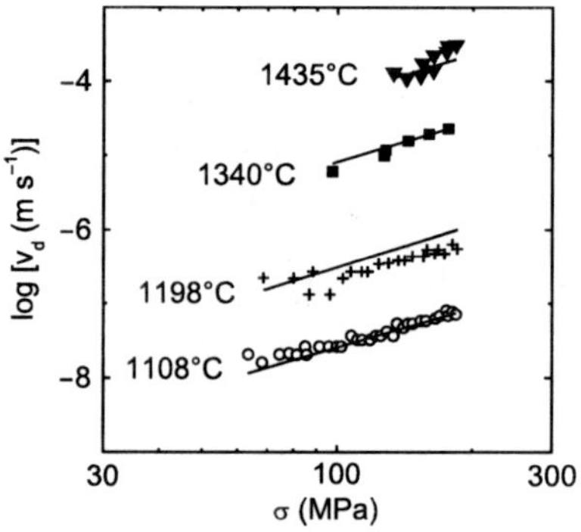

Fig. 7.32. Stress dependence of edge dislocation velocities at different temperatures in ZrO_2–10 mol% Y_2O_3, measured by the stress pulse etching technique. Data from [21]

m' is about 3–5, but at 1,100°C, $m' = 1.4$. The apparent activation energies calculated from Arrhenius plots of dislocation velocities against the reciprocal temperature are of the order of magnitude of 5 eV. The stress range of these measurements touches that of the macroscopic experiments described below only at the high-temperature end. A comparison with the macroscopic data reveals that at high temperatures both the stress exponent and the activation energies are of the same order of magnitude as the data for creep between 1,400 and 1,500°C (e.g., [436]). As it will be described in the following section, these data are interpreted by diffusion-controlled recovery, but not by the dislocation mobility. The low-temperature end of the dislocation velocity data corresponds to the intermediate temperature range discussed below. However, the stress range of the velocity measurements is far below the macroscopic flow stress. Thus, these data are not suited to interpret the macroscopic plasticity data.

7.3.4 Macroscopic Deformation Properties of Cubic ZrO_2

Originally, cubic zirconia single crystals were deformed between 1,200 and 1,500°C (e.g., [437,438]). They were considered brittle below this range. However, careful specimen preparation and a low strain rate of $10^{-6}\,s^{-1}$ enabled the deformation of these crystals in the soft orientation down to 400°C [213, 324] and in the hard orientation down to 500°C [435]. Under confining hydrostatic pressure, the material can be deformed even down to 250°C [214]. In addition to the crystals with 10 mol % Y_2O_3, some experiments were carried out on materials with 15 and 20 mol % Y_2O_3.

The stress–strain curves are mostly smooth except for the materials with higher yttria concentrations where plastic instabilities occur at intermediate temperatures. Some features of the instabilities were discussed before in Sect. 5.3. Particularly at low temperatures, the curves show a strong yield drop effect. The strain hardening coefficient is always low. During almost steady state deformation, the stress–strain curves were interrupted by stress relaxation, strain rate cycling, and temperature change tests. The yield stress σ_y is plotted in Fig. 7.33 as a function of temperature. The yield stresses along $\langle 112 \rangle$ are always lower than those along $\langle 100 \rangle$ except at the highest temperature of 1,400 °C. The 15 and 20 mol% materials, too, have higher yield stresses. The curves show several linear ranges with different temperature sensitivities $\Delta\sigma/\Delta T$. In general, $\Delta\sigma/\Delta T$ is higher for lower temperatures and for the hard $\langle 100 \rangle$ orientation.

Strain rate sensitivities were mostly determined by stress relaxation (SR) tests. The results may be plotted as the logarithm of the relaxation rate vs. the stress. As the relaxation rate is proportional to the plastic strain rate, the reciprocal slope of the curves represents the strain rate sensitivity r, as described above in Sect. 2.1. Normal stress relaxation curves originating from the thermally activated overcoming of obstacles are bent towards the stress axis as plotted in Fig. 7.34a, indicating a decreasing strain rate sensitivity

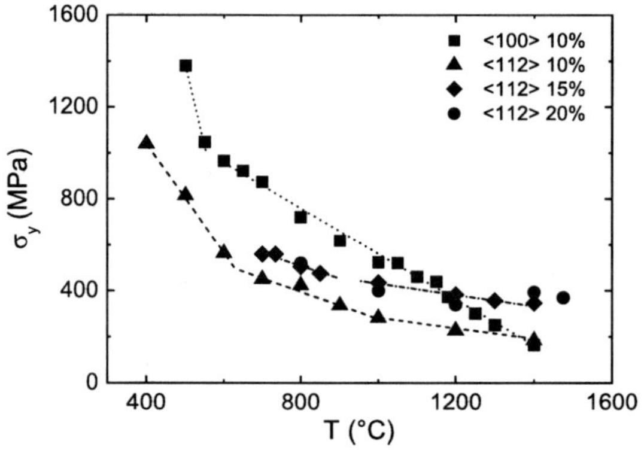

Fig. 7.33. Yield stresses σ_y vs. temperature T. ZrO_2–10 mol% Y_2O_3, $\langle 112 \rangle$ orientation, $\dot{\varepsilon} = 10^{-6}\,s^{-1}$ (*triangles*); $\langle 100 \rangle$ orientation, $\dot{\varepsilon} = 10^{-5}\,s^{-1}$ (*squares*); ZrO_2–15 mol % Y_2O_3, $\langle 112 \rangle$ orientation (*diamonds*); ZrO_2–20 mol% Y_2O_3, $\langle 112 \rangle$ orientation (*circles*). Data from [435]

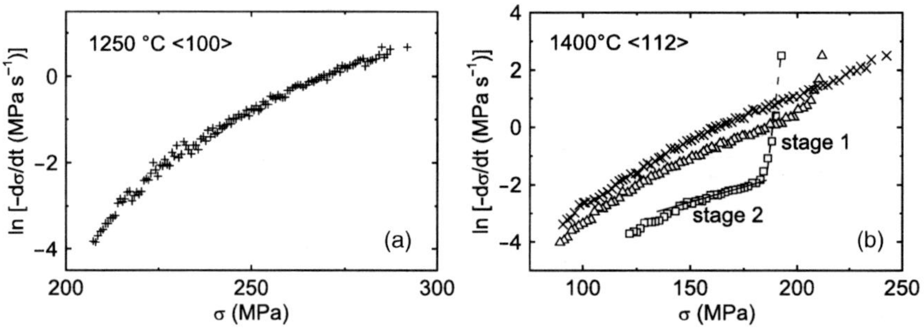

Fig. 7.34. Stress relaxation curves in ZrO_2–10 mol% Y_2O_3. Deformation along $\langle 100 \rangle$ at 1,250°C and $\dot{\varepsilon} = 10^{-5}\,s^{-1}$ (**a**) and along $\langle 112 \rangle$ at 1,400°C and $\dot{\varepsilon} = 10^{-4}\,s^{-1}$ (**b**). $\varepsilon = 0.03\%$ (*squares*), 0.32% (*triangles*), 2.17% (*crosses*). Data from [434, 435]

with decreasing stress. However, in ZrO_2 the curves may also have a different shape being bent in opposite direction. Here, such curves are designated to have an inverse curvature. Frequently, the curves can also be described by two stages, a very steep stage 1 at the beginning at high stress followed by the normal behavior (stage 2) with a much lower slope and usual bending as demonstrated in Fig. 7.34b. Two different processes may cause the inverse curvature. One of them is strain ageing discussed in Sect. 4.11 for range A of Fig. 4.39, and the other one is the action of recovery. The latter takes place in the curves of Fig. 7.34b and will be described in more detail in Sect. 7.3.5. With increasing plastic strain, stage 2 is shifted upwards to higher strain rates so that stage 1 disappears. It disappears also at lower starting strain rates.

The slopes at the beginning of the relaxation curves characterize the strain rate sensitivity corresponding to the strain rate during steady deformation just before the relaxation tests. The results are summarized on a logarithmic scale

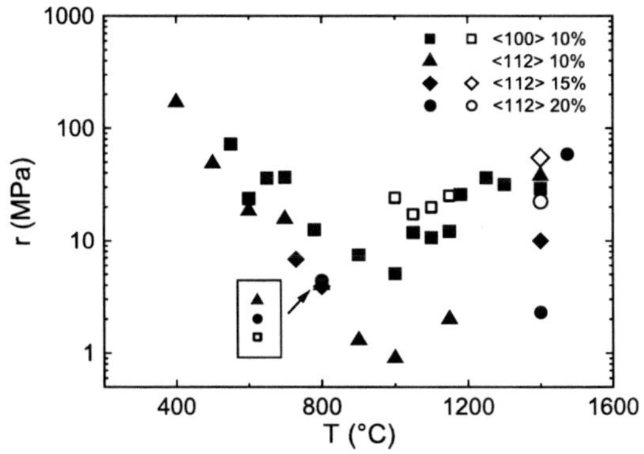

Fig. 7.35. Logarithm of the strain rate sensitivity r of the flow stress as a function of temperature T. Symbols from SR tests as in Fig. 7.33. *Full symbols*: strain rate sensitivity at the beginning of the SR test. *Open symbols*: strain rate sensitivity during stage 2 of the relaxation curves. Data from [435]

in Fig. 7.35. As r depends on the strain, the individual data are extrapolated to a plastic strain of 0.5%. In temperature ranges where the relaxation curves exhibit the two stages, the higher strain rate sensitivities corresponding to the flat stage 2 of the relaxation curves, that is, to strain rates considerably lower than the rate before the relaxation test, are plotted as open symbols. The strain rate sensitivity starts with high values at low temperatures, goes through a minimum near 1,000°C, and increases again at higher temperatures. The data measured along ⟨100⟩ approach those taken along ⟨112⟩ at low and high temperatures. Between about 800 and 1,200°C, they are distinctly higher by a factor of about 5. In the soft ⟨112⟩ orientation, r is very low at 1,000°C, that is, only about 1 MPa. In the instability ranges of the materials with higher yttria concentrations, strain rate sensitivities cannot be determined.

The activation energies calculated from the slopes of the yield stress vs. temperature plots or from temperature change tests combined with strain rate sensitivity data show reasonable values around 3 eV at low temperatures and between 5 and 10 eV at high ones. In the intermediate range around 1,000°C they take very high values, which are not compatible with thermal activation. This results from the low values of the strain rate sensitivity.

7.3.5 Deformation Mechanisms

In the different temperature ranges, the plastic deformation of cubic ZrO_2 single crystals is controlled by a number of mechanisms summarized below.

The Athermal Stress Component

Long-range interactions between the dislocations cause an athermal component τ_i of the flow stress. One contribution to this component is the stress arising from the mutual elastic interaction between parallel dislocations, that is, the Taylor hardening treated in Sect. 5.2.1. Using anisotropic elasticity, the stress contribution τ_p from parallel dislocations can be written as [432]

$$\tau_p = \alpha K b F_m \varrho^{1/2}/(2\pi), \tag{7.13}$$

where α is a numerical constant of about 8, ϱ is the dislocation density, F_m is a dimensionless maximum interaction force, and K is the energy factor of the respective dislocations. K was introduced in (3.40). It can be calculated by anisotropic elasticity theory. In the respective formulae, it replaces the shear modulus. As, except at high temperatures, the dislocations have mostly screw character, K and F_m were selected for screw dislocations. For 700°C, they amount to $K_s = 80.4$ GPa and $F_{ms} = 0.30$ for $\{100\}$ planes ($\langle 112\rangle$ orientation of compression axis) and $F_{ms} = 0.66$ for $\{111\}$ planes ($\langle 100\rangle$ orientation). Using the dislocation densities for ZrO_2–10 mol% Y_2O_3 for the $\langle 112\rangle$ and $\langle 100\rangle$ compression directions mentioned in Sect. 7.3.2, the dependence of τ_p on the temperature is shown in Fig. 7.36. The values are slightly higher for the $\langle 100\rangle$ orientation, owing to the higher F_m value. They are compared with the total (shear) flow stress τ_y calculated from σ_y with the orientation factors of the respective slip systems. In the athermal range around 1,000°C of the $\langle 112\rangle$ specimens, τ_p amounts to about 40 MPa. In situ straining experiments in an HVEM (Sect. 7.3.2) showed that the dislocations move very jerkily, bowing out between large jogs. At 1,150°C, an upper limit of about 75 MPa was estimated of the back stress τ_b resulting from the bowing, which also contributes to the athermal stress component τ_i. Thus, $\tau_i = \tau_p + \tau_b \leq (40 + 75)$ MPa, which just equals the macroscopic flow stress. The athermal nature of the flow stress of ZrO_2–10 mol% Y_2O_3 around 1,000°C along $\langle 112\rangle$ is in agreement with the

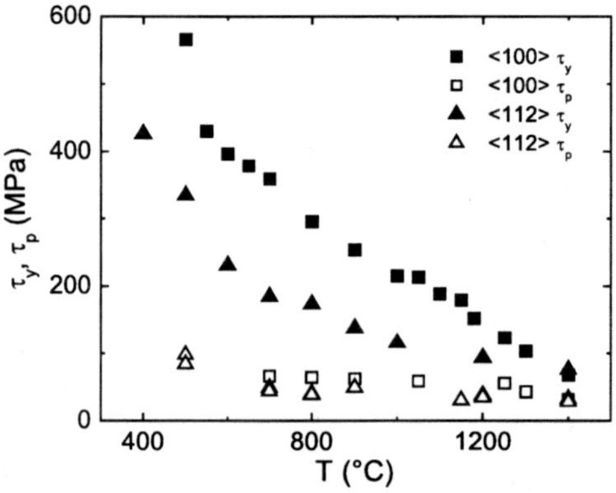

Fig. 7.36. Critical resolved shear stress τ_y (*full symbols*) and athermal flow stress component τ_p from long-range interaction of parallel dislocations (*open symbols*) of ZrO_2–10 mol% Y_2O_3 as a function of temperature. $\langle 112\rangle$ compression axis (*triangles*), $\langle 100\rangle$ axis (*squares*). Data from [435]

low strain rate sensitivity (Fig. 7.35) (or the corresponding large activation volume) and the very high experimental activation energy.

At deformation along ⟨100⟩, in situ straining experiments show dislocations moving continuously on {110} planes also bowing out to similar curvatures as during deformation along ⟨112⟩. It may therefore be concluded that the back stress τ_b and accordingly also the total athermal stress component τ_i assume values similar to those along ⟨112⟩. In this case, however, the (shear) flow stress at 1,000°C is about 200 MPa. Thus, the athermal component amounts to only about 60% of the flow stress. Along the ⟨100⟩ orientation, thermally activated processes influence the flow stress as it is documented by the higher strain rate sensitivity (Fig. 7.35) and the high temperature sensitivity of the flow stress.

It may be concluded that the athermal processes yield an important component τ_i to the flow stress, in particular around 1,000°C. As it increases only slightly with decreasing temperature, it becomes less important at lower temperatures where the total flow stress increases strongly.

Elastic Interactions Between Dislocations and Point Defects

Originally, the dependence of the flow stress of cubic zirconia on the stabilizer content at 1,400°C was interpreted by solution hardening, that is, by the elastic interaction between the dislocations and the yttrium ions, their charge compensating oxygen vacancies or agglomerates of them (e.g., [439]). There is a remarkable difference in the ionic radii of the yttrium and zirconium ions, which may lead to solution hardening owing to the size misfit. Estimates of this contribution to the flow stress applying Mott–Labusch statistics (Sect. 4.5.2) are given in [435]. As shown there, the size misfit interaction may considerably contribute to the flow stress of ZrO_2–10mol% Y_2O_3 over a wide range of temperatures and therefore also to the materials with higher yttria concentrations. Also the order of magnitude of the predicted activation volume fits the range of experimental values. However, this model does not agree with the functional dependencies. The experimental activation volume or the strain rate sensitivity depend on the temperature far more strongly than predicted theoretically. Besides, the strain rate sensitivity should be proportional to the square root of the yttrium concentration instead of being independent of it. In particular, the obstacle distances in the range of 0.1 μm observed by transmission electron microscopy in Sect. 7.3.2 disagree with the very short distances of only a few b between the individual yttrium ions. It may therefore be concluded that direct solution hardening by the solved yttrium ions only weakly contributes to the flow stress.

Agglomerates of yttrium ions and oxygen vacancies form electric and elastic dipoles, which were proved by mechanical loss spectroscopy (e.g., [440]). As the mobility of the oxygen vacancy near the yttrium ion is very high, the relaxation maxima are below 400°C, that is, these defects do not act as fixed obstacles within the range of the present experiments. The respective reorientation processes might lead to the induced Snoek effect in the temperature range of the relaxation maxima, that is, also below the present temperature range. At higher temperatures, the agglomerates dissociate (e.g., [440]).

Precipitation Hardening

Plastic deformation in the temperature range between about 500 and 800°C seems to be controlled essentially by precipitation hardening. The flow stress exhibits a moderate temperature sensitivity (Fig. 7.33). The strain rate sensitivity (Fig. 7.35) corresponds to activation volumes of some tens to some hundreds of b^3. The Gibbs free energy takes reasonable values mostly below 6 eV but its increase with temperature is too strong for a single thermally activated process so that another thermally activated process should be superimposed, probably the Peierls mechanism at low temperatures. The occurrence of precipitation hardening is supported by the bowed-out shape of the dislocations in the respective temperature range (Figs. 7.25, 7.26, and 7.31). The comparison between the activation volume $V = lb\Delta d$ of the order of magnitude of $100\,b^3$ (at about 800°C), determined macroscopically, and the obstacle distance of $l \approx 250\,b$ reveals that the activation distance Δd is a fraction of b. As the precipitates are certainly much larger, this indicates that the effective forces on the obstacles almost reach the obstacle strength so that only the tips of the obstacles are overcome by thermal activation. As shown in Fig. 7.35, the strain rate sensitivity and thus also the activation volume are not influenced by the yttria concentration. Thus, the precipitates should not be connected with the yttria stabilizer. Unfortunately, only a hypothesis can be presented of the nature of the precipitates. According to [441], high concentrations of nitrogen are solved in zirconia crystals owing to their growth in air. At high temperatures, they precipitate to form ZrN particles. As the ZrN precipitates occur at all yttria concentrations studied in [441] and as the crystal material was supplied by the same company as the present one, it is believed that small particles containing nitrogen cause the precipitation hardening in the temperature range between 500 and 800°C.

In [213,324], the temperature dependence of the yield stress and the strain rate sensitivity of the 10 mol% material in the precipitation hardening range are interpreted by a phenomenological obstacle potential suggested in [9]

$$\Delta G(\tau^*) = \Delta G_0 \left(1 - \left(\frac{\tau^*}{\tau_0}\right)^{1/2}\right)^{3/2},$$

with an activation energy at zero stress of $\Delta G_0 \approx 4.2\,\mathrm{eV}$ and of an athermal stress of the obstacle array of $\tau_0 \approx 1,760\,\mathrm{MPa}$. The results for the flow stress are plotted in Fig. 7.37 and for the strain rate sensitivity already in Fig. 4.23. Figure 7.37 contains also low-temperature data from [214] obtained under confining hydrostatic pressure to avoid brittle failure.

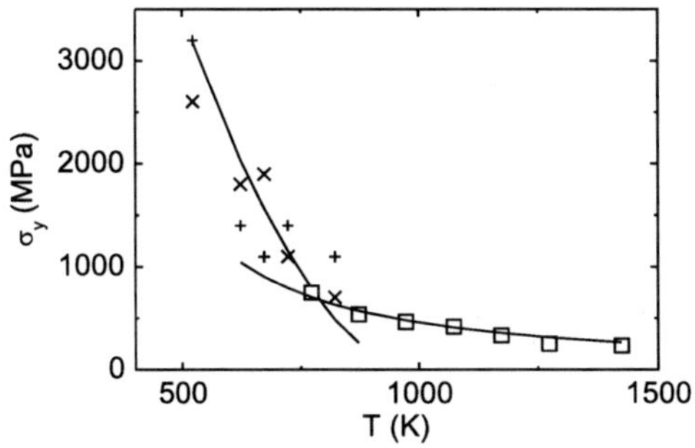

Fig. 7.37. Temperature dependence of the yield stress of cubic ZrO_2 deformed along ⟨112⟩. Experimental data at ambient atmosphere and $\dot{\varepsilon} = 10^{-6}\,s^{-1}$ of ZrO_2–10 mol% Y_2O_3 (*open squares*); at confining hydrostatic pressure and $\dot{\varepsilon} = 2\times10^{-4}\,s^{-1}$ [214], ZrO_2–12.6 mol% Y_2O_3 (*upright crosses*), ZrO_2–10 mol% Y_2O_3 (*tilted crosses*). Data from [213]

The difference in the flow stresses of ZrO_2–10 mol% Y_2O_3 deformed along ⟨100⟩ in comparison to ⟨112⟩ at intermediate temperatures should be related to the dominating thermally activated mechanism, that is, to precipitation hardening, as the strain rate sensitivity is much higher along ⟨100⟩ than along ⟨112⟩. The obstacle distances, however, are practically equal in both directions in agreement with the fact that the dislocations interact with the same obstacles in both orientations. An equal obstacle distance l but a higher strain rate sensitivity along ⟨100⟩ (Fig. 7.35) result in a smaller value of the activation distance Δd. The details of this mechanism are not yet understood well. While for the reference specimens of ZrO_2–10 mol% Y_2O_3 deformed along ⟨112⟩, precipitation hardening ceases in the athermal range around 1,000°C, it still seems to operate at these temperatures for deformation along ⟨100⟩. The contribution of precipitation hardening seems to be independent of the stabilizer concentration. Thus, it does not explain the higher flow stress of the materials with higher yttria concentrations.

The Peierls Mechanism

Below about 500°C, the flow stress of the material with 10 mol% yttria deformed along ⟨112⟩ increases strongly, as demonstrated in Fig. 7.37. This behavior is related to a steep rise of the strain rate sensitivity shown before in Fig. 4.23. In [212, 213], it is interpreted by the transition to the action of the Peierls mechanism. The large strain rate sensitivity below 500°C corresponds to activation volumes below about $10\,b^3$, which are well consistent with the Peierls mechanism. Specimens deformed along ⟨100⟩ also show a steep increase of the flow stress at 500°C, and at low temperatures the activation volumes approach the values in the soft orientation. Thus, the Peierls mechanism is supposed to be also active on the non-cube slip planes at low temperatures.

However, the Peierls mechanism does not seem to be responsible for the difference of the flow stresses for slip on cube or non-cube planes (for the two loading directions) over a wide range of temperatures, as above about 800°C the activation volumes are too large for this mechanism. The specimens with yttria concentrations higher than 10 mol% broke before the range of the Peierls mechanism was reached.

The theory of the transition from the Peierls mechanism to the action of localized obstacles was outlined in Sect. 4.6. It particularly explains the very strong increase in the strain rate sensitivity in the transition region (Fig. 4.23).

Formation of Solute Atmospheres Around Dislocations

While along ⟨112⟩ the ZrO_2–10 mol% Y_2O_3 crystals show an athermal deformation behavior around 1000°C with very low strain rate sensitivities, specimens with 15 and 20 mol% yttria exhibit plastic instabilities discussed in Sect. 5.3. This is usually interpreted by the Portevin–LeChatelier (PLC) effect [346, 347] or dynamic strain ageing, leading to a locking of the dislocations and to an additional contribution to the flow stress (Sect. 4.11).

There are similarities between the load drops of serrations shown in Fig. 5.31 and stress relaxation tests performed in stable ranges near the instabilities. In both cases, deformation takes place at decreasing load with the only difference that during relaxation tests the deformation is driven only by the relaxing elastic deformation of the load train, while at serrated flow the driving rate of the machine is superimposed. In the instability range, stress relaxation curves frequently consist of sudden load drops followed by periods of constant load. At sufficient time resolution, these load drops turn out to be sections of relaxation curves as shown in Fig. 7.38 taken at temperatures around 800°C. When the relaxation rates (slopes of the curves) have reached certain temperature-dependent minimum values marked by the straight lines, the stress remains constant, that is, the relaxation rates turn to zero. In other

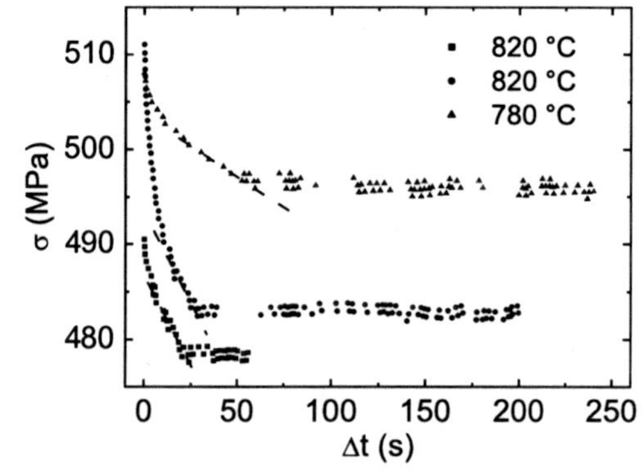

Fig. 7.38. Stress vs. time plots of relaxation tests on ZrO_2–15 mol% Y_2O_3 at $10^{-5}\,s^{-1}$ and 780 and 820°C. Data from [344]

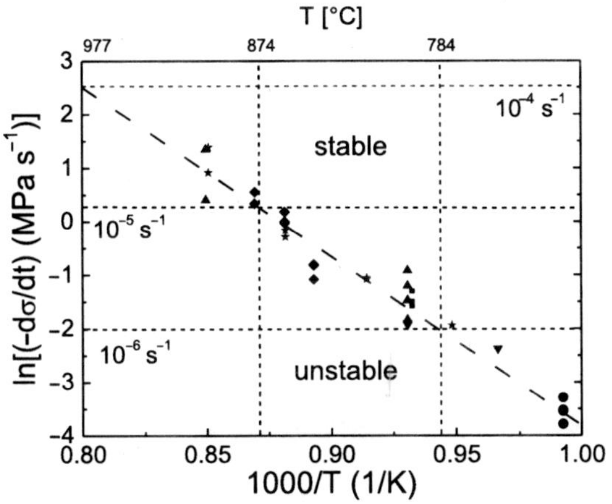

Fig. 7.39. Arrhenius plot of the minimum relaxation rates measured during stress relaxation tests. Horizontal *dashed lines* mark plastic strain rates corresponding to the relaxation rates. Data from [344]

words, below the minimum relaxation rates the dislocations will be totally blocked so that no plastic deformation is possible at lower rates. In the case of serrations, dislocation motion becomes blocked, too, if the strain rate reaches its minimum value, and elastic loading starts again. Figure 7.39 verifies that the minimum strain rates obey an Arrhenius relation with an experimental activation energy of $Q = 2.7\,\text{eV}$. As the minimum deformation rates are the limiting rates below which the deformation is blocked, the Arrhenius line should describe the lower border of the instability range. It actually separates the ranges of stable and unstable deformation.

It is argued in [344] that the plastic instabilities and the yttria concentration dependence of the flow stress between about 800 and 1,200°C are caused by short-range diffusion of unassociated yttrium ions according to the theories of the Cottrell effect and the instabilities outlined in Sects. 4.11 and 5.3. Equations (3.32), (4.99), and (4.100) can be used to estimate the properties of the maximum of the strain ageing curve in Fig. 4.39. Numerical values of the ionic radii of Y^{3+} and Zr^{4+} are 0.102 and 0.084 nm, yielding $(V_s - V_m)/b^3 = 0.0397$, where V_s and V_m are the ionic volumes of the solute and matrix ions. A characteristic dislocation density is $\varrho = 10^{13}\,\text{m}^{-2}$. It may be assumed that the maximum of the curve in Fig. 4.39 is identical with the upper instability border at $T_c \approx 1,400$°C. Then, (4.99) and (4.100) can be used to estimate the necessary diffusion coefficient for the deformation at $\dot{\gamma}_c = 10^{-5}\,\text{s}^{-1}$. It turns out that $D \approx 6 \times 10^{-19}\,\text{m}^2\,\text{s}^{-1}$. This value can be compared with diffusion data, considering that the diffusion coefficient of yttrium should be similar to that of zirconium (e.g., [442]). The value estimated of D fits well that extrapolated between the cation diffusion data from the loop shrinkage study in [443] for ZrO_2–10 mol% Y_2O_3 at lower temperatures and the interdiffusion data [442] at higher ones. Although this does not prove the model, it shows that the diffusion of yttrium ions should be fast enough to cause dynamic strain ageing.

Equation (4.100) gives a rough estimate of the contribution of dynamic strain ageing to the flow stress. Using the same data as above and considering that part of the yttrium ions are associated it follows that $\tau_{\max} \approx 3\,\text{GPa}$. This is one order of magnitude higher than the total flow stress. Thus, defects of much weaker interaction are sufficient to essentially contribute to the flow stress by dynamic strain ageing. The occurrence of dynamic strain ageing may give rise to a flow stress anomaly owing to the maximum in the curve of Fig. 4.39. In zirconia with high yttrium concentrations, the increasing contribution of dynamic strain ageing to the flow stress at temperatures increasing up to 1,400°C may compensate the decreasing contributions of other mechanisms yielding an almost constant flow stress. At temperatures above the upper border of the instability range, fast cation diffusion results in a decrease of the flow stress by both the decreasing action of dynamic strain ageing and the occurrence of dynamic recovery.

The low-temperature border of the instability range is connected with the sharp transition between plastic deformation at a minimum rate and the complete blocking by full ageing of the dislocations during stress relaxation tests. At this transition, the velocity of diffusing solutes becomes equal to the velocity of moving dislocations. Thus, the temperature dependence of the minimum rates demonstrated in Fig. 7.39 is directly connected with the temperature dependence of the underlying diffusion process. There are no cation or solute diffusion data available for temperatures as low as 800°C. The lowest temperature of 1,100°C was reached in the loop shrinkage experiments of [443], yielding an activation energy of 5.3 eV. The substantially lower activation energy of 2.7 eV observed in Fig. 7.39 for the dislocation locking may therefore correspond to short-range diffusion in the dislocation cores. Thus, the character of the diffusion processes may change from short-range diffusion in the dislocation cores at the low-temperature border of the Cottrell effect region to bulk diffusion at the high-temperature one.

Recovery Controlled Deformation at High Temperatures

The deformation of c-ZrO_2 at temperatures around 1,400°C was discussed in detail in terms of recovery-controlled glide in [434] after the original suggestion in [438]. Similar views were also expressed, for example, in [436]. The importance of recovery is documented by the presence of a dislocation network (Fig. 7.30), which may form during multiple slip by the Burgers vector reaction

$$1/2[1\bar{1}0] + 1/2[01\bar{1}] = 1/2[10\bar{1}]$$

and by a decrease of the dislocation density and the corresponding athermal component of the flow stress (Fig. 7.36).

Although in the recovery range the flow stress strongly depends on the temperature, it may consist mainly of two athermal contributions. One is the long-range interaction between parallel dislocations which, after Fig. 7.36, amounts to $\tau_p \approx 30$ MPa for ZrO_2–10 mol% Y_2O_3 deformed along $\langle 122 \rangle$. The other is the back stress τ_b, which the dislocation segments have to overcome to bow out between the nodes of the network, that is, the Frank–Read stress (3.47). With the experimental segment lengths L between 1 and 2.5 µm and the suitable shear modulus, it follows that $\tau_b = 13 \ldots 28$ MPa. Thus, the athermal component of the flow stress of about 50 MPa amounts to about two-thirds of the total flow stress. This is a lower limit as the measured dislocation densities may be underestimated because of the recovery, even for the specimens cooled under load.

The recovery model, in [434] proposed for cubic zirconia, follows a theoretical suggestion by Kocks [444,445] and may particularly explain the two-stage dynamic behavior of the material during stress relaxation tests demonstrated in Fig. 7.34. It is a constitutive model considering both a dynamic law, which controls the mobility of dislocations, and an evolution law of the development of the dislocation structure. The dynamic law can be represented by the power law (4.103) in the form of

$$\dot{\gamma} = \dot{\gamma}_0 \left(\tau/\tau_{\mathrm{th}}\right)^m . \tag{7.14}$$

Here, τ_{th} is a threshold stress, which is related to the long-range stress component $\tau_i = \tau_p + \tau_b$ of the order of magnitude of 50–75 MPa, $\dot{\gamma}_0$ is a constant, and m is the dynamic stress exponent (4.10). It need not be identical with the experimental stress exponent m' defined by (4.12). During straining, the microstructure changes. This is described by an evolution law

$$\mathrm{d}\tau_{\mathrm{th}}/\mathrm{d}\gamma = \Theta_0 \left(1 - R\tau_{\mathrm{th}}\dot{\gamma}^{-1/n}\right), \tag{7.15}$$

where Θ_0 is the initial work-hardening coefficient, and n is the strain-softening exponent. The last term in the parentheses considers dynamic recovery via glide and climb, where R includes the diffusion coefficient. As observed in the loop shrinkage study of [443], the diffusion coefficient strongly depends on the yttria concentration, which explains the dependence of the flow stress on the latter. Under steady state conditions as in creep, τ_{th} is constant so that

$$\dot{\gamma}_{\mathrm{ss}} = \left(R\,\tau_{\mathrm{th}}\right)^n = R'\,\tau^{n'}, \tag{7.16}$$

with $1/n' = 1/n + 1/m$. This is the well known creep law (5.20). In the present experiments at a constant strain rate, the work-hardening is low after the yield point so that (7.16) should also account for the steady state flow stress. Equations (7.14) and (7.15) describe also the transient phenomena as occurring in the stress relaxation experiments. For these experiments, (7.14) and (7.15) can be solved under the (well fulfilled) assumption that the actual stress τ is nearly equal to the threshold stress τ_{th}, yielding

$$\begin{aligned} m' &= \mathrm{d}\ln\dot{\gamma}/\mathrm{d}\ln\tau = \mathrm{d}\ln(-\dot{\tau})/\mathrm{d}\ln\tau \\ &= m - \frac{(m-n)\,A\,(\tau/\tau_1)}{(1-A)\,(\tau/\tau_1)^{m/n} + A\,(\tau/\tau_1)}, \end{aligned} \tag{7.17}$$

with τ_1 being the stress at the beginning of the relaxation and

$$A = \frac{mR\tau_1\Theta_0/S}{(m-n)\,(1+\Theta_0/S)\,\dot{\gamma}_1^{1/n)}}.$$

S is the elastic stiffness defined in Sect. 2.1. At the beginning of relaxation, that is, in stage 1 of the relaxation curve,

$$m' = (\mathrm{d}\ln\dot{\gamma}/\mathrm{d}\ln\tau)_{\tau=\tau_1} = m\,(1-A) + nA, \qquad (7.18)$$

which is close to m for large m (and small A). At the end of the relaxation, that means in stage 2,

$$m' = (\mathrm{d}\ln\dot{\gamma}/\mathrm{d}\ln\tau)_{\tau\ll\tau_1} = n. \qquad (7.19)$$

Thus, the two-stage relaxation curves plotted on a double-logarithmic scale allow one to separate the dynamic dislocation properties from the recovery behavior. Such curves occur also in other materials, for example, in MgO in the high-temperature range. In the following, some features of the deformation behavior described in Sect. 7.3.4 will be discussed qualitatively on the basis of the model.

The stress exponents of the relaxation curves of deformation along ⟨112⟩ at low strains and sufficiently high starting strain rates are about 100 in stage 1 and 4 in stage 2. The high stress exponent in stage 1 approximates the dynamic stress exponent m and describes athermal dislocation motion characteristic also of lower temperatures where the strain rate sensitivity is very low. The strain softening exponent of 4 in stage 2 agrees with the stress exponents derived from incremental load changes performed during creep at small strain rates so that the dynamic stress exponent m could not be observed (e.g., [446]). In the double-logarithmic plot, relaxation curves during deformation along ⟨100⟩ show only a slightly inverse shape with a starting slope of $m' \approx 9$ and a final slope of $m' \approx 6$. This prevents a clear separation of the processes but fits the viscous dislocation motion on non-cube planes.

For deformation along ⟨112⟩, the activation enthalpies at about 1,400°C amount to 5.5–6 eV. This is close to the values of cation self-diffusion in the bulk. For creep processes, diffusion has to take place in both sublattices, but it is controlled by the slower moving species, that is, the cations. Tracer diffusion measurements yielded an activation energy of 4.7 eV [447]. According to the dislocation loop shrinkage study in the lower temperature range between 1,100 and 1,300°C, the activation energy (of climb) is 5.3 eV [443]. Certainly, the latter study fits best the situation of the present experiments. Thus, both the stress exponent and the activation energy of the deformation around 1,400°C are close to the values suggested theoretically for recovery-controlled deformation. While the activation energy of cation diffusion seems to be independent of the stabilizer ions and concentration, the diffusion coefficient decreases by more than one order of magnitude by increasing the yttria concentration from about 10 to about 20 mol% [443, 447]. This dependence explains the increased flow stress of the materials with higher yttria concentrations where recovery is suppressed. At high temperatures, ZrO_2–10 mol% Y_2O_3 exhibits equal flow stresses at the different loading axes as recovery effects are independent of the loading axis.

7.3.6 Summary of Cubic ZrO_2

At all temperatures, part of the flow stress is of athermal nature due to long-range interactions between dislocations and to the back stress of segments bowing out between large jogs or dislocation nodes. This stress part controls the athermal deformation of ZrO_2–10 mol% Y_2O_3 along $\langle 112 \rangle$ around 1,000°C. In all the other situations with higher flow stresses, other mechanisms are superimposed.

At the lowest temperatures where data are available, the Peierls mechanism causes a steep increase of the flow stress and the strain rate sensitivity for glide on both the cube slip planes (activated in the $\langle 112 \rangle$ orientation) and the non-cube slip planes (activated in the $\langle 100 \rangle$ orientation).

Although solution hardening by the yttrium ions may theoretically yield a large contribution to the flow stress, the temperature range between 500 and 800°C with activation volumes up to more than 100 b^3 and dislocations pinned between obstacles about 130 nm apart can best be interpreted by the action of precipitation hardening. According to equal activation volumes of the materials with different yttria contents deformed along the soft $\langle 112 \rangle$ orientation, this mechanism does not cause the dependence of the flow stress on the yttria concentration at intermediate temperatures. It should, however, be responsible for the higher flow stress and the smaller activation volume of non-cube slip at the deformation along $\langle 100 \rangle$.

The formation of yttrium ion atmospheres in the dislocation cores may give rise to the higher flow stress of the materials with higher yttria contents due to strain ageing with serrated yielding at temperatures around 1,000°C.

Around 1,400°C, deformation is controlled by recovery. As the diffusional recovery processes are little influenced by the slip system, the flow stress is equal for the deformation of ZrO_2–10 mol% Y_2O_3 along $\langle 112 \rangle$ and $\langle 100 \rangle$. On the other hand, the decrease of the cation diffusion coefficient with increasing yttria concentration results in the higher flow stress of the materials with higher yttria contents at 1,400°C owing to reduced recovery.

7.3.7 Tetragonal ZrO_2

As outlined in the introduction and in Sect. 7.3.1, materials containing the tetragonal phase of zirconia form a variety of microstructures.

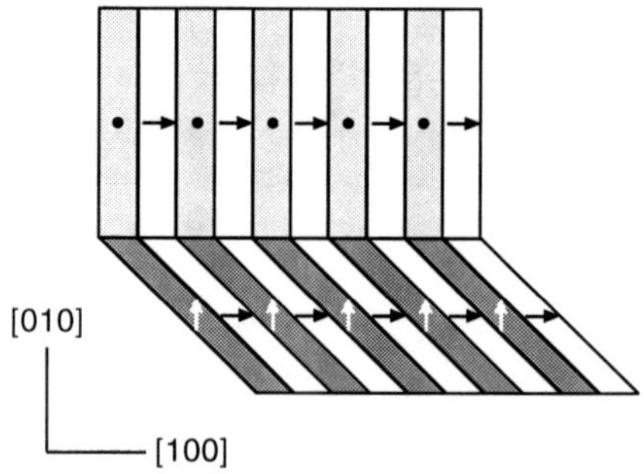

Fig. 7.40. Tetragonal domains with their c axes in different orientation forming two colonies of different composition of domains. The *upper* colony consists of the variants t_1 and t_3, the *lower* one of t_1 and t_2

Microstructure

The tetragonal phase of the different microstructures consists of individual domains with their c axes parallel to the three $\langle 100 \rangle$ axes of the corresponding pseudo-cubic lattice. The domains are designated t_1, t_2, and t_3. Domains of alternating c axes are stacked to form colonies (e.g., [429]) as outlined in the schematic picture of Fig. 7.40. The domain boundaries are the $\{101\}$ twin planes. The colonies grow preferentially in $\langle 111 \rangle$ directions. Owing to the small deviations from orthogonality of the c axes, always three colonies of equal growth direction, consisting of the different combinations of the t_1, t_2, and t_3 domains, form helical arrangements, where the boundaries between the individual colonies are again $\{101\}$ twin planes [448]. The colonies in the metastable t′ zirconia are supposed to have a structure similar to that of the tetragonal precipitates in partially stabilized zirconia. In the $\langle 001 \rangle$ projection of Fig. 7.40, the domain boundaries of the upper colony are inclined by 45° with respect to the image plane, while in the lower colony they are oriented edge-on. Respective micrographs are presented in Fig. 7.41.

Ferroelastic Deformation

Under stress, the tetragonal phase may undergo ferroelastic deformation [427–429, 433, 450]. During tensile loading along a pseudo-cubic axis, domains of c axes different from the loading axis may switch into that type with the c axis in the tensile direction, forming a tetragonal single crystal. In compression, that type of domains with the c axis parallel to the compression axis may disappear. Domain switching in t′-ZrO_2 recorded during in situ straining in an HVEM is illustrated in the following video clip.

Video 7.6. *Domain switching in* t′-ZrO_2 *during ferroelastic deformation at* 1,150°C*:* The video presents two short sequences of domain switching. The untransformed areas appear in strong contrast owing to the electron microscopy contrast

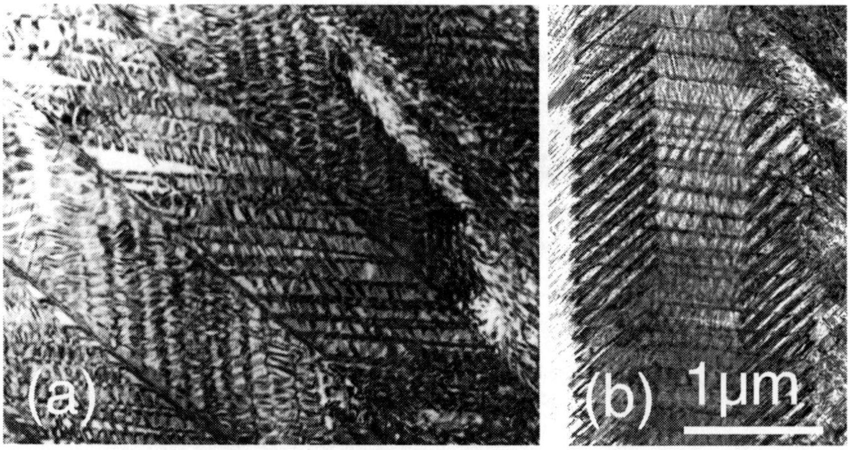

Fig. 7.41. Domain and colony structure in t′-ZrO_2. The edges of the images are parallel to the cube axes. Domain boundaries inclined with respect to the image plane and colony boundaries edge-on (**a**), colony borders inclined (**b**) corresponding to the *lower* colony in Fig. 7.40 rotated by 90°. From the work in [449]

of the domain boundaries and the orientation differences between the domains. The contrast disappears during ferroelastic deformation when tetragonal single crystal regions are formed. The transformation front moves preferentially in the direction of the extension of the colonies rather than by the sidewise growth of the width of the latter. Frame-by-frame analysis reveals that the individual domains transform mostly in time intervals below the resolution of the video recording. This points at an athermal character of the transformation process.

A partially transformed structure is also shown in Fig. 7.42. If the transformation starts simultaneously at different points, the individual transformation fronts will meet leaving behind antiphase boundary-like faults in the oxygen sublattice. These residual defects are frequently located at the former colony boundaries.

During compression, the transformation occurs in such a way that the domains with their c axes parallel to the compression direction always assume the character of their neighboring domains in the same colony. Thus, the neighboring domains become tetragonal single crystal regions, and the high-energy pseudotwin boundaries, where a c direction in one domain has to join an a direction of another one, are generated only along the narrow areas along the former colony boundaries.

The macroscopic behavior of t′-ZrO_2 during compressive loading along a pseudo-cubic axis is demonstrated by the stress–strain curve in Fig. 7.43. The elastic range finishes at a well defined coercive stress σ_t. There follows a range of constant stress where the ferroelastic transformation takes place. As the tetragonality of t′-ZrO_2 amounts to only about 1%, the maximum ferroelastic strain in compression is about 1/3%. After the ferroelastic range, a second

Fig. 7.42. Partially transformed colony structure in t′-ZrO_2 after in situ straining in an HVEM at 1,150°C. From the work in [450]

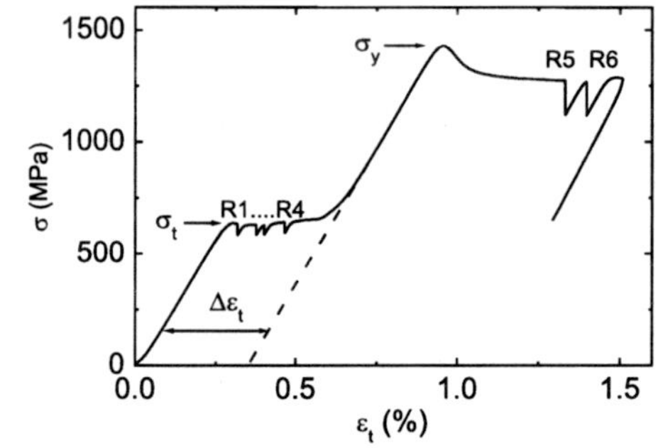

Fig. 7.43. Stress–strain curve of ferroelastic and plastic deformation of t′ zirconia at 500°C. ε_t total strain, $\Delta\varepsilon_t$ transformation strain, σ_t coercive stress, σ_y plastic yield stress, $R_1 \ldots R_6$ stress relaxation tests. Data from [449]

elastic range appears followed by plastic deformation. Below about 1,200°C, the coercive stress is lower than the yield stress of cubic ZrO_2 in the same orientation. At 1,400°C, it is higher. The strain rate sensitivity of the coercive stress is always lower than that of the flow stress of cubic zirconia underlining the athermal character of the transformation.

Plastic Deformation

Dislocations with $1/2\langle 110\rangle$ Burgers vectors as in the cubic phase are complete dislocations in the tetragonal phase only if the Burgers vectors lie in the basal plane. Otherwise, they are partial dislocations that should trail stacking faults in the oxygen sublattice with antiphase boundary-like electron microscopy contrast. Dislocations crossing a domain boundary should create a dislocation in the boundary if their Burgers vector is not parallel to the boundary to fulfil the Burgers vector conservation rule.

As discussed in [449], the microstructure after ferroelastic compressive deformation consists of continuous regions of one tetragonal variant with their c axis perpendicular to the compression axis, in which domains of the size of the former colonies of the respective other variant are embedded. Figure 7.44 shows such regions in bright and dark contrast. After plastic deformation parallel to the c axis of t_2, two sets of dislocations are formed with Burgers vectors parallel to [011] and $[01\bar{1}]$ as illustrated in Fig. 7.44. The dislocations

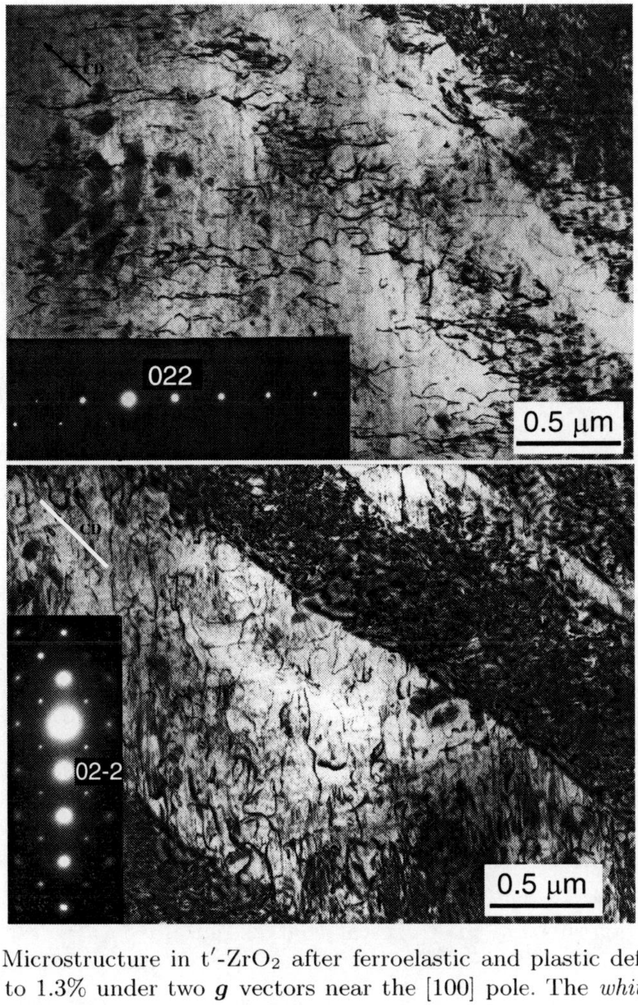

Fig. 7.44. Microstructure in t′-ZrO_2 after ferroelastic and plastic deformation at 1,150°C up to 1.3% under two $\boldsymbol{g}$ vectors near the [100] pole. The *white bar* marks the compression direction. From [449]. Copyright Amer. Ceram. Soc. (2001)

probably extend on {111} planes and are strongly bowed-out with segment lengths of the order of 0.1–0.15 μm. Dislocations of different activated slip systems pin each other by mutual intersections.

The plastic yield stress of t′-ZrO_2 is very high at low temperatures (almost 1,400 MPa at 500°C), decreasing down to about 650 MPa at 1,400°C. Except at 500°C, the strain rate sensitivity is low (10 MPa). The flow stress is interpreted by the sum of the long-range interaction between parallel dislocations according to (7.13) with a dislocation density of $5 \times 10^{13}\,\mathrm{m}^{-2}$, and the back stress of the segments bowing out between dislocation intersections given by the Frank–Read stress (3.47). In agreement with the low strain rate sensitivity, the deformation is mainly of athermal nature. The high flow stress at high temperatures and the accompanying low strain rate sensitivity are explained by recovery being prevented [449].

Partially Stabilized Zirconia

Manifold microstructures may form in partially stabilized zirconia depending on the stabilizer concentration and the thermal treatment. Two structures are shown in Fig. 7.45. The first one consists of medium-sized platelet-like colonies of only two t domain variants of approximately equal thickness embedded in a matrix, which is supposed to be cubic. The other one contains few relatively large precipitation colonies and a high density of small precipitates regularly arranged in a so-called tweed structure. Both elements of the microstructure have a similar morphology on different size scales.

In situ straining experiments in an HVEM on the material of Fig. 7.45b proved that ferroelastic deformation occurs also in partially stabilized zirconia. In general, the untransformed regions show dark contrast under the imaging conditions applied. The figure taken during in situ straining shows these dark untransformed regions and brighter ones of the tweed structure which had already been transformed. The following video demonstrates the process of transformation for both the tweed structure and for individual domains in larger colonies.

Video 7.7. *Domain switching in partially stabilized* ZrO_2 *during ferroelastic deformation at* 1,150°C*:* The video clip contains three sequences of domain switching. The first one shows the partly transformed tweed structure and a few large precipitates with distinct domains. During straining the untransformed dark regions shrink by the motion of the well defined borders between untransformed and transformed regions. Besides, some domains of the large precipitates transform after label A. The second sequence presents again the shrinkage of the untransformed regions with small precipitates, and the third one illustrates the switching of a large domain.

Consequently, in this material under tension the dislocations move after ferroelastic deformation in a crystal that essentially is a tetragonal single crystal.

Video 7.8. *Dislocation motion in partially stabilized* ZrO_2 *after ferroelastic deformation at* 1,150°C*:* After partial transformation the dislocations move in the

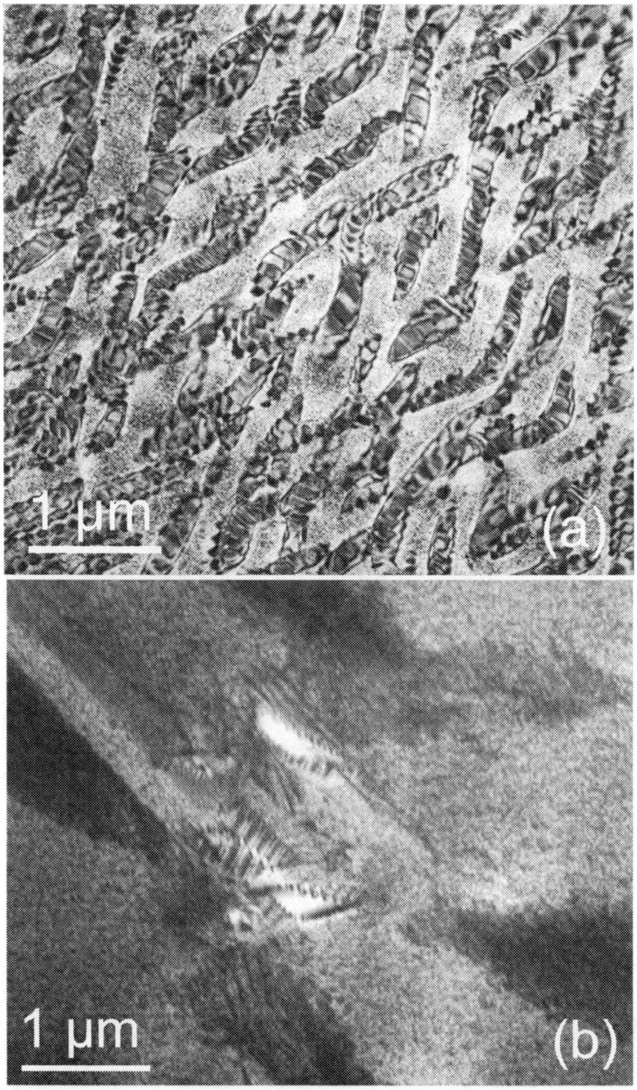

Fig. 7.45. Microstructure in partially stabilized zirconia. Medium-sized t-ZrO_2 precipitate colonies in a cubic matrix (**a**), tweed structure of small precipitates and large domains and colonies (**b**). From the work in [451] (**a**) and [433] (**b**)

transformed regions, in the second sequence of the video near larger precipitates. The motion is quite jerky pointing at athermal dislocation motion as in t′ zirconia.

Models of the strength of partially stabilized zirconia have been developed on the basis of grain boundary hardening. They are reviewed in [433]. These models do not consider the ferroelastic transformation, which may precede plastic deformation. However, ferroelastic deformation has not yet been proved in partially stabilized zirconia with larger precipitates like those in Fig. 7.45a.

Summary of Tetragonal Zirconia

The tetragonal phase of zirconia materials consists of domains of always two of the three tetragonal variants in regular stacking to form larger colonies.

In t′ zirconia and in some microstructures of partially stabilized zirconia, ferroelastic domain switching precedes plastic deformation. As a result, the original domain structure is destroyed, with tetragonal single crystal regions with residual defects being formed, having at least the size of the former colonies. It is not yet clear whether ferroelastic transformation takes place in all microstructures of partially stabilized zirconia, or not.

During subsequent plastic deformation, the dislocations move in an athermal way. The very high yield stress at high temperatures, which is a desired design property, results from the suppression of recovery.

8

Metallic Alloys

The alloys described in this chapter are hardened either by precipitates or by extrinsic particles. As the concentration and size of these obstacles are low, the dislocation motion can be described by the theory of localized obstacles (Sect. 4.5). In the special case where the hardening mechanism inside the particles consists in the formation of antiphase boundaries, as in Al–Li, the matrix dislocations may move in pairs. Impenetrable obstacles like oxide dispersoids are circumvented by the formation of Orowan loops at low temperatures (Sect. 4.7). At high temperatures, they are overcome by climb. However, there may be an attractive interaction between the particles and the dislocations, resulting in a threshold stress for plastic deformation even at relatively high temperatures.

8.1 Precipitation Hardened Aluminium Alloys

Aluminium has the f.c.c. structure with $\{111\}\,1/2\langle 110\rangle$ slip systems as treated in detail in Sect. 3.3. Pure aluminium is very soft at room temperature as the Peierls stress in these close packed metals is important only at very low temperatures. For practical application, aluminium-base alloys are mostly hardened by precipitate particles of other phases (for a review see Nembach [225]). Aluminium alloys are frequently solution treated, that is, they are heated for a few hours at about 500°C to solve the additions, followed by quenching down to room temperature. Afterwards, they are aged at temperatures between room temperature and up to some 200°C to precipitate the particles. During ageing, the particle diameter grows following a $t^{1/3}$ rule, where t is the ageing time. The large particles grow at the expense of the smaller ones. Thus, the concentration of particles decreases during ageing. As the particle strength no longer increases when the particles are surmounted by Orowan looping, the yield stress experiences a maximum at a certain ageing time. This state is called peak-aged. Alloys with lower yield stress owing to shorter ageing times are under-aged, while those with longer ageing times are over-aged.

In several aluminium alloys, the particle microstructures are very well studied so that the yield stresses of the materials can be modeled by the theories discussed in Sects. 4.5 and 4.7. However, by applying conventional methods, one often cannot decide whether the precipitates considered are the relevant obstacles to dislocation glide controlling the flow stress or not. In situ straining experiments in an HVEM, which reveal the morphology and kinematics of the dislocations overcoming the particles, may help answer these questions. The relevant obstacles need not have the density and strength analyzed by other methods. The quantitative data that can be obtained from the in situ experiments are used to check the predictions of the models.

8.1.1 Al–Zn–Mg

The alloy Al-4.54 wt% Zn-1.18 wt% Mg is a material for technical application. For the described experiments [226], specimens were heat treated to grow particles of the η' phase which are large enough to be imaged in the HVEM as small dark dots shown before in Fig. 4.30. These semicoherent particles are supposed to be visible mainly due to structure factor contrast. Under the imaging condition with higher-order reflections, almost all particles should be visible, and the size of the contrast dots should correspond to the extensions of the particles. The particles have an average diameter of $D = 10.5\,\mathrm{nm}$ and an average volume of $V = 1.06 \times 10^{-24}\,\mathrm{m}^3$. They fill a fraction of $f = 0.33\%$ of the volume. Considering the disc shape of the particles, this is by a factor of more than 10 less than the atomic concentration of the Zn and Mg being added. As a measure of the obstacle density, the square lattice distance defined in (4.48) can be calculated by

$$l_{\mathrm{sq}}^{\mathrm{th}} = \sqrt{\pi/(6f)}\,D. \tag{8.1}$$

From the above data, it follows that $l_{\mathrm{sq}} = 170\,\mathrm{nm}$.

The in situ straining experiments (Fig. 4.30) show dislocations pinned at localized obstacles and strongly bowing out between them. These configurations correspond to the so-called stress determining configurations mentioned in the theory in Sect. 4.5. For measuring the obstacle distances, all cusps in the dislocation lines are considered. The distances obey the typical asymmetric shape of the theoretical distribution of Fig. 7.10 with an average value of $l = 62\,\mathrm{nm}$. Effective stresses were determined from the curvature of the bowed-out dislocation segments using the loop fitting method described in Sect. 7.2.2, yielding the reciprocal major half-axis $S = e^{-1} = 16\,\mu\mathrm{m}^{-1}$ of the fitting ellipses as well as the effective stress τ^* according to (7.3). In contrast to the evaluation for MgO, where curvature data were determined from all segments, only clearly imaged segments were selected in the aluminium alloys, because of the reduced quality of the dislocation images due to the

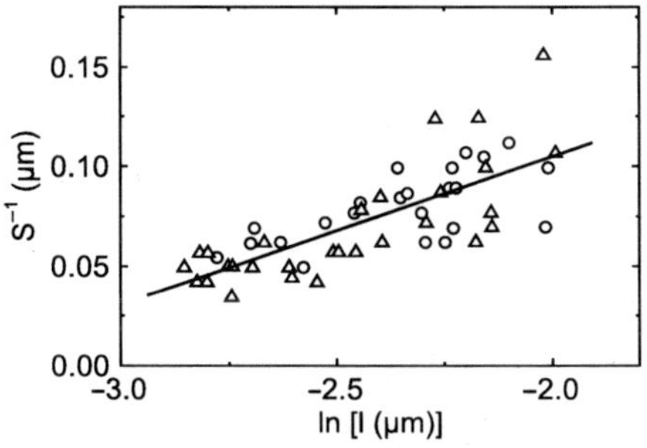

Fig. 8.1. Dependence of the dislocation bowing S on the logarithm of the segment length l in Al–Zn–Mg. Data from [115]

superposition of the contrasts of the precipitates. This selection of segments will certainly prefer strongly bowed segments, leading to higher values of τ^*. The dependence of the dislocation curvature on the logarithm of the segment length, predicted by the theory in Sect. 3.2.7, was proved also for this alloy, as illustrated in Fig. 8.1. The constant in (3.45) amounts to $C = -4.6$, in agreement with the results for MgO. Nevertheless, the theoretical constant $C = -1.61$ was used for the evaluations. These methodical remarks hold also for the other aluminium alloys discussed later.

With (7.6), it follows from l and S that $l_{sq} = 60\,\text{nm} \approx l$. Accordingly, the square lattice distance measured directly at the pinned dislocations l_{sq} is only one third of the value l_{sq}^{th} following from the concentration and size of the precipitates visible in the electron microscope. This means that the dislocations "feel" about ten times more obstacles than the η' particles. The latter grow from Guinier–Preston zones, which are still present in the material. They are strong obstacles due to their stress field but they are not visible in the HVEM micrographs. They were not considered important before the in situ experiments were carried out. As the dislocations move very jerkily over distances long compared to the obstacle distances, most of the dislocation configurations are unstable or $j \approx 1$ in Fig. 4.18. As a consequence, the normalized stress τ' following from the dislocation curvature is only slightly lower than the athermal strength τ_0' of the obstacle array. The overcoming of the obstacles is then more or less of athermal nature. The obstacle data, together with the data of the other Al alloys, are summarized in [115] and in Table 8.1. The values of the effective stress τ^* are partly very high owing to the theoretical value of the constant $C = -1.61$ used in (3.45). It may be concluded from the in situ experiments on the Al–Zn–Mg alloy that the concentration of the precipitates is strongly underestimated from the size and density of particles visible in the electron micrographs. The obstacles are quite strong and are overcome in an athermal way.

The flow stress component from precipitation hardening is certainly not the only contribution to the flow stress of the order of magnitude of $\sigma = 200\,\text{MPa}$. A stress component due to Taylor hardening from long-range elastic interactions between dislocations is estimated via (5.11) to amount to $\tau_i \approx 25$ MPa at a dislocation density of $5 \times 10^{13}\,\mathrm{m}^{-2}$. Further information on the deformation processes can be inferred from the macroscopic deformation properties. Figure 8.2a presents the dependence of the strain rate sensitivity r measured by strain rate cycling tests on the strain, with the strain rate as a parameter. In general, the r values of about 1 MPa are very low resulting in high stress exponents $m' = \sigma/r$, which are characteristic of athermal deformation. The activation volume V of about $450\,b^3$ fits the segment length of $l \approx 220\,b$, yielding an activation distance of $\Delta d = V/(b\,l) \approx 2\,b$. The r values show the usual increase with strain due to work hardening. The interesting point is the inverse dependence on the strain rate, which also holds for

Table 8.1. Parameters characterizing the interaction between dislocations and precipitates in Al alloys

	Al–Zn–Mg	Al–Ag, 6 h	Al–Ag, 9 h	Al–Li, leading disls.
l (nm)	62	111	117	56
S (μm^{-1})	16	11.2	14.9	17
l_{sq} (nm)	60	≈110	≈110	53
l_{sq}^{th} (nm), CTEM	170	≈20	≈20	34
τ'	0.48	0.62	0.87	0.45
τ^*(MPa) with $C = -1.61$	59	45	60	59
estimated τ'_0	0.5	0.65	0.9	0.5
C	−4.6			−5.6

The lines are: segment length, reciprocal major half axis of ellipses fitting bowed-out dislocation segments, square lattice distance from in situ experiments, square lattice distance from conventional transmission electron microscopy (CTEM), normalized effective stress, effective stress, normalized athermal strength of obstacle array, constant in (3.45)

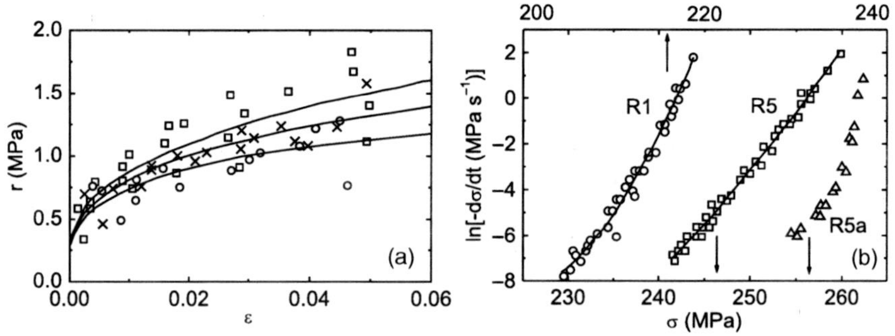

Fig. 8.2. Strain rate sensitivity of the Al–Zn–Mg alloy. Dependence of the strain rate sensitivity measured by strain rate cycling tests on the strain with the strain rate as parameter (**a**). $\dot{\varepsilon} = 4\times10^{-6}\,\mathrm{s}^{-1}$ (*squares*), $4\times10^{-5}\,\mathrm{s}^{-1}$ (*crosses*), $4\times10^{-4}\,\mathrm{s}^{-1}$ (*circles*). Stress relaxation curves (**b**). R5a is a repeated relaxation after R5. Data from [226]

the flow stress itself. This inverse dependence is also obvious from the course of stress relaxation curves plotted in Fig. 8.2b, with their abnormal bending away from the stress axis. As described in Sect. 4.11, the inverse dependence of r on $\dot{\varepsilon}$ may result from diffusion processes in the dislocation cores and corresponds to range A in Fig. 4.39. These effects are also manifest as strain ageing when the deformation machine is stopped and as the strong reduction of the relaxation rate at the repeated relaxation R5a with respect to the original relaxation R5. As described in Sect. 2.1, differences between an original relaxation curve starting from continuous deformation and a repeated one started before steady deformation is reached, again document changes in the microstructure during the first relaxation. According to [452], a great part of the additions remain in solution and can associate with vacancies [453] to form mobile elastic dipoles, which may generate Snoek atmospheres around the dislocations and contribute to the flow stress. Thus, the combined use of microscopic and macroscopic methods is necessary to gain information on the different processes controlling the flow stress.

8.1.2 Al–Ag

Al–Ag alloys are a model system to investigate the decomposition of supersaturated binary alloys and the influence of the resulting obstacle spectra on the mechanical properties. Very detailed studies have been performed on the decomposition kinetics and the influence of the size and density of the precipitates on the yield stress of an Al-1 at% Ag alloy isothermally aged at 413 K after a suitable solution treatment [454, 455]. As a result, the material shows a remarkable anomaly of the ageing kinetics during 10 h, where a primary increase in the flow stress is followed by a drop and a re-increase. As the silver-rich particles have neither a marked lattice constant misfit nor a modulus misfit, stacking fault hardening is considered the main strengthening mechanism [456]. The primary increase was interpreted on the basis of the size and density of precipitates determined by high-resolution electron microscopy [454]. However, in situ straining experiments in an HVEM [283] showed that the basic assumption was not fulfilled that all precipitates visible in the electron microscope represent the relevant obstacle spectrum.

The in situ study was aimed at observing differences in the dislocation behavior between the under-aged state after ageing at 413 K for 6 h and the peak-aged state after 9 h. The basic features of the dislocation structure during in situ straining are shown in Fig. 8.3. Similar micrographs were obtained for both ageing states. As in Al–Zn–Mg, the dislocations exhibit a curly shape, bowing out between localized obstacles. Such micrographs allow the obstacle distances l and l_{sq} to be determined as well as the curvature data S and τ' all listed in Table 8.1. Included in the table are also the square lattice distances $l_{\mathrm{sq}}^{\mathrm{th}}$ calculated from high-resolution TEM. Different from Al–Zn–Mg, the relevant obstacle distances along the dislocations are now about 5 times larger than those predicted from the size and density of contrast dots in the

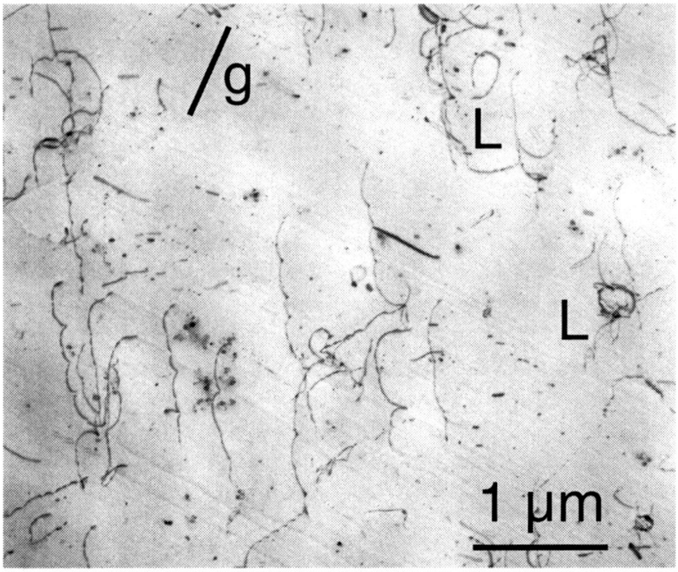

Fig. 8.3. Dislocation structure in an Al-1 at% Ag single crystal aged at 413 K for 9 h, taken during in situ deformation in an HVEM at room temperature. Dislocation loops L. $\boldsymbol{g} = [600]$ parallel to tensile axis, $(0\bar{1}1)$ foil normal. From the work in [283]

high-resolution micrographs. Thus, only part of the precipitates are active as pinning points along the dislocations. While the square lattice distances are almost equal for the two annealing states, the normalized stress for the peak-aged state is clearly higher than that of the under-aged state. These stresses are all very high, indicating that the precipitates are strong obstacles.

The dynamic behavior of the dislocations is demonstrated in the following video clip.

Video 8.1. *Dislocation motion in an* Al-1 at% Ag *single crystal at room temperature:* The clip consists of four short sequences of the same specimen area, demonstrating the pinning of dislocations by the small precipitates. The dislocation motion is very jerky over distances far longer than those between the precipitates. Frequently, dislocation debris and larger dislocation loops are produced very suddenly. The appearance and disappearance of a large loop is indicated by a green circle to the left of the site of the event.

As discussed for Al–Zn–Mg, the jerky motion over larger distances means that τ' is only slightly smaller than the strength of the obstacle array τ_0'.

As mentioned earlier, stacking fault strengthening is widely accepted as the main hardening mechanism in two-phase Al–Ag alloys [456, 457]. Owing to the difference between the stacking fault energies γ_{sf} of aluminium (200 mJ m^{-2}) and silver (20 mJ m^{-2}), a dislocation that encounters a (silver-rich) precipitate will split into two partials enclosing a stacking fault. Following [456], the contribution to the critical resolved shear stress τ_{y} is given by

$$\tau_{\mathrm{y}} = \frac{F_0}{b\,l} = \frac{\Delta\gamma_{\mathrm{sf}}\,d}{b\,l}, \tag{8.2}$$

with the quantities defined in Sect. 4.5 (F_0 maximum interaction force between dislocation and precipitate, l mean distance between obstacles in the slip plane). $\Delta\gamma_{\mathrm{sf}}$ is the difference between the stacking-fault energies of the precipitate and the matrix, and d is the effective length of the dislocation in the particle. The equation holds for the athermal surmounting of the obstacles, which is in agreement with the large slip distances observed in the video clips. The effective length d of the dislocation in the particle depends on the relation between the particle radius and the splitting widths of the dislocation in the matrix as well as in the particle.

The result of $l_{\mathrm{sq}} \approx 5\,l_{\mathrm{sq}}^{\mathrm{th}}$ shows that the dislocations are not interacting with all particles present. It seems likely that a small number of strong obstacles governs the hardening behavior of this alloy. This view is supported by the fact that the curvature of the dislocation segments between the obstacles can explain the whole flow stress. It is well-known that the precipitates consist of ε and η particles (e.g., [458]). Al-Kassab and Haasen [459] applied a field ion microscope-atom probe to investigate an Al-3 at% Ag alloy, isothermally aged at 413 K for different times. They found that, from early to intermediate stages of decomposition, ε and η precipitates coexist with different concentrations of silver atoms of 31 and 57%. It may be assumed that the silver-rich η precipitates control the flow stress. In a mixture of obstacles of different strengths, the mean obstacle distance is determined mainly by the concentration of the strong obstacles (Sect. 4.5.1).

Equation (8.2) can be used to estimate the difference $\Delta\gamma_{\mathrm{sf}}$ between the stacking-fault energies of the matrix and the precipitates. The following parameters are applicable to the specimens aged for 6 h: $d = 2.3$ nm, $l = 110$ nm, and $\tau_{\mathrm{y}} = 17$ MPa from the macroscopic experiments [455]. This results in $\Delta\gamma_{\mathrm{sf}} = 240\,\mathrm{mJ\,m^{-2}}$, which is even larger than the stacking fault energy of $200\,\mathrm{mJ\,m^{-2}}$ for aluminium. On the other hand, the cutting planes of the particles are, on the average, smaller than the particle diameter considered above. Thus, the obstacles observed in the in situ experiments seem to be stronger than the usual models of stacking-fault strengthening can explain. Possible explanations of this discrepancy are given in [283]. Because of the low stacking fault energy in the particles, the stacking fault width should be greater than the particle diameter. In this case, the stacking fault fills the whole particle and the dislocation is fixed to the particle border. This results in an additional elastic contribution to the interaction energy between particle and dislocation [456].

8.1.3 Al–Li

Because of their light weight connected with high elastic stiffness and strength, aluminium–lithium alloys are of considerable technical interest. They harden by precipitates of the ordered δ' phase by analogy with nickel base alloys containing Ni_3Al precipitates. The precipitates have the $L1_2$ structure. A dislocation of the matrix with an $a/2\langle 110\rangle$ Burgers vector produces an antiphase boundary (APB) in the particle. It can be removed when a second dislocation of the same Burgers vector is passing. Thus, the second dislocation moving exactly on the same slip plane is drawn into the particles with antiphase boundaries. This opposes the elastic repulsion between the two dislocations. Accordingly, it is energetically favorable if the dislocations move in pairs leaving particles without faults on the cutting planes. The respective theory [460–462] has originally been developed by Gleiter and Hornbogen and is reviewed in [224]. Evaluations of particle distributions and macroscopic measurements of the mechanical properties are given in [463–466].

Dislocation processes in single crystals grown from the commercial alloy P53 from Alcan were observed in in situ straining experiments in an HVEM [284, 467]. The single crystals contained 8.4 at% Li and were homogenized at 843 K. Afterwards, precipitates of the δ' phase were grown by two-stage annealing. The δ' particles were spherical, having an average diameter of $D = 15.6$ nm and a volume fraction of $f = 0.11$. The alloy is then in the under-aged state.

Figure 8.4 shows a typical dislocation structure. The dislocations move in pairs, in accordance with other microstructural observations (e.g., [468]). The leading dislocations of the pairs strongly bow out between pinning points. The trailing dislocations are less strongly bowed, but also in forward direction. In other parts of the specimens the dislocations move as single dislocations as in Fig. 8.5. They also bow out between obstacles. In general, the dislocations move very jerkily, usually over distances larger than a micrometer, which indicates that the stress is close to the athermal strength of the obstacle array (see Sect. 4.5.1, Fig. 4.18). Most dislocations show a loose arrangement, with each dislocation or dislocation pair moving on an individual slip plane.

Quantitative data from dislocation pairs can be obtained from the micrographs of dislocations under load using the model mentioned above and outlined in Fig. 8.6. The leading dislocation 1 has to create the antiphase boundaries inside the precipitates. It is driven by the applied stress τ and the repulsive internal stress τ_i arising from the trailing dislocation 2. Accordingly, the following force balance holds

$$(\tau + \tau_i)\, l_1\, b - \gamma_{APB}\, D_1 = 0, \tag{8.3}$$

where l_1 is the obstacle distance and D_1 the effective particle diameter along the leading dislocation, and γ_{APB} is the antiphase boundary energy.

The second dislocation is pushed forward by annihilating the antiphase boundaries in the particles and backward by the internal stress. Thus,

$$(\tau - \tau_i)\, l_2\, b + \gamma_{APB}\, D_2 = 0\,. \tag{8.4}$$

The obstacle distance l_2 and the effective particle diameter D_2 generally differ from the respective values of the leading dislocation. According to (3.24) and

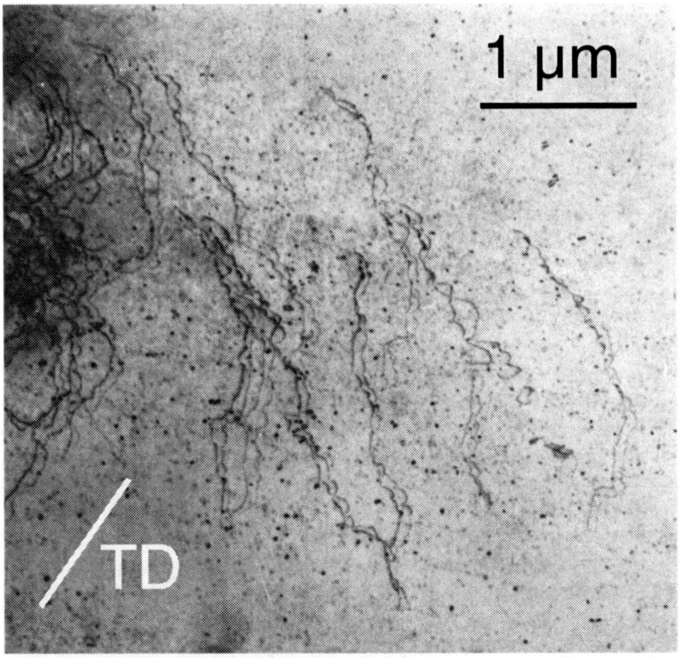

Fig. 8.4. Slip band of dislocation pairs in an Al-8.4 at% Li single crystal during in situ deformation in an HVEM at room temperature. TD tensile direction. From the work in [284]

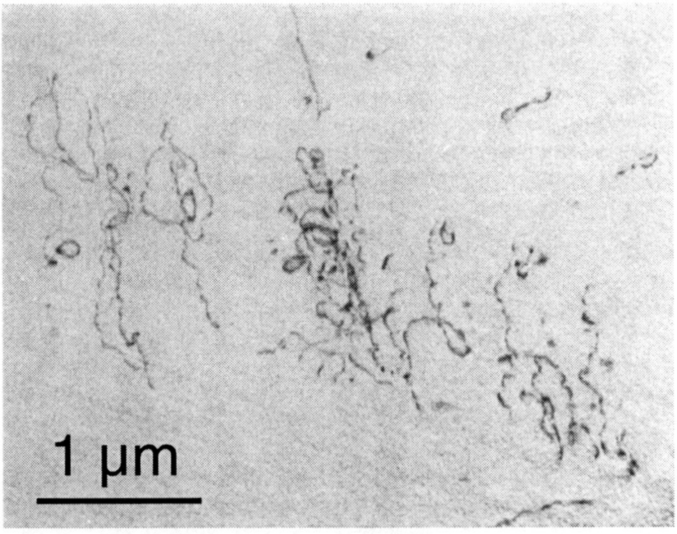

Fig. 8.5. Slip band of unpaired dislocations in an Al-8.4 at% Li single crystal during in situ deformation in an HVEM at room temperature. From the work in [284]

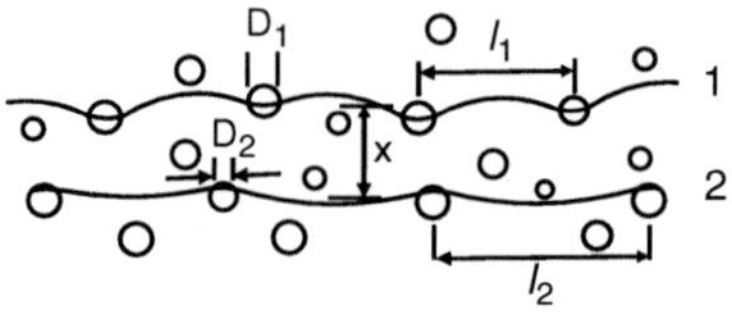

Fig. 8.6. Outline of the model of dislocation pairs in alloys with precipitates of an ordered structure

with $y_0 = 0$, the internal stress between the dislocations of the pairs is given by

$$\tau_i = \mu b/(2\pi x)$$

for screw dislocations, and by

$$\tau_i = \mu b/(2\pi[1-\nu]x)$$

for edge dislocations. Strictly speaking, these formulae hold only for straight dislocations. x is the distance between the dislocations, which can be measured from the micrographs.

Dislocation motion in pairs was simulated in Nimonic PE16, a material strengthened in a way similar to the present Al–Li alloy by particles that have to be cut by forming antiphase boundaries. The precipitates are also of the $L1_2$ structure. The following video presents an example of the simulation by V. Mohles performed in the group of E. Nembach at Münster University.

Video 8.2. *Simulation of the motion of a dislocation pair in* Nimonic PE16*:* The video sequence presents the simulation of the motion of a pair of edge dislocations under conditions very similar to those in the in situ experiments on Al-8.4 at% Li. The volume fraction of the precipitates amounts to $f = 0.089$, the radius of the particles is $r = 24\,b$, and the normalized antiphase boundary energy is $g = \gamma_{\mathrm{APB}}/(b\mu) = 0.015$. The simulation considers the obstacle resistance, the interaction between the dislocations of the pair as well as the self-stress of the bowing dislocation itself. In addition, viscous damping is included. a in the legend is the fraction of the area swept. Details of the simulation method and some results are given in [469]. The video shows the coupled motion of the (red) leading dislocation and the (blue) trailing dislocation. As expected, the leading dislocation strongly bows out while the trailing dislocation is quite smooth. The latter corresponds to the model where the trailing dislocation is pushed forward by removing the antiphase boundaries in the particles cut before by the leading dislocation. These forward forces should be balanced by a slight backward curvature of the connected dislocation segments in the matrix. However, this is in contrast to the experimental observation that the trailing dislocations are bowed-out in forward direction, too. Note that the normalized stress $s = 10^3\tau/\mu$, indicated in the legend, is raised by a factor of almost two during the motion. If the critical (final) stress had acted right from the beginning, the dislocations would have swept most of the area very quickly. Nevertheless, the damped motion visible also in the last part of the video does not correspond to the underdamped jerky motion in the experiments on Al–Li.

Table 8.2. Interaction parameters of leading dislocations in pairs and single screw dislocations with precipitates in Al-8.4 at% Li (Data from [284])

	Disl. pairs		Single screws
	Screws	Edges	
l (nm)	58.9	52.7	53.6
S (μm^{-1})	17.5	16.5	18.5
l_{sq} (nm)	57.5	47.5	51.5
$1/x$(μm^{-1})	19.1	18.3	
τ^*(MPa)	62.7	55.3	68.0
τ_i (MPa)	22.5	33.7	
τ (MPa)	40.2	21.6	

From the micrographs taken during the in situ experiments, the same data as for the other aluminium alloys before were determined for the leading dislocations of the pairs, separately for those of dominating screw and edge character, as well as for unpaired screw dislocations. Details of the evaluation are given in [284]. The results are listed in Table 8.2. In Sect. 8.1.1, the data of Table 8.1 are averages of the values of screws and edges. The present alloy shows also the dependence of the dislocation curvature on the logarithm of the segment length as predicted in (3.45) and proved for MgO and Al–Zn–Mg. The constant C is included in Table 8.1 and fits the values of the other materials.

Using (8.1), a theoretical value of the square lattice distance can again be calculated from the average particle diameter and volume fraction cited above, yielding $l_{sq}^{th} = 34$ nm, which is in relatively good agreement with the measurements from the in situ experiments.

In the force balance (8.3), all parameters are known except the antiphase boundary energy γ_{APB} and D_1. As the dislocations do not cut all the particles at the equator, the effective diameter may be $D_1 = \pi D/4$, where D is the average diameter of the spherical particles. Using the same value for D as above, it follows that $D_1 = 11.8$ nm. For γ_{APB}, this yields 0.090 J m^{-2} for screw dislocations, and 0.071 J m^{-2} for dislocations with a large edge component. These values are remarkably smaller than the literature values of about 0.15 J m^{-2} (e.g., [466]).

The force balance at the trailing dislocation (8.4) can be written as

$$D_2/l_2 = (\tau_i - \tau)\, b/\gamma_{APB}.$$

According to Table 8.2, for the screw dislocations, $\tau_i < \tau$, that is, D_2/l_2 becomes negative. That means, the dislocations of a pair of screws are not so strongly coupled as the model predicts. According to the latter, the trailing dislocation has to bow out in backward direction, which, however, has not been observed in the micrographs of the present study. On the contrary, many trailing dislocations are clearly visible bowing out forward. For the edge dislocations, however, $D_2/l_2 = 0.049$, which is consistent with the model and also with the few micrographs showing edge dislocations.

The disagreement between the behavior of screw dislocations and the simple theory can be explained by assuming that in addition to order hardening a further mechanism contributes to the interaction between dislocations and precipitates, which also acts on the trailing dislocations. It is suggested that this mechanism is modulus hardening. The difference between the shear modulus of the particles and the matrix amounts to 6.7 GPa or about 22% [470]. Estimations on the basis of (4.70) of the model in Sect. 4.7 yield an increase of about 50% of the strength F_0 of the precipitates, but only slight change in the total activation energy. As the particles are overcome in an athermal way, the behavior of the dislocation pairs can be explained by a combination of APB and modulus hardening.

The local effective stress τ^* is approximately equal for the leading screw dislocations in pairs and single screw dislocations. This can be expected if dislocations in both arrangements have to cut the same type of obstacles and to create the antiphase boundaries. The applied stress in the respective regions, however, is drastically different, which is certainly connected with the necessity for the pair mechanism to create dislocations in pairs. This topic will be touched again in Sect. 8.2.

In polycrystals, planar slip frequently occurs. The dislocations can be generated by Franck–Read sources in the grain boundaries. In this case, all dislocations automatically move on exactly the same plane. Planar slip should also be favored for energetic reasons, as the slip resistance of the precipitates is reduced if many dislocations shear the particles on the same plane, thus diminishing the cutting area. In the present study, planar slip has been observed most rarely. An example is given in Fig. 8.7. It shows dislocations emitted apparently from a localized source. In agreement with the model, only the

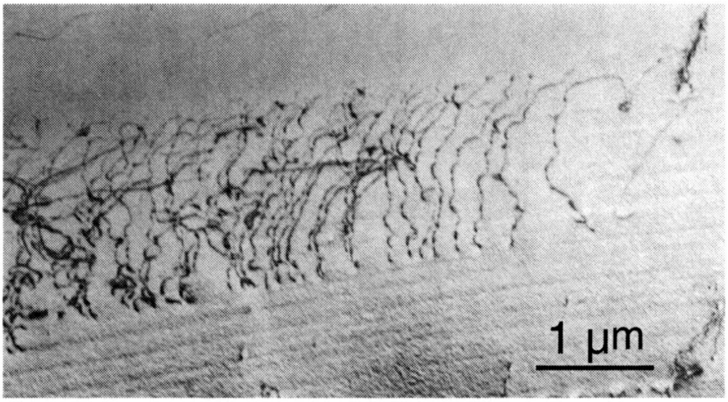

Fig. 8.7. Planar slip of dislocation pairs and single dislocations in Al–Li. From the work in [284]

first dislocations, visible on the left of the figure, move in pairs until the particles are fully sheared. The following single dislocations should not bow out if the particles interact solely by order hardening. This is a further indication of the action of an additional hardening mechanism.

8.1.4 Summary

Three examples were presented of precipitation hardening in aluminium alloys. Only in Al–Li, the obstacle distances agreed satisfactorily with those predicted from the concentration and size of the particles. In Al–Zn–Mg, far more obstacles controlled the dislocation motion than those considered before based on conventional methods. The opposite case occurred in Al–Ag, where only part of the precipitates visible in conventional and high-resolution TEM impede the dislocation motion.

As demonstrated in Table 8.1, the normalized effective stress τ' is always high in these alloys, between 0.45 and 0.87, where the lower values belong to the under-aged alloys. The high values indicate that the situation is close to the Orowan case. It should be remembered that in a line tension analogon Orowan looping appears for $\tau' < 1$ if the self-stress of the interacting dislocation segments is taken into account.

A problem is the calculation of the effective stress τ^* from the curvature measure S, which requires the knowledge of the dislocation line energy. There is great uncertainty because of the constant C in formulae like (3.45), (5.17), or (5.18). The measurements of the dependence of the curvature of dislocation segments on their length in Al–Zn–Mg and Al–Li confirm the value of C discussed in Sect. 3.2.7 for MgO, leading to very low line energies. Nevertheless, the values of τ^* quoted in this section on aluminium alloys were based on the theoretical value of C. This may be kept in mind if some of the effective stress data appear unrealistically high.

8.2 Dislocation Generation in Metals

This section presents several video sequences of the creation of new dislocations in metals, thus supplementing the experimental observations of Sect. 5.1.2. In many metals, the dislocations are generated by the double-cross slip mechanism. A further example from Al–Ag described in Sect. 8.1.2 is given in the following video.

Video 8.3. *Loop generation and opening in an* Al–Ag *single crystal:* The video clip shows the very sudden generation of large dislocation loops and their opening resulting in the generation of new dislocations moving away. The places of events are marked by green circles. After the second labeling, the shift of the upper loop downwards is caused by a reaction with a dislocation moving very quickly from the lower center of the image to the left, thereby annihilating parts of the loop. All processes take place very quickly.

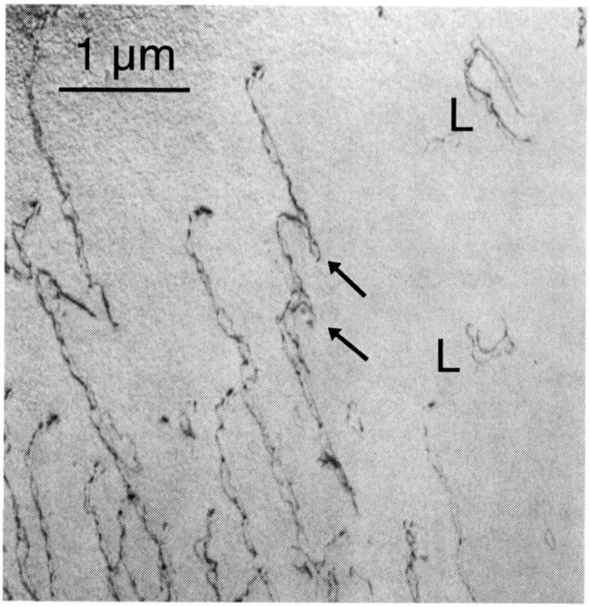

Fig. 8.8. α-like dislocation configurations indicating dislocation generation by the double-cross slip mechanism in Al–Li. From the work in [284]

The very sudden appearance of the loops and their opening underline that dislocation multiplication is controlled mainly by long-range dislocation processes. It is therefore of athermal character.

A special situation occurs in the Al–Li alloy treated in Sect. 8.1.3, where the dislocations move in pairs and are spread over relatively wide slip bands as in Fig. 8.4. This loose arrangement again hints at the operation of the double-cross slip mechanism. But, to multiply in pairs, both dislocations of the pair have to cross slip together onto a cross slip plane and afterwards to cross slip again together back to a plane parallel to the original one. This is astonishing in so far as the dislocations of the pair do not attract but repel each other, and as they are not coupled together by a continuous planar fault. The APB faults exist only inside the precipitates. Nevertheless, the dislocations multiply in pairs as documented by the α-like dislocation configurations marked by arrows in Fig. 8.8. There are also small loops and half-loops L consisting of dislocation pairs.

A material in which dislocations are created during plastic deformation in a dispersed way by the double-cross slip mechanism as well as in localized repeatedly operating (Frank–Read) sources is a titanium alloy with 6 wt% aluminium. The α-phase has the h.c.p. crystal structure. The dislocations with a-type Burgers vectors $\boldsymbol{b} = a/3\langle 11\bar{2}\rangle$ move preferentially on prismatic and basal planes. The flow stress depends on annealing treatments. The responsible solution hardening is supposed to result from the formation of short-range order [471, 472]. The order is partially destroyed by the moving dislocations. This requires energy, which causes a friction stress for the dislocations. Further, dislocations moving on the same plane move in a less ordered region and consequently experience a lower friction stress. This favors planar slip. The specimens shown in the following video sequences were produced in cooperation with M. Mills and T. Neeraj. The crystals were all homogenized at 900°C for 24 h and then either air-cooled or annealed at 600°C for another 24 h followed by ice water quenching (step-annealed). Both treatments cause the formation of different degrees of short-range order. The first two videos present several examples of the creation of dislocations by the double-cross slip mechanism.

Video 8.4. *Creation and opening of dislocation loops in air-cooled* Ti-6 wt% Al *at room temperature:* This video consists of three sequences, in which many events show the creation of dislocation loops and their opening to new dislocations as well as the annihilation of screw dislocations of opposite sign. In the first sequence, a small loop L is annihilated by a dislocation gliding nearby. The second sequence shows the opening of several loops to form pairs of dislocations. The dislocation at J acquires a jog, which trails a dipole growing to a loop and finally also to two new dislocations. In the third sequence, at D two screw dislocations of opposite sign trap each other to form a dipole. A dislocation loop expands at L to form two new dislocations. The right one annihilates with another dislocation of opposite sign so that only a small debris remains, which later on grows to a loop. At A, the left dislocation moves to the screw dislocation dipole formed before and annihilates one dislocation of the dipole so that only a single dislocation remains. This underlines the dynamic character of dislocation multipoles as discussed in Sect. 5.1.3.

Video 8.5. *Dislocation generation at a jog in step-annealed* Ti-6 wt% Al *at room temperature:* A jog in a screw dislocation trailing a short dipole is labeled J. The dipole opens to a loop, which is growing until the edge segments emerge at the surface so that two new dislocations are formed.

The next video shows several localized dislocation sources.

Video 8.6. *Localized (Frank–Read) sources in step-annealed* Ti-6 wt% Al *at room temperature:* In contrast to the preceding videos, here dislocations are generated by localized (Frank–Read) sources. One source is marked by a green circle. It consists of a half-loop, which is pinned by a high jog J at the lower end, as shown in Fig. 8.9. The jog causes a shift of the emergence points of the two dislocation branches through the surface with respect to the trace of the slip plane marked by the dashed line. One arm of the dislocation rotates around the jog, thereby always emitting one dislocation moving to the right and another one moving to the left. Two further sources are located at the left edge of the frame. The dislocations emitted from these sources are pinned temporarily at a dislocation of opposite sign forming a screw dislocation dipole. All the emitted dislocations moving to the right form a pile-up, the back stress of which blocks the sources. They operate again only after the dislocations of the pile-up have moved far enough away. This behavior was described before in Sect. 5.1.2 and is typical of all localized dislocation sources.

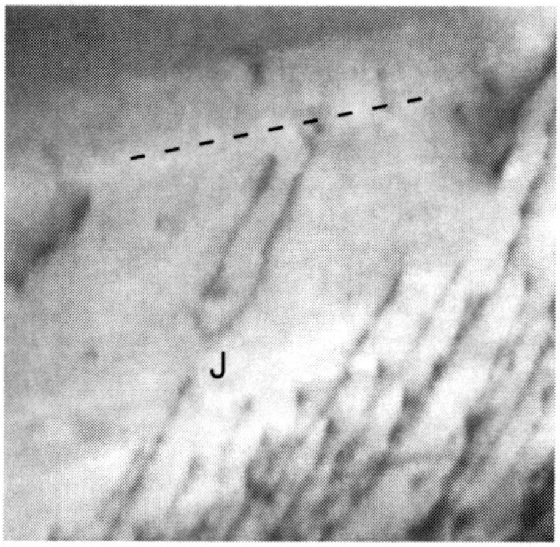

Fig. 8.9. Frank–Read source in Ti-6 wt% Al consisting of a half loop with a high jog. From a cooperation with M. Mills and T. Neeraj

As a consequence, dislocations with a high screw component are effectively pinned by high jogs. These jogs may trail a dipole, which opens to a dislocation loop, or one branch rotates around the jog, thus emitting a greater number of dislocations. The occurrence of localized dislocation sources in Ti-6Al is a prerequisite to planar slip taking place in this material. The slip localization is favored by the strain softening due to the destruction of the short-range order.

A two-phase material with different mechanisms of dislocation generation in the two phases is duplex steel. It consists of grains of b.c.c. (α) ferrite with $\{110\}\,1/2\langle 111\rangle$ slip systems and of f.c.c. (γ) austenite with $\{111\}\,1/2\langle 110\rangle$ slip systems. The Burgers vectors in the two phases correspond to each other. While the dislocations in the ferrite grains multiply by the double-cross slip mechanism, localized (Frank–Read) sources operate in the austenite grains, as demonstrated by the two following video clips.

Video 8.7. *Dislocation multiplication in a ferrite grain during the deformation of duplex steel at room temperature:* In the two-phase material, slip has to be transferred between grains of the same phase and of different phases. In the present example, slip in a grain in the middle of the left side of the frame is transferred to the bright grain on the right side. Thus, many dislocations emerge from the phase or grain boundary. In addition, the clip contains a number of α-like dislocation configurations, which are characteristic for dislocation multiplication and which develop to loops and later on to two isolated dislocations. At J, a jog trailing a short debris dipole is shifted conservatively along the dislocation until it disappears at the surface. The α-like configuration at L opens to a loop that expands to two dislocations.

Video 8.8. *Frank–Read source in an austenite grain during the deformation of duplex steel at room temperature:* The video clip presents the operation of a localized dislocation source in an austenite grain. The nature of the pinning agent is not clear.

A connection of the source dislocation with the dislocation network is not obvious. It can be extinguished under the present contrast conditions, or the pinning agent is a high jog in the dislocation as in the preceding example of a localized source. The moving branch rotates quite fast because of the athermal character of the dislocation motion. After one dislocation is emitted, the source stops due to the back stress of the emitted dislocations, which pile up against the grain or phase boundaries. When the outer dislocation breaks through the boundaries, part of the back stress is released and a new dislocation can be generated.

In summarizing, it may be stated that both mechanisms of dislocation generation are quite similar and may operate simultaneously in a single material. If cross slip is easy, many sources may be created, the jogs may move conservatively along the multiplying screw dislocations after one new dislocation loop has been created. Thus, slip spreads easily over the crystal volume. If cross slip is restricted, the jogs may remain in their positions, and the sources may operate repeatedly. Localized sources emitting many dislocations always occur when the pinning agents are nodes of the dislocation network.

8.3 Oxide Dispersion Strengthened Materials

Hard incoherent particles can be incorporated into a metal matrix to improve the creep resistance of alloys at high temperatures. Frequently, the particles are oxides, leading to oxide dispersion strengthened (ODS) alloys. At low temperatures, such particles have to be bypassed by Orowan looping (Sect. 4.7). At high temperatures, the dispersoids are easily overcome by climb [473]. Nevertheless, there exists a well-defined threshold stress for creep, below which the ODS alloys exhibit very low creep rates. The interpretation of the threshold stress is based on an attractive interaction, which causes the pinning of the moving dislocations on the departure side of the particles [474]. It was stated that if the interfaces between the particles and the matrix do not transfer shear strains, the stress field of the dislocations may partly relax at the interface leading to a reduced dislocation energy [475, 476], which causes the attractive interaction. Most experimental data on the creep behavior of ODS alloys are interpreted in terms of the respective model by Arzt and Rösler [477] or its variant including the thermally activated detachment from the particles [478] (Arzt–Rösler model). Recently, Xiang and Srolovitz [479] simulated the overcoming of penetrable and impenetrable particles without and with misfit by combined glide and climb applying linear relations between the glide and climb velocities and the respective stress components. It turned out that during overcoming the particles, the dislocations may assume a great variety of configurations depending on the actual parameters. There is some transmission electron microscope evidence of dislocations pinned to particles (e.g., [480, 481]), but there was no direct proof that the dislocations spend most of their time in the respective configurations on the departure side of the particles. In situ straining experiments in an

HVEM were performed to study the dislocation dynamics and the interaction between dislocations and particles in the ODS alloy INCOLOY MA956 [482] and the ODS superalloy INCONEL MA754 [483]. The studied alloys consist of Fe-19.9Cr-4.6Al-0.38Ti-0.51Y-0.019C-0.014N-0.19O and Ni-20Cr-0.3Al-0.5Ti-1Fe-0.05C-0.6Y_2O_3 (wt%). The activated slip planes were determined from the traces the moving dislocations generated on the specimen surfaces. The Burgers vectors are of $a/2\langle 111\rangle$ type for MA956 and $a/2\langle 110\rangle$ for MA754.

8.3.1 Microscopic Observations in Oxide Dispersion Strengthened Alloys

Experimental results are mainly described for MA956. They are supplemented by those of MA754. At room temperature, slip was observed only on $\{110\}$ planes. At the beginning of the in situ experiments, slip occurs in relatively narrow slip bands, as shown in Fig. 8.10. The slip plane is inclined by about 65° to the image plane so that the dislocations appear only in a narrow projection. Resting dislocations are mainly of screw character, being quite straight then. Only edge or mixed segments bow out strongly. With increasing strain, the dislocations cross slip onto another $\{110\}$ plane, with the slip bands broadening until the whole volume is filled with dislocations. Many small dislocation loops occur in the slip bands.

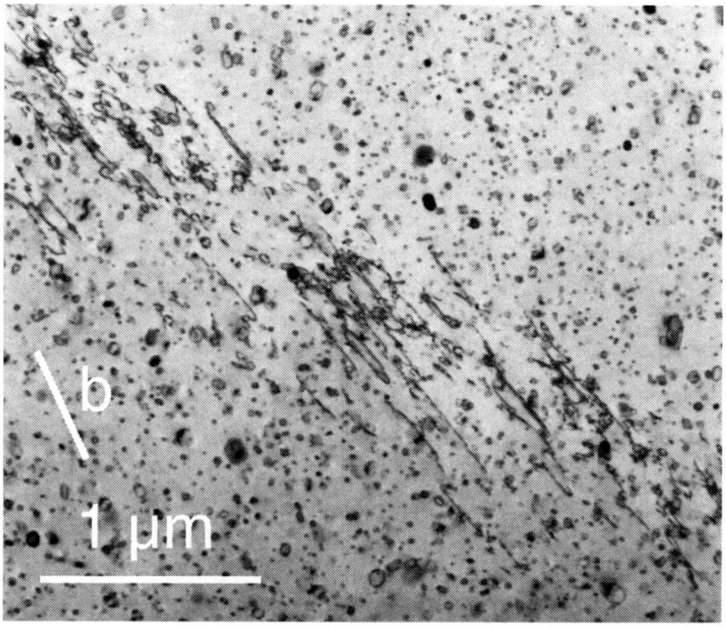

Fig. 8.10. Dislocations moving on $\{011\}$ slip planes during in situ deformation of MA956 at room temperature. From the work in [482]

The dynamic behavior of dislocations at room temperature is shown in the video.

Video 8.9. *Dislocation motion in* INCOLOY MA956 *at room temperature:* At room temperature, the dislocations move in a very jerky way over distances longer than those between the oxide particles.

The jerky motion hints at the athermal character of the deformation at room temperature.

At elevated temperatures, slip occurs in MA956 on {110}, {112}, and {123} planes often including cross slip between these planes. Most experiments were made at 700°C. A typical example is presented in the following video.

Video 8.10. *Dislocation motion in* MA956 *at* 700°C*:* This video clip consists of two sequences. The dislocations move in a smooth and viscous way. At the particles, a cusp develops, with the adjoining dislocation segments bowing out. When the maximum bowing is reached, that is, when the maximum force acts, the particles are mostly overcome without long waiting times in the equilibrium positions.

In the second video sequence, dislocations move on a {112} plane. The slip plane is inclined by about 25° with respect to the image plane so that the width of the slip plane is about 1 μm. The average dislocation velocity is 10–15 nm s^{-1}, which is slow enough to resolve different stages of overcoming the dispersoids. The dislocation velocity was approximately equal for screw and edge dislocations. The first part of this sequence shows several stages of surmounting a single strong obstacle, marked by a green dot above it. Selected frames are presented in Fig. 8.11. In the first frame taken at a time of 0 s, the obstacle is first contacted. Afterwards, the dislocation moves from the arrival side to the departure side of the obstacle in about 7 s forming only a weak cusp at the obstacle. It then slowly bows out reaching its equilibrium bowing after a total time of 21 s. In this configuration, the dislocation forms a sharp cusp on the departure side. It remains in this position for less than 1 s. At 21.6 s, the dislocation starts to detach from the obstacle. At 22.7 s, it has left the particle completely, the strong bowing has disappeared, and the dislocation moves in a viscous way. While traveling to the next particle, which takes about 10 s, most of the dislocation is quite straight. Certainly, the particle that had been overcome during the first 22 s was a particularly large one. Nevertheless, the process is very much the same also for smaller particles, which are not imaged in the HVEM, but which cause the formation of cusps along the dislocation line.

To increase the particle size, some specimens were annealed. An example is presented in the following video.

Video 8.11. *Dislocation motion in annealed* MA956 *at* 700°C*:* This clip of the annealed alloy with particles of larger size confirms the observation made above that the dislocations detach from the pinning centers without long waiting times in the equilibrium configurations. The dislocation moving upwards undergoes cross slip onto a plane with an inclined trace.

In conclusion, one can say that around 700°C the four following stages of dislocation motion in an array of incoherent particles can be distinguished.

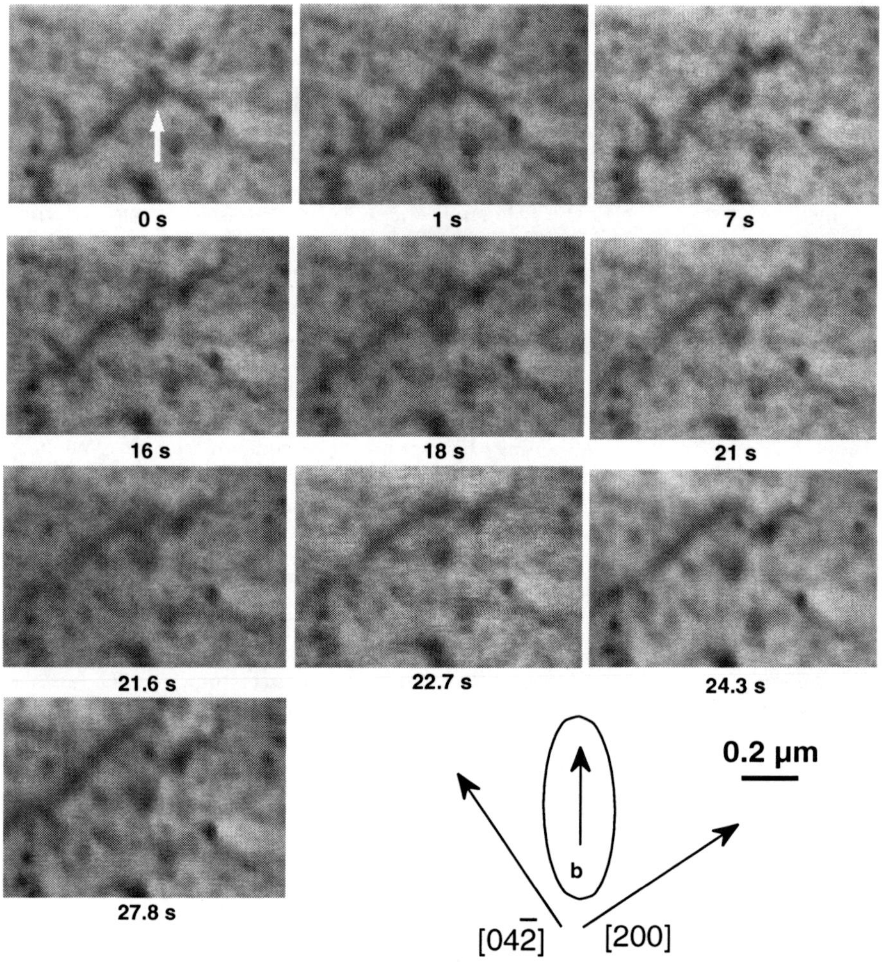

Fig. 8.11. Sections of a video sequence showing a dislocation in MA956 moving at 700°C on a {112} plane overcoming a large oxide particle. The plane is inclined by about 25° with respect to the image plane. From the work in [482]

1. The dislocations move in a viscous way between configurations in which they are locally pinned. This points to a frictional force acting on the moving dislocations.
2. After contacting a particle, the dislocation moves to the departure side in rather a short time, with a relatively small force acting.
3. The dislocation gets pinned on the departure side of the particle while the adjacent segments in the matrix slowly bow out until they reach their elastic equilibrium configurations. This process of bowing out between the particles takes a relatively long time. The back stress of the bowed segments reduces the stress available for the viscous motion, causing a

lower velocity before the equilibrium bowing is reached. This back stress, which reduces the applied stress temporarily, is supposed to be the main strengthening effect of the particles around 700°C.

4. After a short lifetime of the equilibrium configuration, the dislocation detaches from the particle, straightens, and glides to the next obstacle as described under 1.

Dislocation configurations at other temperatures resemble those at 700°C. The dislocation motion at lower temperatures is viscous between the stable configurations. The slip distances after overcoming the particles are not much greater than the particle distances. A different situation holds for higher temperatures as verified by the following two video clips.

Video 8.12. *Dislocation motion in* MA956 *at* 1010°C*:* At this high temperature, the dislocation motion becomes quite jerky.

Video 8.13. *Collective dislocation motion in* MA956 *at* 1010°C*:* The video clip comprises two sequences taken from a narrow slip band. In the band, the dislocations move in a collective way where the dislocations are approximately equally spaced. The dislocation motion is quite jerky at this temperature. Many dislocations are pinned at the same particles so that successive configurations resemble each other.

In the second clip, many dislocations pass a very large particle almost in the center of the frame. The course of the dislocations while passing the particle can be followed.

The jerky motion at high velocities between the particles suggests that the frictional forces, which act at temperatures around 700°C, disappear at higher temperatures. In addition, the slip distances become larger than a micrometer. Near 1000°C, the dislocations mostly move in a collective way. This suggests that long-range elastic interactions between the dislocations play an important role.

Quantitative data on the dislocation dynamics can be derived from the micrographs and video recordings. Some efforts were made to measure the obstacle distance l at 700°C. First, the distances were determined between the cusps in all bowed-out positions that were recorded during a 3-minute observation of the second part of the dynamic Video 8.10. In a second approach, video frames were selected randomly for determining the pinning point distances of a larger number of segments. Finally, l can be estimated from the size and density of the dispersoids. It follows from all three measurements that the segment length between such obstacles which effectively impede the dislocation motion at about 700°C amounts to about 200 nm. Similar estimates were made for MA754. The results are summarized in Table 8.3.

The stress acting on individual bowed-out dislocation segments was measured from the dislocation curvature determined by matching calculated dislocation shapes with the dislocation images as described in Sect. 5.2.3 and by applying (5.18). The elastic constants for MA956 were taken from [484] for the alloy PM 2000, which should be similar to INCOLOY MA956. As no

Table 8.3. Interaction parameters between moving dislocations and oxide dispersoids in the ODS alloys MA956 and MA754 (Data from [482, 483])

Mat., T (°C)	l (nm)	y_0 (nm)	τ_b (MPa)	$\tau_{OR,s}$ (MPa)	$\tau_{OR,e}$ (MPa)	$\sigma_{OR,aver}$ (MPa)
MA956, 23	141[a]	111	56	107	56	204
MA956, 700	200	141	25	77	33	138
MA956, 880				68	28	120
MA754, 745	150	78	90	123	78	252

[a]This value was obtained from long segments selected for measuring τ_b

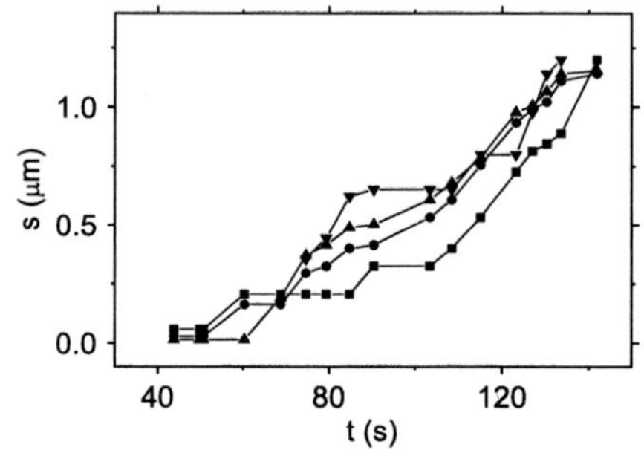

Fig. 8.12. Plots of dislocation displacement s vs. time t in MA956. The four curves correspond to different points marked along the dislocation line observed in Video 8.10 and shown in Fig. 8.11. Data from [482]

values for INCONEL MA754 were available, the constants of Ni, which is the main constituent of MA754, were applied. In MA956, there is a large ratio between the line energies of edge and screw dislocations, which is reflected in the strongly elongated shape of the dislocation loops as plotted in the inset of Fig. 8.11. A high value of the line tension of screw dislocations results in their low curvature at room temperature (Fig. 8.10). In (5.18), the theoretical constant $C = -1.61$ was used, indicating that the values of τ_b should be upper limits. Average values of the observed back stresses are listed in Table 8.3.

To characterize the dynamic behavior of dislocations at high temperatures, the displacement s of the dislocations was plotted vs. the time t. Figure 8.12 shows an example, with four curves originating from different positions along the single dislocation observed in the second part of Video 8.10. Other moving dislocations show a similar behavior. This plot demonstrates that only parts of the dislocations are pinned temporarily while others are still moving, leading to a relatively homogeneous motion of the dislocations as a whole, whereby the times of resting and moving are of the same order of magnitude, as concluded above qualitatively.

The microscopic observations on MA956 are essentially confirmed by the in situ deformation experiments on INCONEL MA754 as illustrated by the succeeding video clip.

Video 8.14. *Dislocation motion in the ODS alloy* INCONEL MA754 *at* 655, 745, *and* 790°C*:* The video clip consists of three sections taken at different temperatures. At the low temperature of 655°C, dislocations move quite smoothly. The interaction with the particles does not strongly influence the dislocation motion. At the intermediate temperature of 745°C, the motion is still quite viscous. In the first two sequences, the interaction with large particles marked by green dots is prominent. In the first case, the dislocation detaches immediately after full bowing is reached. In the second case, the dislocation stays in its pinned configuration for a long time. The third sequence shows viscous motion without remarkable interaction with the particles. At the highest temperature of 790°C, the dislocations detach from the large particle again immediately.

In summarizing the microscopic observations it may be stated that in the relevant temperature range the dislocations move in a viscous way, where the waiting times for detachment from the departure side of the particles do not dominate the average dislocation velocity, in contrast to the predictions of the Arzt–Rösler model.

8.3.2 Macroscopic Deformation Properties

Two samples of MA956 were tested, one at room temperature up to 600°C and the second between 600 and 900°C. Figure 8.13 presents the dependence of the flow stress σ on the temperature. It shows an initial decrease above room temperature, perhaps a plateau at 300–400°C, and again a decrease at higher temperatures.

The strain rate sensitivity of the flow stress was measured by stress relaxation tests. Figure 8.14a shows the strain rate sensitivity r as a function of temperature. Data points with full symbols are calculated from the slope at the beginning of the relaxation tests, corresponding to the strain rate of the continuous deformation before the relaxation tests. At intermediate temperatures, that is, at 400 and 500°C, the relaxation curves have an "inverse" curvature as discussed in Sect. 4.11 and in more detail in Sect. 9. In this range,

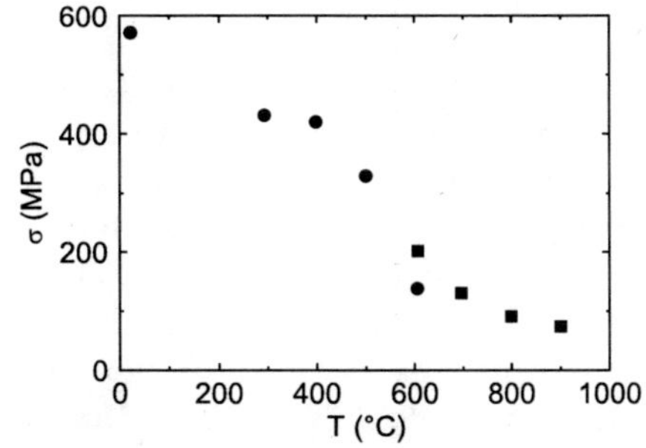

Fig. 8.13. Temperature dependence of the critical flow stress σ. Data from two samples deformed at different temperatures (*squares and circles*). The flow stresses were corrected for work hardening. Data from [482]

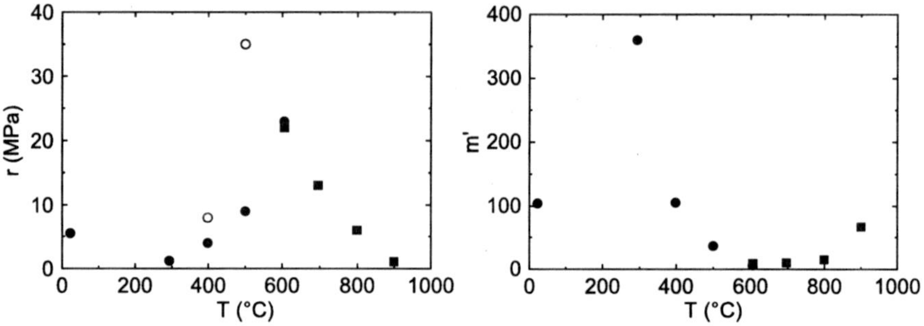

Fig. 8.14. Temperature dependence of the strain rate sensitivity of stress r (**a**) and the experimental stress exponent m' (**b**). Values of r taken at the beginning of the relaxation tests are plotted as full symbols and those determined at intermediate relaxation rates as open ones. *Squares* and *circles* correspond to different samples. Data from [482]

r values plotted as open symbols are determined also for intermediate relaxation rates. The figure shows that the strain rate sensitivity decreases from about 5.5 MPa at room temperature down to 1.2 MPa at 300°C. Afterwards, the initial values exhibit a maximum at about 600°C, decreasing again above 700°C down to a very small value at 900°C. The strain rate sensitivity at intermediate relaxation rates (open symbols) shows a strong maximum at about 500°C. Below, the strain rate sensitivity will be discussed also in terms of the stress exponent $m' = \mathrm{d}\ln\dot{\varepsilon}/\mathrm{d}\ln\sigma = \sigma/r$. The temperature dependence of m' is plotted in Fig. 8.14b.

8.3.3 Deformation Mechanisms

In the following, processes which may control the flow stress of the two ODS alloys at different temperatures will be discussed. This is mainly based on the data of INCOLOY MA 956.

Long-Range Dislocation Interactions

Mutual elastic interactions between dislocations may contribute to the flow stress by Taylor hardening depending on the dislocation density ϱ and being described by (5.11) in Sect. 5.2.1. For a rough estimate, the numerical constant α is chosen equal to π and $\mu = 4\pi E_s/b^2$. Dislocation density data of deformed specimens of MA956 are rare and slightly controversial [485, 486]. The dislocation density seems to be constant between 400 and 600°C at $\varrho \approx 5 \times 10^{13}\,\mathrm{m}^{-2}$, yielding $\tau_i \approx 40$ MPa at some intermediate temperature. At high temperatures, ϱ decreases down to $2.5 \times 10^{13}\,\mathrm{m}^{-2}$ [485]. At 400°C, deformation is localized in shear bands having a very high dislocation density of $3.5 \times 10^{15}\,\mathrm{m}^{-2}$ [486]. This yields the very high internal stress contribution of $\tau_\mathrm{i} \approx 300\,\mathrm{MPa}$. As to compare the calculated stresses with macroscopic flow stress data (Fig. 8.13), an orientation factor of $m_\mathrm{s} = 0.4$ may be considered for the single-crystal like macroscopic specimens. It may therefore be concluded that Taylor hardening contributes essentially to the critical flow stress of MA956 at all temperatures. Unfortunately, no dislocation density data are available for room temperature. In [483], the athermal stress component for MA754 is estimated to be $\tau_\mathrm{i} \leq 52\,\mathrm{MPa}$.

Orowan Stress

The array of oxide particles in MA956 can be characterized by an average particle diameter of $D = 24\,\mathrm{nm}$ and a volume fraction of $f = 0.0145$ [486]. The volume fraction was estimated from the distribution of the diameters and the total density of the particles via a stereological formula. From these data, the square lattice distance l_sq on the slip plane can be calculated in the most simple way by (8.1), yielding $l_\mathrm{sq} \approx 150\,\mathrm{nm}$. This is consistent with the average obstacle distance of $l \approx 200\,\mathrm{nm}$ estimated above from the in situ straining experiments. At zero temperature, these obstacles can be bypassed only if the applied shear stress exceeds the Orowan stress, given by (4.71) and (4.72). Calculated Orowan stresses τ_OR for screw and edge dislocations, based on the data of D and l_sq quoted above, are listed in Table 8.3. The last column in the table contains the average value between screw and edge dislocations of the normal Orowan stress $\sigma_\mathrm{OR,aver}$.

Both long-range dislocation interactions and the Orowan stress are of athermal character. It is therefore proposed that the flow stress plateau between about 300 and 400°C in Fig. 8.13 is controlled by these two mechanisms. In this case, the back stress τ_b estimated above should approximately be equal to the Orowan stress. In Table 8.3, the value of the back stress of $\tau_\mathrm{b} = 56\,\mathrm{MPa}$ at room temperature was measured along edge dislocations. It fits the Orowan stresses for edge dislocations, but not the average Orowan stress. This discrepancy may illustrate the accuracy (or the errors) of the experimental and theoretical methods used. As demonstrated in Fig. 8.14a, the strain rate sensitivity is very low at 300°C, in agreement with the athermal character of the deformation in this temperature range. The drastic decrease of the flow stress above 400°C should then be due to the thermally activated overcoming of the oxide particles but also to a decreasing contribution of Taylor hardening due to recovery. This view is supported by the fact that the particle-free matrix material Kanthal shows the same plateau as well as the decrease of the flow stress above 400°C [487]. Both processes leading to the flow stress decrease require diffusion. Above about 500°C, the flow stress should be lower than the Orowan stress. This is indicated in Table 8.3, where the back stress τ_b at 700°C is less than half the average Orowan stress. The situation is similar for MA754.

The Thermally Activated Detachment Model

In the high-temperature range, the flow stress reduced by the athermal component from long-range dislocation interactions is lower than the Orowan stress. Only then can the theory of Arzt and Rösler [473–478] be applied. This theory assumes that after overcoming the particles by climb, the dislocations become pinned on their departure side. In the athermal case, the stress for the detachment of the dislocations from the particles is given by

$$\tau_{\mathrm{d}} = \tau_{\mathrm{OR}}(1 - \kappa^2)^{1/2} \,. \tag{8.5}$$

κ is a parameter describing the relaxation of the stress field of the dislocation on the particle surface. At a finite temperature, the dislocations can detach from the particles at stresses lower than τ_{d}. They spend most of their time at the departure side of the particles waiting for the thermally activated detachment so that this process controls the average dislocation velocity. The in situ experiments described in Sect. 8.3.1 show clearly that this does not happen at the relevant temperatures (700°C for MA956). Figure 8.12 demonstrates that mostly only parts of a dislocation are resting while others are still moving, and that the traveling times between the stable positions are approximately equal or even longer than the lifetimes in the pinned positions. As discussed in connection with the second part of Video 8.10 and Fig. 8.11, the dislocations spend a long time for bowing between the particles before they reach their elastic equilibrium positions. The time of awaiting the detachment is only a small fraction of the total time. However, frictional forces surely impede the dislocation motion and result in their viscous motion, which will be discussed below. No frictional stress is considered in the thermally activated detachment model.

As pointed out in Sect. 8.3.1, above 700°C, the dislocation motion in MA956 becomes jerky, indicating that the frictional stress is no longer important while the dislocations are still pinned to the particles. This can be explained by the Orowan process being still active at high temperatures, but it is also consistent with the thermally activated detachment model.

For the Orowan process, the temperature dependence of the flow stress is only due to the temperature dependence of the energy factor E_0 of the dislocations in (4.71). Respective values of the Orowan stress are listed in Table 8.3 for screw and edge dislocations at different temperatures. Accordingly, the temperature sensitivity of the stress contribution of the Orowan process amounts to $\mathrm{d}\sigma_{\mathrm{OR}}/\mathrm{d}T \approx -0.13\,\mathrm{MPa\,K^{-1}}$ for screw dislocations and to about $-0.07\,\mathrm{MPa\,K^{-1}}$ for edge ones at temperatures above 700°C. Between 700 and 800°C, the experimental value is $\mathrm{d}\sigma/\mathrm{d}T \approx -0.4\,\mathrm{MPa\,K^{-1}}$. Thus, the flow stress in MA956 above 700°C depends on the temperature far more strongly than predicted by the Orowan mechanism. The reason is the operation of a thermally activated mechanism, for instance the thermally activated detachment model.

According to [478], in this model the stress exponent can be expressed by

$$m = [3\Gamma D/(2kT)]\,(1-\kappa)^{3/2}\,(1-\tau/\tau_{\mathrm{d}})^{1/2}\,\tau/\tau_{\mathrm{d}}.$$

Γ is the line tension. For a rough estimate, an average value of the energy factors of screw and edge dislocations can be used for Γ. κ is not known very well. According to [488], $k = 0.66$ for PM 2000. In [483], the same value is estimated for MA754. The stress acting locally is given by the back stress τ_{b}. Using the data listed in Table 8.3 for 700°C, yields $\tau_{\mathrm{b}}/\tau_{\mathrm{OR}} \approx 0.45$ or, with (8.5), $\tau_{\mathrm{b}}/\tau_{\mathrm{d}} \approx 0.6$. With these data, the theoretical value of m turns out to be about 60 at this temperature. It decreases with increasing temperature, also due to the decreasing τ_{b}. The experimental values of m' in Fig. 8.14b increase from about 10 at 700°C up to about 70 at 900°C. While the experimental values of the stress exponent are in the correct order of magnitude of the estimated one, their temperature dependence contradicts the theoretical dependence of the thermally activated detachment model so that the deformation between about 700 and 900°C should not be controlled by this model. Above 900°C, m' is reported to decrease [488] so that the model may be active in this high-temperature range. However, the high values of m' at high temperatures may also hint at an increasingly athermal character of the deformation, which should be controlled mainly by τ_{i}.

Solution Hardening

The additions to the alloys in lower concentrations (Ti, perhaps also a fraction of the Al and the trace elements in MA956 as well as the trace elements in MA754) may give rise to solution hardening. Here, solution hardening is understood as the interaction between dislocations and nondiffusing solutes. The order of magnitude of the parelastic interaction energy between the solutes and the dislocations can roughly be estimated from the maximum interaction force (3.34) and an activation distance of b, which together with estimated quantities yields a fraction of an electron-volt. Such weak obstacles may give rise to the low-temperature increase of the flow stress of about 150 MPa between 300°C and room temperature in Fig. 8.13 as well as to the simultaneous increase of the strain rate sensitivity in Fig. 8.14. However, these obstacles are easily overcome at high temperatures.

From the strain rate sensitivity r, the activation volume V can be calculated by (4.9). At room temperature with $r = 5.5$ MPa, V amounts to about $120\,b^3$. The square lattice distance l_{sq} between the solutes can roughly be estimated by (4.48). With the concentration c being a fraction of one percent, l_{sq} will be of the order of magnitude of $15\,b$. This relation between V and l_{sq} is well consistent with the Mott–Labusch theory of solution hardening (Sect. 4.5.2). Thus, a small part of the flow stress near room temperature can consistently be explained by solution hardening. It follows from the jerky motion of dislocations at room temperature that this part is only small. In particular, solution hardening will not be responsible for the high-temperature friction mechanism causing the viscous dislocation motion observed in the in situ experiments at high temperatures.

Diffusional Point Defect Drag

At high temperatures, the solutes may diffuse giving rise to a dynamical point defect drag. The physical origins and the respective theories are described in Sect. 4.11. It is suggested here that the point defect drag controls the velocity of the viscously moving dislocations in the respective temperature ranges (around 700°C in MA956 and at all temperatures investigated in MA754). It is estimated in Sect. 4.11 and experimentally proved on different materials as reviewed in [489] that some 1000 ppm of a solute are necessary for a remarkable contribution to the flow stress. Such concentrations should be available in the alloys investigated.

As pointed out in Sect. 4.11, point defect drag may cause an inverse dependence of the strain rate sensitivity on the strain rate, that is, a decreasing strain rate sensitivity with increasing strain rate or stress. As mentioned in Sect. 8.1.1 and illustrated in Fig. 8.2, this may be observed as an inverse curvature of the stress relaxation curves. In MA956, the inverse strain rate behavior occurs at 400 and 500°C as described in Sect. 8.3.2 and shown by the open symbols in Fig. 8.14a. Accordingly, the point defect drag may cause the whole maximum of the strain rate sensitivity between about 300 and 800°C as well as a contribution to the flow stress, showing a maximum, too. This process is superimposed on a very rapid high-temperature decrease of the flow stress above 400°C. Nevertheless, the dynamical point defect drag should be responsible for the viscous motion of dislocations observed in the in situ experiments. Thus, this mechanism should represent the thermally activated process during the high-temperature deformation of MA956 up to 900°C. The diffusional drag in ODS materials has been considered before, for example, in [481, 487, 490].

8.3.4 Summary

Oxide dispersion strengthened materials are substantially hardened by the inclusion of the oxide dispersoids. From room temperature up to temperatures where diffusion sets in, the particles cause an Orowan stress contribution which, together with the long-range stress contribution from the elastic interaction between dislocations, constitutes the main part of the flow stress. Near room temperature, solution hardening by trace elements is superimposed. At high temperatures, the particles can be surpassed by climb. The dislocations may then be pinned at the departure side of the dispersoids by an elastic interaction. The detachment from the pinned configurations may be supported by thermal activation. In situ straining experiments in an HVEM have shown that the dislocations move in a viscous way, which may be controlled by diffusional point defect drag. The back stress resulting from the interaction with the dispersoids slows down the viscous dislocation motion. This may be the main impeding action of the particles. At very high temperatures, the dislocations move in an athermal collective way.

8.4 Plastic Deformation During Fracture of Al_2O_3/Nb Sandwich Specimens

High-strength materials are frequently so-called composites, where hard particles or fibres are embedded in a soft matrix. Their mechanical properties and in particular their fracture parameters depend sensitively on the strength of the interfaces between the hard and soft components. As both constituents of these composites, for example, ceramics and metals, have quite different elastic and plastic properties, the fracture processes along or near the interfaces may considerably differ from those in homogeneous materials. The fracture along the interfaces is controlled by the work of adhesion W_{ad}. However, the total fracture energy J_c, the so-called J integral, may be more than three orders of magnitude higher than the work of adhesion. For a review, see [491]. This difference results from the occurrence of plastic deformation both near the crack, which increases the fracture toughness, as well as far from it.

If the stress intensity factor at the crack K, introduced in Sect. 7.2.6, is high enough, new dislocations can be nucleated at the crack tip. The Rice–Thomson model for homogeneous solids [492] has been extended to bi-materials in [493]. Finally, dislocation nucleation at an interface crack is described by a Peierls-type model [494]. In addition to the direct dislocation nucleation, dislocations can also be generated in the plastic zone of a crack by dislocation multiplication as in homogeneous materials during plastic deformation (Sect. 5.1.1). Although the role of plastic deformation for interface fracture is recognized, experimental studies on dislocation creation and motion during deformation processes are only rarely reported.

To study the fracture behavior of metal/ceramic interfaces, α-Al_2O_3/Nb sandwich specimens were produced by ultra-high vacuum diffusion bonding with geometrically and chemically well defined interfaces. They have been thoroughly investigated (e.g., [495–497]) and are therefore well suited to serve as a model system for interface fracture studies. In situ deformation experiments inside an HVEM were performed to investigate the processes of plastic deformation in connection with fracture near the interface [498].

The micro-tensile specimens were cut from tri-crystal bars consisting of two α-Al_2O_3 single crystal pieces enclosing a 2 mm thick niobium sheet. The two interfaces between the Nb sheet and the Al_2O_3 bars had different orientation relations:

OR A: $(110)_{\mathrm{Nb}} \parallel (0001)_{\mathrm{Al2O3}}$, $[1\bar{1}0]_{\mathrm{Nb}} \parallel [01\bar{1}0]_{\mathrm{Al2O3}}$

OR B: $(110)_{\mathrm{Nb}} \parallel (0001)_{\mathrm{Al2O3}}$, $[1\bar{1}0]_{\mathrm{Nb}} \parallel [2\bar{1}\bar{1}0]_{\mathrm{Al2O3}}$.

The micro-tensile specimens used for the results below always had the orientation named 2 with a $[001]_{Nb}$ foil normal.

Niobium has the b.c.c. crystal structure with Burgers vectors of $1/2\langle 111\rangle$ type and $\{110\}$ and $\{112\}$ slip planes [499]. Two of the four $1/2\langle 111\rangle$ Burgers vectors are parallel to the interface planes. The latter are perpendicular to the tensile direction so that the orientation factors of the corresponding slip systems are zero. For the present specimens with foil normal 2, the traces of the $\{110\}$ slip planes comprise angles of 45° with the tensile direction. The traces of the $\{112\}$ planes are either perpendicular to the tensile direction or form angles of 18.4° with it. Thus, the planes can be identified by the orientations of the traces only. In some cases, also slip systems with macroscopically zero orientation factors were observed.

During in situ straining, the cracks moved either along the interface or within the niobium, in contrast to macroscopic fracture measurements, where the crack formed mostly in the brittle sapphire [14]. Nevertheless, plastic deformation occurred only in the soft niobium. Figure 8.15 presents a frame from a video showing a crack growing along the interface. The upper part of the figure shows the transparent sapphire crystal with the interface perpendicular to the tensile direction TD. The lower part displays the deformed niobium with a high density of dislocations. The crack has become blunted in the niobium.

From the width of the blunted crack tips, the so-called crack opening displacement (COD) δ_c can be measured. It ranges between about 70 and 360 nm for different cracks grown during in situ straining. In homogeneous materials, the crack opening displacement allows one to determine the energy

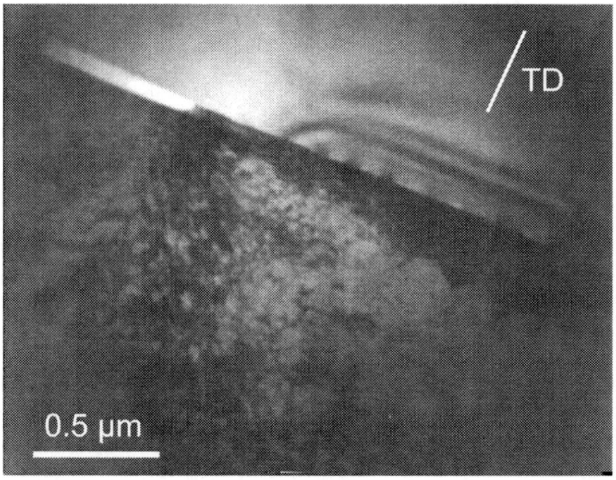

Fig. 8.15. Video frame of a crack in an Al_2O_3/Nb sandwich specimen growing slowly along the interface of OR A2. TD tensile direction. From [498]. Copyright (2004) Carl Hanser Verlag, München

release rate G_{cCOD} defined in Sect. 7.2.6 [500]

$$G_{\mathrm{cCOD}} = m\,\delta_{\mathrm{c}}\sigma_{\mathrm{y}},$$

where m is a numerical factor with $1 < m < 3$ and σ_{y} is the yield stress of the material. In a first approximation, it is assumed that this relation holds also for fracture along or near the interface. The yield stress of Nb was measured along the $\langle 110 \rangle$ direction characteristic of the present loading conditions and with the same thermal history. It amounts to $\sigma_{\mathrm{y}} = 84\,\mathrm{MPa}$ [14]. With this value and an intermediate value of $m = 2$, G_{cCOD} assumes values between about 12 and $60\,\mathrm{J\,m^{-2}}$.

These energy release rates can be compared with the work of adhesion W_{ad} and with the macroscopically measured fracture energies. W_{ad} amounts to only about $0.5\,\mathrm{J\,m^{-2}}$ for OR A and to $0.95\,\mathrm{J\,m^{-2}}$ for OR B (quoted in [491]). The present G_{cCOD} values are one to two orders of magnitude higher due to the plastic deformation near the crack tip. Average values of G_{c} determined from the stress intensity factors amount to 313 and $426\,\mathrm{J\,m^{-2}}$ [14]. Thus, macroscopic specimens are stronger than it is indicated by the crack opening during in situ straining experiments. The average total work of fracture J_{c} is of the order of magnitude of 1,200–1,700 $\mathrm{J\,m^{-2}}$ [14]. Consequently, a large amount of the plastic work is expended far from the crack tip and is not directly increasing the critical stress intensity factor. This plastic deformation leads to the formation of voids in front of the main crack, so that void formation and coalescence represents an essential mechanism of crack extension in the sandwich samples.

The dislocation structure of a specimen with OR A2 is displayed in Fig. 8.16. The dislocations belong to a slip plane, which has traces parallel to the trace of the interface, that is, the (112) or $(11\bar{2})$ plane with the maximum orientation factor of 0.471. The Burgers vectors point out of the interface plane. Other dislocations in this specimen move on the $(01\bar{1})$ plane with $m_{\mathrm{s}} = 0.408$. The dislocations exhibit long straight segments in accordance with the action of the double-kink mechanism in the b.c.c. material.

The dynamic behavior of the dislocations is illustrated in the following video clip.

Video 8.15. *Dislocation motion in the plastic zone of cracks in* Al_2O_3/Nb *sandwiches:* The video presents four sequences, the first three in A2 orientation, and the last one in B2 orientation. The series starts with dislocations moving on (121) planes, which do not contain the Burgers vectors pointing out of the interface plane, that is, owing to the complex stress state near the crack these dislocations move in spite of their macroscopic zero orientation factor. The straight segments move smoothly but at a relatively high speed over distances comparable with the spacings between the dislocations. Thus, these jumps are controlled by the long-range internal stresses due to other dislocations. The second sequence displays a dislocation multiplication. It starts from a debris dipole at a jog marked by a green arrow. With the dislocation moving further, the dipole opens and the dislocation assumes the shape of a φ. Later on, the arc of the φ grows and is terminated to become a full elongated loop, before

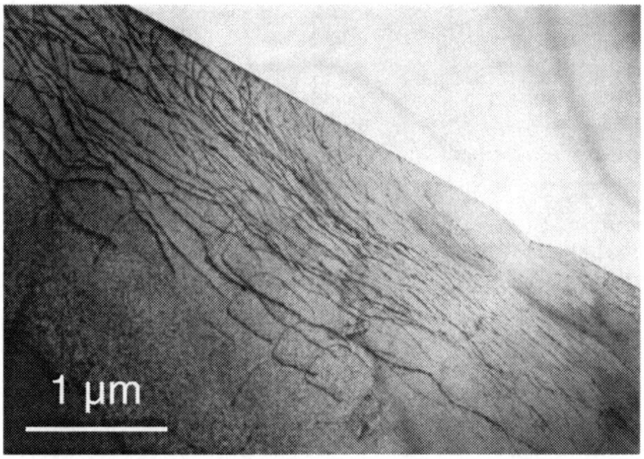

Fig. 8.16. Dislocations moving in a specimen with OR A2 on a plane with a trace parallel to the interface. From [498]. Copyright (2004) Carl Hanser Verlag, München

the original dislocation moves away. Finally, two new dislocations move away. The last two sequences present dislocations moving away from the crack. In the third part, the crack shortly appears at the upper edge of the frame. In the last sequence, the crack is located outside the image below the lower right corner.

To summarize the above, extensive plastic deformation occurs near the crack tip but also quite far away from it. Many dislocations move on planes of {112} type with high macroscopic orientation factors. Near the interface, dislocations also move on slip systems with zero macroscopic orientation factors, indicating a complicated stress state. New dislocations are created by the double-cross slip multiplication mechanism far from the crack. However, many dislocations originate also from regions near the crack tip. It cannot be decided whether these dislocations are emitted from the crack or multiply in regions of high stress near the crack tip.

9

Intermetallic Alloys

Intermetallic alloys are a special class of materials with mechanical properties that bridge the gap between metals of relatively low yield stress and good plasticity, and ceramics of high yield stresses even at high temperatures but strong brittleness at low ones. Many properties of intermetallic alloys result from their lower crystal symmetry. They increasingly gain structural applications at high temperatures because of their good strength-to-weight ratios.

9.1 Introduction

Ordered intermetallic compounds are alloys with two or more metal constituents, the atomic concentrations of which are close to fixed stoichiometric ratios. These constituents may form long-range order so that the different species of atoms occupy differently designated lattice sites in typical superlattice structures, which are based on the three principal metal structures, i.e., the f.c.c., b.c.c., and h.c.p. structures. Between the different atom species there may exist strong and also frequently directional bonds. The materials may have different degrees of atomic disorder resulting from low ordering energies or from deviations from the stoichiometric compositions. These deviations are realized by vacancies in specific sublattices or by antisite defects, i.e., by atoms occupying the sites of other constituents. General properties of intermetallic compounds are summarized, e.g., by Sauthoff [501] and by Westbrook and Fleischer [502].

In intermetallic compounds, the dislocations have complete Burgers vectors which are mostly multiples of the Burgers vectors in the corresponding simple metal structures. These superdislocations dissociate into superpartial dislocations including planar faults like antiphase boundaries (APBs) and different kinds of stacking faults. The dislocations may have complicated three-dimensional core structures, which impede the dislocation motion. The properties of dislocations and their relation to the characteristics of macroscopic deformation were summarized by Yamaguchi and Umakoshi [503].

Many intermetallic alloys show a flow stress anomaly, i.e., an anomalous increase of the flow stress with increasing temperature within a certain temperature range, mentioned earlier in Sect. 4.11 (Fig. 4.40). The anomaly is a desired property for structural applications since it retains a high strength up to high temperatures. The physical origins may be manifold and are frequently connected with transitions of the dislocation cores into an immobile state. On the other hand, the occurrence of the anomaly is often quite unspecific of the particular material. It is observed also at deformation by ordinary dislocations with simple cores. In these cases, the anomaly may be caused by diffusion processes as discussed in Sect. 4.11.

The basic features of the structure of intermetallics will be introduced on materials of the A_3B composition like Ni_3Al. These alloys frequently crystallize in the $L1_2$ structure shown in Fig. 9.1a. The B atoms occupy the corners of the underlying f.c.c. lattice, whereas the A atoms are situated in the face centered positions. The arrangement of the atoms on the (111) slip plane is illustrated in Fig. 9.1b. The complete Burgers vector is of type $\langle 110 \rangle$, e.g., from the B atom at 0 to the B atom at β. It is twice as long as the Burgers vector in the f.c.c. structure (see, e.g., [504]). The dislocations can dissociate into two superpartial dislocations with Burgers vectors of $1/2\langle 110 \rangle$, e.g., one from 0 to α and the other one from α to β. The shift of the a layer and all above it by the superpartial Burgers vector $\boldsymbol{b}_{\mathrm{A}} = 0\alpha = 1/2[\bar{1}10]$ creates an antiphase boundary (APB) as a planar fault. The fault may be annihilated by the second superpartial with the Burgers vector $\alpha\beta$. Superlattice dislocations enclosing an APB were first observed in [505] and have been mentioned before for cutting the particles with the $L1_2$ structure in an Al–Li alloy (Sect. 8.1.3).

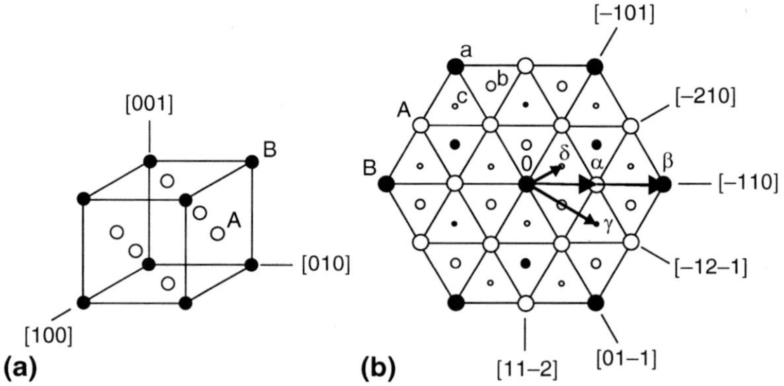

Fig. 9.1. Unit cell (**a**) and atomic arrangement on the slip plane (**b**) of an A_3B compound of the $L1_2$ structure. Three layers of atoms a, b, c are stacked in the ...$abcabc$... sequence and are drawn by circles of decreasing size

Other possible shift vectors on the octahedral plane are $\boldsymbol{b}_{\mathrm{S}} = 0\gamma = 1/3[\bar{1}2\bar{1}]$, which creates a so-called superlattice intrinsic stacking fault (SISF), and $\boldsymbol{b}_{\mathrm{C}} = 0\delta = 1/6[\bar{2}10]$ producing a complex stacking fault (CSF). The SISF on the $\{111\}$ plane causes an $\ldots abab \ldots$ (h.c.p.) stacking not violating the nearest neighbor relations. It corresponds to a usual stacking fault in the f.c.c. structure. The CSF also produces an h.c.p. stacking, however, violating the nearest neighbor relations. A method of studying the stability of the different planar faults is the construction of the so-called γ surface developed by V. Vitek for b.c.c. metals [506]. A planar fault is created by cutting the crystal along a plane and shifting both parts with respect to each other by a fault vector $\boldsymbol{f}$. The fault energy γ is determined by suitable atomistic methods (pair potential, embedded atom, first principles, or other methods) and plotted against $\boldsymbol{f}$. The minima in this plot, the γ surface, are the positions of the stable fault vectors. Calculations of the γ surface for the $L1_2$ structure with different central force potentials [507] showed that ABPs with the above fault vector $\boldsymbol{b}_{\mathrm{A}}$ are always stable on $\{001\}$ planes since the nearest neighbor relations are not violated. APBs on $\{111\}$ planes may be stable or unstable depending on the applied potential. The fault vector may have a component out of the fault plane. The SISFs are always stable on the $\{111\}$ plane. Similar planar faults exist also in the other crystal structures of intermetallics. Theoretical calculations of the γ surfaces and the core structure of intermetallic alloys are reviewed by Vitek and Paidar [508].

The different mechanisms explaining the flow stress anomaly in intermetallics can be classified into three groups. The details of the models will be described in the sections of the particular materials:

1. Formation of sessile dislocation locks due to particular nonplanar core configurations. These mechanisms are specific for particular crystal and dislocation core structures of intermetallics. Such sessile cores can result from cross slip of the leading superpartial from the $\{111\}$ slip plane in the $L1_2$ structure onto a $\{100\}$ cross slip plane to reduce the total energy of the dislocation. The cross slip is supposed to be thermally activated so that the rate of formation of the locks increases with increasing temperature. The locked configurations may be unlocked by the formation of double-kinks which smoothly spread along straight dislocations as for the simple Peierls mechanism (Sect. 4.2.2) or by unlocking long segments leading to a jerky motion (locking–unlocking mechanism, Sect. 4.4).
2. Cross slip processes not specific for the particular dislocation core structure. Segments of a screw dislocation may cross slip onto another slip plane and bow out on this plane so that the dislocation contains segments bowing out on different planes. The cusps between the segments are then not glissile in forward direction, they pin the dislocation. A variant of this model is the usual double-cross slip mechanism described in Sect. 5.1.1, where the jogs formed by double-cross slip are not glissile in forward

direction of the whole screw dislocation. The anomaly again results from higher cross slip frequencies at higher temperatures.
3. Various mechanisms involving point defects and diffusion. In the most simple case, thermal vacancies with their concentration increasing with increasing temperature are considered fixed localized obstacles to the dislocation motion. Besides, diffusing intrinsic and extrinsic point defects may form point defect atmospheres around the dislocations, which cause a diffusional drag as described in Sect. 4.11. Finally, dislocation cores may decompose into components with nonplanar Burgers vectors. The whole dislocation can then move only by the combination of glide and conservative climb, i.e., a point defect flow from one component dislocation to the other.

All these mechanisms explain the occurrence of a flow stress anomaly. However, they have different implications on the shape of the dislocations and their kinematic behavior, the stability and the activation parameters of deformation, and on the orientation dependence of the flow stress. Although a wealth of experimental data has been collected on the deformation of the different intermetallic materials, an unequivocal interpretation of these data is frequently not yet possible. The following materials are discussed mainly with respect to the flow stress anomaly.

9.2 Ni_3Al

The $L1_2$ ordered γ' Ni_3Al phase is the strong constituent of the γ/γ' nickel base superalloys. These alloys combine high strength at high temperatures with a good oxidation resistance. In single crystal form, they are employed, e.g., as turbine blades in the hot parts of aero jet engines. For a review, see [509]. The γ' phase has been studied for about 50 years. It precipitates in the form of cubes regularly arranged in a disordered γ phase matrix. The matrix forms narrow channels between the precipitates. Plastic deformation starts in these soft channels. The dislocations with $1/2\langle 110\rangle$ Burgers vectors become superpartial dislocations when they enter the hard γ' phase. Ni_3Al shows a flow stress anomaly with a peak temperature of about 750°C, which is responsible for the high strength of the Ni base superalloys at high temperatures. The anomaly is connected with a high work-hardening rate and a low strain rate sensitivity of the flow stress.

9.2.1 Microscopic Observations and Dislocation Dynamics

Ni_3Al may be considered a prototype material of intermetallics. Its dislocation properties are reviewed in several papers (e.g., [510]) and will therefore only briefly be reported here. The crystal structure and possible planar faults were described above. The slip system dominating the anomaly range of the flow stress is $\{111\}\langle 110\rangle$. Several dissociation schemes have been discussed. Those most important for the octahedral plane are (e.g., [503])

$$[\bar{1}10] = 1/2[\bar{1}10] + \text{APB} + 1/2[\bar{1}10] \tag{9.1}$$

$$= 1/6[\bar{2}11] + \text{CSF} + 1/6[\bar{1}2\bar{1}] + \text{APB} + 1/6[\bar{2}11] + \text{CSF} + 1/6[\bar{1}2\bar{1}]. \tag{9.2}$$

In [511], the fault energies were determined by weak-beam TEM. As mentioned earlier, the APB energy may be lower on the cube plane than on the octahedral one so that the APB extends also on this plane, depending on the ratio between the energies on both planes. This will be discussed in more detail in the next section. Dislocation configurations have been investigated in many studies mainly by TEM, which is not described here. The dislocations are then in a static configuration which need not be characteristic of mobile dislocations.

In situ experiments in a TEM were performed by Molénat and Caillard [512, 513]. From this work, a movie sequence of dislocation motion at room temperature is presented in the following video.

Video 9.1. *APB jumps in* Ni_3Al-0.25at%Hf *at room temperature (from the work in [512]). Courtesy of D. Caillard:* The weak-beam images of the dislocations are straight in screw orientation. The dislocations are dissociated into superpartials connected by an APB. Further dissociation of the superpartials cannot be resolved during the in situ experiments. Within the time resolution of the video recordings, the dislocations move in jumps. The width of the jumps can be quite large, e.g., a new dislocation appears suddenly in the upper part of the frame. Frequently, however, the dislocations jump over distances comparable with the width of the dislocations. This is shown in Fig. 9.2. The contrast feature marked by a triangle may be taken as a reference point. Such jumps over the distance of the dissociation width are called APB jumps. Only part of the upper dislocation in the movie undergoes an APB jump. The moved and un-moved parts are connected by macrokinks, as will be discussed in the next section. A detailed analysis shows that the dislocations in the mobile configuration extend mainly on the octahedral slip plane. Only dislocations at rest are dissociated on the cube cross slip plane. These features have been revealed only by in situ experiments. It was only later, that this mechanism was observed by conventional TEM, too [514].

A similar behavior is observed up to about 300°C [513]. Above this temperature up to the peak temperature, dislocation motion on $\{111\}$ planes during in situ experiments takes place in avalanches with large slip distances on many slip planes. After motion, the dislocations are locked by dissociation on the cube plane. Under shear stress, they may bow out on this plane.

Ni_3Al is probably the only intermetallic material in which dislocation velocities have been measured as a function of stress and temperature. Figure 9.3 presents the results of a study by Nadgorny and Iunin [515] using the stress pulse-double etching method (Sect. 2.2). The dislocations were created by a scratch. The velocities were measured on groups of dislocations

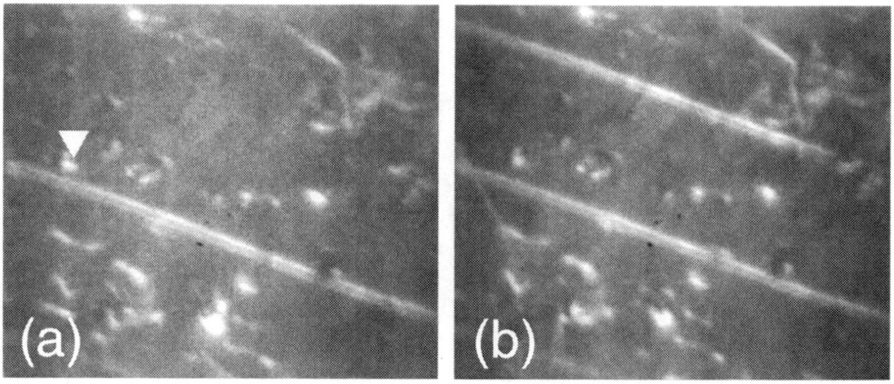

Fig. 9.2. Two stages of an APB jump in Ni_3Al-0.25at%Hf at room temperature. From the work in [512]. Courtesy of D. Caillard

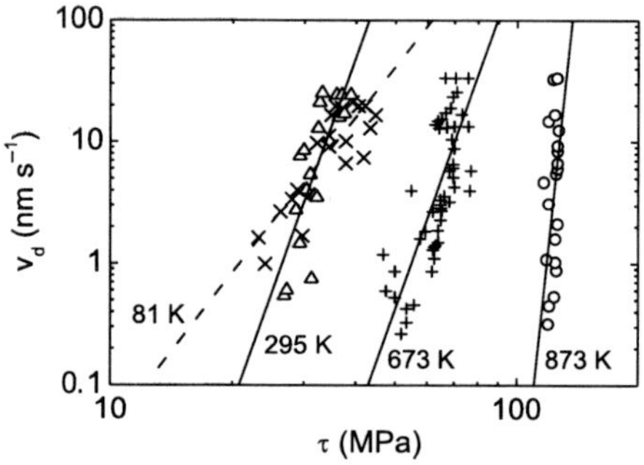

Fig. 9.3. Dependence of the dislocation velocity on the resolved shear stress τ in Ni-22.9at%Al single crystals at different temperatures. Data from [515]

having moved out of the scratch. The stress necessary to reach a certain dislocation velocity strongly increases with increasing temperature, which corresponds to the flow stress anomaly. At room temperature, the stress to attain a velocity of $3\,\mathrm{nm\,s^{-1}}$ is equal to the macroscopic flow stress. At high temperatures, it is lower, probably due to an increased athermal contribution of long-range dislocation interactions. The slope of the curves corresponds to the stress exponent m according to (4.11). The values of m are 4.2, 9.6, 9.5, and 31 for the temperatures of 81, 295, 673, and 873 K. The low values are consistent with the dislocation velocity being controlled by a thermally activated process. The high value of 31 at 873 K agrees with the usually observed low strain rate sensitivity in the anomaly range. These measurements are the only direct proof that the flow stress anomaly is caused by a reduction of the dislocation mobility and not by collective dislocation processes like work-hardening although the anomaly is mostly connected with high work-hardening rates.

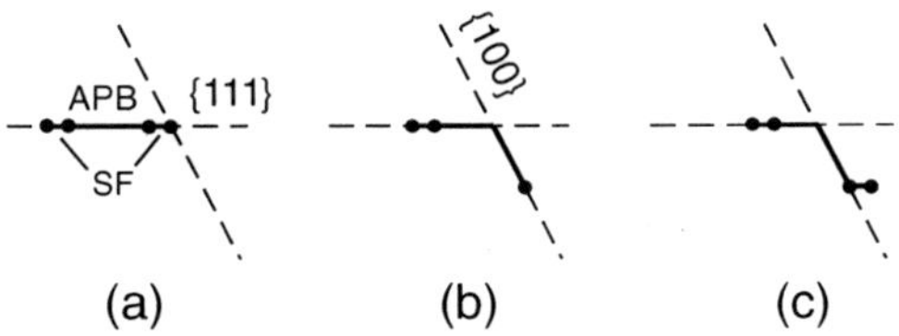

Fig. 9.4. Locking of a dislocation in the $L1_2$ structure by cross slip (Kear–Wilsdorf lock)

9.2.2 Models of the Flow Stress Anomaly

The basic model of explaining a flow stress anomaly by special properties of the dislocation cores in intermetallic materials is the formation of the so-called Kear–Wilsdorf locks after the paper by Kear and Wilsdorf [516]. As shown in (9.1), the superdislocation in the $L1_2$ structure can dissociate into two superpartials enclosing an APB. The superpartials can further split including CSFs as outlined in Fig. 9.4a. One superpartial may cross slip onto the $\{100\}$ cross slip plane (Fig. 9.4b) if the APB energy is lower on this plane. The cross-slipped superpartial may then dissociate again on a plane different from $\{100\}$ making the whole dislocation immobile. A first quantitative theory of the flow stress of intermetallics with the $L1_2$ structure including the flow stress increase with rising temperature in the anomaly range and the following high-temperature decrease was given by Takeuchi and Kuramoto [517]. The dislocations are assumed to form localized Kear–Wilsdorf locks by thermally activated cross slip from the $\{111\}$ glide plane onto the $\{100\}$ cross slip plane. The thermal activation is assisted by the shear stress on the cross slip plane so that an equation like (4.6) holds for the rate of activation. For the dynamic breakaway from these locked configurations, the shear stress on the glide plane has to exceed a critical value, and for the steady state motion, a condition has to be fulfilled similar to the Friedel criterion for localized obstacles (Sect. 4.5.1) that one new obstacle has to be formed after the dislocation has moved to the critical configuration for breakaway. The flight motion follows the viscous law (4.74). Because of the thermally activated nature of the formation of Kear–Wilsdorf locks by cross slip, the flow stress increases with increasing temperature. Above the peak temperature, strongly temperature depending slip on the cross slip plane leads to a rapidly decreasing flow stress.

A considerable extension of the theory was given by Paidar, Pope and Vitek [518]. It is usually quoted as the PPV model. In this theory, the formation of a kink pair of the leading superpartial on the $\{100\}$ cross slip plane is considered as depicted in Fig. 9.5. The energy for the formation of the double-kink is written as

$$H = W + 2Ch - \Delta EL - \tau_c bLh - Mh^2/2L \,.$$

Here, the first term is the energy of the two constrictions which have to be formed on the leading superpartial before cross slip can start. The second

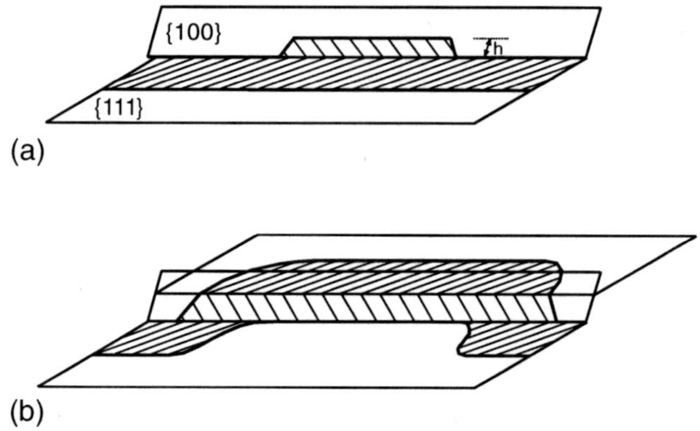

Fig. 9.5. Formation of a kink pair in the $L1_2$ structure from the {111} glide plane onto the {100} cross slip plane (**a**) and cross slip back to a parallel {111} plane (**b**). The dissociation of the superpartials into Shockley partials is not shown. After [520]

term is the energy of the kink pair on the cross slip plane with C being the specific kink energy per length, and h is the kink height (or width of the cross-slipped segment). The third term describes the gain in energy due to the change of the APB from the octahedral slip plane onto the cube cross slip plane. It is proportional to the length L of the cross-slipped segment. ΔE is positive only if $\gamma_0 < \gamma_1/\sqrt{3}$. γ_0 and γ_1 are the APB energies on the {100} and {111} planes. The fourth term is the work done by the external stress τ_c on the cross slip plane. Finally, the last term represents the interaction energy between the two kinks in a screw dislocation (4.23). The activation energy for the cross slip process is given for the saddle point configuration where the length L takes a critical value. The kink height is either $b/2$ or b. In the PPV model, the activation energy for cross slip, i.e., that for locking, is controlled by four effects: the difference in the APB energies γ_0 and γ_1, the shear stress on the cross slip plane, the elastic energy difference between the stress induced compression or dilatation of the superpartial pair before a jump and the equilibrium configuration afterwards, and finally, the nature of the dissociation of the superpartials on the octahedral planes. For the unlocking process, the same assumptions are made as for the model of Takeuchi and Kuramoto [517]. The PPV theory explains the occurrence of a flow stress anomaly as well as its orientation dependence and a tension–compression asymmetry, which all have been observed experimentally.

The models described so far do not consider certain microstructural observations, in particular the occurrence of the APB jumps in the lower temperature range of the flow stress anomaly (Video 9.1). D. Caillard and V. Paidar have extended the theory to include these features [519, 520] by assuming that the locking cross slip process is of short range and occurs only over widths of b or a few b to form incomplete Kear–Wilsdorf locks. Unlocking takes place by a second cross slip back to a plane parallel to the original

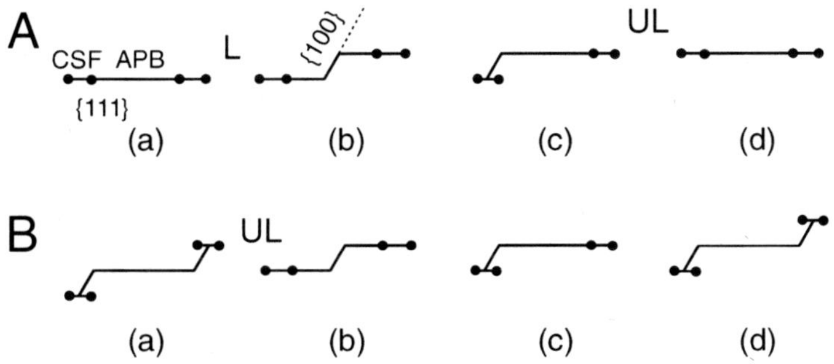

Fig. 9.6. Dislocation jumps over variable distances (*line A*) and APB jumps (*line B*) in intermetallics with the $L1_2$ structure. *L* locking, *UL* unlocking. After [519]

octahedral slip plane, similar to the double-cross slip mechanism for dislocation multiplication (Sect. 5.1.1). As illustrated in Fig. 9.6, a dislocation in the mobile configuration (a) in line A locks (L) itself by cross slip of the leading superpartial onto a cube plane (b). In (c), it is still locked by the trailing superpartial. When the latter removes the APB on the cube plane by gliding to the plane of the leading superpartial, the dislocation is unlocked (UL) and the mobile configuration is restored (d). The dislocation can then glide over relatively long variable distances. This process is one case of the locking–unlocking mechanisms briefly described in Sect. 4.4. In line B of Fig. 9.6, the sequence starts with a dislocation locked at both superpartials (a). The dislocation can unlock to take the same semi-locked configuration (b) as in line A. If the leading superpartial in (c) cross slips before the trailing one is unlocked, the whole dislocation is completely blocked again (d). The slip distance now corresponds to the width of the APB in the locked configuration. Since the cross slip heights are small in both cases, the macroscopic plane of motion does not deviate remarkably from the octahedral glide plane. With increasing temperature, locking becomes more complete, i.e., the dislocations spread on the cross slip plane. These complete Kear–Wilsdorf locks unlock only at high temperatures.

In [519], all forces acting on the long and straight superpartial dislocations are calculated, similar to [521,522]. The ranges of the existence of incomplete Kear–Wilsdorf locks in plots of the normalized stress on the octahedral glide plane on the cross slip height are then established for a fixed value of $\gamma_0/\gamma_1 = 0.8$ and different values of stress on the cross slip plane. The extension of the model [520] takes into account the thermally activated nature of the cross slip events. It also distinguishes between a process similar to the kink pair formation with a cross slip height $h = b$ as shown in Fig. 9.5a with a low activation energy to form incomplete Kear–Wilsdorf locks, and cross slip over larger distances, which is controlled by a critical bow out in the cross slip plane. This process requires a higher activation energy and leads to incomplete Kear–Wilsdorf locks of different strengths and to complete locks. In both cases, by cross slip the leading partial returns onto a plane parallel to the original slip plane as illustrated in Fig. 9.5. The model explains most of the experimental results, including the strong work-hardening in the anomaly range.

A quite different approach to the motion of a superdislocation with kinks was taken in the simulations by Mills and Chrzan [523]. Dislocations are pinned locally by random events with their probability depending on the local orientation of the dislocation line. The distribution of the pinning agents turns out not to be random but to consist of strongly pinned near-screw segments connected by mobile superkinks. The dislocation moves by sidewise spreading of the kinks at high velocities. Owing to fluctuations in the density of the kinks, whole segments may become immobile. Thus, the jerky motion observed experimentally is not a consequence of locking and unlocking of long segments but of the nonrandom distribution of the macrokinks. Several versions of such models have been discussed in [524]. Petukhov [525], too, gave a statistical treatment of the formation of pinning agents by cross slip and subsequent unpinning.

In conclusion one can say, a number of refined models have been developed for estimating the flow stress of intermetallic compounds with the $L1_2$ structure, particularly of the flow stress anomaly. These models are based on the dissociation of the superdislocations into superpartials connected by an APB. The energy of the APB being lower on the cube cross slip plane than on the octahedral glide plane, together with a shear stress component acting on the cross slip plane lead to localized cross slip, which locks the cross-slipped dislocation segments. Such models are exemplary also for intermetallics with other crystal structures. However, as will be discussed in the following sections, the flow stress anomalies may have many other origins, too.

9.3 γ-TiAl

Titanium aluminides are prospective materials for structural applications at intermediate temperatures owing to their light weight, their relatively good corrosion resistance and, particularly, to their high flow stress up to temperatures above 650°C. The latter is due to the occurrence of a flow stress anomaly, at least a constancy of the flow stress with increasing temperature over a wide range. Materials of technological interest based on Ti–Al are mostly two-phase alloys consisting of the soft TiAl γ-phase having the $L1_0$ structure and the harder Ti_3Al α_2-phase with the $D0_{19}$ structure. Single-phase γ-alloys have some major disadvantages regarding their room temperature formability, creep and also strength properties. However, in two-phase alloys, the major part of the plastic deformation occurs in the γ-phase so that the single-phase γ-alloys may serve as a model material of the plastic constituent of other TiAl based materials. Comprehensive reviews on γ-TiAl based materials are given, e.g., in [526–528].

The single-phase γ-alloys form on the Al-rich side of the stoichiometric TiAl composition, whereas the two-phase ($\alpha_2 + \gamma$) alloys occur on the Ti-rich side. In the latter, different microstructures may develop depending on further alloying elements and the thermo-mechanical treatment. These microstructures frequently involve a lamellar structure consisting of relatively wide twin-related lamellae of the γ-phase and thin α_2 lamellae. These microstructures essentially influence the macroscopic deformation properties of the technical alloys. However, the present section is restricted to the dislocation behavior and plastic properties of the γ-phase, in particular to the deformation at room temperature and in the anomaly range. In Ti_3Al, dislocation dynamics was studied by in situ straining experiments in a TEM in [529, 530].

9.3.1 Crystal Structure and Slip Geometry

The $L1_0$ structure of the γ-phase of TiAl is an f.c.c. based structure with a small tetragonality of $c/a \approx 1.02$. It is characterized by lattice planes alternately occupied by Al and Ti atoms as demonstrated in Fig. 9.7a. Consequently, the {111} slip planes consist of separate rows of Al and Ti atoms in ⟨110⟩ directions (Fig. 9.7b). The shift vector $\boldsymbol{b}_o = 0\alpha = 1/2\langle 110\rangle$ does not produce a planar fault, it corresponds to complete dislocations. Frequently, the vectors in this tetragonal structure are also written as $1/2\langle 110]$ to indicate that the third index may not be mixed up with the other two. However, this is self-evident in the tetragonal structure. Dislocations with $1/2\langle 110\rangle$ Burgers vectors are called simple or ordinary dislocations. The shift vector $\boldsymbol{b}_{sp} = 0\beta = 1/2\langle 101\rangle$ causes an APB and corresponds to a superpartial dislocation. The complete superdislocation has the Burgers vector $\boldsymbol{b}_s = 0\gamma = \langle 101\rangle$.

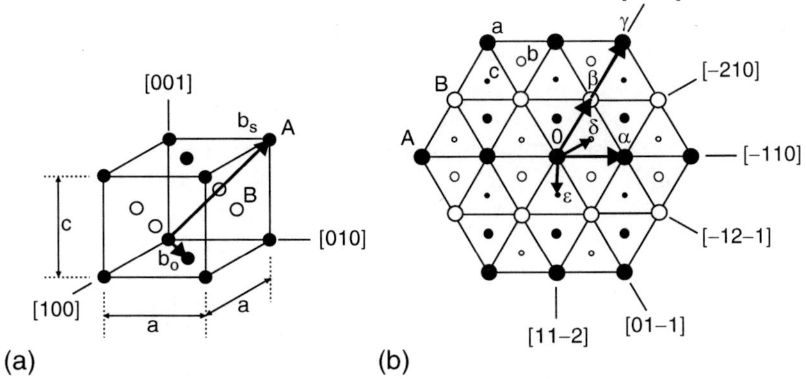

Fig. 9.7. Unit cell with Burgers vectors $\boldsymbol{b}_o$ and $\boldsymbol{b}_s$ of ordinary and superdislocations (**a**) and atomic arrangement on the {111} slip plane (**b**) of an AB compound with the $L1_0$ structure. Three layers of atoms *a*, *b*, *c* are stacked in the ...*abcabc*... sequence and are drawn by circles of decreasing size

In Fig. 9.7a, the Burgers vectors of ordinary and $\langle 101 \rangle$ superdislocations are indicated. There exist also superdislocations with $\boldsymbol{b}_{s2} = 1/2\langle 112 \rangle$, which are rare and are not considered further. The fault vectors 0δ and 0ϵ produce a CSF and an SISF. In addition to dislocation motion, γ-TiAl may also be deformed by twinning of $\{111\}\langle 11\bar{2} \rangle$ type.

9.3.2 Microscopic Observations

The following observations are based on in situ straining experiments in an HVEM on coarse-grained single-phase γ-Ti-52at%Al containing 750 at ppm of oxygen [286], and on two-phase $(\alpha_2 + \gamma)$ Ti–47Al–2Cr–0.2Si (at%) [531] with a so-called near-γ microstructure consisting of large γ grains with the α_2-phase located in the grain boundary pockets. In situ experiments in a conventional TEM in [532, 533] yielded similar results.

Room Temperature

The room temperature deformation is mainly carried by dislocations, mostly ordinary ones and occasionally $\langle 101 \rangle$ superdislocations as demonstrated in Fig. 9.8 for the near-γ alloy. The figure shows the same specimen area under different imaging conditions so that in Fig. 9.8a only the ordinary dislocations are imaged, and in Fig. 9.8b only the superdislocations. The majority of dislocations are of screw character and move on $\{111\}$ planes. Most dislocations move in slip bands. The dislocation density within these bands is of the order of magnitude of 1×10^{13} to $3 \times 10^{13}\,\mathrm{m}^{-2}$.

The ordinary dislocations are of curly shape indicating that the dislocation motion is impeded by obstacles between which the dislocation segments bow out under stress. In most cases, the obstacle sites are not aligned accurately in screw direction, instead they are distributed over the slip plane. To determine the spatial arrangement of the bowed-out dislocation segments, some stereo pairs of the dislocation structure were taken under load and in the unloaded state after the in situ experiments. They reveal that most of the segments bow out on a single slip plane, or on parallel planes, respectively. These observations are important for identifying the obstacle mechanism. The deep cusps in the dislocation line are clearly due to jogs, some of which are marked by J in Fig. 9.8a. The jogs trailing dipoles (dislocation debris) are marked by D. Most of the dipoles are not arranged in edge direction. If the jogs are high enough, both arms of the dipole may pass each other as at M, leading to dislocation multiplication. There are only a few micrographs showing edge dislocation segments. Under load, edge dislocations exhibit the same curly shape as screw dislocations do, as is demonstrated in Fig. 9.9. Figure 9.8b illustrates that superdislocations S are pinned like the ordinary ones. This is shown more clearly in Fig. 9.10 at the head of a slip band of superdislocations. However, their bowing is always less strong. Slip traces prove that these dislocations had moved during the in situ experiment. In conclusion, it may be

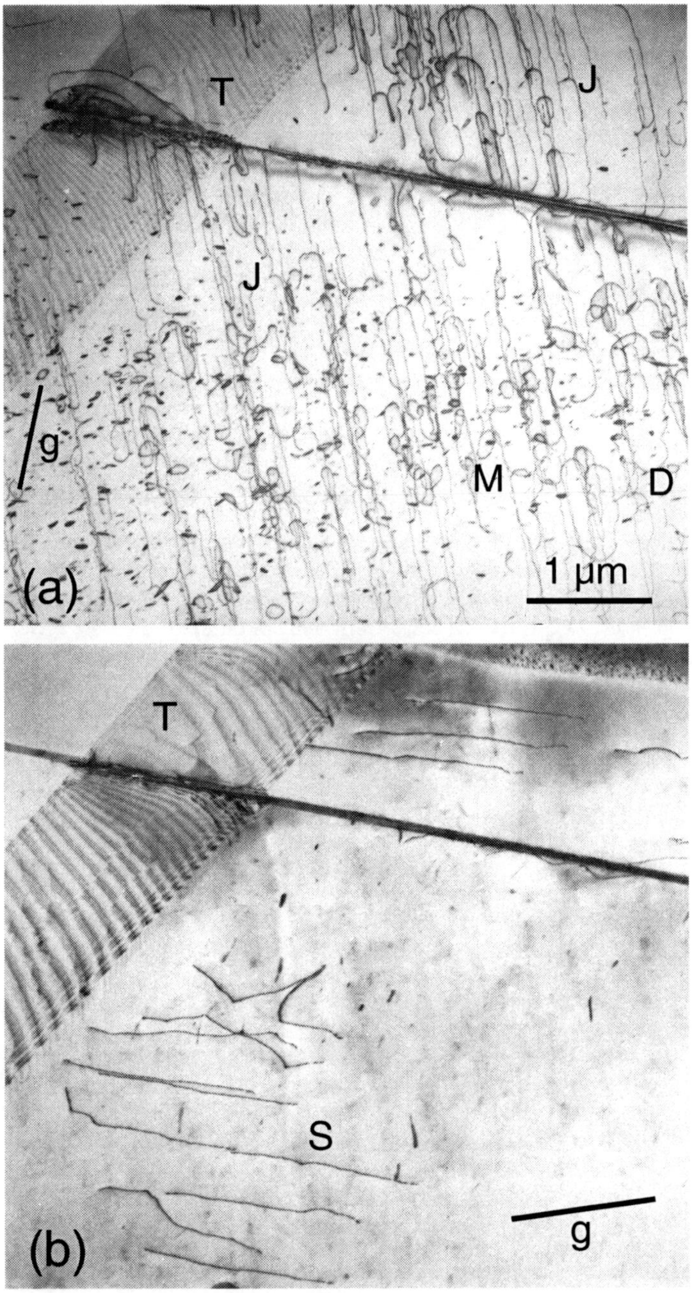

Fig. 9.8. Dislocations and a twin T under load during the in situ deformation of near-γ TiAl at room temperature. Image normal [101]. (**a**) Ordinary dislocations with $\boldsymbol{b} = 1/2[\bar{1}\bar{1}0]$ imaged with $\boldsymbol{g} = [\bar{1}\bar{1}1]$. Superdislocations with $\boldsymbol{b} = [0\bar{1}\bar{1}]$ are extinguished. (**b**) The superdislocations are imaged with $\boldsymbol{g} = [\bar{1}11]$, whereas the ordinary dislocations are out of contrast. From the work in [531]

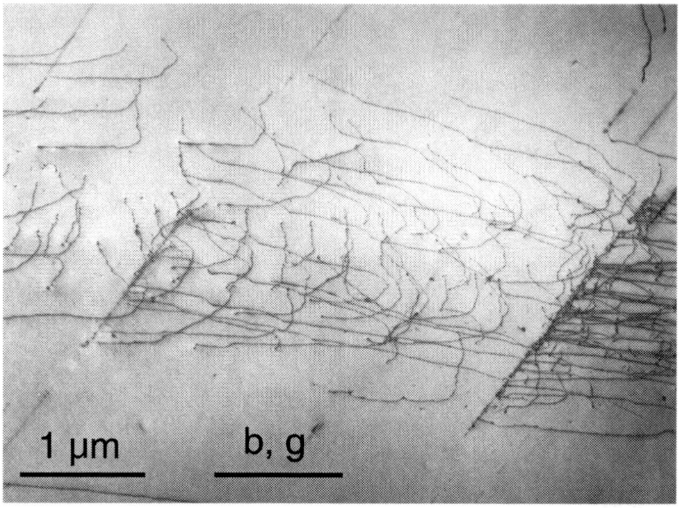

Fig. 9.9. Slip band taken during in situ deformation of γ-TiAl at room temperature showing ordinary mixed and edge dislocations. Diffraction vector $\boldsymbol{g} = [1\bar{1}0]$. $\boldsymbol{b} = 1/2[1\bar{1}0]$ is the only possible Burgers vector of ordinary dislocations at the [110] pole. The traces in $[1\bar{1}\bar{2}]$ directions are from defects on $(1\bar{1}1)$ planes, probably twins. The ordinary dislocations are created at these defects. From the work in [286]

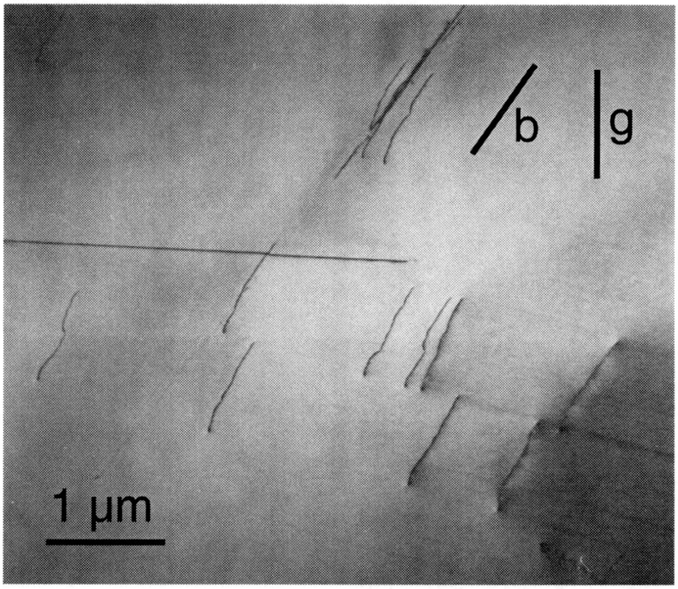

Fig. 9.10. Head of a slip band of superdislocations during in situ deformation of γ-TiAl at room temperature. $\boldsymbol{b}$ projection of [101] Burgers vector, $\boldsymbol{g} = [001]$ diffraction vector near the [110] pole. From the work in [286]

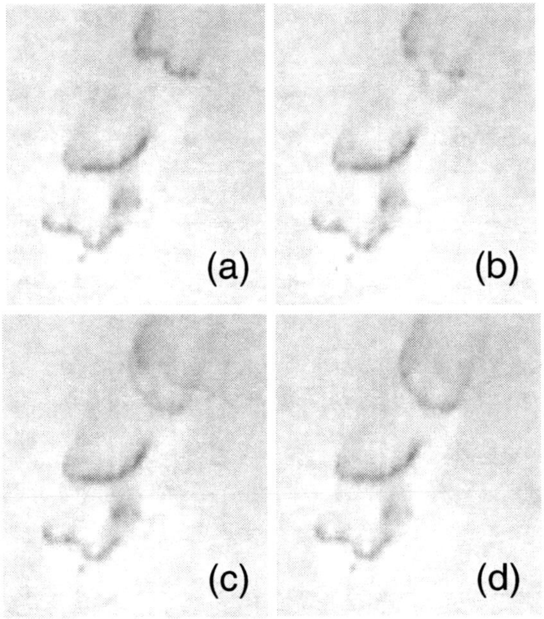

Fig. 9.11. Four successive frames from the first part of Video 9.2 of the motion of ordinary dislocations in γ-TiAl

pointed out that obstacles impede the motion of both ordinary dislocations and superdislocations, independent of their character, i.e., screw or edge. Figure 9.8 shows also a twin lamella T. Below, a remark will be made about the twinning dislocations.

Video recordings reveal the kinematic behavior of the moving dislocations.

Video 9.2. *Motion of ordinary dislocations in* γ-TiAl *at room temperature:* This clip consists of two sequences showing the motion of individual ordinary screw dislocations. In the first clip, the dislocations move by jumps of individual segments over areas comparable with the square of the obstacle distance. This is demonstrated by the sequence of successive frames in Fig. 9.11. In (**b**) and (**c**) several positions are superimposed due to the afterglow of the luminescence screen of the recording system. In the second one, dislocations frequently move as a whole over distances in the order of magnitude of the obstacle distance. They are curved also in the intermediate states.

In slip bands of a higher density of dislocations, the dislocations frequently move in a collective way, as demonstrated by the following video sequences.

Video 9.3. *Motion of ordinary dislocations in slip bands in* γ-TiAl *at room temperature:* The video presents three clips of ordinary dislocations moving in slip bands:

1. A narrow band originates from a crack. The dislocations move in a jerky mode over distances large compared to the obstacle distance along the dislocations, frequently in a collective way.
2. This clip shows the head of a slip band mostly consisting of screw dislocations. Partially, the leading dislocations overcome individual obstacles. Some of them are jogs which, along the screw dislocations, are driven toward the edge components. The edge segments are also locally pinned. Later on, dislocations are driven by the stress field of the succeeding dislocations and move over larger distances.
3. In the interior of the band in 2., many dislocation segments are of near edge character. Their mode of motion resembles that of the screw dislocations.

These sequences illustrate that the kinematic behavior of mixed and edge ordinary dislocations is similar to that of screws, and that the dislocations in dense bands move in a collective way, indicating the importance of long-range elastic interactions between the dislocations.

Superdislocations mostly move in jumps over long distances. Only once their motion has been recorded on video tape.

Video 9.4. *Motion of* $\langle 101 \rangle$ *superdislocations in* γ-TiAl *at room temperature:* The video is from the same experiment as Fig. 9.10 is. The image normal is [110]. $1/2[\bar{1}10]$ is the only possible Burgers vector direction of ordinary dislocations, which is perpendicular to the [001] diffraction vector and the image normal. There are no (screw) dislocations visible in this orientation. Thus, all dislocations in this video are superdislocations. A few dislocations move almost simultaneously right of the circular marks. The left ones belong to the $(\bar{1}11)$ slip plane, which is oriented edge-on. The right one glides on the (111) plane, which is also the slip plane of Fig. 9.10. Since the specimen normal is not identical with the image normal, the traces of the slip plane in the video and in the figure are not perpendicular to the [001] $\boldsymbol{g}$ vector. These superdislocations move over shorter distances in a jerky way.

Part of the plastic deformation of γ-TiAl is also carried by the formation of deformation twins. This mechanism was briefly introduced in Sect. 3.3.3. The motion of twinning dislocations was recorded in the following video.

Video 9.5. *Motion of twinning dislocations in* γ-TiAl *at room temperature:* Between obstacles, the twinning dislocations bow out like ordinary and superdislocations. However, owing to their low line tension, the bowing is very strong. The jerky motion of the twinning dislocations is most similar to that of the ordinary ones, including the collective behavior in the twinning lamella.

It may be concluded in a qualitative sense that all types of dislocations, ordinary, super, and twinning dislocations, are impeded by similar obstacles and that they move in a jerky way by jumps over different distances, depending on the average dislocation velocity.

To obtain quantitative data on the obstacle mechanism, the distances l between the pinning points were measured along the dislocations. For ordinary dislocations of all characters as well as for screw superdislocations the average values are always around 100 nm. The histogram of the obstacle distances along ordinary screw dislocations is of the asymmetric shape typical of localized obstacles as discussed in Sects. 4.5.1 and 7.2.2 ($H(v)$ in Fig. 7.10b). The effective stress τ^* corresponding to an equilibrium curvature of the bowed-out dislocation segments was determined by fitting the shape of the segments calculated by the line tension theory of elastically anisotropic materials to the shape in the micrographs as described in Sect. 5.2.3 and by (5.18). The line tension data used were calculated by Yoo [534]. While ordinary dislocations are of almost circular shape, superdislocations are elliptic, with their screw segments showing a very low curvature. Ordinary and superdislocations indicate the same value of the effective stress of $\tau^* \approx 40$ MPa. It is important to note that the dislocation bowing only slightly relaxes after unloading.

High Temperatures

Similarly to room temperature, deformation at intermediate temperatures occurs by the motion of ordinary and superdislocations as well as by twinning. Figure 9.12 shows ordinary dislocations moving during the deformation at elevated temperatures of 360 and 575°C. The different widths of the slip traces in Fig. 9.12a, e.g., of dislocations A and B, indicate that the dislocations must contain large jogs. They cause a wide variation of the width of the slip trails because of the flat inclination of the slip planes. In spite of the jogs, the dislocations are quite smooth, i.e., the extent of bowing is drastically reduced compared to that at room temperature so that the cusps in the dislocation line corresponding to the obstacles are scarcely visible. Nevertheless, the average obstacle distance is still approximately equal to its room temperature value. Most of the dislocations are additionally pinned at the surfaces of the specimen, probably owing to an oxide layer. The kinematic behavior of the dislocations is similar to that at room temperature. At 575°C (Fig. 9.12), the number of cusps is strongly decreased. The dislocations are pinned by only a few jogs about 0.5–1 μm apart.

Compared to lower temperatures, in the high-temperature range, the shape as well as the dynamic behavior of dislocations have changed strongly. The obstacles, which affect the motion of all dislocations at room temperature and at intermediate ones, no longer act above about 600°C. As demonstrated in Fig. 9.13, the ordinary dislocations with $1/2[1\bar{1}0]$ Burgers vectors are smoothly bent with large radii of curvature. The spatial arrangement of the dislocations was proved by stereo pairs taken after the in situ experiments at room temperature. The observed nonplanar arrangement shows that their motion is not restricted to their slip planes. In addition to the curved dislocation segments being predominant, some segments are straight and arranged in $[10\bar{1}]$ direction. They are the only segments lying on the original (111) slip plane.

The dynamic behavior of dislocations at high temperatures drastically differs from that at low and intermediate temperatures. On loading the

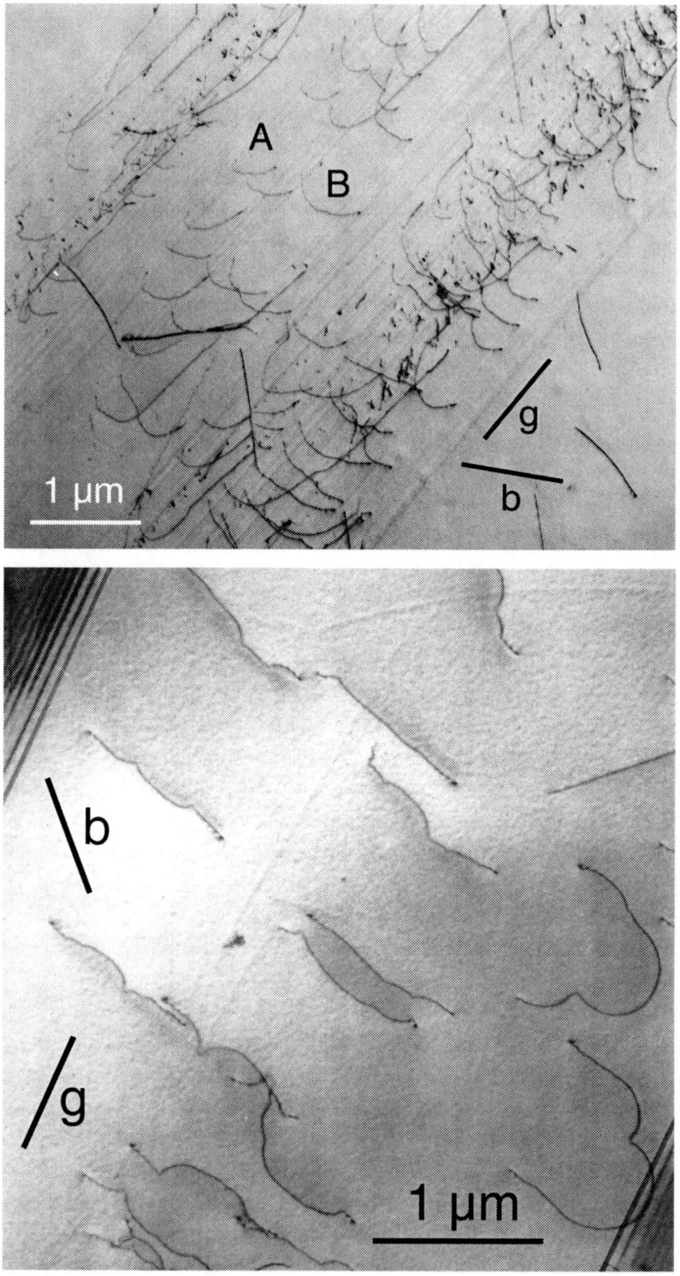

Fig. 9.12. Ordinary dislocations of mainly screw character moving during in situ straining of γ-TiAl at elevated temperatures. (**a**) 360°C, $\boldsymbol{g} = [101]$ diffraction vector, $\boldsymbol{b}$ projection of 1/2[110] Burgers vector. (**b**) 575°C, $\boldsymbol{g} = [1\bar{1}0]$ diffraction vector, $\boldsymbol{b}$ projection of $1/2[1\bar{1}0]$ Burgers vector. From the work in [286]

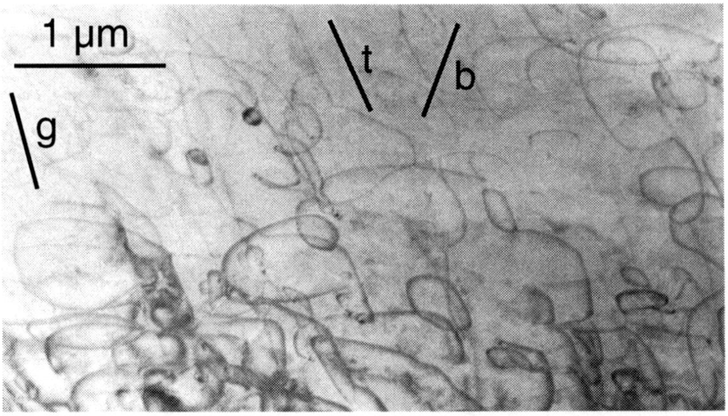

Fig. 9.13. Ordinary dislocations under load during in situ deformation of γ-TiAl at 655°C. $\boldsymbol{g}$ = [111] diffraction vector, $\boldsymbol{b}$ projection of $1/2[1\bar{1}0]$ Burgers vector, $\boldsymbol{t}$ projection of $[10\bar{1}]$. From the work in [286]

specimens for the first time during the in situ experiments above about 600°C, large numbers of ordinary dislocations and some superdislocations usually appear instantaneously after pronounced load drops like those demonstrated in Fig. 9.39 for $MoSi_2$ single crystals. This avalanche-like multiplication has never been recorded. If ordinary dislocations again start to move, their motion is no longer jerky but smooth in a viscous way as illustrated by the following video.

Video 9.6. *Motion of ordinary dislocations in* γ-TiAl *at* 655°C*:* In the temperature range of the flow stress anomaly, smoothly curved dislocations move in a viscous way.

Superdislocations with ⟨101⟩ Burgers vectors occur less frequently at elevated temperatures than simple dislocations do. Figure 9.14 shows that they are characterized by long screw segments linked by superkinks. The movement of these superdislocations has never been recorded. Obviously, they jump over larger distances, adopt the screw orientation and get locked.

9.3.3 Macroscopic Deformation Parameters

The temperature dependence of the yield stress of γ-TiAl is demonstrated in Fig. 9.15, together with two NiAl materials which will be described later. In contrast to the NiAl materials, which do not show a flow stress anomaly, γ-TiAl exhibits a weak anomaly, with the yield stress slightly increasing up to about 700°C.

The strain rate sensitivity of the flow stress of γ-TiAl was studied over a wide range of strain rates by stress relaxation experiments, where the dependence of the strain rate sensitivity r on the strain rate $\dot{\varepsilon}$ can be followed along

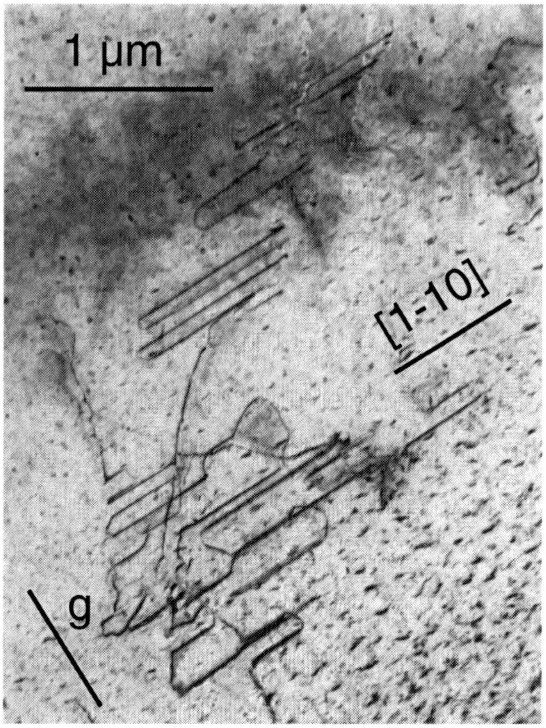

Fig. 9.14. Superdislocations in γ-TiAl under load which had moved at 655°C. $\boldsymbol{g}$ = [001] diffraction vector. From the work in [286]

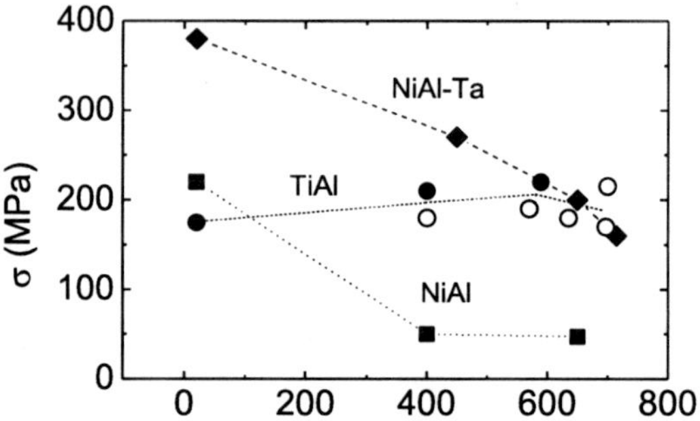

Fig. 9.15. Temperature dependence of the yield stress of the intermetallic alloys γ-Ti-52at%Al, NiAl and NiAl-0.2at%Ta. The NiAl single crystals are deformed along the soft ⟨421⟩ orientation. Data from [535]

a single relaxation curve (see Sect. 2.1). These relaxation curves cover more than three orders of magnitude of the strain rate. r is low at room temperature as listed in Table 9.1. At and above 400°C, the relaxation curves show the typical "inverse" curvature with a low strain rate sensitivity at high stress (or

Table 9.1. Dependence of the strain rate sensitivity r and activation volume V in γ-TiAl on the temperature

T (°C)	r (MPa)	V (nm^3)
21	0.9	13.5
400	2.27	12.3
570	4.2	8.3
630	7.55	4.95
700	12.7	3.17

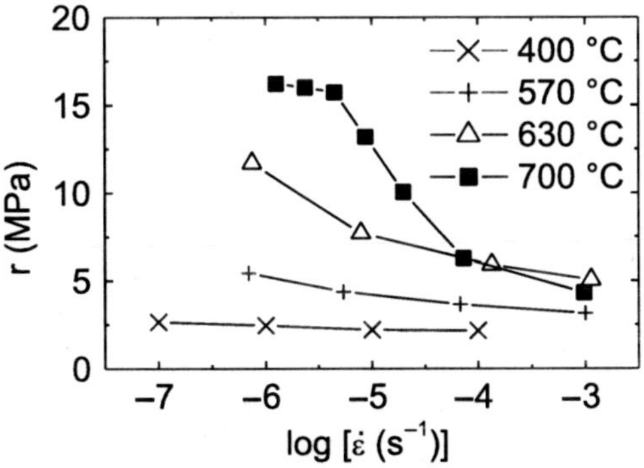

Fig. 9.16. Dependence of the strain rate sensitivity of the stress in γ-Ti-52at%Al on the logarithm of the strain rate with the temperature as parameter. Data from [535]

strain rate), and with a higher strain rate sensitivity at lower stresses, like the curve R1 in Fig. 8.2b measured on Al–Zn–Mg. Strain rate sensitivities were determined at different points along the relaxation curves and then plotted versus the actual strain rate. The latter was obtained from the stress rate using the stiffness of the sample and fixtures during unloading at the end of the experiments according to (2.7). Data from experiments at different temperatures are plotted in Fig. 9.16. Table 9.1 presents the values at $\dot{\varepsilon} = 10^{-5}\,\mathrm{s}^{-1}$ together with the corresponding activation volumes V from (4.9), based on an average orientation factor of $m_s = 1/3$. The activation volumes are constant up to 400°C before they decrease. At 400°C, the strain rate sensitivity decreases only weakly with increasing strain rate. With increasing temperature, r increases, but its decrease with increasing strain rate becomes very pronounced. In general, only a minor part of the total flow stress (about 30% at 630°C) relaxes during usual relaxation times (about 10 min).

9.3.4 Deformation Mechanisms

Though the studied single-phase γ-TiAl shows only a slightly increasing flow stress and a decreasing activation volume over the whole temperature range investigated, the dislocation processes differ greatly between room temperature and high temperatures. Nevertheless, the microstructure is always characterized by the presence of ordinary dislocations, superdislocations and microtwins with a clear dominance of ordinary dislocations. The difference in the role of the three types of dislocations with respect to the results in [536] may be due to the lower aluminium content of the present material.

Room Temperature up to About 430°C

The essential experimental observations in the low-temperature range are as follows:

- Ordinary dislocations bowing out between obstacles about 100 nm distant dominate the deformation.
- The cusps in the dislocation line forming at the obstacles are aligned not exactly along the screw orientation but are spread on the slip plane.
- Neighboring segments of ordinary dislocations bow out on the same $\{111\}$ plane, or on parallel planes.
- Under stress, the bowing decreases with increasing temperature.
- Superdislocations with $\langle 101 \rangle$ Burgers vectors show the same bowed-out shape as the ordinary ones do.
- All types of dislocations show the same kind of jerky motion.

Ordinary dislocations dominate the deformation in the described experiments at all temperatures. In contrast to [537] but in agreement with [538], they have a high mobility. The theory predicts that ordinary dislocations have extended cores leading to a low mobility [539–541]. Experimental observations, however, show that the cores of ordinary dislocations should be rather compact [542]. Nevertheless, theoretical estimates suggest a high Peierls stress (e.g., [543]), which agrees with the observation that dislocation segments bowing out under stress relax only little after unloading.

A number of transmission electron microscope studies have been carried out to assess the contributions of the different dislocation types at different temperatures. While some authors mainly observe superdislocations at room temperature (e.g., [537]) others find a dominance of ordinary dislocations at all temperatures (e.g., [544]). Apparently, the activation of the different types of dislocations depends on the concentration of oxygen impurities within the specimens [544–546] as well as on the content of aluminium [536]. In the experiments described above, superdislocations do not play an important role. Thus, the following discussion is based mainly on the properties of ordinary dislocations.

The literature offers three models to explain the curly shape of ordinary dislocations, the first two of which are also thought to cause the flow stress anomaly at higher temperatures:

1. Louchet and Viguier [547] and Viguier et al. [548] assume that neighboring segments of ordinary screw dislocations cross slip on different $\{111\}$ planes and bow out on them as schematically drawn in line a of Fig. 9.17.

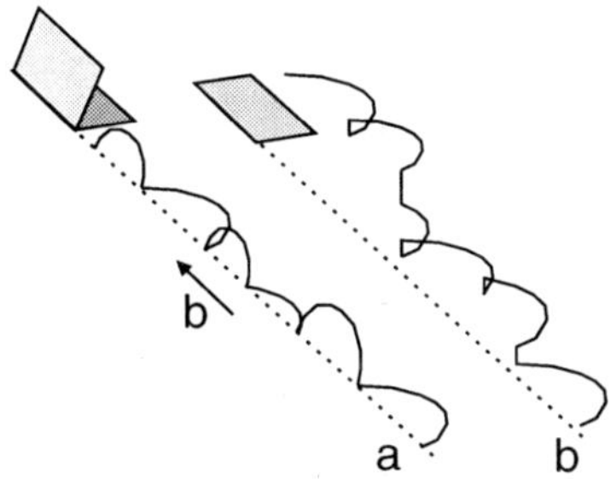

Fig. 9.17. Models of pinning by cross slip in TiAl. *Line a*: Cross slip of individual segments on different planes according to [547, 548]. *Line b*: Double-cross slip as described in Sect. 5.1.1, Fig. 5.5

The points linking these segments are aligned almost exactly in screw orientation. They cannot glide together with the dislocations so that they form obstacles. The dislocations move very jerkily by pushing the pinning links along the dislocations in the direction of the Burgers vector to form long mobile segments, which spread very quickly. This mechanism is also called the local pinning unlocking (LPU) model. The present observations are not in accordance with this model as the obstacles are aligned not exactly along the screw orientation. Besides, obstacles occur along all types of dislocations, including edge dislocations. While jogs in edge dislocations may impede the dislocation motion by their lattice friction, too, leading to bowed-out segments, edge dislocations cannot cross slip, which is a prerequisite to the formation of pinning agents according to the LPU model. Besides, stereo images do not show neighboring segments bowing out on different crystallographic planes but mostly on parallel ones.

2. Sriram et al. [549] carefully investigated the spatial arrangement of the ordinary dislocations by post-mortem electron microscopy. They found the individual segments bowing out on parallel planes and the cusps not strictly aligned in screw direction as illustrated by line b in Fig. 9.17. The authors assume that this structure of the dislocations is formed by the double-cross slip mechanism described in Sect. 5.1.1, i.e., all cusps in the dislocations correspond to jogs with a spectrum of heights ranging from the atomic height up to a maximum height which is determined by the dipole opening criterion. These experimental observations are in accordance with most of the features of the microstructure of the resting dislocations at low temperatures. Deep cusps or the trailing of dipoles suggest that the ordinary screw dislocations have many jogs. However, cusps were also observed in mixed and edge dislocations as proved in Fig. 9.9. As discussed earlier, edge dislocations cannot cross slip to form these jogs. The kinematic behavior of dislocations presented in the Video 9.2 and in Fig. 9.11 does not agree with jogs being the dominating obstacles on screw dislocations. The jogs can glide along the direction of the Burgers

vector. This motion leads to long segments with very large bow-outs, before the configuration gets unstable. This mechanism was identified in in situ straining experiments on NiAl single crystals at room temperature [287] and will be described in Sect. 9.4. Such long segments have not been observed in TiAl. Thus, while the double-cross slip mechanism may explain most of the features of the resting dislocations as also observed in post-mortem studies, it fails to explain the kinematic behavior of dislocations in the in situ experiments at room temperature.

3. Almost all experimental results can be interpreted consistently if it is assumed that during the deformation between room temperature and about 430°C, in addition to deep cusps at high jogs, the weak cusps in the dislocations result from an interaction between dislocations and extrinsic spatially localized obstacles following the theory described in Sect. 4.5.1, i.e., probably small precipitates. The main arguments are the following. The cusps in the resting ordinary dislocations are lined up not exactly along the screw orientation but are distributed on the slip plane (Fig. 9.8). The cusps appear in all types of dislocations, also in edge and mixed ordinary dislocations (Fig. 9.9), superdislocations (Fig. 9.10) and in twinning partial dislocations (Video 9.5, see also [532]). The latter dislocations are not able to cross slip on $\{111\}$ planes to form jogs. The obstacle distance l is independent of the character and type of dislocations and of the temperature, and the activation volume is independent of the temperature too (Table 9.1). The dynamic behavior of dislocations is compatible with precipitation hardening.

Some quantitative estimates are following regarding the precipitation hardening mechanism. It is argued in [286] that the precipitates are small oxide particles about 4 nm wide. Applying Friedel statistics, the activation volume V measured in macroscopic tests is related to the obstacle distance l and the size by (4.55) and (4.4). Using $l \approx 100$ nm and $V = 13.5\,\text{nm}^3$ from above, the activation distance Δd turns out to be about 0.7 nm or $2.5\,b$. This fits well the estimated obstacle width of 4 nm. The small value of Δd indicates that the equilibrium position of the dislocations waiting for thermal activation is near the tip of the interaction potential, or that the stress is not much smaller than the athermal flow stress of the obstacle array. This corresponds to the kinematic dislocation behavior partly with jumps over larger distances (Fig. 4.18).

The stress $\tau^* \approx 40$ MPa estimated above from the dislocation bowing can be compared with the macroscopic flow stress in Fig. 9.15. It is important to note that the bowing of the dislocation segments only slightly relaxes after the specimens are unloaded. This indicates that a friction stress τ_p acts on the dislocations, which prevents the relaxation and which is only slightly smaller than τ^*. Friction stresses were calculated for ordinary dislocations, e.g., in [541] using the embedded-atom method. These stresses are high for screw dislocations, intermediate for 60° dislocations, and low for 30° and edge dislocations. This may be a reason why edge dislocations are more mobile than screw dislocations so that the latter dominate the dislocation structures.

In addition to the stress components τ^* and τ_p, a long-range internal stress component τ_i contributes to the flow stress. An estimation via the Taylor hardening model (5.11) and an intermediate dislocation density of $\varrho = 2 \times 10^{13}\,\mathrm{m}^{-2}$, quoted earlier, yields $\tau_i \approx 44\,\mathrm{MPa}$. Thus, the total resolved yield stress estimated from the microstructure

$$\tau_y = \tau^* + \tau_p + \tau_i \approx 2\tau^* + \tau_i$$

will be about 125 MPa. With the stress value from Fig. 9.15, the macroscopic resolved yield stress $\tau_y = m_s\sigma_y$ is only about 60 MPa. Thus, the stress components resulting from the dislocations bowing out between obstacles, the friction stress and the long-range athermal stress overestimate the macroscopic yield stress. There may be several reasons, for instance using an incorrect line tension in estimating τ^* (see the discussion in Sect. 5.2.3). With rising temperature, thermal activation becomes more and more important. This leads to a decreasing contribution of τ^* to the flow stress and, accordingly, to a weaker curvature of the dislocation segments between the precipitates, as observed in Fig. 9.12. Finally, the precipitates are overcome spontaneously and no longer act as obstacles. A very similar behavior of dislocations was observed in two-phase Ti–Al materials [531].

Flow Stress Anomaly

Several attempts have been made to explain the flow stress anomaly of TiAl. As reviewed by Vitek [550], the core structure remarkably influences the properties of dislocations in ordered noncubic phases. First explanations of the flow stress anomaly in TiAl were made on the basis of $\langle 101 \rangle$ superdislocations by Hug et al. [551] in analogy to the Kear–Wilsdorf locks in $L1_2$ alloys [516]. The energy of the planar faults on the $\{010\}$ plane is lower than on the $\{111\}$ plane. Therefore, dissociated $\langle 101 \rangle$ superdislocations may cross slip from the $\{111\}$ slip plane to the $\{010\}$ cross slip plane where they become locked. In a more general way, the cores of the dislocations may either extend on their slip plane resulting in a glissile configuration, or they may assume a three-dimensional structure leading to a low mobility (e.g., [539,540]). While, at room temperature, superdislocations with $\langle 101 \rangle$ Burgers vectors are flexible under load and not locked (Fig. 9.10), at high temperatures they show straight segments in screw orientation, which are connected by superkinks as in Fig. 9.14, in accordance with the theory. However, the motion of superdislocations has never been recorded in in situ experiments. Although the different behavior of superdislocations at room temperature and at high temperatures may explain the flow stress anomaly, their small number suggests that they do not considerably influence the high-temperature deformation under the conditions of the present in situ experiments. Consequently, the flow stress anomaly has to be explained on the basis of ordinary dislocations only.

The first two models by Louchet and Viguier [547] and Sriram et al. [549] introduced in the foregoing section to interpret the curly shape of ordinary dislocations in TiAl at room temperature were also used to explain the flow stress anomaly. Both models assume that cross slip leading to the formation of pinning centers is thermally activated so that the number of pinning points increases with increasing temperature. However, above about 430°C, the dominating ordinary dislocations no longer bow out between localized obstacles. In the in situ experiments, the ordinary dislocations are created instantaneously in an avalanche, which had already been observed in earlier in situ experiments, as reviewed in [552]. If they move again, they are smoothly bent, moving in a viscous way as shown in Fig. 9.13 and Video 9.6, similarly to NiAl described in Sect. 9.4. This had not been observed before in TiAl. It clearly contradicts the cross slip models of the flow stress anomaly. First, the dislocation segments no longer show the curly shape, and second, the dislocations move continuously, but not in the proposed pinning–unzipping mode. As discussed earlier, the dislocations are not confined to their slip planes, indicating that climb is involved. Smoothly curved dislocations not lying on a slip plane were recently observed also in a post-mortem study [553]. This view is supported by the formation of helical dislocation structures in two-phase Ti–Al during in situ heating experiments [554]. Climb requires lattice diffusion of both atomic species. The formation and migration energies of vacancies in TiAl are between 1 and 1.6 eV [555,556]. Thus, these vacancies are present and mobile in the temperature range around the flow stress maximum. However, climb usually facilitates dislocation motion and results in recovery so that it does not induce additional friction.

Since the low-temperature mechanism of precipitation hardening ceases to operate above 575°C, an additional process must cause the increase or at least the constancy of the flow stress in the range of the anomaly, associated with the viscous type of dislocation motion. Viscous motion at high temperatures is not consistent with simple models of a superposition of long- and short-range obstacles to dislocation motion, as outlined in Sect. 5.2.2 and Fig. 5.20. In the high-temperature (athermal) case, the dislocation motion becomes very jerky. It is a major aim of the intermetallics chapter of this book to stress that diffusion-controlled processes may give rise to additional friction impeding the motion of dislocations in intermetallic alloys. Locking mechanisms and diffusion-controlled processes are compared in [557]. These diffusion processes may have quite different origins as will be discussed in the present and following sections. γ-TiAl probably is a relatively simple case where point defect atmospheres form around the dislocations as described in Sect. 4.11. The atmosphere formation may be superimposed with climb, which follows from the dislocations not being restricted to their glide planes at high temperatures. The atmospheres around dislocations may form from solute impurities as proposed in [552] to cause strain ageing effects. Such effects have been observed between 150 and 350°C [558], i.e., at temperatures where the precipitation hardening ceases to contribute to the flow stress. The amplitude of strain ageing increases when the Al content decreases below 50%. It is therefore concluded that the responsible point defects are antisite defects which have noncentrosymmetrical stress fields. They may associate with vacancies which change their position by diffusional jumps producing an induced Snoek effect. In range B of Fig. 4.39, the strain ageing takes place. It is associated with a negative strain rate sensitivity. A negative steady-state strain rate sensitivity (see Fig. 2.2) was observed at 350°C [559]. The maximum of the stress contribution from the atmospheres is located left of range B, i.e., at a higher temperature.

By means of the theory of the Cottrell effect in Sect. 4.11 it was estimated from (4.100) that several hundred p.p.m. of the diffusing defects are necessary to cause a remarkable contribution to the flow stress. The concentration of the antisite defects is in the percent range and the concentration of vacancies is also high at temperatures of the anomaly. Applying the same theory, the strain rate $\dot{\gamma}_c$ at the maximum stress can be related to the diffusion coefficient D of the diffusing defects by (4.99). Using appropriate data of $\dot{\gamma} = 10^{-5}\,\mathrm{s}^{-1}$, $\varrho_m = 2 \times 10^{13}\,\mathrm{m}^{-2}$, and the temperature of the flow stress maximum of $T = 730°\mathrm{C}$, the diffusion coefficient of the defects moving in the atmospheres should be about $D = 2.4 \times 10^{-19}\,\mathrm{m}^2\,\mathrm{s}^{-1}$. This value fits very well the extrapolated diffusion data of Ti in TiAl [556,560]. It is higher than the diffusion coefficient of Al [556]. On a semi-quantitative level, the present estimates show that intrinsic point defects in TiAl causing strain ageing effects may considerably contribute to the flow stress anomaly.

The assumption that point defect atmospheres control the dislocation mobility in the range of the flow stress anomaly is consistent with the inverse dependence of the strain rate sensitivity r on the strain rate $\dot{\varepsilon}$ in Fig. 9.16. The data belong to the stable range A in Fig. 4.39. The viscous motion of the dislocations in the in situ experiments corresponds to this range of positive strain rate sensitivity. At high dislocation velocities, the strain rate sensitivity becomes negative in range B leading to plastic instabilities, which may occur in TiAl at intermediate temperatures (e.g., [561]). In the in situ experiments, the instantaneous formation and motion of ordinary dislocations during the first loading belong to range B. It will be discussed in the following sections that in the respective materials, the occurrence of a flow stress anomaly is always accompanied by the inverse dependence of the strain rate sensitivity on the strain rate or stress.

As briefly outlined in Sect. 9.3.3, only a small part of the stress relaxes during the stress relaxation tests. This is a qualitative indication that a remarkable part of the flow stress is of athermal nature, which seems to contradict the view of the dislocation motion being controlled by the thermally activated processes of point defect atmospheres. However, it was pointed out already by Hirth and Lothe [12] that, according to the Orowan equation (3.5), a reduced dislocation mobility increases the mobile dislocation density ϱ_m, which, in turn, results in an increased athermal component of the flow stress. Unfortunately, to the authors' knowledge, there are no representative data available of dislocation densities in TiAl and many other intermetallic alloys in the different temperature ranges to estimate the temperature dependence of the athermal stress component.

The in situ experiments in the intermediate temperature range demonstrate that between about 430 and 600°C the low-temperature mechanism of the thermally activated overcoming of obstacles changes into the high-temperature mechanism involving lattice diffusion. There is probably a wide range when both mechanisms are superimposed.

9.3.5 Summary

In γ-TiAl at room temperature, all types of dislocations, i.e., ordinary dislocations of all orientations, superdislocations, and twinning dislocations, bow out between spatially distributed pinning points and move in a jerky way. This can best be explained by a precipitation hardening mechanism controlling the dislocation motion between room temperature and about 430°C. At higher temperatures, the dominating ordinary dislocations are smoothly bent and move either in an unstable or in a viscous way. This may be interpreted by the diffusion-controlled formation of Snoek atmospheres from antisite defects and vacancies, which cause an additional friction by strain ageing effects, thus being responsible for the flow stress anomaly. The occurrence of the anomaly is connected with an inverse dependence of the strain rate sensitivity on the strain rate or stress.

9.4 NiAl

The intermetallic compound NiAl has the B2 crystal structure. It shows a strong plastic anisotropy. As demonstrated in Fig. 9.15, the material does not exhibit a flow stress anomaly along a soft deformation axis. Along the hard axis, however, the flow stress is several times higher than in the soft orientations. Besides, it shows a weak anomaly, i.e., the flow stress decreases only slightly between room temperature and about 500°C, as reviewed by Miracle [562]. For applications at high temperatures, NiAl is alloyed with metals like Nb and Ta to form two-phase alloys with strengthening precipitate particles and grain boundary layers containing ternary phases, e.g., Laves phases, as described by Sauthoff [563, 564].

9.4.1 Crystal Structure and Slip Geometry

Figure 9.18 presents the unit cell of an AB compound with the cubic B2 structure. One atom species A occupies the corners of the cube and the other atom B the body-centered positions. The possible Burgers vectors are $\langle 100 \rangle$, $\langle 011 \rangle$, and $\langle 111 \rangle$. Dislocations with the latter Burgers vectors are superdislocations. The critical stress to move the dislocations is much lower for dislocations

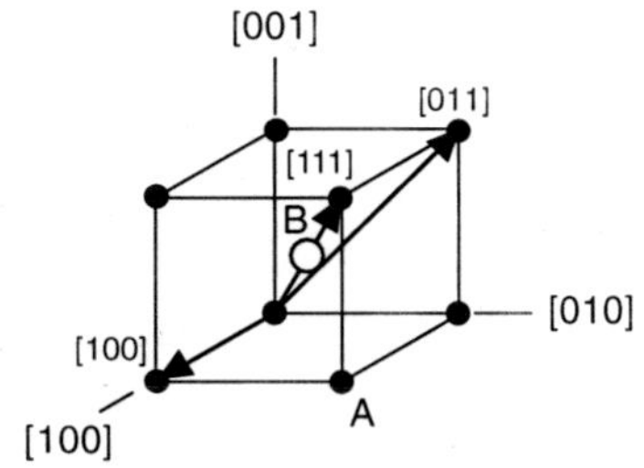

Fig. 9.18. Unit cell of the B2 structure with [100], [011], and [111] Burgers vectors of the $(0\bar{1}1)$ slip plane

with $\langle 100 \rangle$ Burgers vectors than for the other two. Therefore, middle-oriented specimens with reasonable orientation factors for $\langle 100 \rangle$ dislocations have a low flow stress. These orientations are called soft orientations. For the deformation along the hard $\langle 100 \rangle$ directions, the $\langle 100 \rangle$ dislocations have zero orientation factors so that the two other types of dislocations are activated. The flow stress is then much higher.

9.4.2 Microscopic Observations

The following experimental results are from in situ straining experiments either on stoichiometric NiAl single crystals with (110) foil surfaces strained along the soft $[1\bar{1}1]$ or $[2\bar{2}1]$ tensile directions [287], or on coarse-grained stoichiometric polycrystals containing 0.2at% Ta [565]. The single crystals are deformed exclusively by dislocations with the short $\langle 100 \rangle$ Burgers vectors. The same holds for the NiAl-0.2at%Ta polycrystals at room temperature. At higher temperatures, in NiAl-0.2at%Ta the dislocations with $\langle 110 \rangle$ Burgers vectors were activated in the special orientation of the large grains during the in situ straining experiments. Video 9.10 originates from the in situ fracture study in [566].

At room temperature, the $\langle 100 \rangle$ dislocations with a large screw component in both materials are strongly pinned as shown in Fig. 9.19. The average distance between the pinning agents amounts to $l = 76$ nm in NiAl and to 150 nm in NiAl-0.2at%Ta, corresponding to about $260\,b$ and $500\,b$, respectively. Between the pinning centers, the bowed-out segments are of angular shape. The main features of this shape can be explained in terms of the line tension model in anisotropic elasticity, which was described in Sect. 3.2.7. Figure 9.20 shows the equilibrium shape of loops on a $\{110\}$ plane, calculated by (3.41) with elastic constants from [567] for determining the dislocation energy factor $K(\beta)$ and its derivative in the framework of anisotropic elasticity. As pointed out earlier in [568], screw dislocations with $\boldsymbol{b} = \langle 100 \rangle$ are unstable on $\{100\}$ planes. In screw orientation, the line energy E has a maximum in contrast to elastic isotropy where it shows a minimum. At the same time, the line tension Γ is negative, resulting in the instability. The figure shows that the

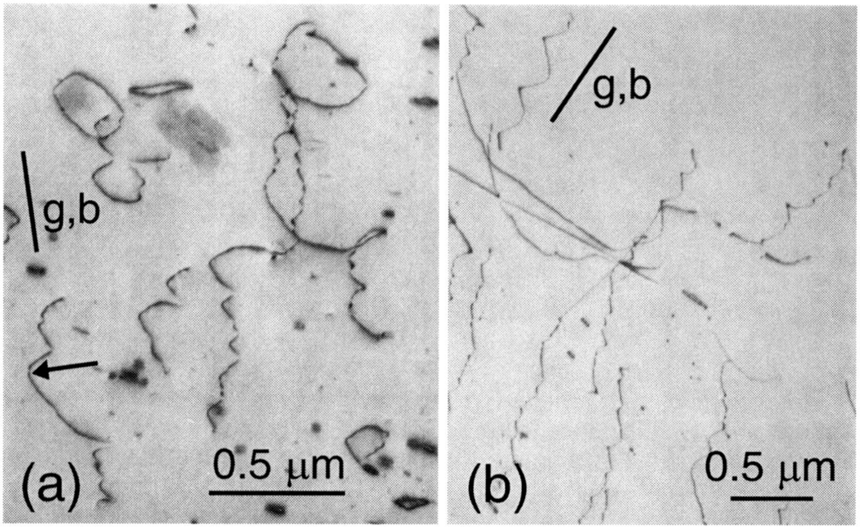

Fig. 9.19. Dislocations with ⟨100⟩ Burgers vectors on {010} planes under load during in situ deformation at room temperature. (**a**) NiAl single crystal in a soft orientation. (**b**) NiAl-0.2at%Ta polycrystal. Image normal (110), diffraction vector $\boldsymbol{g} = [\bar{1}10]$, $\boldsymbol{b}$ projection of [010] Burgers vector. From the works in [287, 565]

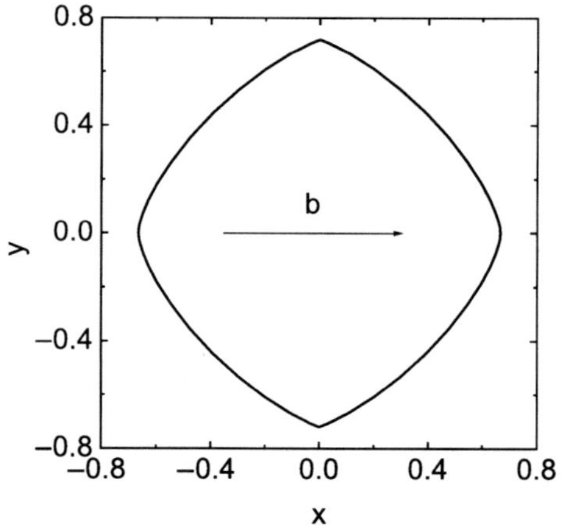

Fig. 9.20. Equilibrium shape of a dislocation with [100] Burgers vector on a (011) plane in units of $E_s/(\tau\, b)$ calculated within the framework of the line tension model using anisotropic elasticity. Elastic constants for room temperature from [567]

instability occurs on {110} planes too. It also holds for other planes, leading to the sharp "knee" at the screw parts and to the dominance of mixed dislocation characters as in Fig. 9.19. At about 45°, E has a minimum, whereas Γ exhibits a maximum. The local effective shear stress τ^* can be estimated from the size of the calculated loops fitting the images of bowed-out dislocation segments by the method described in Sect. 5.2.3 using (5.18). The average value of the minor half axis x_0 of the loops, taken from relatively long bowed-out segments, is 95 nm, which has to be considered a rough estimate because of the angular shape of the segments. For NiAl-0.2at%Ta, the size x_0 could not be estimated unambiguously. A lower limit is $x_0 = l/2 = 75$ nm.

The detailed analysis of the video recordings enabled the obstacles impeding the motion of screw dislocations to be identified as jogs. This is demonstrated in the following video clip.

Video 9.7. *Motion of dislocations with* $\langle 100 \rangle$ *Burgers vectors in an* NiAl *single crystal at room temperature:* The clip consists of three parts. The first one shows the typical movement of dislocations with large screw components by shifting jogs along the dislocation line. Selected frames are reproduced in Fig. 9.21. The dislocation moves on a $\{210\}$ plane as indicated by the $[\bar{1}12]$ trace t on the (110) surface. Some major positions of the dislocation are summarized in Fig. 9.22. These positions are not identical with those of Fig. 9.21. At 0 s, the dislocation under consideration is anchored by two jogs, J_1 and J_2. Because of the greater length of the central segment, the jogs are subjected to an outward tangential force, initiating the outward motion of the jogs. At 6.16 s, the central segment bows out extraordinarily wide. In the next frame of the figure, two positions of the lower part of the dislocation appear in weak contrast due to a quick motion during the exposure of the frame. Thereby, the dislocation produced a debris D and acquired a new jog J_3 by cross slip. Later on, J_3 is shifted along the dislocation in the Burgers vector direction. The most prominent feature of this kind of motion is the formation of long segments by the lateral spreading of the limiting jogs leading to very large bow-outs as at 6.16 s. These large bow-outs cannot be explained by the action of localized obstacles like precipitates, since the moving dislocation segment has a high probability to contact the next obstacle after sweeping an area of the square of the square lattice distance (see Sect. 4.5.1). The steps in the outer traces in the compilation of positions in Fig. 9.22 reveal that the dislocation undergoes cross slip during its motion.

Between their stable positions, the dislocations move in a viscous way at a velocity that can be resolved by the video recording. This suggests that a lattice friction mechanism is also active, in agreement with the fact that the segments bowed-out under load do not remarkably relax when the specimen is unloaded.

In the second video sequence, part of a dislocation annihilates with another one, which is arranged on another plane, as indicated by the different lengths of both dislocations. The third part presents the motion of edge segments. They have a zig-zag shape, too, owing to the elastic equilibrium shape demonstrated in Fig. 9.20.

Video 9.8. *Motion of dislocations with* $\langle 100 \rangle$ *Burgers vectors in* NiAl-0.2at%Ta *at room temperature:* Dislocations move in a slip band originating at a crack. The elementary processes of motion are not well resolved but are similar to those in pure NiAl.

The activated slip planes are identified from slip traces, from points where the dislocations emerge through the surface and from the path of these points

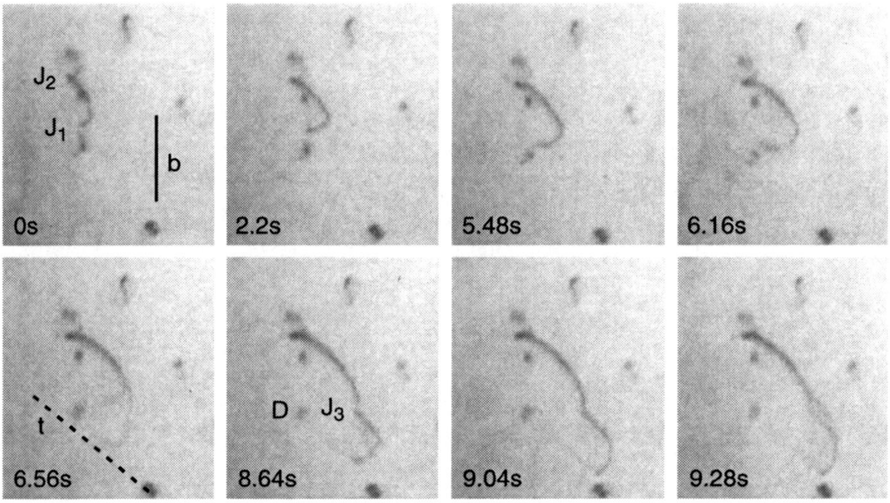

Fig. 9.21. Sequence of frames from Video 9.7 showing the motion of a dislocation with a ⟨100⟩ Burgers vector with the projection $\boldsymbol{b}$ by shifting the jogs along the screw direction. The line $\boldsymbol{t} = [\bar{1}12]$ is the trace of the {210} slip plane. From the work in [287]

Fig. 9.22. Schematic drawing of different stages of the motion of the dislocation in NiAl shown in Fig. 9.21. The stages are not the same as those in Fig. 9.21

during motion, recorded in the videos. According to that, the dislocations glide on {100}, {110}, and {210} planes, with frequent cross slip events between these planes. That cross slip is easy was observed earlier in [569,570]. Accordingly, many short dislocation dipoles are created forming the debris shown earlier in Fig. 5.19. The dislocation mobility does not remarkably differ between the different planes.

At elevated temperatures, the concentration of jogs on dislocations with ⟨100⟩ Burgers vectors is drastically reduced. Figure 9.23 demonstrates that the shape of the dislocation segments is still controlled by the line tension

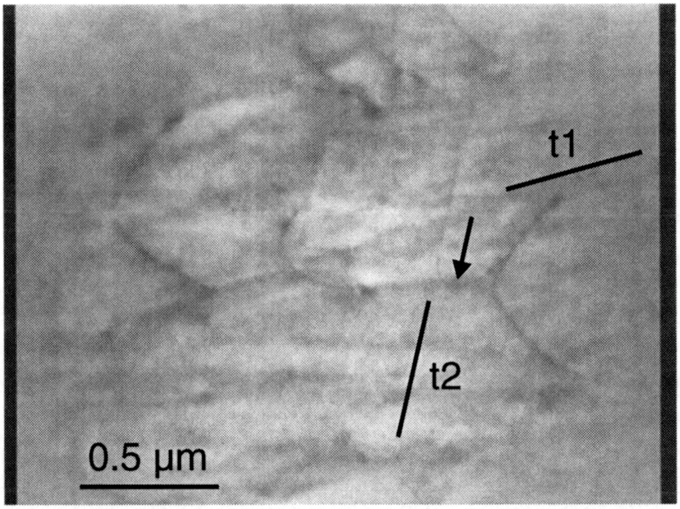

Fig. 9.23. Dislocation motion in an NiAl single crystal during in situ deformation at 475° C. The dislocation marked by an arrow has cross slipped from a {100} plane with trace t1 onto a {110} plane with trace t2. From the work in [287]

with the typical knee in the dislocation line at the elastically unstable screw orientation. The average size of the fitting theoretical loops amounts to about $x_0 = 560$ nm. The following video is characteristic also of higher temperatures.

Video 9.9. *Motion of dislocations with* ⟨100⟩ *Burgers vectors in an* NiAl *single crystal at* 475° C*:* At elevated temperatures, the dislocations of mixed character near screw orientation move viscously at different velocities depending on the long-range interactions with other dislocations. During the motion, the shape of dislocations governed by the line tension is preserved. Dislocations frequently cross slip from a {100} plane onto a {110} plane as marked in Fig. 9.23. Two of such events are labeled by violet circles just before the dislocations emerge at the right edge of the frame. The slip traces are mostly relatively straight, but sometimes also serrated as indicated by CS, which also hints at cross slip. The short second part of the video presents a dislocation moving quickly on the (vertically running) plane with frequent cross slips.

At 475°C, slip appears preferentially still on {100} and {110} planes with frequent cross slips between these planes. At higher temperatures, the slip trails become noncrystallographic so that probably climb, too, contributes to the dislocation motion. Some features of the motion of edge dislocations are presented in Video 9.10.

Video 9.10. *Motion of dislocations with* ⟨100⟩ *Burgers vectors in an* NiAl *single crystal at* 565° C*:* The band of dislocations with dominating edge character is emitted at a crack tip. The dislocations are slightly bowed. The dislocation right of the violet label drags a prismatic half-loop behind. Another prismatic half-loop at P is moved forward and backward on its glide cylinder by the interaction with the passing dislocations of the band.

In NiAl-0.2at%Ta, dislocations with $\langle 110 \rangle$ Burgers vectors were activated in grains in the hard orientation. Their shape strongly depends on their glide plane. Figure 9.24a shows such dislocations marked A, moving on $\{111\}$ planes and trailing slip traces. Segments parallel to the surfaces along $[0\bar{1}1]$ seem to be quite immobile. The moving parts show preferred orientations along $[12\bar{1}]$ directions, which do not correspond to the edge and screw orientations. They can also not be explained by the elastic equilibrium shape and may therefore be determined by an anisotropic glide resistance. Video records show that these dislocations move in a continuous viscous way. Figure 9.24a also exhibits a large number of dislocations like B of the same type of Burgers vectors which are not located on $\{111\}$ slip planes and which are difficult to interpret.

Figure 9.24b demonstrates that the morphology of dislocations with $\langle 110 \rangle$ Burgers vectors greatly differs when they move on $\{110\}$ planes. They consist of very straight segments in $\langle 111 \rangle$ orientations, i.e., they are neither of screw nor of edge character. This observation was confirmed by post-mortem TEM of Ni-44at%Al single crystals deformed in the hard orientation [571]. The straight shape can also not be interpreted by the elastic equilibrium shape, which is quite smooth without instabilities. The segments are connected by small arcs. The following video exhibits the motion of these dislocations.

Video 9.11. *Motion of dislocations with* $\langle 110 \rangle$ *Burgers vectors on* $\{110\}$ *planes in* NiAl-0.2at%Ta *at* 475°C*:* At elevated temperatures, the straight dislocations with $\langle 110 \rangle$ Burgers vectors move in a viscous way on $\{110\}$ planes.

The core structure of edge dislocations with $\langle 100 \rangle$ and $\langle 110 \rangle$ Burgers vectors was studied by Mills and Miracle [572] by high-resolution TEM. According to that, the cores of the $\langle 100 \rangle$ dislocations are compact but have large elastic strains corresponding to the relatively large Burgers vectors. The $\langle 110 \rangle$ dislocations may dissociate into two partial dislocations with $1/2\langle 111 \rangle$ Burgers vectors including an APB, or they decompose into two dislocations with $\langle 100 \rangle$ Burgers vectors. The decomposition was proved also for dislocations along $\langle 111 \rangle$ directions with a decomposition distance of only 2 nm [571]. For edge dislocations, it is larger. The consequences of such a kind of decomposition will be discussed in Sect. 9.4.4.

9.4.3 Macroscopic Deformation Parameters

As shown in Fig. 9.15, stoichiometric NiAl single crystals strained along a soft orientation, which are deformed by $\langle 100 \rangle$ dislocations, do not exhibit a flow stress anomaly. The NiAl-0.2at%Ta polycrystals not only show much higher flow stresses but also a normal temperature dependence. In contrast to the in situ straining experiments above, where at high temperatures only grains in the hard orientation were observed, mainly dislocations with $\langle 100 \rangle$ Burgers

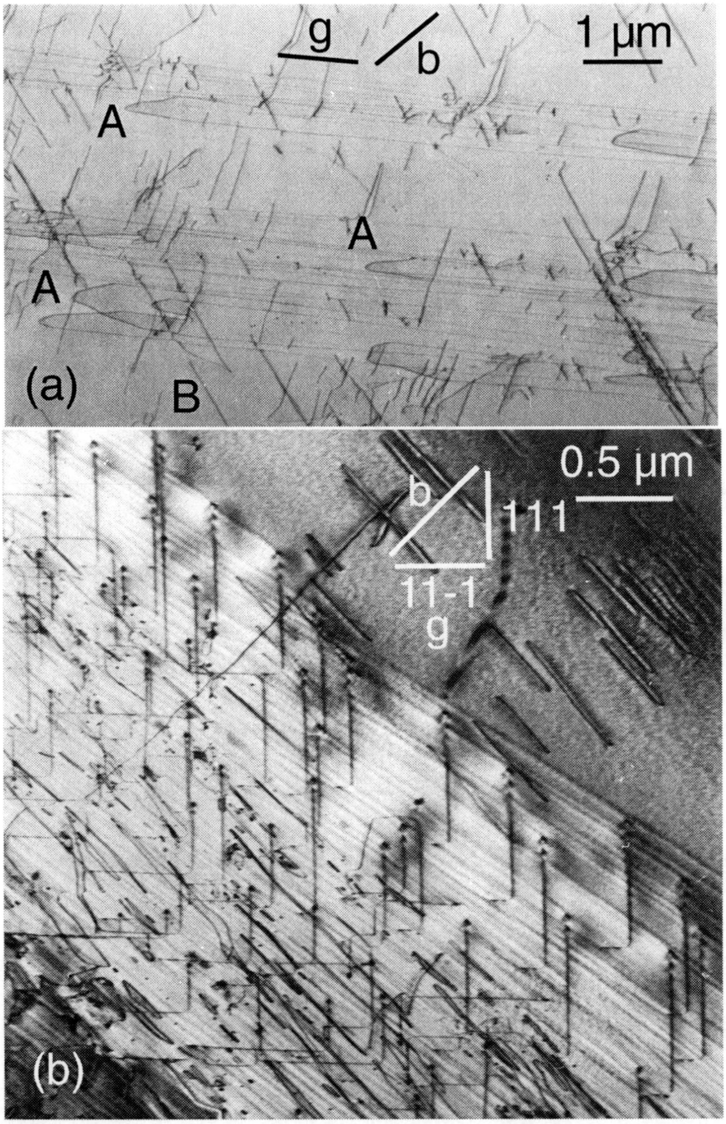

Fig. 9.24. Dislocations with ⟨110⟩ Burgers vectors during in situ deformation of NiAl-0.2at%Ta at 475°C. (**a**) Dislocations A moving on {111} planes and trailing slip traces and those moving on other planes B. Diffraction vector $\boldsymbol{g} = [0\bar{1}1]$, $\boldsymbol{b}$ projection of $[\bar{1}10]$ or $[\bar{1}\bar{1}0]$ Burgers vectors. (**b**) Dislocations moving on {110} planes. Indicated are the projections of the $[11\bar{1}]$ and [111] directions and of the [110] Burgers vector onto the image plane and the imaging vector $\boldsymbol{g} = [0\bar{1}1]$. From the work in [565]

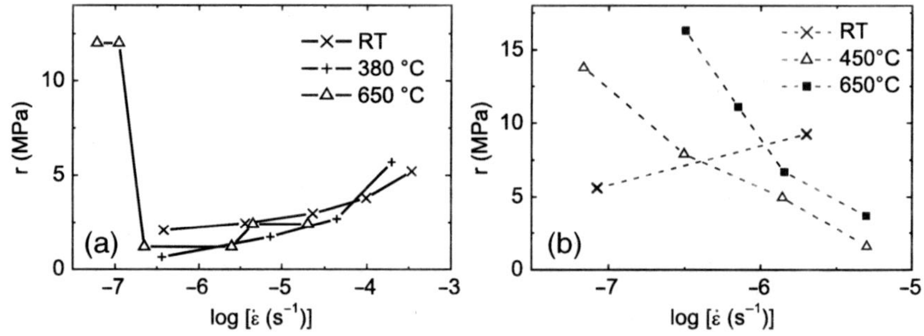

Fig. 9.25. Dependence of the strain rate sensitivity of the flow stress on the logarithm of the strain rate in (**a**) NiAl single crystals deformed along ⟨421⟩ and in (**b**) NiAl-0.2at%Ta polycrystals. Data from [573]

vectors will carry the deformation. As reviewed in [562], crystals deformed in the hard orientation show a weak flow stress anomaly with a plateau, where a transition occurs between glide on the {112}⟨111⟩ system below about 325°C and {110}⟨110⟩ above that temperature. Thus, the in situ straining experiments with ⟨110⟩ dislocations belong to the range where these dislocations control the deformation in the hard orientation.

The dependence of the strain rate sensitivity r on the logarithm of the strain rate $\dot{\varepsilon}$ is plotted in Fig. 9.25a for the NiAl single crystals. At all the three temperatures selected, there is a positive dependence corresponding to normal obstacle behavior. An exception is the highest temperature where r is high at very low strain rates due to recovery. In NiAl–Ta, however, the behavior of the strain rate sensitivity is normal only at room temperature (Fig. 9.25b). At elevated temperatures, this material exhibits an inverse dependence of r on $\log \dot{\varepsilon}$.

9.4.4 Deformation Mechanisms

Like in TiAl, different mechanisms control the dislocation mobility in NiAl at room temperature and at high temperatures. At room temperature, the dislocations have configurations with bowed-out segments and are moving in a jerky way. At high temperatures, they are smooth and move viscously.

Room Temperature

The static configurations and especially the details of the mode of forward motion described earlier (Fig. 9.19, Video 9.7) suggest that the obstacles which pin the dislocations with ⟨100⟩ Burgers vectors at room temperature are jogs. Thus, these dislocations are clearly an example of the jog mechanism suggested by Sriram et al. [549] for γ-TiAl. However, as proved in NiAl in soft orientations, the jog mechanism does not cause a flow stress anomaly. The jogs

Table 9.2. Deformation data of NiAl materials at room temperature. Data rom [287, 565]

Material	σ_y (MPa)	r (MPa)	V/b^3	l/b	x_0 (nm)	τ^* (MPa)
NiAl ⟨421⟩	220	2.5	215	260	95	90
NiAl-0.2at%Ta	380	9.3	57	520	>75	<110

are created by double-cross slip as proposed for dislocation multiplication in the Koehler–Orowan mechanism described in Sect. 5.1.1. The high frequency of cross slip in NiAl is quite strange since only screw segments can cross slip. However, this orientation is unstable and does not exist in equilibrium configurations of the dislocations. Before cross slip, first a screw segment has to be created, the energy of which is about 9% higher than that of the stable branches adjoining the knee at the screw orientation. Apparently, stress components on the cross slip plane assist this process.

As several times before in this book, via (5.18) the (shear) back-stress can be calculated resulting from the bowing-out of dislocation segments pinned between obstacles. Table 9.2 lists the results together with other deformation data, using the values of the minor half-axis x_0 of the bowed-out segments and the relevant constants. These values are large fractions of the total yield stresses σ_y if average orientation factors are considered. Thus, the jog mechanism contributes essentially to the total flow stress at room temperature.

Comparing the activation volume V following from the strain rate sensitivity r with the obstacle distance l from the in situ straining experiments yields for NiAl an activation distance Δd of about b if (4.4) and the data of Table 9.2 are used. Unfortunately, there does not exist a theory of the activation parameters of the motion of screw dislocations by the sidewise shifting of jogs so that this value cannot be interpreted microscopically. For NiAl-0.2Ta, Δd roughly equals only about $0.1\,b$. The corresponding small activation volume should result from the solution hardening of the Ta additions. In [565], based on the theory of Mott statistics described in Sect. 4.5.2 it is argued that solution hardening may explain both the increased flow stress of NiAl-0.2Ta with respect to pure NiAl as well as the corresponding decrease of the activation volume. A further contribution to the flow stress certainly originates from the Peierls mechanism. If there were no lattice friction, the jogs would easily glide along the screw direction so that long segments became unstable and the dislocations would easily proceed forward.

High Temperatures

For the ⟨100⟩ dislocations in NiAl at 475°C, the effective stress $\tau^* \approx 20$ MPa measured from the dislocation bowing is only about 20% of the macroscopic flow stress. Viscous motion of dislocations at these temperatures was described before for TiAl (Sect. 9.3). It cannot be explained by the Peierls mechanism based on the data of the low-temperature increase of the flow stress. According to standard theory, thermally activated processes controlling the dislocation mobility at low temperatures should not be active above specific temperatures. Apparently, like in TiAl, diffusion-controlled processes are important also in NiAl. At high temperatures, climb is indicated by noncrystallographic slip trails. Because of the low mobility of vacancies, characterized by a migration enthalpy of 2.1 eV [560], diffusion processes may become evident only in the temperature range of the viscous dislocation motion.

As estimated in [565], the solution hardening of the Ta additions in NiAl-0.2Ta will rather quickly decrease with increasing temperature, in contrast to the weak normal temperature dependence of the yield stress in Fig. 9.15. It may therefore be concluded that the Ta additions contribute to the flow stress at higher temperatures through the formation of impurity atmospheres around the dislocations. This is in agreement with the inverse dependence of the strain rate sensitivity on the strain rate, which is observed in NiAl–Ta but not in pure NiAl, as documented in Fig. 9.25. Corresponding strain ageing effects were observed in NiAl materials containing additions of Ti, Zr, and Hf [574].

The formation of point defect atmospheres can explain the viscous motion of dislocations at high temperatures and also the occurrence of a flow stress anomaly. However, it does not explain that the dislocations with $\langle 110 \rangle$ Burgers vectors arrange in very straight configurations along the mixed $\langle 111 \rangle$ orientation as in Fig. 9.24 and Video 9.11. Apparently, in NiAl single crystals deformed in the hard $\langle 100 \rangle$ orientation the situation is quite complex. Dislocations with $\langle 100 \rangle$ Burgers vectors are not activated because of their zero orientation factor. Below a certain transition temperature, the material is deformed by dislocations with $\langle 111 \rangle$ Burgers vectors (e.g., [568, 575]) at a very high resolved shear stress of 600 MPa. Above the transition temperature, the deformation occurs either by kinking via $\langle 100 \rangle$ dislocations [576] or by slip of $\langle 110 \rangle$ dislocations [577]. In [578, 579], for instance, it is suggested that the slip transition is achieved by decomposition of the $\langle 111 \rangle$ dislocations according to

$$[111] \rightarrow [110] + [001].$$

This reaction is energetically favorable for dislocations with near-edge orientation on $(\bar{1}10)$ planes [580]. By atomistic calculations, Srinivasan et al. [581,582] have shown that the $\langle 111 \rangle$ dislocations may occupy a metastable state along the mixed $\langle 111 \rangle$ orientation. These dislocations can move as a whole but if the components are separated wide enough, they may decompose completely. All reaction products are then arranged parallel to the $\langle 111 \rangle$ mixed orientations of the parent dislocations. The components with $\langle 100 \rangle$ Burgers vectors experience no shear stress. Thus, the $\langle 110 \rangle$ components have to move away. This is possible only at sufficient mobility, i.e., if the temperature is high enough.

As mentioned in Sect. 9.4.2, edge dislocations with $\langle 110 \rangle$ Burgers vectors are observed to have a decomposed core after the reaction [572]

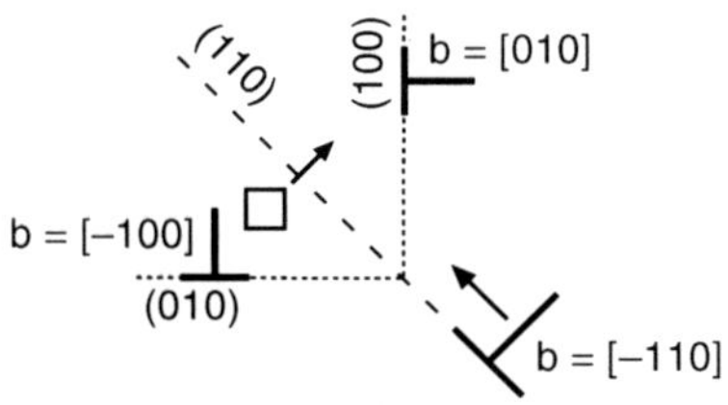

Fig. 9.26. Decomposition of an edge dislocation of $[\bar{1}10]$ Burgers vector as suggested in [583]

$$[110](1\bar{1}0) \rightarrow [100](010) + [010](100). \tag{9.3}$$

The indexes in parentheses mark the slip planes of the respective dislocations. Based on the microstructural observations, the reaction is studied by elastic and atomistic calculations in more detail by Mills, Srinivasan and Daw [583]. It may occur by glide of the product dislocations on their (010) and (100) glide planes as outlined in Fig. 9.26. It is favored for edge dislocations since their line orientation is contained in the glide plane of the parent dislocation as well as in low-index glide planes of both product dislocations. From calculations of the elastic energy, the reaction is neutral, i.e., there is no radial force between the product dislocations. Taking the elastic interaction energy into account, the reaction yields a gain in energy. Atomistic calculations by the embedded atom method (EAM) show that the decomposition takes place under a shear stress on the parent dislocation. In its decomposed state, the dislocation cannot glide. However, it can move in upward direction indicated by the arrow at the parent dislocation in Fig. 9.26 by a vacancy flux from the [100] dislocation to the [010] dislocation as indicated by the open square with the arrow. This process requires only short-range diffusion, but no lattice diffusion over larger distances. It is called conservative climb.

In Fig. 9.24, Video 9.11 and in the TEM study on Ni-44at%Al in [571], the dislocations with $\langle 110 \rangle$ Burgers vectors on $\{110\}$ planes are mostly oriented along the mixed $\langle 111 \rangle$ directions. Line tension calculations within the framework of anisotropic elasticity yield a smooth equilibrium shape of $\langle 110 \rangle$ dislocations without elastic instabilities. Besides, static relaxation and molecular dynamics calculations by the EAM at 0 K do not show higher Peierls stresses for particular orientations [571]. Thus, the straight oriented course of the mixed dislocations may be explained by the same decomposition reaction as (9.3), which now reads

$$[110](\bar{1}10) \rightarrow [100](0\bar{1}1) + [010](10\bar{1}). \tag{9.4}$$

The $\{110\}$ planes on which the decomposition may take place are also low-index planes, they are preferred slip planes of the $\langle 100 \rangle$ dislocations. Besides, the $\langle 111 \rangle$ preference orientation of the dislocations is contained in the slip planes of both the parent dislocation and of the reaction products.

Table 9.3. Decomposition of dislocations with [110] Burgers vectors on the $(1\bar{1}0)$ plane in NiAl

Reaction	(9.3)		(9.4)	
Total disl.				
Line orient., ϕ	$[00\bar{1}]$, 90°		[111], 35.3°	
E_{0t} (10^{-9} N)	1.201		0.945	
Comp. disls.				
BV	[100]	[010]	[100]	[010]
Plane, ϕ	(010), 90°	(100), 90°	$(0\bar{1}1)$, 54.7°	$(10\bar{1})$, 54.7°
E_{0c} (10^{-9} N)	0.666	0.666	0.694	0.694
Energy gain				
$E_{0t} - \sum E_{0c}$ (10^{-9} N)		−0.13		−0.44

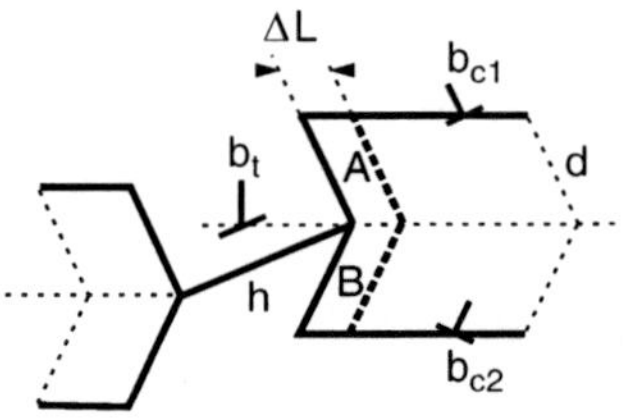

Fig. 9.27. Motion of a macrokink in a dislocation with ⟨110⟩ Burgers vector decomposed according to (9.3). After [583]

The geometry and energy balance of both reactions are summarized in Table 9.3. The energies are the prelogarithmic factors E_0 calculated in the framework of anisotropic elasticity. According to the table, the reaction (9.4) along ⟨111⟩ is slightly unfavorable in terms of the elastic energy factors of the dislocations involved. The interaction energy is not considered.

Mills et al. [583] present a model of the diffusional motion of an edge dislocation decomposed according to (9.3) by the spreading of constricted macrokinks. They assume that the dislocation forms a constricted macrokink of height h as outlined in Fig. 9.27. b_t is the ⟨110⟩ total Burgers vector, and b_{c1} and b_{c2} are the ⟨100⟩ Burgers vectors of the components. To shift the macrokink by a distance ΔL, the area A has to be swept by the dislocation with b_{c1} acting as a vacancy source while the area B is swept by absorbing these vacancies. Under very simplifying assumptions, the change in energy by shifting the macrokink by a distance dL is

$$\mathrm{d}W = (\tau b_t h + \sigma b_{c1} d + E_{110} - 2E_{100})\,\mathrm{d}L. \tag{9.5}$$

Here, τ is the shear stress acting on the macrokink with ⟨110⟩ Burgers vector. This kink is of screw character. σ is the normal stress along a ⟨100⟩ direction causing the dislocation with b_{c1} to climb, whereas the dislocation with b_{c2} does not experience a climb force from the external stress. Finally, d is the width over which the component dislocations have decomposed. The first term in

(9.5) is the gain in energy by shifting the macrokink, the second one is the loss in energy to form the vacancies on the dislocation with b_{c1}. As described above, the change in the self-energy by the decomposition $E_{110} - 2E_{100}$ is zero for the decomposed edge dislocation.

During the motion of the macrokink by $\mathrm{d}L$, a number of vacancies $\mathrm{d}N$ have to be transferred from the vacancy source at the border of the area A of the dislocation with Burgers vector b_{c1} to the vacancy sink at B of the dislocation with b_{c2}, which equals

$$\mathrm{d}N = \frac{d}{a^2}\,\mathrm{d}L,$$

where a is the lattice parameter. The change in the elastic energy per transferred vacancy is

$$w = -\frac{\mathrm{d}W}{\mathrm{d}L}\frac{\mathrm{d}L}{\mathrm{d}N} = (\tau b_t h - \sigma b_{c1} d)\,\frac{a^2}{d}.$$

Applying (4.83) with $\Delta\left(\Delta G_f\right) = w$, the vacancy supersaturation at the source dislocation is

$$\Delta c_v = c_v - c_{v0} = c_{v0}\left[\exp\left(\frac{w}{kT}\right) - 1\right],$$

with c_{v0} being the thermal equilibrium concentration of vacancies. In the limit of $\Delta c_v \ll c_{v0}$, there is

$$\Delta c_v \approx \left(\frac{c_{v0} w}{kT}\right).$$

In a simple model, the vacancy flux from the source at A having a vacancy concentration $c_{v0} - \Delta c_v$ to the sink at B with a concentration $c_{v0} + \Delta c_v$ occurs by pipe diffusion along the component dislocations. Applying Fick's law (4.91) and considering that $\tau = \sigma/2$ in the hard orientation, finally yields

$$v_k = \frac{4a^4 D_p \left(h/\sqrt{2} - d\right)\sigma}{d^3\,kT} \tag{9.6}$$

for the velocity of the macrokink. D_p is the pipe diffusion coefficient. The macrokink velocity exhibits a strong temperature dependence via the pipe diffusion coefficient. It depends on the geometric parameters of the kink and decomposition width, where the latter is particularly important. This parameter will further be considered in Sect. 9.5. Mills et al. [583] observed a good agreement with experimental data by applying the geometric parameters of the microstructural investigations and macroscopic data at a temperature of 600°C, and by assuming that the activation energy of pipe diffusion is half of that of bulk diffusion. For the mixed dislocations oriented along ⟨111⟩, the motion occurs by combined glide and conservative climb. The character of the climb-controlled motion is not jerky, but viscous.

In a later model [571], it was considered that most dislocations orient along ⟨111⟩ and that macrokinks have not been observed. According to the revised model, the movement of the dislocation consists of a succession of climb and glide of the ⟨100⟩ component dislocations, as sketched in Fig. 9.28. The forces acting on the product dislocations are the components of the external stress, the elastic interaction between the products, a chemical force due to the point defect super- or subsaturation and, in case of bowed segments, the line tension. In the hard orientation, the external tensile or compressive stress causes a pure climb force on one component dislocation and no force on the other. This would lead to an increase in the decomposition width. The elastic interaction between the components tries to arrange them perpendicular to the glide plane of the total dislocation and then cause a radial attraction. In addition, the dislocation climbing under the external stress changes the point defect concentration away from the equilibrium, which results in a chemical force on the other product dislocation. If the point defect flux generated at one component is totally annihilated at the other one (conservative climb), the dislocation couple will move along the glide plane of the nondecomposed parent dislocation. The deviation of the dislocation along ⟨111⟩ from the edge orientation results in a superposition of glide of the components on their low-index glide planes. These processes may occur on an atomic scale as assumed in the scheme of Fig. 9.28.

Srinivasan et al. [571] estimate the dislocation velocity by assuming the climb to be diffusion-controlled (see Sect. 4.10.4). They arrive at a formula very similar to (4.92). Instead of the cut-off radius R in the logarithmic term there appears the decomposition width d. Accordingly, the velocity depends strongly on the temperature via the diffusion coefficient and weakly on the decomposition width. Inserting data for the deformation at about 600°C, the (bulk) diffusion coefficient of Ni at that temperature, an experimental decomposition width of 2 nm for the dislocations aligned along ⟨111⟩, and a characteristic dislocation density of $2 \times 10^{13}\,\mathrm{m}^{-2}$ yields a dislocation velocity in the $10^{-5}\,\mathrm{s}^{-1}$ range, only one order of magnitude lower than that of the flow stress data. This

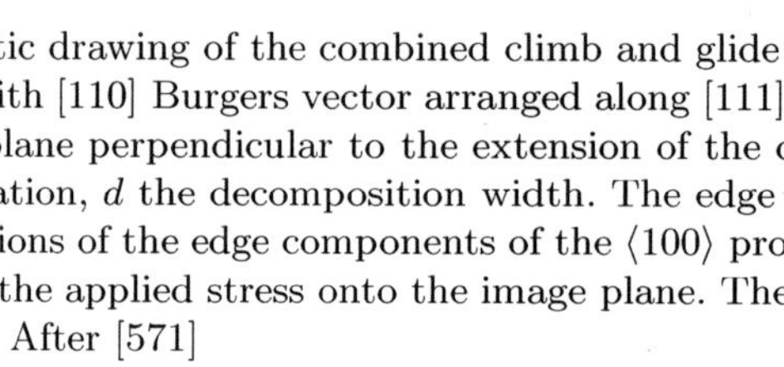

Fig. 9.28. Schematic drawing of the combined climb and glide motion of a decomposed dislocation with [110] Burgers vector arranged along [111] in NiAl. The image plane is the (111) plane perpendicular to the extension of the dislocation. P is the direction of propagation, d the decomposition width. The edge dislocation symbols indicate the projections of the edge components of the ⟨100⟩ product dislocations. σ is the projection of the applied stress onto the image plane. The open square marks a diffusing vacancy. After [571]

estimation is certainly no proof of the model. In particular, the assumption of diffusion-controlled climb is questionable since it requires the characteristic diffusion length to be great compared to the jog distance. This is certainly not fulfilled at the small separation of the component dislocations. Besides, the coordinated diffusion during the conservative climb in the elastic stress field of the composite dislocation may be controlled rather by pipe diffusion than by bulk diffusion. Thus, a comprehensive theory of the motion of these dislocations still has to be established.

9.4.5 Summary

The dislocations with $\langle 100\rangle$ Burgers vectors in NiAl, which are activated by the deformation near room temperature along soft orientations, are pinned by high jogs in the screw components. The jogs are created by double-cross slip. They are stabilized against lateral motion and annihilation by lattice friction. This is exactly the model suggested by Sriram et al. [549] for TiAl. It does not lead to a flow stress anomaly. At elevated temperatures, the $\langle 100\rangle$ dislocations move in a viscous way pointing at the operation of a diffusion-controlled mechanism. In the hard orientation, dislocations with $\langle 111\rangle$ and $\langle 110\rangle$ Burgers vectors are active at high temperatures. When moving on $\{110\}$ planes, the $\langle 110\rangle$ dislocations may be very straight and oriented along the edge and mixed $\langle 111\rangle$ directions. As suggested by Srinivasan et al. [571] and in the present book, the straight oriented arrangement may result from a decomposition into dislocations with $\langle 100\rangle$ Burgers vectors. The preferred orientations are the intersection lines between the glide planes of the parent dislocation and low-index glide planes of the product dislocations. The decomposition may occur by glide and is therefore different from the climb dissociation observed by H. Saka et al. in several materials (e.g., in β brass [584]). The decomposed dislocations cannot move by glide alone but by the aid of conservative climb, i.e., a diffusional flow of vacancies from one product dislocation to the other. It will be discussed in the following sections that this model need not be restricted to dislocations with $\langle 110\rangle$ Burgers vectors in NiAl, but may be more general.

9.5 FeAl

Like NiAl, FeAl has the B2 structure with $\langle 111\rangle$ and $\langle 100\rangle$ Burgers vectors. The dominating slip planes are of $\{110\}$ type. FeAl single crystals show a flow stress anomaly with a maximum at about 550°C as shown in Fig. 4.40.

9.5.1 Microscopic Observations

The following data originate from experiments on an Fe-43at%Al single crystal [8], which was annealed at 400°C for 1 week followed by slow cooling to room temperature to anneal out the excess point defects originating from crystal growth. Specimens for in situ straining experiments were cut with a [753] tensile axis and $(1\bar{2}1)$ foil planes. Samples for macroscopic compression tests were prepared with the same loading axis and a $(13\,\bar{50}\,53)$ side face about 10° off $(0\bar{1}1)$. This orientation favors slip on the $(101)[11\bar{1}]$ system. Notable orientation factors also have $(011)[11\bar{1}]$, (101)[010], and (011)[100].

In FeAl, there is a transition of the dominating Burgers vector from ⟨111⟩ to ⟨100⟩ [585,586] just at the temperature of the flow stress peak, which might cause the anomaly [587]. However, rapid quenching after deformation showed that dislocations with ⟨111⟩ Burgers vectors dominate during deformation at the high strain rate of $5 \times 10^{-2}\,s^{-1}$ even at the high temperature of about 650°C [588]. Similarly, it was observed that the deformation at the peak temperature of 550°C started by dislocations with ⟨111⟩ Burgers vectors, whereas those with ⟨100⟩ Burgers vectors appeared only after a few percent of plastic strain [589]. Conventional TEM in the high-voltage electron microscope of macroscopically deformed specimens was performed to secure that the deformation of the present material is carried by dislocations of the same Burgers vector up to the flow stress maximum. In Fig. 4.40, the temperatures at which dislocation structures were studied are marked by dashed grey vertical lines. The microstructure with dislocations of ⟨111⟩ Burgers vectors after deformation at 495°C was presented in Fig. 3.22. Dislocations with these Burgers vectors prevail up to the flow stress maximum. Stereo image pairs show that most dislocations are not confined to their slip planes indicating the importance of climb. Further deformation at 564°C activates dislocations with ⟨100⟩ Burgers vectors as demonstrated before in Fig. 1.4.

Due to elastic instabilities, many dislocation lines of both Burgers vectors exhibit an angular shape with a sharp knee. Some of them are indicated in the figures. If the overall orientation of a dislocation lies in the unstable range, the dislocation decomposes into stable segments as labeled by A in Fig. 3.22. In [590], a comprehensive study was given of the elastic properties of dislocations with all relevant Burgers vectors in B2 Fe–Al of different composition. According to that, dislocations with ⟨111⟩ and ⟨100⟩ Burgers vectors are elastically unstable in certain orientation ranges. The insets of Figs. 3.22 and 1.4 present the calculated shapes of dislocation loops on the (101) plane using the line tension model in anisotropic elasticity with elastic constants from [591]. The line tension maxima and minima of the ⟨111⟩ dislocations are not located at the screw and edge orientations. They differ by a factor of about 8, resulting in the elongated shape of the equilibrium loop. Between about 38° and 55°, i.e., for orientations close to $[10\bar{1}]$, the dislocations are unstable resulting in edges in the dislocation line. These knees and preferential orientations are evident in Fig. 3.22, which also shows many mixed dislocations having decomposed into a zig-zag shape, e.g., at A. The energy of the [010] dislocation shows an anomalous decrease with increasing edge character with its minimum at about 125°. The shape of this dislocation is symmetric around the Burgers vector direction with relatively flat segments at the two ⟨111⟩ orientations. The screw

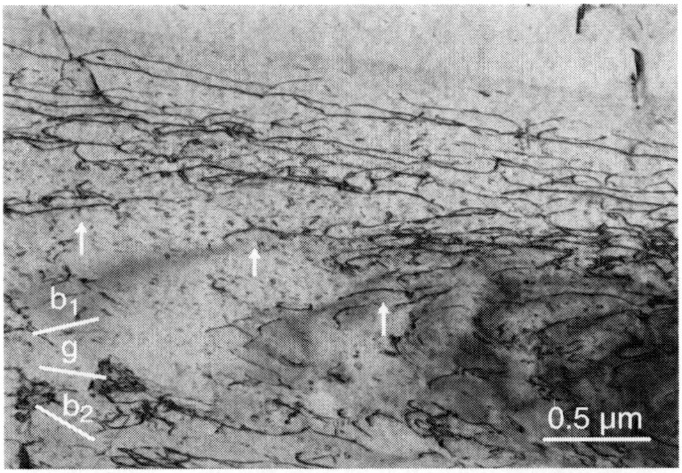

Fig. 9.29. Dislocations in Fe-43Al moving during an in situ straining experiment at room temperature. $\boldsymbol{g} = (10\bar{1})$ near the $[1\bar{2}1]$ pole. $\boldsymbol{g}$ is parallel to the trace of the (101) primary slip plane. $\boldsymbol{b}_1$ and $\boldsymbol{b}_2$ are the projections of the $[11\bar{1}]$ and $[\bar{1}11]$ Burgers vectors. From [8]. Copyright Elsevier Ltd. (2005)

orientation is unstable in the range between about 14° and −14°. The instability of the screw direction and the preferential orientations are obvious from Fig. 1.4. Apparently, the edge orientation seems to be another preferential orientation which is not described by the elastic dislocation properties.

In Fig. 4.40, the temperatures at which in situ straining experiments were performed in the HVEM are indicated by black vertical lines. At room temperature, dislocation motion was observed in the plastic zone of a crack. Figure 9.29 shows the dislocation structure during deformation. The projections of the Burgers vectors are marked. Moving near-screw dislocations with Burgers vector $\boldsymbol{b}_1$ are labeled by arrows. In the upper part of the figure, long dislocations of 30° character have developed. All dislocations bow out between cusps in the dislocation lines. The distances l between the obstacles are in the range of 0.1 μm.

The dynamic behavior of the dislocations is demonstrated in the following video sequences.

Video 9.12. *Motion of dislocations with* ⟨111⟩ *Burgers vectors in* Fe-43at%Al *at room temperature:* The clip consists of four sequences. Compared to the others, the first one is taken at a lower magnification. The dislocations move in a jerky way over distances longer than the segment length between the pinning points. The long dislocations are of near-screw character.

In situ experiments at 363°C, i.e., when the flow stress begins to increase, show straight slip traces belonging to glide on {110} planes, with cross slip events between different such planes. The dislocations with ⟨111⟩ Burgers vectors consist of straight segments of two orientations. One is consistent with

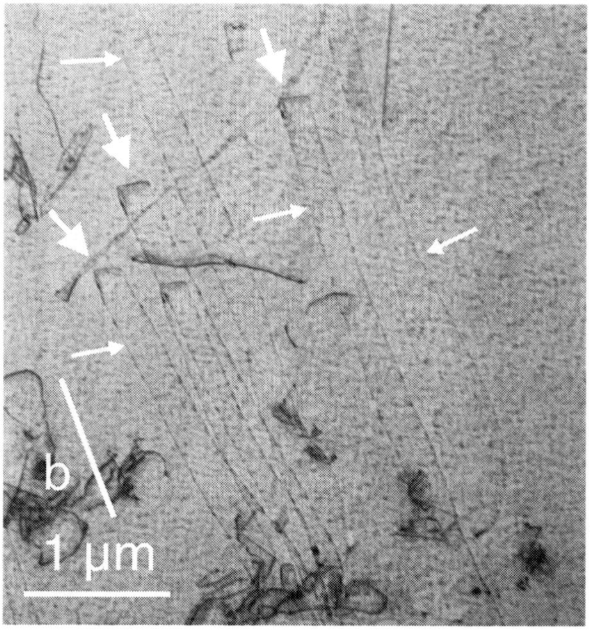

Fig. 9.30. Slip traces and dislocations with ⟨111⟩ Burgers vectors moving during in situ straining in Fe-43at%Al at 530°C. $\boldsymbol{g} = (10\bar{1})$ near the $[1\bar{2}1]$ pole. From [8]. Copyright Elsevier Ltd. (2005)

the screw direction, whereas the other one corresponds to a mixed orientation with a large edge component, which may be ⟨111⟩ on the respective {101} plane. At 530°C, i.e., in the anomaly range of macroscopic deformation tests, the slip traces (thin arrows in Fig. 9.30) are curved pointing at the superposition of climb or cross slip in addition to glide. Nevertheless, the traces are close to those of the (011) planes. The dislocations (thick arrows) imaged with the $(10\bar{1})$ reflection belong to the primary $[11\bar{1}]$ Burgers vectors. The projections of the straight segments are again consistent with the screw orientation and the 60° character on the (011) plane. The next video shows the motion of such dislocations.

Video 9.13. *Motion of dislocations with ⟨111⟩ Burgers vectors in* Fe-43at%Al *at* 530°C*:* The dislocations of angular shape move very slowly in a viscous way.

In addition, slip bands occur with dislocations that are extinguished at $\boldsymbol{g} = (10\bar{1})$. These dislocations can be imaged with a (211) $\boldsymbol{g}$ vector at the $[1\bar{3}1]$ pole as demonstrated in Fig. 9.31 taken at 447°C. They are also invisible with $\boldsymbol{g} = (200)$ so that their Burgers vector is [010]. The orientation of the slip traces suggests again (101) slip planes. These dislocations are straight, with their projections corresponding to ⟨111⟩ directions, i.e., the angle between the Burgers vectors and the dislocation lines is 54.7°. Their motion is recorded in

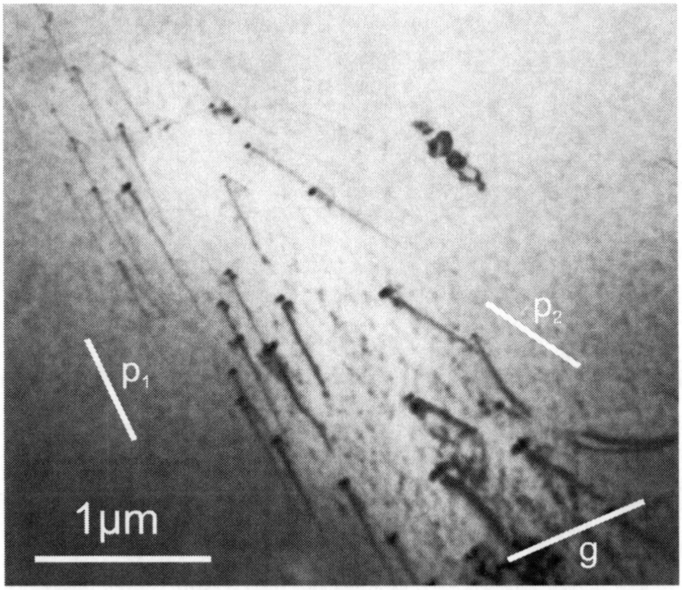

Fig. 9.31. Dislocations with ⟨100⟩ Burgers vectors moving during in situ deformation of Fe-43Al at 447°C. $\boldsymbol{g} = (112)$ near the $[1\bar{3}1]$ pole. Projections $\boldsymbol{p}_1 = [71\bar{4}]$ of $[\bar{1}11]$ direction and $\boldsymbol{p}_2 = [\bar{4}17]$ of $[11\bar{1}]$ direction. From [8]. Copyright Elsevier Ltd. (2005)

Video 9.14. They move and multiply in bands separated from the bands with ⟨111⟩ dislocations.

Video 9.14. *Motion of dislocations with* ⟨100⟩ *Burgers vectors in* Fe-43at%Al *at* 447 *and* 530°C*:* The dislocations of a straight oriented shape move viscously. Even at a temperature lower than in Video 9.13, the ⟨100⟩ dislocations move faster than those with ⟨111⟩ Burgers vectors. After unloading, they relax into a curved shape.

At high temperatures far above the flow stress maximum (near 700°C), the dislocations show a totally different behavior, which is documented in the following video.

Video 9.15. *Dynamic formation and annihilation of dislocations in* Fe-43at%Al *at* 700 °C*:* Smoothly curved dislocations move viscously in a very irregular way pointing at a combination of glide with rapid climb. New dislocation length is created continuously by certain segments bowing out. On the other hand, dislocation segments coming across each other frequently annihilate, so that the dislocation density remains constant on a low level over a long time of observation. At L near the end of the sequence, a small loop is created by a moving dislocation, which shortly afterward collapses again. The loop is also shown in Fig. 9.32. When the dislocations stop moving, they form straight segments, the projections of which are consistent with the three ⟨100⟩ line directions.

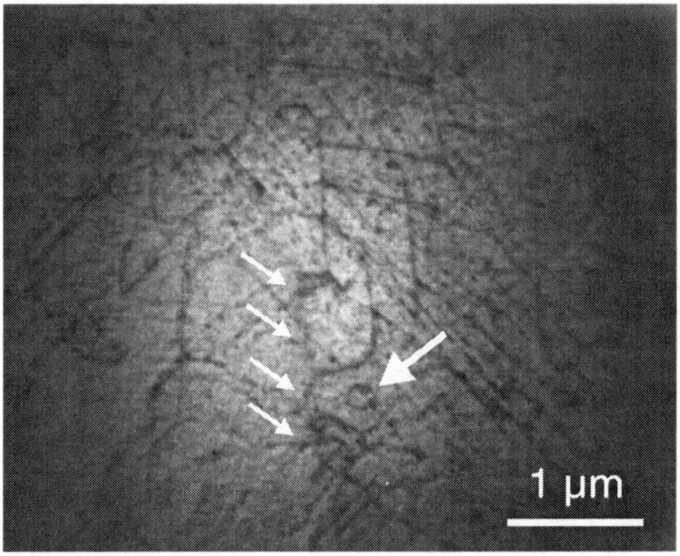

Fig. 9.32. Frame from Video 9.15 showing the blurred image of a moving dislocation marked by *small arrows* and a dislocation loop indicated by a *thick arrow*. From [8]. Copyright Elsevier Ltd. (2005)

The irregular nonplanar motion of the dislocations suggests that, at this high temperature, climb is not only a process of recovery but also a mode of dislocation motion.

9.5.2 Macroscopic Deformation Parameters

Figure 9.33 shows two stress–strain curves at 550°C with sections taken at the different strain rates indicated. The curves are either smooth or serrated depending on the strain rate and, in general, also on the temperature. Smooth deformation curves occur only at room temperature for strain rates above $10^{-6}\,s^{-1}$, and above about 550° for strain rates less than $10^{-3}\,s^{-1}$. Below 550°C, optical microscopy shows single slip on localized bands corresponding to the (101) primary slip plane. Above that temperature, traces are revealed belonging to the (011) plane.

As mentioned earlier, the temperature dependence of the yield stress was presented before in Fig. 4.40. The curve consists of three ranges, a low-temperature range of a normal temperature dependence, i.e., with a slightly decreasing flow stress between room temperature and about 350°C, a range of an anomalous increase between about 350 and 550°C, and a high-temperature range above 550°C with again a normal decrease of the flow stress. In this range, very strong yield points occur. Figure 4.40 presents the lower yield

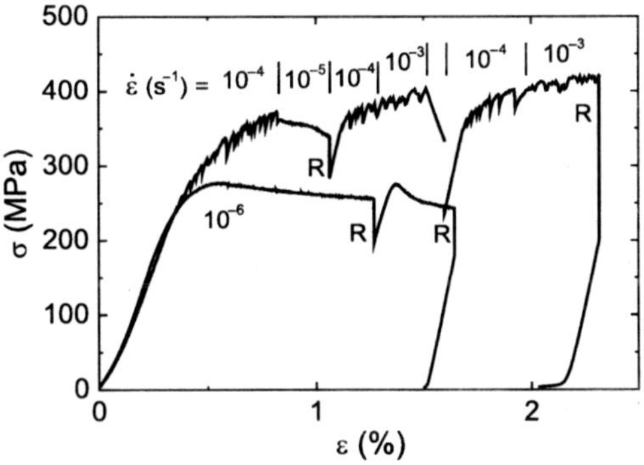

Fig. 9.33. Stress–strain curves of Fe-43at%Al deformed at 550°C at different strain rates $\dot{\varepsilon}$. R are stress relaxation tests. Data from [8]

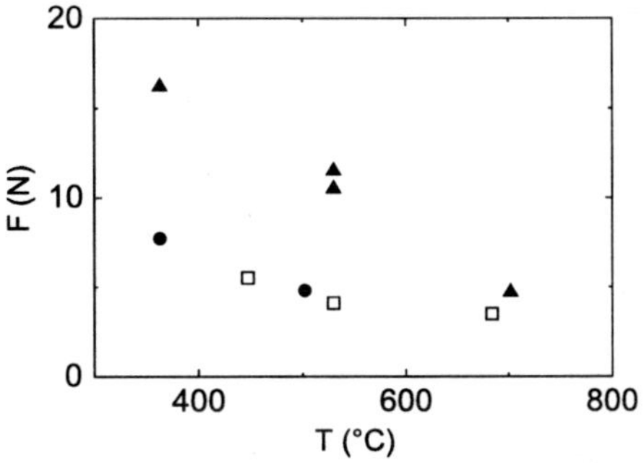

Fig. 9.34. Dependence of the load to achieve dislocation velocities suitable for observation in the HVEM for three in situ experiments including temperature changes. Each symbol belongs to one experiment. Data from [8]

points. The upper yield point of 342 MPa at 594°C has almost the same value as the stress maximum at 550°C.

The tensile stage for the in situ straining experiments in the HVEM allowed the measurement of the force acting on the micro-specimens. During the experiments at higher temperatures, several times a single specimen was deformed successively at different temperatures. The straining stage was controlled in such a way that dislocation velocities were achieved in the range that can well be recorded (about 10–1,000 nm s^{-1}). In Fig. 9.34, these loads are plotted versus the temperature for three in situ experiments. Each symbol corresponds to one specimen deformed at different temperatures. In all experiments, there is a continuous decrease of the loads with increasing temperature, i.e., the anomalous increase of the flow stress observed in macroscopic experiments did not occur during the in situ tests.

The strain rate sensitivity of the flow stress was studied by stress relaxation tests. The relaxation curves are quite straight or have the usual curvature

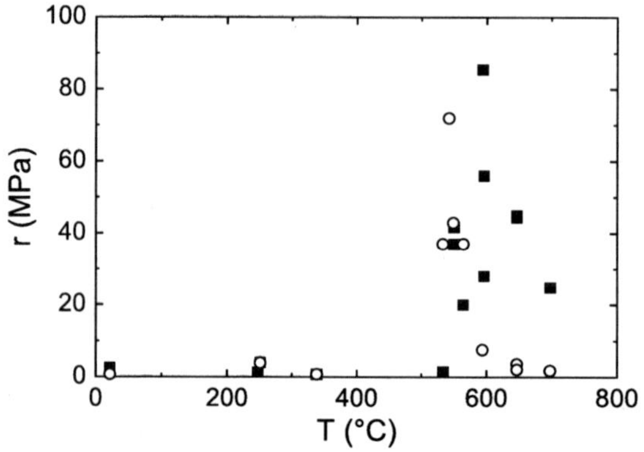

Fig. 9.35. Temperature dependence of the strain rate sensitivity r measured from the slope of stress relaxation curves at relaxation (strain) rates of $\dot{\varepsilon} = 10^{-5}\,\mathrm{s}^{-1}$ (*filled squares*) and $10^{-7}\,\mathrm{s}^{-1}$ (*open circles*). Data from [8]

towards the stress axis, corresponding to most obstacle mechanisms. For a narrow range within the anomaly range (at 533 and 542°C), the curves have a typical two-stage shape with an inverse curvature bending away from the stress axis. As outlined in Sect. 4.11, the inverse curvature points at a mechanism with the dynamics of the Cottrell effect for the flow stress anomaly. Values of the strain rate sensitivity r of the flow stress shown in Fig. 9.35 were evaluated from the slope of the stress relaxation curves at strain rates of 10^{-5} and $10^{-7}\,\mathrm{s}^{-1}$. They are very small up to about 530°C and show a maximum at about 600°C, i.e., above the flow stress peak. In the anomaly range, the values measured at the strain rate of $10^{-7}\,\mathrm{s}^{-1}$ are higher than those at $10^{-5}\,\mathrm{s}^{-1}$, in agreement with the inverse curvature of the relaxation curves. In the high-temperature range, the behavior is normal as expected from the normal temperature dependence of the flow stress. Strain rate change experiments over a wide range of strain rates between 5×10^{-7} to $10^{-4}\,\mathrm{s}^{-1}$ show zero strain rate sensitivity at room temperature, 250 and 533°C.

9.5.3 Deformation Mechanisms

The following discussion of the processes controlling the deformation of FeAl is divided according to the temperature ranges, room temperature, the range of the flow stress anomaly, and the range above the peak temperature.

Room Temperature

The in situ straining experiments show dislocations with $\langle 111 \rangle$ Burgers vectors slightly bowing out between obstacles of an average distance of $l \approx 0.1\,\mu\mathrm{m} = 200\,b$ (Fig. 9.29) and moving in a jerky way (Video 9.12). This is a mechanism typical of low temperatures also in other intermetallic alloys. In principle, the obstacles may be jogs as in NiAl (Sect. 9.4.4) or small precipitates as in TiAl (Sect. 9.3.4). In situ straining experiments on FeAl at room temperature were performed earlier, yielding similar results [592].

At room temperature, the strain rate sensitivity r exhibits a normal dependence on the strain rate. Using (4.9), the value of $r = 2.6\,\text{MPa}$ at $\dot{\varepsilon} = 10^{-5}\,\text{s}^{-1}$ (Fig. 9.35) is related to an activation volume of $V \approx 3.7 \times 10^{-27}\,\text{m}^3 = 30\,b^3$. In microscopic terms, the activation volume is given by $V = \Delta d\,lb$ (4.4) yielding an activation distance of $\Delta d \approx 0.15\,b$. This is a normal value for dislocation motion controlled by localized obstacles like small precipitates (see the similar estimates for ZrO_2 (Sect 7.3.5), Al–Zn–Mg (Sect. 8.1.1) or TiAl (Sect. 9.3.4)). The very jerky motion over large distances indicates that the dislocation motion is of almost athermal nature (Fig. 4.18), i.e., that the Peierls mechanism, which frequently acts at low temperatures, is not important.

Flow Stress Anomaly

As discussed in [593], a number of models were established to explain the flow stress anomaly in B2 ordered FeAl crystals. Since macroscopic deformation between room temperature and the peak temperature is carried by dislocations with the $\langle 111 \rangle$ Burgers vector (Sect. 9.5.1), these dislocations must be responsible for the anomaly. The models include:

- Cross slip of dislocations with $\langle 111 \rangle$ Burgers vectors from their $\{110\}$ slip plane onto $\{112\}$ planes where they get pinned, by analogy with Kear–Wilsdorf locks in Ni_3Al [594]
- Local decomposition of dislocations with $\langle 111 \rangle$ Burgers vectors into partials with $\langle 001 \rangle$ and $\langle 110 \rangle$ Burgers vectors, which impedes the dislocation motion [595]
- Localized climb of the antiphase boundary-coupled partials [596] according to a model originally proposed to explain the flow stress anomaly in β brass [584]
- Hardening owing to the high concentration of thermal vacancies at the temperatures of the flow stress anomaly (vacancy hardening model) [593] based on the observation that the anomaly occurs only in specimens in which the high excess vacancy concentration originating from the crystal preparation is annealed out by long-time annealing at relatively low temperatures [597]

The in situ straining experiments described earlier show a viscous motion of straight oriented dislocations. Such a behavior is not considered in the literature on deformation of B2 Fe–Al. The viscous motion contradicts all pinning/unpinning mechanisms with jerky dislocation motion like the first three models listed above.

From the literature, it is obvious that during the deformation of FeAl at high temperatures point defects play an essential role, e.g., the structural defects originating from deviations from stoichiometry [598]. The effects of thermal vacancies are considered in the vacancy hardening model of George and Baker [593], mentioned above, and in other related papers [599–601]. In this model, individual thermal vacancies appearing in the range of the flow stress anomaly are supposed to pin the dislocations as localized nondiffusing obstacles to be treated by the theory of localized obstacle hardening as described in Sect. 4.5.1. There are several experimental observations supporting the importance of thermal vacancies in connection with the flow stress anomaly. In experiments with fast up-heating to the deformation temperature, the flow stress anomaly occurs only after an incubation time of the order of some minutes. This time equals the time necessary to establish the equilibrium concentration of thermal vacancies [602]. Besides, after quenching down from temperatures in the anomaly range, the specimens show the same flow stress at room temperature as those deformed at the annealing temperatures [600]. However, vacancies should not act as fixed localized obstacles at the high temperatures in the range of the anomalous increase of the flow stress as will be discussed below.

It is difficult to estimate the interaction strength between a vacancy and a dislocation. Considering the vacancy as an inclusion of zero bulk modulus, the maximum interaction energy is given by (e.g., [603])

$$\Delta G_0 = \frac{9K_\mathrm{e}V}{16\pi^2}\,\frac{1+\nu}{1-\nu},$$

where V is the volume of the vacancy taken as $V = a^3/2$, and ν is Poisson's ratio ($\nu = 0.384$ [604]). K_e is the energy factor of the edge dislocation with $\langle 111\rangle$ Burgers vector on the $\{110\}$ plane. It replaces the shear modulus. K_e is calculated using the anisotropic elastic constants from [591] mentioned above. With these data, $K_\mathrm{e} = 108\,\mathrm{GPa}$ and $\Delta G_0 = 1.03\,\mathrm{eV}$ are obtained. Assuming, for simplicity, a triangular interaction profile of the width of the absolute value of the Burgers vector, the resulting maximum force is given by

$$F_0 = \Delta G_0/(0.5\,b).$$

This yields $F_0 = 6.6 \times 10^{-10}\,\mathrm{N}$. To decide whether Friedel or Mott statistics has to be applied, the parameter ξ_0 can be estimated from (4.52). For the low concentration of vacancies of $c \approx 3\times10^{-4}$ at the flow stress peak [605] and with the width of the obstacles $x_0 \approx b$, the square lattice distance $l_\mathrm{sq} = ac^{-1/2}/2$, and the energy factor of the edge dislocation mentioned above, it follows that $\xi_0 \approx 0.07 \ll 1$ so that Friedel statistics can be applied. Then, the contribution of the obstacles to the flow stress at zero temperature is given by (4.56). Thus, $\sigma_0^* = \tau_0^*/m_\mathrm{s} = 170\,\mathrm{MPa}$ with the orientation factor $m_\mathrm{s} = 0.44$ for the $(101)[11\bar{1}]$ slip system. In terms of the Friedel model, the contribution of solution hardening to the flow stress shows a strong temperature dependence with a limiting temperature

$$T_0 = \frac{\Delta G_0}{k\,\ln(\dot\gamma_0/\dot\gamma)},$$

above which the contribution is small. k is again Boltzmann's constant, $\dot{\gamma}$ the (shear) strain rate and $\dot{\gamma}_0$ the pre-exponential factor in the Arrhenius equation of the strain rate (4.8). With a characteristic value of $\ln(\dot{\gamma}_0/\dot{\gamma}) \approx 25$, T_0 becomes about 200°C. Consequently, vacancies of the estimated concentration at the flow stress peak may contribute to the flow stress at low temperatures, e.g., quenched-in vacancies at room temperature, but they should not cause the flow stress anomaly at much higher temperatures. The fact that such small and relatively weak obstacles as individual vacancies are easily surmounted at high temperatures by the aid of thermal activation is generally neglected in the literature of the vacancy hardening model of the flow stress anomaly in FeAl. Vacancy agglomerates, which may act also at higher temperatures, usually occur at much lower concentrations than the individual vacancies do.

Several experimental results can well be interpreted by dynamic strain ageing. Effects of strain ageing were observed in FeAl in several experiments (e.g., [606–608]), but not in all. In the experimental material above, this concerns, in particular, the occurrence of the flow stress anomaly itself, the inverse behavior of the strain rate sensitivity (Fig. 9.35), and the appearance of serrated yielding (Fig. 9.33) connected with slip localization. According to [608], serrated yielding depends on the orientation of the loading axis. The present orientation is in the range of unstable slip. Serrated yielding is certainly favored in single crystals in single slip orientation where slip localization is promoted. It is also supported in a stiff testing machine used here, where sudden strain increments cause large stress drops. As described in Sect. 5.3, the so-called strain rate softening instabilities do not require a negative strain rate sensitivity. It is sufficient if the strain rate sensitivity is small, which is fulfilled in the present experiments.

It was discussed in Sect. 4.11 that diffusion processes in the dislocation cores like the Cottrell effect may cause a flow stress anomaly. This idea was applied in Sect. 9.3.4 to the anomaly in γ TiAl. The respective formulae are (4.100) for the maximum contribution of this effect to the flow stress, and (4.99) for the relation between the critical strain rate and the diffusion coefficient and the temperature. It may be assumed that the diffusing defects are the Fe vacancies. The interaction strength β is given by (3.32) for solutes with a hydrostatic stress field. By comparing this formula with (14–42) in [12] it may be concluded that for vacancies $\Delta\Omega$ is given by the relaxation volume of the vacancy. Since it is not known, it is assumed that it amounts to 30% of the atomic volume, i.e., $\Delta\Omega = 0.3a^3/2$. Using again K_e for the shear modulus, the maximum flow stress contribution is only $\sigma_{max} = 8.6$ MPa. With a characteristic dislocation density of $\varrho_m = 2 \times 10^{13}\,m^{-2}$, the diffusion coefficient D necessary to obtain the flow stress maximum at the specific temperature $T_{max} = 550$°C and strain rate $\dot{\varepsilon}_{max} = 10^{-5}\,s^{-1}$ turns out to be $D = 10^{-19}\,m^2\,s^{-1}$. This is exactly the diffusion coefficient of Fe at 550°C [609]. Thus, the Cottrell effect model of dynamic strain ageing can qualitatively interpret the dynamic deformation properties. It is consistent with the diffusion coefficient of Fe vacancies, but it fails to describe the amount of the anomalous flow stress increase.

Dynamic strain ageing does also not explain the straight crystallographic shape of the dislocations moving during the in situ experiments (Fig. 9.30 and Video 9.13). In general, the straight appearance of the dislocations can be explained by the Peierls mechanism. However, it was shown above that the Peierls mechanism is no more important at room temperature. Nevertheless, the straight arrangement of the moving dislocations is obviously a property of the dislocation cores. There exist only a few theoretical studies of the dislocation core configurations in FeAl (e.g., [610, 611]). They suggest that dislocations with both $\langle 111 \rangle$ and $\langle 100 \rangle$ Burgers vectors dissociate into partial dislocations with Burgers vectors close to $1/2\langle 111 \rangle$ which, however, slightly point out of the $\{110\}$ slip plane. Thus, these dislocations cannot move conservatively in the dissociated configuration. The critical stresses to move the dislocations on $\{110\}$ planes at zero temperature are between 8 and 320 MPa. They are very different for screw and edge dislocations, but it seems that these values are not relevant for the dislocation motion in the range of the flow stress anomaly.

It is suggested in [8] to explain the straight crystallographically oriented shape of dislocations by a decomposition model similar to that for the motion of dislocations with $\langle 110 \rangle$ Burgers vectors during deformation of NiAl single crystals in the hard $\langle 100 \rangle$ orientation described in Sect. 9.4.4. The proposed model is fundamentally different from the climb dissociation model [584], which was applied to FeAl in [596]. In FeAl, almost all preferential orientations of the dislocations during motion are of the type $\langle 111 \rangle$, i.e., of the screw orientation and a mixed orientation for dislocations with $\langle 111 \rangle$ Burgers vectors as well as of a mixed orientation for the $\langle 100 \rangle$ dislocations. Only the case of the $\langle 111 \rangle$ screw dislocation was treated in the core structure calculations in [610, 611]. Possible decomposition schemes in analogy to the model for NiAl are listed in Table 9.4. According to that, the screw dislocation on a (101) plane with a $[11\bar{1}]$ Burgers vector can decompose by glide into three components with complete $\langle 100 \rangle$ Burgers vectors

$$[11\bar{1}](101) \rightarrow [100](0\bar{1}\bar{1}) + [010](101) + [00\bar{1}](\bar{1}10) \qquad (9.7)$$

(reaction 1 in Table 9.4). All the slip planes involved are of the type $\{110\}$, which seems to be the preferred slip plane in FeAl. The slip plane of the second component is the same as that of the total dislocation so that this component can move by glide. The two other components have to move by combined glide and conservative climb as discussed for NiAl. Table 9.4 lists also the character ϕ of all dislocations. The prelogarithmic factors of the energy $E_{0\mathrm{t}}$ of the starting dislocation and those of the components $E_{0\mathrm{c}}$ on the respective slip planes were calculated again within the framework of anisotropic elasticity theory as in Table 9.3. The last column of Table 9.4 contains the energy gain $E_{0\mathrm{t}} - \sum E_{0\mathrm{c}}$ by the reactions. It is strongly negative for the considered reaction 1 (9.7) so that this reaction is not driven by the elastic energy. Nevertheless, it

Table 9.4. Suggested decomposition of dislocations with $[11\bar{1}]$ Burgers vectors on (101) planes in FeAl

Reaction	1			2		
Total disl.						
line orient.	$[11\bar{1}]$			$[\bar{1}11]$		
ϕ	0°			109.5°		
E_{0t} (10^{-9} N)	1.005			2.039		
Comp. disls.						
BV	[100]	[010]	$[00\bar{1}]$	[100]	[010]	$[00\bar{1}]$
Plane	$(0\bar{1}\bar{1})$	(101)	$(\bar{1}01)$	$(01\bar{1})$	(101)	(110)
ϕ	54.7°	54.7°	54.7°	125.3°	54.7°	125.3°
E_{0c} (10^{-9} N)	0.593	0.593	0.593	0.593	0.593	0.593
Energy gain (10^{-9} N)			−0.774			0.26

may occur owing to a special core configuration although this reaction is not suggested by the calculations in [610, 611]. On the other hand, the near-edge dislocation with a $[11\bar{1}]$ Burgers vector with a very high energy can decompose with a considerable gain in elastic energy (reaction 2 in Table 9.4).

In the original model for NiAl [571], a dependence of the decomposition width d on the deformation parameters is not explicitly considered. This leads to a monotonously increasing dislocation mobility with increasing temperature. It is now suggested to extend the model by taking into account a possible dependence of the width of the decomposed dislocations on the dislocation velocity and the temperature. Since no detailed theory of this process is yet available, it is assumed that the decomposition width plays a role similar to the width of the diffusing point defect cloud around the dislocations in the theory of the Cottrell effect described in Sect. 4.11. This dependence results in the dynamic dislocation behavior of Fig. 4.39, leading to the occurrence of a flow stress anomaly, the inverse dependence of the strain rate sensitivity on the stain rate or stress, and to plastic instabilities, phenomena which are all observed in the present material. Since both processes, the Cottrell effect and the conservative climb motion of a decomposed dislocation require diffusion processes in the dislocation cores over similar distances, (4.99) should give a reasonable estimation of the temperature of the flow stress anomaly. Climb in the dislocation cores in FeAl is quite probable since general climb was observed in the same temperature range.

High-Temperature Range

The transition from the range of the flow stress anomaly to the high-temperature range with a strongly decreasing flow stress may be caused by two processes. As first suggested in [587], the dominating Burgers vectors change from ⟨111⟩ to ⟨100⟩. A comparison between Videos 9.13 and 9.14 from in situ straining experiments had shown that dislocations with ⟨100⟩ Burgers vectors are much more mobile than those with ⟨111⟩ ones. In addition, recovery processes become important.

In the in situ straining experiments, there is no indication of dislocations with ⟨100⟩ Burgers vectors being formed by the decomposition of dislocations with ⟨111⟩ Burgers vectors. Instead, the ⟨100⟩ dislocations moved in locally separated bands and were generated within these bands by multiplication. During their motion, the ⟨100⟩ dislocations are also straight and oriented along ⟨111⟩ directions (Fig. 9.31), which may partly be due to the line tension. However, the straight shape can also result from dissociation reactions, which are not discussed in detail.

As demonstrated by Video 9.15 taken at 700°C in the high-temperature range, irregular dislocation motion points at rapid climb as a mode of dislocation movement which also causes very fast recovery so that the dislocation density during in situ straining remains on a low level. As the work-hardening is low, the high-temperature deformation can conveniently be treated in terms of an equation for steady state creep (5.20), as suggested in [593]. Depending on the particular creep model, stress exponents are expected between about 4 and 6. Experimental stress exponents m' determined from the strain rate sensitivity r according to (4.12) are equal to about 5 in the whole high-temperature range. These experimental values fit the range of the theoretical values for steady state creep. Thus, the strain rate sensitivity data also suggest the importance of recovery during the high-temperature flow stress decrease.

9.5.4 Summary

Between room temperature and the temperature of the maximum of the flow stress anomaly, Fe-43at%Al single crystals in a single slip orientation are deformed by dislocations with ⟨111⟩ Burgers vectors. At room temperature, the dislocations bow out most probably between localized obstacles. The flow stress anomaly is accompanied with an inverse dependence of the strain rate sensitivity of the flow stress on the strain rate or the stress in a narrow temperature range and with serrated flow. Curved slip traces during in situ straining experiments in an HVEM and a nonplanar arrangement of dislocations after macroscopic deformation indicate the importance of climb in the range of the flow stress anomaly. In the in situ experiments, viscously moving dislocations are straight and crystallographically oriented. These phenomena are interpreted by a decomposition of the cores of the dislocations with ⟨111⟩ Burgers vectors into three component dislocations with ⟨100⟩ Burgers vectors where, similar to the model for NiAl described in Sect. 9.4.4, two of the component dislocations have to move by combined glide and conservative climb. This model can be extended by considering a dependence of the decomposition width on the temperature and the dislocation velocity resulting in a dynamic behavior like the Cottrell effect and causing dynamic strain ageing. Above the peak temperature, dislocations with ⟨100⟩ Burgers vectors dominate. The deformation includes general climb and can be described by an equation for steady state creep.

9.6 Molybdenum Disilicide

Materials based on molybdenum disilicide with its high melting point of 2,030°C are promising for structural applications at high temperatures because of their relatively high flow stress and good oxidation resistance above 1,000°C [612]. However, like other high-temperature materials $MoSi_2$ is very brittle at room temperature and suffers from a low creep resistance at temperatures of ductility. It has the tetragonal $C11_b$ structure with a high c/a ratio. This implies strong plastic anisotropy. In the first studies on single crystals, a number of slip systems were observed, but plastic deformation was achieved only above 900°C [613–616]. Some of the slip systems have zero orientation factors for loading along [001] so that this orientation is a hard orientation. In a thorough study of the orientation dependence of the plastic properties, it was found that $MoSi_2$ can be deformed plastically even down to room temperature, and lower, along orientations away from [001] [617]. In these soft orientations, a flow stress anomaly occurs in different ranges between about 600 and 1,100°C, depending on the orientation. The temperature dependence of the critical resolved shear stress of most of the slip systems was gained from macroscopic deformation data and from studies of the slip lines on the surfaces and from the transmission electron microscope determination of the Burgers vectors.

9.6.1 Crystal Structure and Slip Geometry

As mentioned above, stoichiometric $MoSi_2$ has the $C11_b$ structure up to the melting point with an axial ratio c/a of almost exactly $\sqrt{6}$ as shown in Fig. 9.36. The structure consists of three b.c.c. cells, which are slightly compressed and stacked along the c axis. This results in a pseudo-hexagonal symmetry of the $\{110\}$ planes with almost identical lengths of the vectors $\langle 110 \rangle$ and $1/2\langle 111 \rangle$ as well as $1/2\langle 331 \rangle$ and $\langle 001 \rangle$. However, crystallographically, these directions are not equivalent as demonstrated in Fig. 9.37, which shows the arrangement of two neighboring layers of atoms on the (110) plane by black and grey circles. Among numerous slip systems possible, $\{110\}1/2\langle 111 \rangle$, $\{011\}\langle 100 \rangle$, $\{010\}\langle 100 \rangle$, $\{110\}1/2\langle 331 \rangle$, $\{013\}1/2\langle 331 \rangle$, and $\{011\}1/2\langle 111 \rangle$ were experimentally proved [613–616]. The first four systems have zero orientation factors for loading along [001]. The latter two have high flow stresses, which makes [001] a hard orientation. The following experimental data are restricted to the soft $(1\bar{1}0)1/2[111]$ and $(011)[100]$ slip systems included in Fig. 9.36a and b.

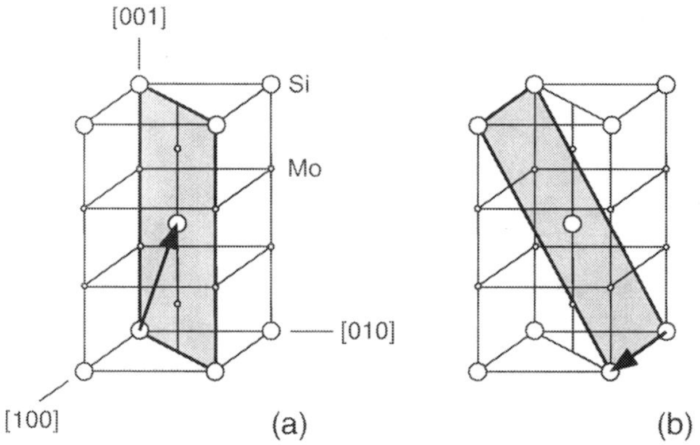

Fig. 9.36. Crystal structure of $MoSi_2$ with $(1\bar{1}0)1/2[111]$ (**a**) and $(011)[100]$ slip systems (**b**)

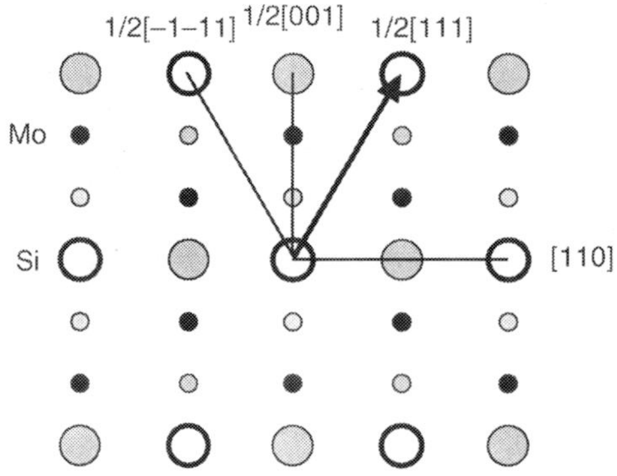

Fig. 9.37. Structure of the $(1\bar{1}0)$ slip plane in $MoSi_2$ with $1/2[111]$ Burgers vector. Black circles are the atoms of the plane, grey ones those on the neighboring layers

9.6.2 Microscopic Observations

To activate the mentioned slip systems, $MoSi_2$ single crystals were deformed along the [201] and [110] directions. The experimental data are collected from [15, 288, 618–622].

Deformation Along ⟨201⟩

At this deformation axis, mainly dislocations with $1/2\langle 111\rangle$ Burgers vectors are activated. Figure 9.38a illustrates the dislocation structure from a macroscopic deformation experiment at 300°C. As will be shown in Sect. 9.6.3, this temperature corresponds to the low-temperature decrease of the flow stress. Under different imaging directions, most dislocations follow a straight course. Via the contrast extinction rule (Sect. 2.3), their Burgers vector has

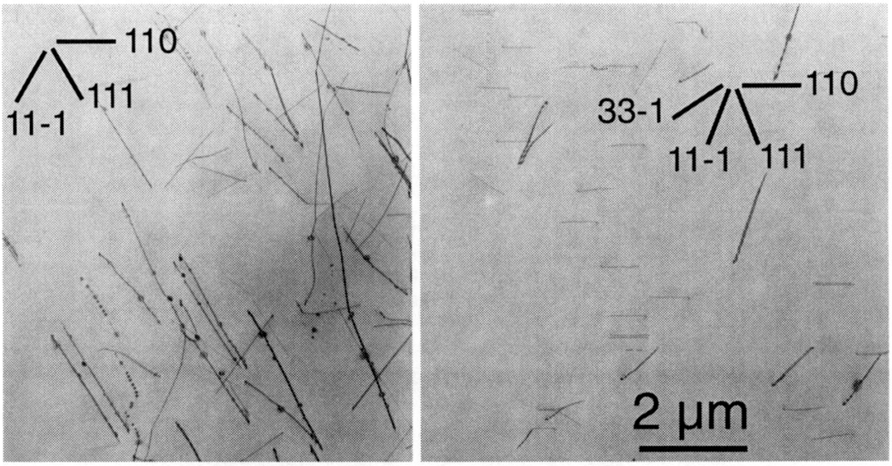

Fig. 9.38. Dislocation structures in $MoSi_2$ single crystals deformed along [201] taken in an HVEM. (**a**) 300°C, plastic strain 0.37%, beam direction $\boldsymbol{n} \approx [1\bar{1}0]$ perpendicular to $(1\bar{1}0)$ slip plane, Burgers vector $\boldsymbol{b} \parallel [11\bar{1}]$, $\boldsymbol{g} = (110)$. (**b**) at first 500°C, then 850°C, 1.5%, $\boldsymbol{n} \approx [010]$, $\boldsymbol{b} \parallel [111]$, $\boldsymbol{g} = (200)$. Black bars mark the indicated directions in the slip plane in (**a**) and their projections onto the image plane in (**b**). From the work in [620]

been determined to be $1/2[11\bar{1}]$. The figure shows a view perpendicular to the slip plane so that the line orientations can directly be determined. According to that, the majority of dislocations is arranged parallel to [111]. Thus, they are of 60° character. In addition, also 30° and edge dislocations occur. Very few dislocations have ⟨100⟩ Burgers vectors.

In the temperature range of the flow stress anomaly, slip becomes localized in narrow slip bands. The dislocations in Fig. 9.38b have 1/2[111] Burgers vectors. They are mostly straight again in 60° and edge orientation. However, in contrast to the low-temperature range, the 60° dislocations are now parallel to [110]. As mentioned earlier and illustrated in Fig. 9.37, the 60° directions parallel to ⟨111⟩ and ⟨110⟩ are crystallographically not equivalent. Thus, between the low-temperature range and the anomaly range, the dominating line orientation of straight 60° dislocations with 1/2⟨111⟩ Burgers vectors exhibits a transition from ⟨111⟩ to ⟨110⟩.

At temperatures between about 900 and 1,000°C, a few in situ straining experiments in the HVEM were successful with deformation along [201] demonstrating the dynamic dislocation behavior in the anomaly range. Similar to γ-TiAl, dislocations are generated suddenly during elastic loading connected with load drops LD demonstrated in the load time diagram of an in situ straining experiment in Fig. 9.39. As in macroscopic experiments, slip is concentrated in narrow slip bands. In all cases analyzed, the dislocations had 1/2⟨111⟩ Burgers vectors and were moving on {110} planes. Since these dislocations were formed during the sudden load drops, they must have moved at

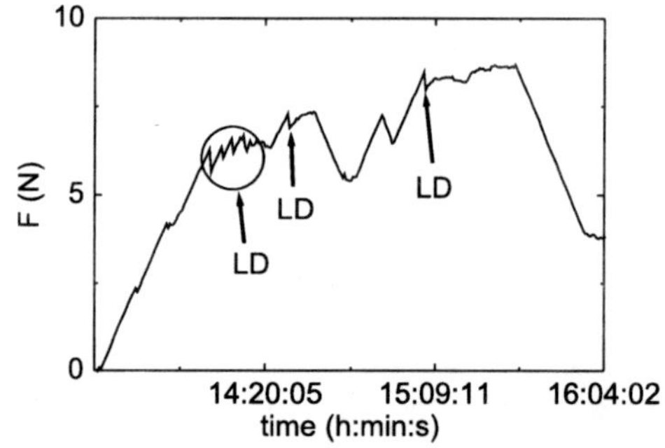

Fig. 9.39. Load–time recording of the in situ straining experiment on $MoSi_2$ of Videos 9.16 and 9.17 with sudden load drops LD. Data from [619]

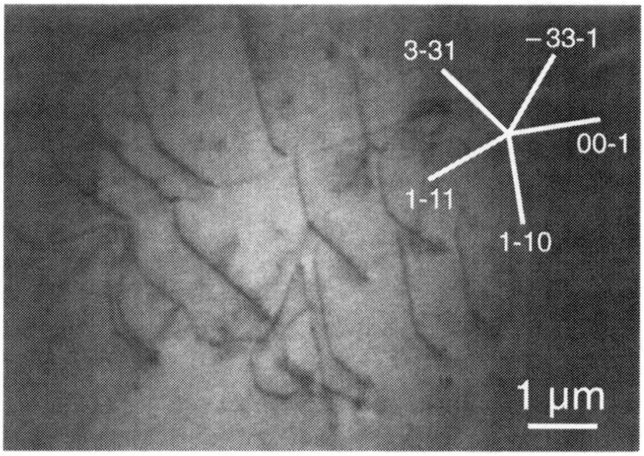

Fig. 9.40. Section of a frame from Video 9.16 showing dislocation motion in $MoSi_2$ at 990°C. The c axis denoted by $[00\bar{1}]$ lies in the image plane, the other bars mark the projections of the indicated directions onto the image plane. The Burgers vector is $1/2[1\bar{1}1]$. From the work in [288]

very high velocities. Later on, during continuous deformation, the dislocations moved in a viscous way as demonstrated in the following video clip taken at almost 1,000°C.

Video 9.16. *Dislocation motion and generation in* $MoSi_2$ *deformed at* 990° *C along* [201]*:* The dislocations are mostly straight along crystallographic orientations, which are indicated in Fig. 9.40. The preferred orientations of the dislocations with an angular shape are $[1\bar{1}0]$ and $[3\bar{3}1]$, i.e., 60° and edge at the $1/2[1\bar{1}1]$ Burgers vector, as in the macroscopic experiments. During the motion in direction of the slip traces imaged in weak contrast, curved segments and superkinks form, which move along the dislocations as marked by M in the video. At the dislocation labeled K in Fig. 9.41, several superkinks are visible. When moving at higher velocities, the dislocations prefer the 60° character.

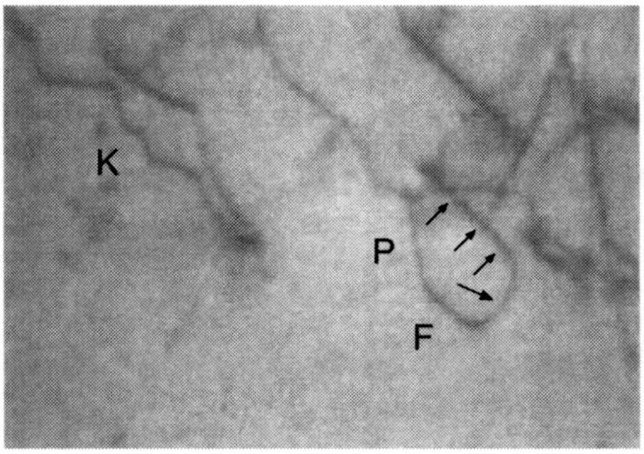

Fig. 9.41. Section of a frame from Video 9.16 showing the localized dislocation source

At S in the lower part of the video, a one-ended localized dislocation source as described in Sect. 5.1.1 and Fig. 5.1b generates new dislocations. The source is also shown in Fig. 9.41. It consists of one arm P lying on the primary (110) slip plane, and of a fixed segment F on another plane, therefore having a low mobility.

In $MoSi_2$, cross slip of dislocations with $1/2\langle 111\rangle$ Burgers vectors seems to be very restricted. It has never been observed in the in situ experiments that these dislocations cross slip from their $\{110\}$ slip planes onto the $\{011\}$ planes, which contain the same Burgers vector. The movable arm on a plane parallel to that of the segment P is marked by arrows. With each revolution of the moving branch, one dislocation is emitted to the right and another to the left. While the dislocations moving to the left leave the frame of the recording, the dislocations emitted to the right pile up in the stress field of other pre-existing dislocations. As described in Sect. 5.2.1, the piled-up dislocations cause a back stress which shields the source. Later on, these dislocations move away and the source operates again. The generation of several tens of dislocations has been observed.

Usually, intrinsic pinning agents which cause dislocation generation are jogs in a screw dislocation, which have edge character and cannot glide in the forward direction of the dislocation motion. The observed source in $MoSi_2$ is insofar particular as the pinning segment is located in a 60° dislocation, where its pinning is due to a low mobility on a plane away from the easy $\{110\}$ slip plane.

Some details of the motion of dislocations with $1/2\langle 111\rangle$ Burgers vector are also shown in the following video.

Video 9.17. *Dislocation motion in* $MoSi_2$ *deformed at* 990° *C along* [201]*:* In this video sequence, the segments extending along $[3\bar{3}1]$ develop several superkinks and move faster than those along $[1\bar{1}0]$ so that the latter extend. Later on, the lower part of the leading dislocation bows out. The bowing segment is bordered by a jog labeled J. The cusp in the dislocation line forming at the jog causes line tension forces on the jog driving it in upward direction along the $[1\bar{1}0]$ segment. As the jog itself has edge or mixed character, it cannot glide in this direction. Therefore climb must have contributed to the sidewise motion of the jog. Later on, part of the dislocation emerges through the foil surface.

In conclusion, dislocations with 1/2⟨111⟩ Burgers vectors arrange mostly in straight crystallographically oriented configurations, both within the range of the low-temperature decrease of the flow stress as well as in the anomaly range. However, a transition occurs from the dominating 60° dislocations along ⟨111⟩ at low temperatures to those along ⟨110⟩ at high ones.

At temperatures above the flow stress maximum, i.e., above 1,000°C, in addition to the 1/2⟨111⟩ dislocations those with ⟨100⟩ Burgers vectors are observed more frequently than at lower temperatures. They occur preferentially as Frank loops.

Deformation Along ⟨110⟩

In this orientation, dislocations with 1/2⟨111⟩ Burgers vectors have zero orientation factors. Therefore, dislocations with ⟨100⟩ Burgers vectors are activated. An example of the microstructure produced in the low-temperature range at 380°C was shown in Fig. 2.6 including the analysis of the ⟨100⟩ Burgers vectors using the contrast extinction rule (Sect. 2.3). Figure 9.42 presents the dislocation structure of a specimen cut parallel to the (110) cross-sectional face, taken from the same deformation experiment. In this projection, all four

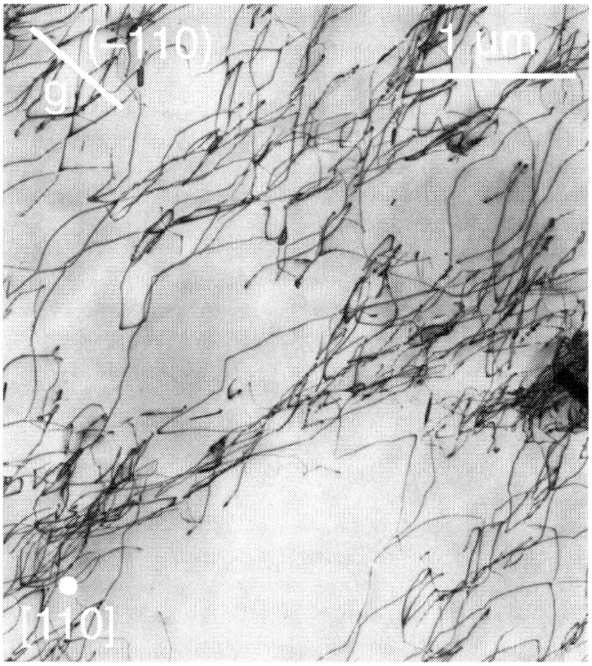

Fig. 9.42. Dislocation microstructure of $MoSi_2$ deformed along [110] at 380°C from the same deformation experiment as Fig. 2.6. Foil plane parallel to the (110) cross section plane. From [15]. Copyright Wiley-VCH (2005)

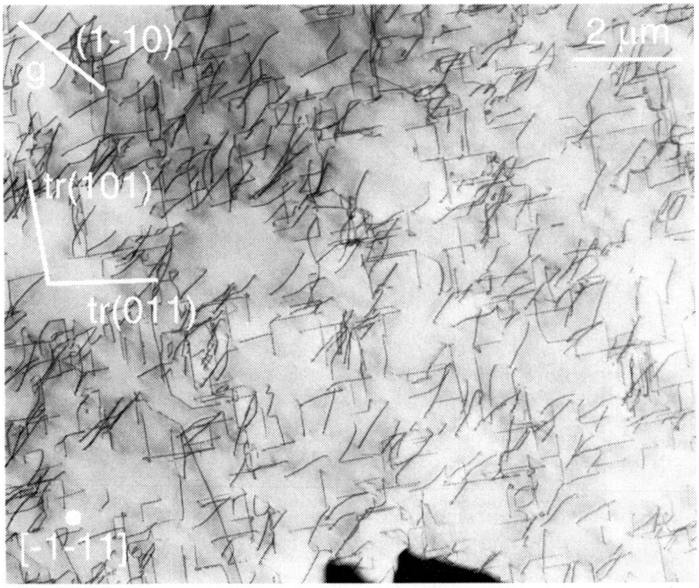

Fig. 9.43. Microstructure in $MoSi_2$ after deformation along ⟨110⟩ at 606°C by 0.2%. Foil plane parallel to the (001) side face of the compression specimen. The *straight lines* marked tr indicate the directions of the traces of the respective planes imaged edge-on. From [15]. Copyright Wiley-VCH (2005)

{101} slip planes are inclined to the foil plane (in contrast to Fig. 2.6b and d, where two of the four slip planes are imaged edge-on). It is obvious that the dislocations are curved at this temperature at the end of the normal low-temperature range of the flow stress. The dislocation density was estimated to amount to about $2.8 \times 10^{13}\,\mathrm{m}^{-2}$.

The dislocation structures in the lower part of the anomaly range are documented in Figs. 9.43 and 9.44 of a specimen deformed at 606°C. The foil is cut parallel to the (001) side face of the compression sample. In Fig. 9.43, the specimen is imaged near the $[\bar{1}\bar{1}1]$ pole with $\boldsymbol{g} = (1\bar{1}0)$ so that all dislocations with ⟨100⟩ Burgers vectors are visible. Many dislocations are straight and oriented parallel to the traces of the (101) and (011) planes, which are imaged edge-on in the projection of the figure. Thus, it may be assumed that these dislocations extend on these slip planes. Figure 9.44 presents a specimen area close to that of Fig. 9.43 imaged with $\boldsymbol{g} = (1\bar{1}0)$ near the [001] pole where all dislocations are visible. These dislocations have again ⟨100⟩ Burgers vectors. In the projection of the figure, most dislocations are also straight and oriented parallel to either the $[1\bar{1}0]$ direction, indicated by the $\boldsymbol{g}$ vector, or to the perpendicular [110] direction. Thus, these dislocation lines extend along ⟨111⟩ directions. It is argued in [15] that most of the dislocations which are not parallel to the traces in Fig. 9.43 are also aligned parallel to ⟨111⟩ directions.

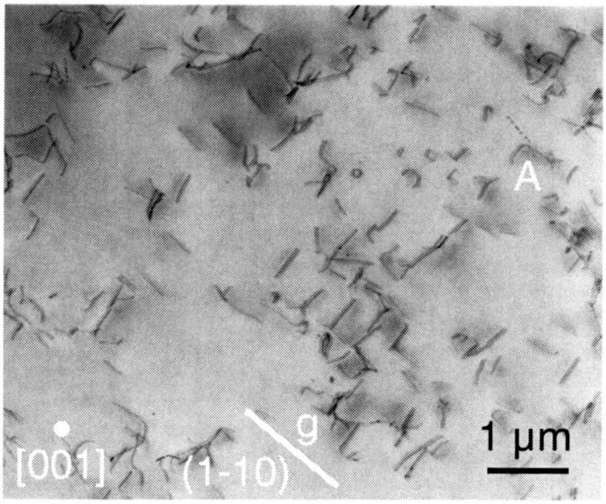

Fig. 9.44. Microstructure of the same specimen as Fig. 9.43 taken near the [001] pole. From [15]. Copyright Wiley-VCH (2005)

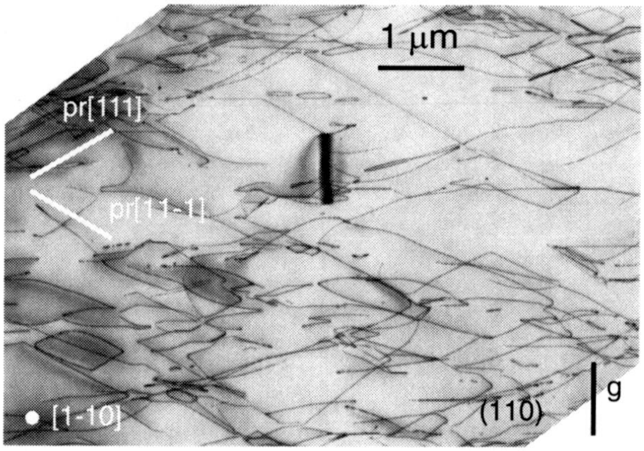

Fig. 9.45. Dislocation structure after deformation at 672°C by 0.25%. The *straight lines* marked pr indicate the directions of the projections of the indicated orientations. From [15]. Copyright Wiley-VCH (2005)

The dislocation structures formed at a small strain at a higher temperature within the anomaly range (672°C) are displayed in Fig. 9.45 taken near the $[1\bar{1}0]$ pole. The compression axis is parallel to the (110) $\boldsymbol{g}$ vector. All four $\{101\}$ slip planes are inclined to the foil plane at equal angles. The images show that the dislocations have angular shapes with segments appearing in the projections of the $\langle 111 \rangle$ directions. Burgers vector and stereometric analysis have shown that practically all dislocations are parallel to the projections of the four $\langle 111 \rangle$ directions. It may therefore be concluded that in the range

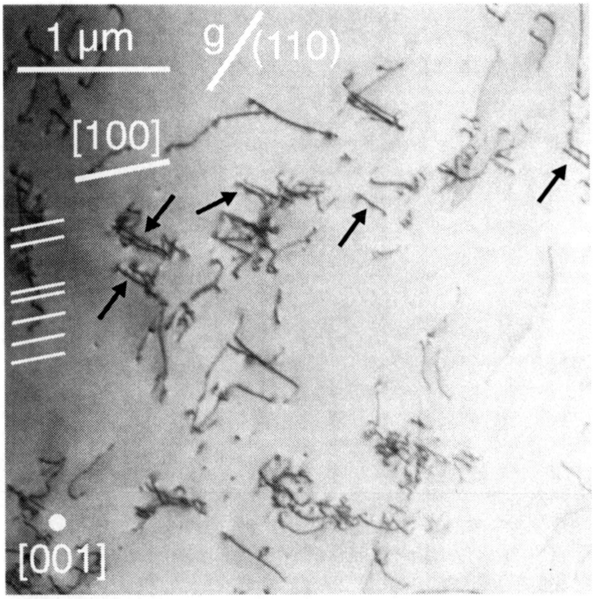

Fig. 9.46. Dislocations during in situ straining in an HVEM at 475°C along ⟨110⟩ of a specimen pre-deformed at 606°C. From [15]. Copyright Wiley-VCH (2005)

of the flow stress anomaly, the dislocations with ⟨100⟩ Burgers vectors show distinctly preferred orientations along ⟨111⟩.

In situ straining experiments were tried on pre-deformed specimens at different temperatures. Only one test at 475°C near the flow stress minimum was successful. Figure 9.46 presents the microstructure obtained. Most dislocations originate from the pre-deformation. Dislocations having moved during the in situ loading have produced slip traces, which are visible as weak bright lines. Some of them are marked by short white lines near the left edge of the figure. At the [001] pole, the slip traces run in [100] directions, i.e., they belong to the (011) and $(0\bar{1}1)$ slip planes. Some of the dislocations which had moved are marked by black arrows. These dislocations are straight. Their projections follow the $[\bar{1}10]$ direction. Thus, the dislocation lines should be oriented again along ⟨111⟩ directions. Accordingly, this single successful in situ experiment confirms the result of the above post-mortem studies that the active dislocations are straight and oriented along ⟨111⟩ at intermediate temperatures.

In the high-temperature range above the flow stress peak, the dislocations are no longer straight and crystallographically oriented but smoothly bow out. At high dislocation densities, they form dislocation bands by the interaction between dislocations on (101) and $(\bar{1}01)$ planes, which both contain the [010] Burgers vector.

Stacking Faults

In $MoSi_2$, stacking faults on (001) habit planes were described in a number of papers, e.g., [623–625]. Taking into account the results on faults observed after deformation at high temperatures, specific climb dissociation reactions were considered, e.g., in [616,626], leading to APBs and CSFs on (001) planes. As a second process, the formation of grown-in condensation faults on (001) planes is considered due to a loss of Si during the growth of $MoSi_2$ single crystals at temperatures up to 1,600°C [624,625]. These faults are bordered by Frank type dislocations having a Burgers vector parallel to [001]. During the in situ straining experiments described above, faults on (001) planes formed instantaneously. These faults were studied in detail in [627].

The faults were created principally during in situ straining or heating in the HVEM at temperatures between 440 and 1,000°C and only if dislocations with $1/2\langle 111\rangle$ Burgers vectors were present. These dislocations may originate either from in situ loading or from a pre-deformation in macroscopic experiments. Figure 9.47 presents an example of the formation of such faults during in situ annealing at 630°C. The faults are generated so quickly that the process itself has never been recorded.

The displacement vector $\boldsymbol{R}_\mathrm{F}$ of the faults and the Burgers vectors of the bordering dislocations have been analyzed by the methods of TEM diffraction contrast, e.g., the contrast extinction rule (2.9) in Sect. 2.3. An example of a fault bounded by dislocations A and B is presented in Fig. 9.48. The details of the analysis are described in [627] including further figures. The analysis clearly showed that the displacement vector $\boldsymbol{R}_\mathrm{F}$ equals either 1/6[001] or 1/2[001], which both characterize intrinsic stacking faults, and that the Burgers vector of one dislocation bordering each fault is parallel to [001] but that the other dislocation does not have the same Burgers vector, see in particular Fig. 9.48g where dislocation B is extinguished, but A is not. The Burgers vector of the second dislocation should be parallel to [110]. Although the whole analysis in [627] is not unambiguous, it is argued that the Burgers vectors of the bordering dislocations are 1/2[001] and 1/2[110]. Thus, these defects cannot be stacking faults, and produced by the condensation of vacancies due to a loss of Si, which on all edges should be surrounded by Frank partial dislocations. Besides, the instantaneous formation of the defects contradicts the creation by a diffusion process. Since the generation of the defects is bound to the presence of dislocations with $1/2\langle 111\rangle$ Burgers vectors, the best explanation of their formation is the dissociation of these dislocations according to

$$1/2[111](1\bar{1}0) \rightarrow 1/2[001](1\bar{1}0) + 1/2[110](001). \qquad (9.8)$$

The reaction leads to faults with a fault vector equal to the absolute value of the Burgers vector of the Frank-type partial with $\boldsymbol{R}_\mathrm{F} = -1/2[001]$. The faults can be created either by climb of the Frank-type dislocation with a 1/2[001] Burgers vector, or by glide of the partial dislocation with the 1/2[110]

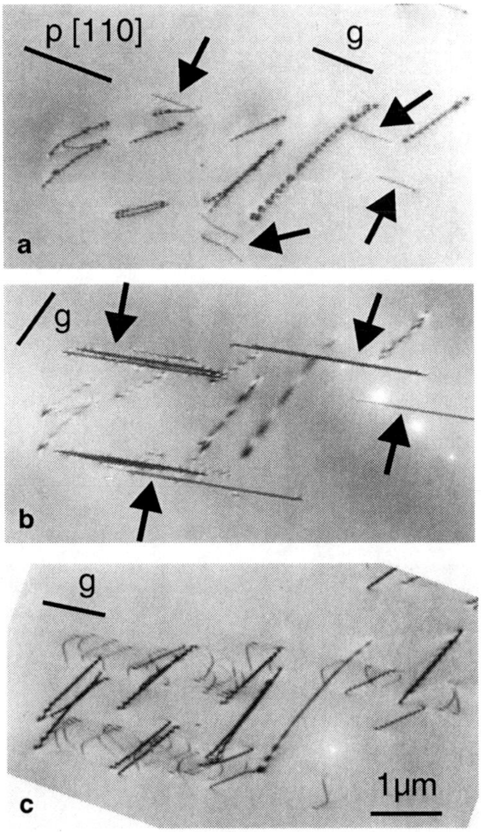

Fig. 9.47. Formation of planar faults on (001) planes on dislocations with a 1/2[111] Burgers vector during an in situ annealing experiment on an $MoSi_2$ single crystal at 630°C. (**a**) Microstructure with 1/2[111] dislocations after macroscopic compression. Beam direction near $\boldsymbol{n}$ = [010], $\boldsymbol{g}$ = (200). p is the projection of the [110] direction onto the image plane. Dislocations oriented along this direction are marked by *arrows*. (**b**) Sudden formation of faults labeled by *arrows* during in situ annealing at about 630°C. $\boldsymbol{g}$ = $(\bar{1}01)$. The faults are imaged edge-on. (**c**) The faults after post-mortem analysis at $\boldsymbol{n}$ = $[0\bar{3}\bar{1}]$ and $\boldsymbol{g}$ = (200). After [627]. Copyright Taylor & Francis Ltd. (2002)

Burgers vector, both on the (001) plane. The indexes in parentheses mark the slip planes of the respective dislocations. Their ⟨110⟩ intersection line is the preferred orientation of the initial dislocation for dissociation. The fault vector corresponds to a removal of three (001) layers, two Si layers and one Mo layer, without a loss of stoichiometry.

The planar faults are not created on all dislocations with 1/2⟨111⟩ Burgers vectors. The width of the faults scatters over a wide range. However, faults with a large width appear particularly during in situ straining experiments

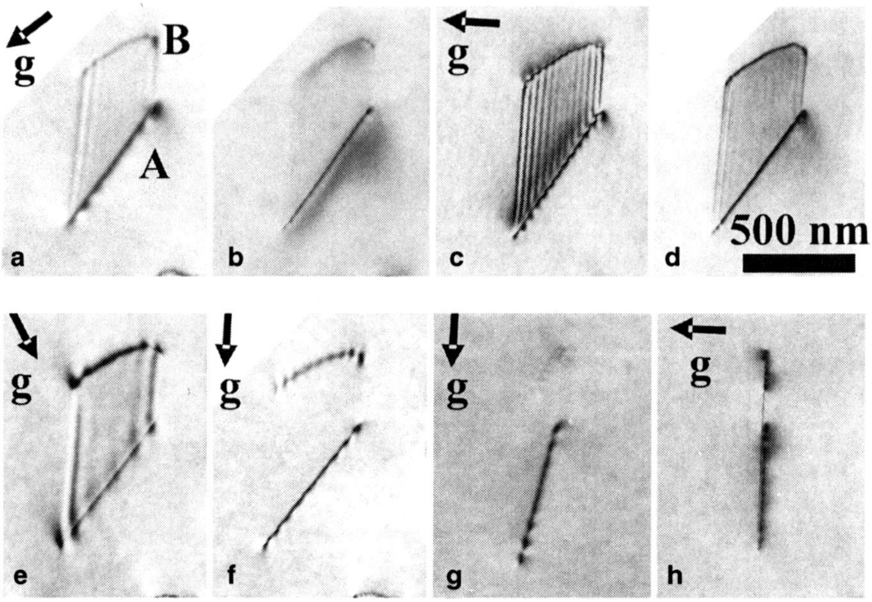

Fig. 9.48. Contrast analysis of a planar fault on a (001) plane and its bordering partial dislocations A and B in $MoSi_2$. (**a**) $\boldsymbol{g} = (213)$ at $\boldsymbol{n} = [03\bar{1}]$. (**b**) $\boldsymbol{g} = 2 \times (213)$ at $\boldsymbol{n} = [03\bar{1}]$. (**c**) $\boldsymbol{g} = (013)$ at $\boldsymbol{n} = [03\bar{1}]$. (**d**) $\boldsymbol{g} = 3 \times (013)$ at $\boldsymbol{n} = [03\bar{1}]$. (**e**) $\boldsymbol{g} = (2\bar{1}\bar{3})$ at $\boldsymbol{n} = [03\bar{1}]$. (**f**) $\boldsymbol{g} = (200)$ at $\boldsymbol{n} = [03\bar{1}]$. (**g**) $\boldsymbol{g} = (200)$ at $\boldsymbol{n} = [010]$. (**h**) $\boldsymbol{g} = (008)$ at $\boldsymbol{n} = [010]$. After [627]. Copyright Taylor & Francis Ltd. (2002)

under the applied load. The formation of the planar faults on $1/2\langle 111\rangle$ dislocations arranged along $\langle 110\rangle$ suggests that the cores of these dislocations are unstable and tend to dissociate. The same dislocations dominate the plastic deformation in the range of the flow stress anomaly. The consequences will be described in Sect. 9.6.4.

9.6.3 Macroscopic Deformation Parameters

Figure 9.49 displays typical stress–strain curves up to small plastic strains of $MoSi_2$ single crystals deformed along the soft [201] direction. The yield stress shows the typical three temperature ranges, a region of the normal decrease below 500°C, the anomaly range up to 990°C and a normal high-temperature range above that temperature. The stress–strain curves are smooth with a yield drop effect, except in the anomaly range where plastic instabilities (serrated yielding) occur. For deformation along [110], the curves are smooth at all temperatures with one exception. An example of the deformation at 738 and 726°C was shown in Fig. 2.2b.

The temperature dependence of the yield stress is presented in Fig. 9.50 for the two loading axes. Both curves show the three characteristic ranges. In general, the flow stresses for loading along $\langle 110\rangle$ are higher than those

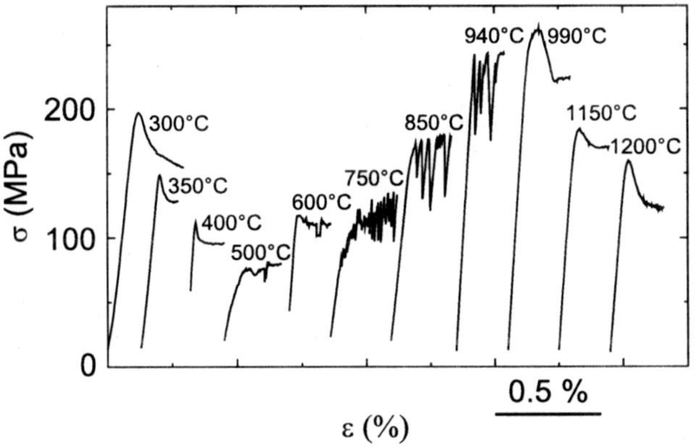

Fig. 9.49. Stress–strain curves of $MoSi_2$ single crystals deformed along the soft ⟨201⟩ direction at different temperatures. Data from [288]

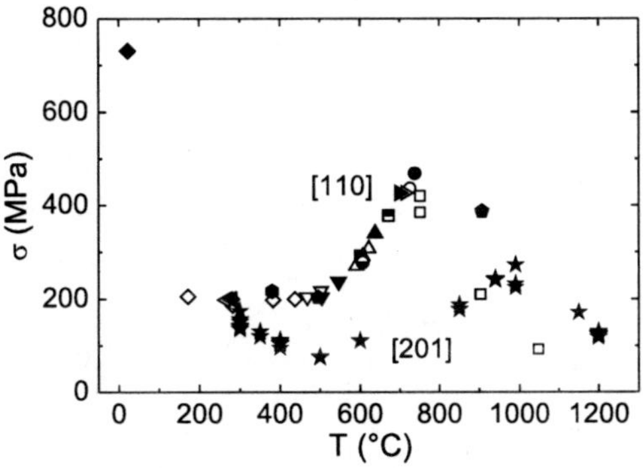

Fig. 9.50. Temperature dependence of the yield stress of $MoSi_2$ single crystals deformed along ⟨201⟩ (*fivefold stars*) and ⟨110⟩ (*full symbols* for data from the first loading of a specimen, *open symbols* for data taken after temperature changes corrected for work-hardening). Data from [15]

for ⟨201⟩, and the flow stress peak is shifted to a lower temperature. (The orientation factors are similar, $m_s = 0.4$ for ⟨201⟩ and $m_s = 0.46$ for ⟨110⟩.)

The strain rate sensitivity r of the flow stress was again measured by stress relaxation tests. Some examples of the deformation along ⟨110⟩ are shown in Fig. 9.51. At room temperature, the curve exhibits the normal bending toward the stress axis. At increasing temperature (281°C), the curves with normal curvature become very steep, corresponding to low values of r. In the anomaly range at 502°C, the curve is still very steep, but it starts to show an inverse curvature with increasing strain rate sensitivity at decreasing strain rate or stress. At higher temperatures in the anomaly range (638°C, 698°C), the curves are steep at the beginning but then clearly show the inverse curvature. Only at low strain rates, the curves bend back to the normal curvature

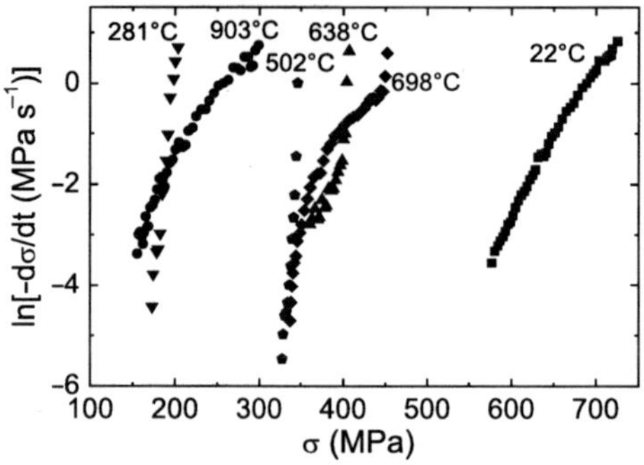

Fig. 9.51. Stress relaxation curves of $MoSi_2$ single crystals deformed along ⟨110⟩. Data from [15]

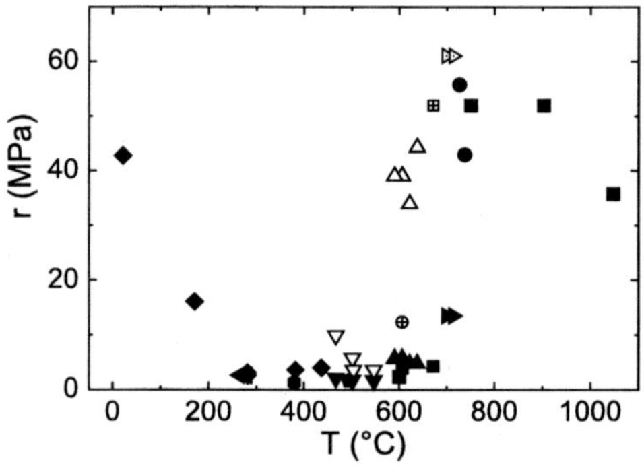

Fig. 9.52. Temperature dependence of the strain rate sensitivity of $MoSi_2$ single crystals deformed along ⟨110⟩. *Filled symbols*: data from the beginning of the relaxation tests belonging to the strain rate prior to the tests. *Open symbols*: for relaxation curves with inverse curvature, data belonging to a lower strain rate of about $10^{-7}\,s^{-1}$ (corresponding to about $\ln[-\dot{\sigma}(\mathrm{MPa\,s^{-1}})] = -3$). Data from [15]

as demonstrated for 698°C. At high temperatures, the curves assume their normal curvature again.

The dependence of the strain rate sensitivity belonging to the strain rate before the tests measured from the beginning of the relaxation curves is plotted in Fig. 9.52 as full symbols. The curve has exactly the same shape as that for FeAl in Fig. 9.35 with very low values in the anomaly range. For relaxation curves with an inverse curvature, r values measured at the lower strain rate of about $10^{-7}\,s^{-1}$ are plotted as open symbols. Accordingly, these values are greater than those corresponding to the initial higher strain rate.

For the deformation along ⟨201⟩, the temperature dependence of the strain rate sensitivity is similar, the minimum is located at about 800°C. At 850°C, strain rate cycling experiments yield negative values of the steady state strain

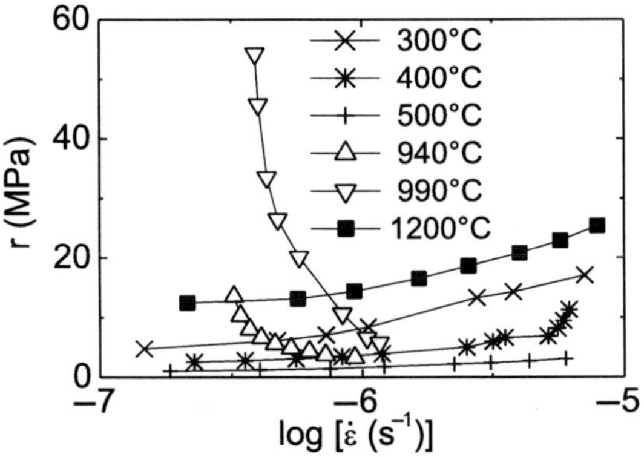

Fig. 9.53. Strain rate sensitivity of the flow stress of $MoSi_2$ single crystals deformed along ⟨201⟩ in dependence on the logarithm of the strain rate. Parameter is the temperature. Data from [621]

rate sensitivity. The dependence of the strain rate sensitivity r on the strain rate $\dot{\varepsilon}$ is demonstrated in Fig. 9.53. This dependence is normal for all temperatures except the two temperatures of 940 and 990°C within the range of the flow stress anomaly.

As described in Sect. 5.2.3, stress relaxation tests can also yield information on the effective stress component τ^*. In Fig. 9.51, relaxation curves with normal curvature are bent towards the stress axis, i.e., they converge against certain stress values. In this case, the difference between the starting stress of the relaxation test and the final stress may be called the relaxing stress $\Delta\sigma_{rel}$. Values of $\Delta\sigma_{rel}$ cannot be obtained from short-time stress relaxation curves as that for room temperature, or from curves with inverse curvature in the anomaly range as that for 638°C. In the high-temperature range, the internal stress can also relax owing to recovery. Nevertheless, some values of $\Delta\sigma_{rel}$ for the deformation along ⟨110⟩ in the low-temperature range are plotted in Fig. 9.54 as a function of temperature. The relaxing stress decreases with increasing temperature. For the deformation along ⟨110⟩, a few temperature change tests yielded activation enthalpies of about 0.4 eV at room temperature and 2.5 eV at 800°C.

9.6.4 Deformation Mechanisms

Different mechanisms control the plastic deformation of $MoSi_2$ in the three temperature ranges. For a better understanding of the dislocation behavior, the elastic properties of the dislocations of the two slip systems studied were calculated in [15]. According to this, the elastic anisotropy of $MoSi_2$ does not lead to elastic instabilities as, e.g., in NiAl (Fig. 9.20). The elastic equilibrium shapes of the dislocations under stress are smooth, almost circular for the {011}⟨100⟩ slip system.

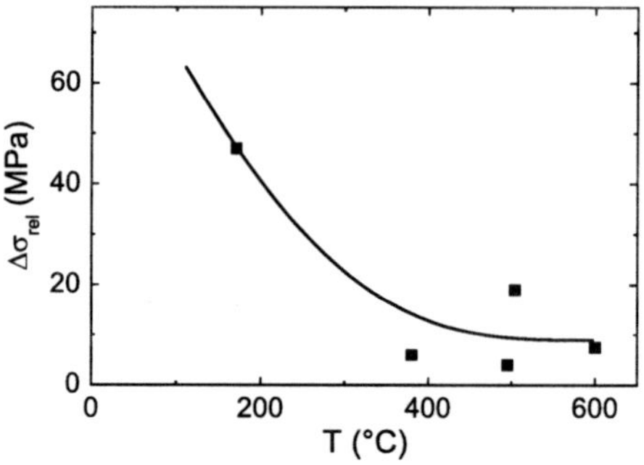

Fig. 9.54. Temperature dependence of the relaxing stress in $MoSi_2$ deformed along $\langle 110 \rangle$. Data from [15]

Low Temperatures

As in other materials, a remarkable part of the total flow stress is controlled by long-range dislocation interactions leading to an athermal stress component τ_i as described in Sect. 5.2.2. It may be estimated from the dislocation density ϱ using the Taylor hardening model (5.11). For the numerical constant, a value of $\alpha = \pi$ is used. The shear modulus μ is replaced by the anisotropic K value for the screw dislocation, $K_s = 4\pi E_{0s}/b^2 = 198.4\,\text{GPa}$ as introduced in (3.40). With the dislocation density of $\varrho = 2.8 \times 10^{13}\,\text{m}^{-2}$ from Sect. 9.6.2 for the deformation at 380°C along $\langle 110 \rangle$, the athermal stress component becomes $\tau_i \approx 170\,\text{MPa}$. At this temperature, the engineering flow stress σ is only about 200 MPa corresponding to a (shear) flow stress τ of only 92 MPa. Thus, the estimation with the Taylor model greatly overestimates the athermal stress component τ_i. Nevertheless, this estimation shows that the athermal stress component represents probably a large part of the total flow stress and cannot be neglected. It may be assumed that the effective stress τ^* can completely relax during long-time stress relaxation experiments. In this sense, the relaxing stresses plotted in Fig. 9.54 may be considered lower limits of the effective stress. Converted to shear stresses, τ^* may amount to about 40 MPa at room temperature and decreases down to only about 5 MPa between 400 and 600°C. This is 12% or 5%, respectively, of the total shear flow stress. Thus, in the low-temperature range, the athermal stress component is the dominating part of the flow stress. In specimens deformed along $\langle 201 \rangle$, the dislocation densities and accordingly also the internal stress component are lower [619].

As demonstrated in Fig. 9.38a, dislocations with $1/2\langle 111 \rangle$ Burgers vectors are straight and oriented along $\langle 111 \rangle$ directions in the low-temperature range. In the studies of the author and his coworkers, no micrographs are available for dislocations with $\langle 100 \rangle$ Burgers vectors, but in [617] it is described that these dislocations also tend to align along $\langle 111 \rangle$ at room temperature. These features are explained by the action of the double-kink or Peierls mechanism in the low-temperature range of the deformation of $MoSi_2$.

According to Sect. 4.6, the dependence of the Gibbs free energy of activation on the effective stress τ^* for the double-kink mechanism can be expressed by (4.66). From this potential, there follows the activation volume

$$V(\tau^*) = -\frac{\partial \Delta G}{\partial \tau^*} = \frac{\pi \Delta F_{\text{fk}}}{4\tau_{\text{p}}} \ln \frac{16\tau_{\text{p}}}{\pi \tau^*}, \quad (9.9)$$

with ΔF_{fk} being the formation energy of a single kink, and τ_{p} being the Peierls stress at $T = 0\,K$. As discussed above, the athermal stress component is large but not well-known so that the analysis of V vs. τ^* is not possible. Therefore, (4.66) and (9.9) together with the Arrhenius equation (4.8) were solved for T. The result is

$$T = T_0 \left[1 - 2\exp(-\chi V)(\chi V + 1)\right], \quad (9.10)$$

with

$$T_0 = \frac{2\Delta F_{\text{fk}}}{k \ln(\dot{\gamma}_0/\dot{\gamma})} \quad \text{and} \quad \chi = \frac{4\tau_{\text{p}}}{\pi \Delta F_{\text{fk}}}.$$

T_0 has a similar meaning as in Sect. 4.5.2. Figure 9.55 presents the respective plots for temperatures where the Peierls mechanism is supposed to operate. The data of the regression analysis are collected in Table 9.5. The kink formation energies are calculated using $\ln(\dot{\gamma}_0/\dot{\gamma}) = 20$. The kink energies assume reasonable values. Using ΔF_{fk} and τ_{p} from the table, (4.66) and (4.8) yield an activation energy is 0.5 eV for the deformation along ⟨110⟩, which is consistent with the activation enthalpy of 0.4 eV determined from the temperature change test (Sect. 9.6.3). The higher value for the deformation along ⟨201⟩ results from the larger absolute value of the Burgers vector. In accordance with the higher temperature of the flow stress minimum, T_0 is also higher.

According to the elastic theory of the Peierls mechanism, the Peierls stress τ_{p} is related to the Peierls energy W_{p} by (4.21), and this, in turn, to the kink

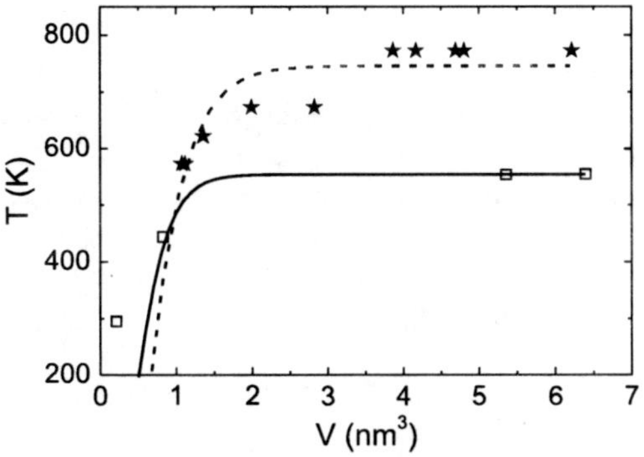

Fig. 9.55. Plot of temperature T versus activation volume V for deformation along ⟨110⟩ with data below 300°C (*squares*) and deformation along ⟨201⟩ at and below 500°C (*fivefold stars*). Theoretical curves after (9.10) and the data in the text. In the regression curve for the deformation along ⟨110⟩, the room temperature point is not considered. Data from [15]

Table 9.5. Parameters of the regression curves in Fig. 9.55 characterizing the Peierls mechanism in $MoSi_2$. Data from [15]

Loading axis	T_0 (K)	χ (nm^{-3})	ΔF_{fk} (eV)	τ_p (MPa)
$\langle 201 \rangle$	746	3.25	0.64	263
$\langle 110 \rangle$	554	4.71	0.48	282

energy $\Delta F_{fk} = W_{fk}$ by (4.26). The width of the Peierls valleys may be set $h = b$. The line energy is $E_d = \ln(R/r_0)E_0 = 6.37E_0$, if $1/\sqrt{\varrho}$ is set for the outer cut-off radius R of the dislocations with $\varrho = 2.8 \times 10^{13}\,m^{-2}$ from Sect. 9.6.2 and $r_0 = b$ for the inner one. With these data and the value of the Peierls stress for the deformation along $\langle 110 \rangle$ from Table 9.5, the kink energy would be $\Delta F_{fk} = 0.55\,eV$. This value is close to that from the regression analysis of the activation volume. That means that the parameters of the Peierls mechanism, ΔF_{fk} and τ_p, determined earlier are consistent within the framework of the elastic theory of this mechanism.

The relatively low values of τ_p (263 and 282 MPa) with respect to the flow stresses at room temperature, $\tau = 336\,MPa$ for the $\langle 110 \rangle$ deformation axis, emphasize the large contribution of the long-range internal stress to the total flow stress, as discussed above. If the dislocation multiplication and consequently the dislocation density are triggered by the effective stress as it is discussed in Sect. 5.1.1, a Cottrell–Stokes like behavior may result, i.e., the increasing effective stress from the Peierls mechanism with decreasing temperature causes also an increasing athermal stress component, which may be even greater than the thermal one.

Flow Stress Anomaly

It is suggested in [620] and [15] that the flow stress anomalies in the slip systems studied here are caused by the same mechanism as developed by Mills et al. [583] for the dislocations with $\langle 110 \rangle$ Burgers vectors in NiAl (Sect. 9.4.4). This mechanism was applied above to FeAl (Sect. 9.5.3), comprising the dissociation of the dislocations along preferred orientations by glide and the motion of the dissociated dislocations by combined glide and conservative climb. The importance of diffusion processes in connection with the flow stress anomaly in $MoSi_2$ was noted before not only in [618] but also in [628]. For the dislocations with $1/2\langle 111 \rangle$ Burgers vectors the dissociation reactions may read

$$1/2[111](1\bar{1}0) \rightarrow 1/6[331](1\bar{1}0) + 1/6[3\bar{3}1](013) + 1/6[\bar{3}31](103) \qquad (9.11)$$

for the preferred $\langle 331 \rangle$ edge orientation and

$$1/2[111](1\bar{1}0) \rightarrow 1/6[331](1\bar{1}0) + 1/6[3\bar{3}1](\bar{1}16) + 1/6[\bar{3}31](1\bar{1}6) \qquad (9.12)$$

for the preferred mixed $\langle 110 \rangle$ orientation. Most of the references (e.g., [617, 629]) cite the dissociation of dislocations with $1/2\langle 111 \rangle$ Burgers vectors according to

$$1/2[111] \rightarrow 1/4[111] + 1/4[111].$$

The results in [617] confirm this reaction for screw dislocations but show a dissociation outside the {110} plane for dislocations with an edge component. The three slip planes involved in (9.11) and (9.12) intersect one another always along the preference directions of the dislocations. In both reactions, the first partial can glide on the glide plane of the parent dislocation. The {013} planes of the product dislocations in (9.11) are considered before as slip planes in $MoSi_2$ [617]. The dissociation according to (9.12) does not contradict the weak beam images of Fig. 1 in [629] and of Fig. 7 in [617], which resolve two partials of different contrast, but it does not agree with Fig. 10b in [616]. Calculations of the prelogarithmic factors of the dislocation energy using anisotropic elasticity show that the dissociation is accompanied by a 3% gain in elastic energy. The faults between the partial dislocations are APBs. They have high energies [630] so that the dissociation width should be small. The latter two partials cannot glide on the $(1\bar{1}0)$ glide plane of the total dislocation. Thus, the motion of the whole dislocation requires conservative climb in the dislocation core, which is possible solely at sufficiently high temperatures. The latter partials in (9.12) are pure edge components for the 60° dislocations along [110], which may explain that these dislocations experience the highest glide resistance leading to their dominance in the dislocation structure in the anomaly range.

For the dislocations with $\langle 100 \rangle$ Burgers vectors oriented along $\langle 111 \rangle$ a dissociation reaction is suggested in [631]

$$[100](0\bar{1}1) \rightarrow 1/6[301](12\bar{3}) + 1/6[30\bar{1}](1\bar{4}3). \qquad (9.13)$$

The three slip planes involved intersect along the [111] preference direction of the dislocations. The reaction is accompanied with an energy gain of about 5%. The total dislocation can again move on its $(0\bar{1}1)$ slip plane only by a vacancy flux from one partial to the other. As suggested for FeAl in Sect. 9.5.3, the dependence of the dissociation width on the dislocation velocity and temperature is necessary to explain the occurrence of a flow stress anomaly.

The following experimental observations favor the proposed model:

- Within the anomaly range, dislocations tend to orient along specific crystallographic directions as described earlier. The crystallographic orientation is not a consequence of an anisotropy of the line tension. The elastic equilibrium shapes are smoothly curved for both slip systems. The dislocations are curved and not oriented at high temperatures above the anomaly range as well as at the flow stress minimum below the anomaly range (Fig. 9.42).
- Dislocations with $1/2\langle 111 \rangle$ Burgers vectors arranged along $\langle 110 \rangle$ are unstable and may dissociate according to (9.8) to form stacking faults on (001) planes.

- The dynamic properties of dislocations with rapid multiplication at high dislocation velocities at first loading and viscous motion afterwards agree with Cottrell effect-like behavior.
- The strain rate sensitivity is small in the anomaly range (Fig. 9.52), showing an inverse dependence on the strain rate, i.e., the strain rate sensitivity is higher for smaller strain rates or stresses (c.f., Fig. 9.53).
- At least for the deformation along ⟨201⟩, slip is localized in narrow bands with plastic instabilities (serrated yielding) occurring (Fig. 9.49).

Since a theory of the conservative climb model with velocity and temperature depending dissociation width is not yet available, some estimations are made regarding the position and height of the flow stress maximum using the theory of the Cottrell effect as it is described in Sect. 4.11 and applied several times in this book for other intermetallics. The relations between the temperature $T_{\max}$ of the flow stress maximum, the diffusion coefficient D, the elastic strength of the point defects β from (3.32), and the contribution to the flow stress $\tau_{\max}$ are given by (4.99). The shear modulus μ is replaced by the anisotropic K value of the edge dislocation K_e. With elastic constants for 650°, there follows $K_e = 187.2\,\text{GPa}$. Poisson's ratio for dislocations with ⟨100⟩ Burgers vectors is very small in agreement with their almost circular equilibrium shape. By comparing (14–47) and (14–42) of [12], it may be concluded that the relaxed volume of a vacancy should enter (3.32) for $\Delta\Omega$ if the diffusion takes place by the motion of vacancies. It is assumed here that it equals about 30% of the atomic volume, i.e., $\Delta\Omega = 0.3a^2c/6$. a and c are the lattice constants. Using these data yields $\beta \approx 2.8 \times 10^{-29}\,\text{N}\,\text{m}^2$. Then, it follows from (4.99) that the diffusion coefficient should be about $D \approx 5 \times 10^{-19}\,\text{m}^2\,\text{s}^{-1}$ to obtain the flow stress maximum at a strain rate of $\dot{\gamma} = 10^{-5}\,\text{s}^{-1}$ and a temperature of $T_{\max} = 1010\,\text{K}$ for the deformation along ⟨110⟩ (Fig. 9.50). Diffusion coefficients for both Mo and Si in $MoSi_2$ single crystals are published in [632,633]. According to these measurements, the diffusion coefficients are strongly anisotropic with higher diffusivities perpendicular to the c axis than parallel to it. Besides, in the temperature range considered, only Si diffuses. Comparable diffusion coefficients for Mo are observed only above about 1,500 K. The Si diffusion coefficients at 1,010 K are about $10^{-16}\,\text{m}^2\,\text{s}^{-1}$ for the diffusion perpendicular to the c axis and $3 \times 10^{-18}\,\text{m}^2\,\text{s}^{-1}$ for diffusion parallel to it. Thus, the experimental diffusion coefficients of Si are considerably higher than those required for the diffusion in the dislocation core for the Cottrell effect. However, climb-controlled motion of the dislocation requires the diffusion of both constituents, i.e., Si and Mo, so that the slower species should control the total rate. Mo bulk diffusion is far too slow so that the model requires strongly enhanced diffusion in the dislocation core.

The proposed mechanism should explain the different positions of the flow stress maximum for the deformation along ⟨110⟩ and ⟨201⟩. For dislocations with 1/2⟨111⟩ Burgers vectors on {110} planes, the value $K_e = 213.5\,\text{GPa}$ is only slightly higher than the respective value for the {011}⟨100⟩ slip system.

The remarkably larger Burgers vector cancels if (3.32) is inserted into (4.99). However, an essential difference may arise from the anisotropy of the diffusion coefficient. In each case, the transport of matter has to occur in the direction of motion of the dislocation. For the dislocations with ⟨100⟩ Burgers vectors and ⟨111⟩ line directions, the transport occurs mainly perpendicular to the c axis. This is the direction of a high diffusion coefficient. Alternatively, the direction of motion of dislocations with 1/2⟨111⟩ Burgers vectors aligned along ⟨110⟩ has a large component parallel to the c axis. In this direction, the diffusion coefficients of both Si and Mo are considerably lower than perpendicular to it which, according to (4.99), results in a higher temperature T_{max} of the flow stress maximum as observed in Fig. 9.50. The maximum contribution to the flow stress τ_{max} in (4.99) is proportional to b^{-3}. This may lead to a smaller value of τ_{max} for the deformation along ⟨201⟩ under otherwise similar conditions.

In addition to the mechanism causing the flow stress anomaly, there is certainly also a large athermal flow stress component due to long-range dislocation interactions. As described earlier, it dominates at the flow stress minimum between 300 and 500°C. It may even increase in the anomaly range triggered by the increasing anomalous stress contribution.

High-Temperature Range

Above the flow stress peak, the flow stress decreases in the normal way for both deformation axes (Fig. 9.50). As shown in [15], for the ⟨110⟩ compression axis, the dislocations are no longer straight and oriented. Like in other materials, the high-temperature deformation can conveniently be treated in terms of a power law equation for steady state creep (5.20). From a single experiment in the high-temperature range, it follows that the experimental stress exponent m' defined in (4.12) decreases from about 7.5 at 750°C down to 2.5 at 1,050°C. In other materials, the creep exponents scatter between 2 and 6 (Sect. 5.2.4). Thus, the m' values are roughly in this range. The activation enthalpy should correspond to the activation enthalpy for self-diffusion. The experimental value is 2.5 eV at 800°C (Sect. 9.6.3) and the diffusion enthalpies are 1.93 and 2.33 eV for the diffusion of Si, and about 5.4 and 6.2 eV for the diffusion of Mo, perpendicular and parallel to the c axis each [632,633]. Accordingly, the situation is similar to that of the Cottrell effect interpretation of the anomaly range: the diffusion data required for the respective model are between the data for the diffusion of Si and Mo. At present, a convincing interpretation of this fact cannot be given. For other intermetallic materials, there is no such great difference between the diffusion data of the different constituents.

9.6.5 Summary

The low-temperature increase of the flow stress of the two soft slip systems {110}1/2⟨111⟩ and {011}⟨100⟩ in $MoSi_2$ single crystals is most probably caused by the Peierls mechanism. In addition, a very great athermal part of the flow stress originates from long-range dislocation interactions. Both slip systems show an anomalous increase of the flow stress at intermediate temperatures. For the $\{110\}1/2\langle 111\rangle$ slip system, the transition to the anomaly range is accompanied with a change of the preferred orientation of straight dislocations from the 60° $\langle 111\rangle$ orientation to the 60° $\langle 110\rangle$ orientation, which are crystallographically not equivalent. The instantaneous formation of stacking faults during in situ experiments in an HVEM from the dissociation of dislocations with $1/2\langle 111\rangle$ Burgers vectors oriented along $\langle 110\rangle$ indicates that these dislocations are not very stable. Their dynamic behavior is characterized by rapid multiplication and glide at very high velocities during first loading and later on by viscous motion. The occurrence of the flow stress anomaly is explained by the same model as suggested for FeAl, i.e., a dissociation of the dislocations oriented along crystallographic directions by glide out of the glide plane of the total dislocation and the motion in the dissociated state by a combination of glide and conservative climb. Respective dissociation reactions are proposed for dislocations with both Burgers vectors studied. To obtain the anomaly, it has to be assumed that the dissociation width depends on the dislocation velocity and the temperature. The normal high-temperature decrease of the flow stress is interpreted by recovery-controlled deformation.

9.7 Conclusions on Intermetallics

All the materials presented in this chapter show a normal temperature dependence of the flow stress at low temperatures. The mechanisms controlling the dislocation motion are the same as in other materials classes, i.e., the Peierls mechanism (in $MoSi_2$), localized obstacles like small precipitates (in γ-TiAl and probably also in FeAl) or jogs resulting from cross slip (for the $\{011\}\langle 100\rangle$ slip system in NiAl). Thus, they are not specific of the special crystal structures in intermetallic alloys.

Except NiAl, all the materials exhibit a flow stress anomaly. In some materials, the anomaly is caused by special properties of the superdislocations like the formation of Kear–Wilsdorf locks or similar reactions as in Ni_3Al. In TiAl, superdislocations also contribute to the plastic deformation. However, at all temperatures ordinary dislocations dominate so that the anomaly has to be explained by processes connected with ordinary dislocations. In several materials discussed here, the properties of the anomaly are not very specific to the particular crystal structure. But the temperature of the flow stress maximum seems to correlate with diffusion properties. This correlation becomes obvious from the Cottrell effect-like behavior. The dislocations may move in a viscous or unstable manner. The viscous motion is not typical of the high-temperature deformation controlled by long-range dislocation interactions as in pure metals where the dislocations move in a jerky way over distances of the spacing between dislocations. It is also not characteristic of all locking–unlocking mechanisms, which are proposed for the deformation of intermetallic materials. A particular feature of the Cottrell effect-like behavior is the inverse dependence of the strain rate sensitivity r on the strain rate $\dot{\gamma}$ itself, i.e., decreasing r with increasing $\dot{\gamma}$ in range A of the strain ageing curve in Fig. 4.39. The inverse behavior is observed in all materials in which the anomaly seems to be connected with diffusion, i.e., in TiAl, FeAl, and $MoSi_2$.

Strain ageing with Cottrell effect-like behavior may cause the viscous motion of ordinary dislocations in γ-TiAl where associates of antisite defects and vacancies may give rise to the induced Snoek effect. Point defect atmospheres around the dislocations may arise also from other existing lattice defects like impurities or additions in low concentrations. The theory describes the interaction of these defects with the dislocations, e.g., [634] for NiAl. Binding energies are calculated by determining the difference between the energy of a dislocation containing the defect in its core and the sum of the energies of the dislocation without the defect and the formation energy of the defect. In most cases, positive or negative binding energies are obtained below 1 eV. However, in the particular case of an Ni antisite defect in the core of a dislocation on $\{01\bar{1}\}$ with a $\langle 111\rangle$ Burgers vector, the dislocation reconstructs locally, and the binding energy turns out to be remarkably higher than the formation energy of the defect. This means that the energy of the dislocation including the defect in its core is lower than the energy of the dislocation without the defect. It is suggested in [618] that generally in intermetallics, the lowest energy state of a dislocation may be a configuration containing a certain concentration of intrinsic point defects in the dislocation core, independent of further chemical or structural disorder of the material, i.e., the arrangement of atoms near the core may differ from the regular arrangement in the ordered structure. During motion, this low-energy state can follow the dislocation only via diffusional processes including a frictional force, in close analogy to the formation of atmospheres of existing defects like the Cottrell atmospheres. Molecular dynamics simulations allowing for atomic mobility should be carried out to prove the possibility of such a mechanism. The process is of intrinsic nature, suggesting a viscous motion of the dislocation as observed during in situ straining experiments on several intermetallic materials.

A special feature of dislocations with $\langle 110\rangle$ Burgers vectors in NiAl and of both Burgers vectors studied in FeAl and $MoSi_2$ at high temperatures is the occurrence of straight dislocations arranged along preferred crystallographic directions and moving viscously. The origin cannot be the usual lattice friction as this operates at low temperatures, as observed for the $\langle 100\rangle$ dislocations in $MoSi_2$. They are straight at low temperatures owing to the Peierls mechanism, curved at intermediate temperatures and straight again at temperatures of the flow stress anomaly. Besides, the low activation energy characteristic of the Peierls mechanism is observed only at low temperatures. For the intermetallic alloys with obstacle mechanisms at low temperatures, the Peierls mechanisms is expected to operate at even lower temperatures. Thus, the straight oriented arrangement of the dislocations at high temperatures has to be a particular core property.

A suitable model to describe the straight dislocation configurations is the glide decomposition model originally suggested by Mills et al. [583] to explain the behavior of ⟨110⟩ dislocations in NiAl as described in Sect. 9.4.4. Similar processes may operate also in the other intermetallics with straight dislocations listed earlier. Thus, this mechanism may perhaps be of more general nature. For NiAl, it is proved by weak beam and high-resolution TEM. For the other materials it is still speculative, though for several of the reactions the calculations of the prelogarithmic factors of the elastic energy of the dislocations involved using anisotropic elasticity indicate a gain in energy. The characteristic feature is the decomposition or dissociation by glide, where the product dislocations do not lie on the glide plane of the parent dislocation, thus making the dislocation core non-planar. The slip planes of all components, i.e., the parent dislocation and the reaction products, intersect along a single line defining the preferred orientation of the dislocations. Since most of the product dislocations cannot glide on the glide plane of the total dislocation, these partials have to move by involving climb. The point defects produced by one partial are then absorbed by the other, which is called conservative climb. At present, there are two models discussed in Sect. 9.4.4 for estimating the dislocation velocity. They predict very different dependencies of the dislocation velocity on the dissociation width. To explain the flow stress anomaly, the models have to be extended by assuming that this width depends on the dislocation velocity and the temperature in a way similar to the width of the Cottrell atmospheres. The model is still hypothetic but it should be worth spending more theoretical and experimental work on it.

At temperatures above the flow stress peak, most intermetallic materials behave in a way similar to other metals. The flow stress is controlled by long-range dislocation interactions where the dislocation structures are influenced by recovery. In addition, climb seems to be a mode of dislocation motion.

In intermetallic materials, the contribution of athermal processes to the total flow stress is mostly an open question. Unlike in pure metals where the dislocation substructures and densities are studied in great detail, very few experiments are available giving information on the athermal part of the flow stress. For $MoSi_2$ deformed along ⟨110⟩, it was estimated that the athermal part of the flow stress in the low-temperature range may amount to 80%. A similar situation may hold for the range of the anomalous increase of the flow stress. Thus, the increasing thermal component of the flow stress may trigger the dislocation generation thus influencing also the athermal stress component as it was discussed in Sect. 5.2.2. More experimental results are desirable in this respect.

10

Quasicrystals

In 1984, Shechtman, Blech, Gratias, and Cahn [635] detected diffraction patterns of rapidly solidified aluminium manganese alloys with sharp peaks having three-dimensional icosahedral symmetry. These diffraction patterns enclose six symmetry axes with fivefold symmetry. The observation was very surprising since fivefold symmetry violates the rules of crystallography based on crystal structures formed by translational periodicity of a unit cell decorated with the atoms. The fact that the diffraction patterns show sharp reflections suggests that the materials have long-range order. This order combined with missing translational symmetry can be explained by a new class of materials with so-called quasiperiodic order, in addition to the crystalline and amorphous states of solid matter. The new class of quasicrystals with ideal atomic structures was introduced by Levine and Steinhardt [636] in the same year as was the detection of the diffraction patterns with fivefold symmetry, after early mathematical works on quasiperiodic structures and the observation of modulated structures and incommensurate modulated phases.

Quasiperiodic structures can be formed by a number of different procedures. One consists in filling the plane or space by arranging two "unit cells" called tiles according to so-called matching rules. In the two-dimensional case the tiles are rhombi of equal edge length with angles of 36° and 72° and one or two arrows at their edges. The matching rules require that adjacent tiles meet with the same number and direction of the arrows. The resulting structures were described by Penrose as early as in 1974 [637]. The quasiperiodic structures can be derived also by projection from a periodic higher-dimensional hyperspace as will be described below. An introduction to these methods and to the basic properties of quasicrystals is given by Janot [638].

Quasicrystals exhibit the same kinds of defects as crystals, i.e., point defects like vacancies, dislocations, grain boundaries, etc., which may be connected with elastic strains. In quasicrystals, the latter are called phonon strains. In addition, the quasicrystalline order may be disturbed by violations of the matching rules. These kinds of defects are specific to quasicrystals and are termed phason strains. Dislocations in quasicrystals were first observed by Hiraga and Hirabayashi [639]. The dislocation strain field includes both phonon and phason strains. In a three-dimensional quasicrystal, it can be described by a six-dimensional Burgers vector. Because of the missing periodicity, the quasicrystalline structure is not intact after a dislocation has moved, but a layer of phason faults is trailed if the temperature is not high enough for annealing out these defects. Quasicrystals were therefore regarded as intrinsically brittle.

Nevertheless, quasicrystals can be deformed plastically at high temperatures. The plastic deformation parameters were first investigated on polyquasicrystals, e.g., [640–643]. The discovery of thermodynamically stable ternary quasicrystalline phases [644] and the resulting possibility of growing large single quasicrystals [645] formed the basis for studying the intrinsic plastic properties of quasicrystals. The observation of a drastic increase of the dislocation density during the high-temperature deformation of Al–Pd–Mn single quasicrystals pointed at the operation of a dislocation mechanism of plastic deformation [646]. This view was confirmed by the direct evidence of dislocation motion during high-temperature in situ deformation experiments in an high-voltage electron microscope (HVEM) [647]. Materials containing quasicrystalline phases find first applications because of their particular surface properties (e.g., [648]), the hydrogen storage capacity (e.g., [649]) or a high strength up to high temperatures of alloys containing quasicrystalline phases (e.g., [650, 651]).

In the following sections, the properties of dislocations in quasicrystals and their role in the plastic deformation will be described. The treatment is confined to the two materials where experimental data of the author and his coworkers are available, i.e., the stable phases Al–Pd–Mn with the icosahedral structure and of Al–Ni–Co with the decagonal structure. To help understanding the properties of dislocations, first the quasiperiodic structures and their construction by the cut and projection method are described.

10.1 Structure of Quasicrystals

Janner and Janssen [652] were the first to present an analytical description of incommensurate structures. It was later on extended by Duneau and Katz [653] and Elser [654] to construct quasiperiodic structures by projecting a higher-dimensional space onto the physical space. The method is illustrated in Fig. 10.1 for the generation of a one-dimensional quasiperiodic lattice by the projection from a two-dimensional square hyperlattice. In the origin of the hyperlattice with the coordinate axes h_1 and h_2, a second coordinate system is placed having the coordinates $x_\parallel$ and $x_\perp$. The $x_\parallel$ axis defines the physical space or, in quasicrystal terminology, the parallel space $E_\parallel$. The $x_\parallel$ axis is rotated against the h_1 axis of the hyperspace by the angle α, which represents an irrational slope $\tan\alpha$. Thus, except the origin the $x_\parallel$ axis does not meet any other lattice point. The $x_\perp$ axis defines the one-dimensional perpendicular or orthogonal space $E_\perp$. The one-dimensional quasiperiodic lattice in the physical space is generated by projecting the lattice points of the hyperspace onto

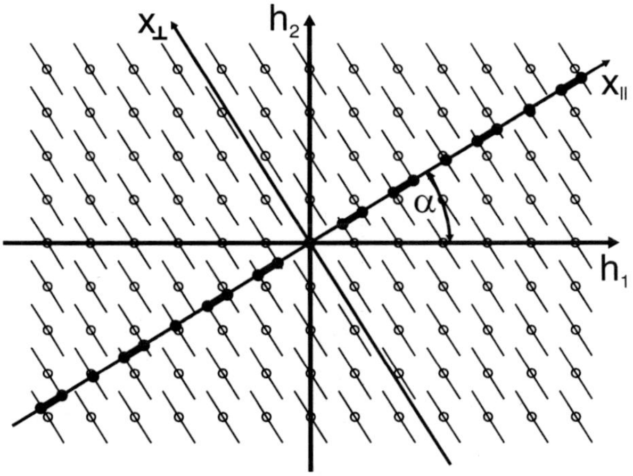

Fig. 10.1. Construction of a one-dimensional quasiperiodic point lattice from a two-dimensional periodic lattice by the cut and projection method

the $x_\parallel$ axis within a strip having the cross section parallel to $x_\perp$ equal to that of the square unit cell. This is equivalent to decorating the hyperlattice points with segments of the length of the strip and generating the lattice points in real space at the intersections between the segments and the $x_\parallel$ axis. In Fig. 10.1, the lattice points in real space are indicated by full black circles. The distances between the lattice points consist of two unit distances, short and long ones characterized by thick and thinner lines along $x_\parallel$. These distances follow a so-called Fibonacci sequence, which is not periodic, and which can be generated by a recursive deflation rule where at each recursion step the segments are replaced according to *short* $\rightarrow$ *long* and *long* $\rightarrow$ *long* + *short*. A three-dimensional quasiperiodic structure is generated by the projection of a six-dimensional hyperlattice onto the three axes in parallel space. Quasicrystals can be quasiperiodic either in all three dimensions as the icosahedral structure discussed below, or in two dimensions and periodic in the third as the following decagonal structure, or in one dimension only. If the slope $\tan\alpha$ in Fig. 10.1 has a rational value, the matter is crystalline. A large periodicity length corresponds to a large lattice constant. The structure is then called an approximant. The respective alloys may be called structurally complex intermetallic phases.

As will be shown below, the atoms in quasicrystals are arranged along crystallographic directions and on crystallographic planes. Although these structures lack periodicity, they produce sharp diffraction spots. The Fourier transform of the point lattice in the six-dimensional hyperspace is again a point lattice in the six-dimensional reciprocal space. It can be projected onto the three-dimensional physical reciprocal space to yield the diffraction pattern. As in this procedure all lattice points are projected and not only those within a strip, there appears an infinite number of diffraction spots. Since the sensitivity of the recording medium is restricted, only sufficiently strong spots are recorded.

10.1.1 Quasicrystals with Icosahedral Symmetry

To obtain the icosahedral symmetry from a projection of the six-dimensional hyperspace, the irrational slope of the axes of the parallel space with respect to those of the hyperspace is set equal to the reciprocal value of the Golden Mean $\tan\alpha = \tau^{-1}$, where $\tau = \frac{1}{2}\left(\sqrt{5}+1\right) = 1.61803\ldots$. The icosahedron implies six fivefold axes, 10 threefold ones and 15 twofold ones as demonstrated in Fig. 10.2. The reciprocal space exhibits icosahedral symmetry, too. Therefore, the diffraction diagrams cannot be indexed by integers of three basis vectors only. To index the three-dimensional parallel reciprocal space by integers, six basis vectors are necessary, according to the integral indexing of the six-dimensional hyperlattice. Depending on the choice of the basis vectors, different indexing systems have been proposed. The projection of the six principal axes of the hyperspace yields the six fivefold directions $\boldsymbol{e}_1$ to $\boldsymbol{e}_6$ of the icosahedron. Taking the latter as the basis vectors conserves the indexing of the six-dimensional reciprocal hyperspace.

Most frequently, icosahedral quasicrystals are indexed by applying the system introduced by Cahn, Shechtman and Gratias [655]. It is based on the three principal axes x, y, z of the cubic system of Fig. 10.2 parallel to three orthogonal twofold axes. To denominate the irrational coordinates with integer indexes, the coordinates along the axes are written as $h + \tau h'$, etc. The three components of the parallel subspace $E_{\|}$ are related to the six coordinates of the six-dimensional hyperspace n_1 to n_6 by

$$\begin{aligned} h + \tau h' &= (n_1 - n_4) + \tau(n_2 + n_5) \\ k + \tau k' &= (n_3 - n_6) + \tau(n_1 + n_4) \\ l + \tau l' &= (n_2 - n_5) + \tau(n_3 + n_6). \end{aligned} \tag{10.1}$$

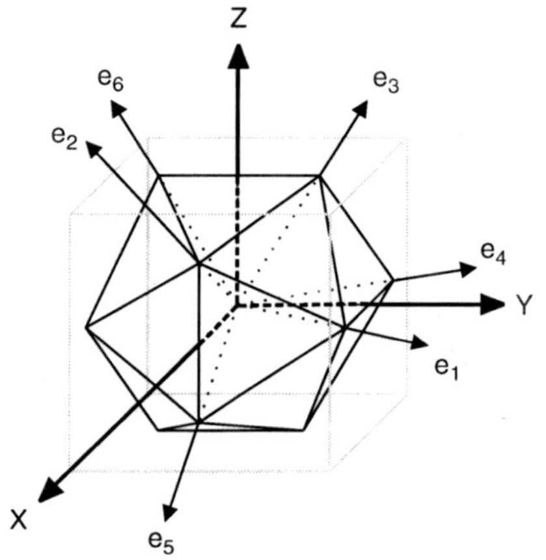

Fig. 10.2. Icosahedron with its three orthogonal twofold axes x, y, and z and the six 5-fold axes $\boldsymbol{e}_1$ to $\boldsymbol{e}_6$

The components of the orthogonal subspace $E_\perp$ are given by

$$\begin{aligned} h' - \tau h &= (n_2 + n_5) + \tau(n_1 - n_4) \\ k' - \tau k &= (n_1 + n_4) + \tau(n_3 - n_6) \\ l' - \tau l &= (n_3 + n_6) + \tau(n_2 - n_5). \end{aligned} \tag{10.2}$$

The vector $(h + \tau h',\ k + \tau k',\ l + \tau l')$ is abbreviated by $(h/h',\ k/k'\ l/l')$ and the vector $(h' - \tau h,\ k' - \tau k,\ l' - \tau l)$ by $(h'/\bar{h},\ k'/\bar{k}\ l'/\bar{l})$. To obtain physical $\boldsymbol{g}$ vectors in reciprocal parallel and perpendicular spaces, the index vectors have to be multiplied by $\left(a_0\sqrt{2(2+\tau)}\right)^{-1}$, where a_0 is the lattice constant of the six-dimensional cubic primitive hyperlattice. For the stable i-Al–Pd–Mn phase, $a_0 = 0.645\,\text{nm}$. The threefold symmetry axes result from the addition of three neighboring fivefold directions, e.g., $\boldsymbol{e}_1 + \boldsymbol{e}_3 + \boldsymbol{e}_4$, and the twofold ones from the sum of the two neighboring directions, e.g., $\boldsymbol{e}_1 + \boldsymbol{e}_2$. In addition, there are two other distinctive types of poles, the 30 pseudo-twofold poles between two neighboring twofold poles and 30 poles in the middle between two orthogonal twofold poles, which correspond to the $\langle 110 \rangle$ poles of the cubic crystal structure. Figure 10.3 illustrates experimental diffraction patterns of these highly symmetric orientations.

To derive the real atomic structure of a quasicrystalline alloy, the lattice points in the hyperspace are decorated with three-dimensional hypersurfaces with an extension only in the perpendicular space $E_\perp$. They consist of different shells with each shell corresponding to a particular chemical element. The intersection of the cutting face of the projection with these shells determines

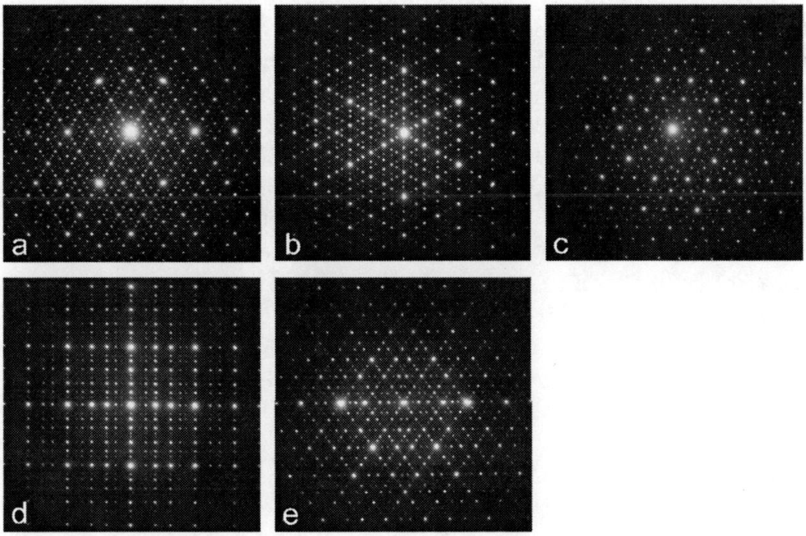

Fig. 10.3. Electron diffraction patterns of an i-Al–Pd–Mn single quasicrystal along the main symmetry axes. (**a**) twofold, (**b**) threefold, (**c**) fivefold, (**d**) pseudo twofold, (**e**) "$\langle 110 \rangle$" direction

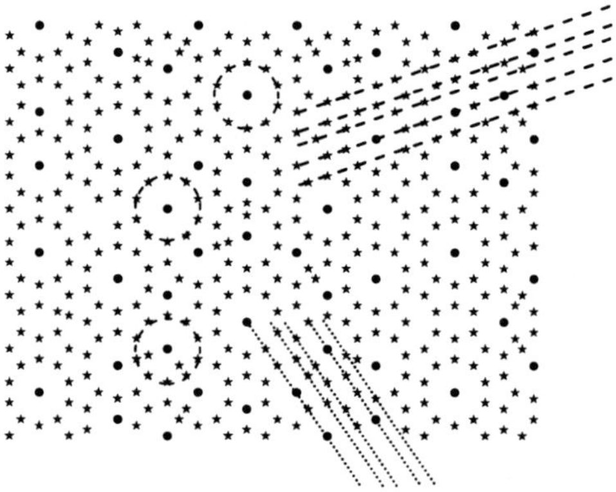

Fig. 10.4. Atom plane perpendicular to a fivefold direction in i-Al–Pd–Mn after the cluster model of Boudard et al. [656]

the element to be positioned at the respective site in the physical space $E_{\|}$. The structure models are experimentally studied mostly by X-ray and neutron diffraction, evaluating the intensities of hundreds of reflections. Figure 10.4 presents the arrangement of atoms on a plane perpendicular to a fivefold axis in i-Al–Pd–Mn, according to the study by Boudard et al. [656]. The figure shows that the structure can be considered an arrangement of partly penetrating clusters. Some of them are marked by circles. In the plane, they consist of a central Mn atom and a ring of 10 Al atoms. Recent studies [657, 658] describe these structures by the decoration of a three-dimensional Penrose tiling with so-called Mackay and Bergman type clusters with different shells of different symmetry, i.e., dodecahedral, icosahedral, and icosidodecahedral. These clusters contain up to 50 atoms and have a diameter of about 1 nm. They are believed to be particularly strongly bound.

In addition to the cluster structure, Fig. 10.4 reveals that the atoms in the quasicrystal structure are arranged in rows along crystallographic directions marked by dashed and dotted lines, and these rows are located on crystallographic atomic planes, where the distances between the rows and between the planes are not periodic. This crystallographic aspect was studied in [659]. Both views on the same quasiperiodic structure have been discussed in [660]. They are both important with respect to the dislocation properties to be treated in the following sections.

10.1.2 Quasicrystals with Decagonal Symmetry

Decagonal quasicrystals are two-dimensional quasicrystals consisting of quasiperiodic atomic planes with decagonal symmetry, which are stacked in a periodic order. The respective diffraction patterns were first observed in 1985 [661,662]. The stable phase Al-15at%Ni-15at%Co was first synthesized by Tsai et al. [663]. The diffraction patterns may be indexed by a linear combination of multiples (H_i) of five reciprocal basis vectors pointing at the corners of a regular pentagon ($\boldsymbol{e}_2^*$ to $\boldsymbol{e}_6^*$) and a sixth one in the direction of the periodic axis (e_1^*). Since of the five vectors in the quasiperiodic plane only four are linearly independent, one of them can be omitted so that the diffraction vector can be written as

$$\boldsymbol{g} = (H_1, H_3 - H_2, H_4 - H_2, H_5 - H_2, H_6 - H_2)\,.$$

After the system by Steurer [664], the decagonal quasilattice is derived from a five-dimensional hyperspace with the physical (parallel) space based on the three orthogonal basis vectors $\boldsymbol{v}_1$, $\boldsymbol{v}_2$, $\boldsymbol{v}_3$ and the perpendicular space with two orthogonal basis vectors $\boldsymbol{v}_4$, $\boldsymbol{v}_5$. The basis vectors of the five-dimensional reciprocal lattice are then

$$\begin{aligned} \boldsymbol{d}_\mathrm{i}^* &= a^* \left(\cos\frac{2\pi i}{5},\ \sin\frac{2\pi i}{5},\ 0,\ \cos\frac{6\pi i}{5},\ \sin\frac{6\pi i}{5}\right), \quad i = 1, \ldots, 4 \\ \boldsymbol{d}_5^* &= a_5^*\,(0,\ 0,\ 1,\ 0,\ 0), \end{aligned} \tag{10.3}$$

and those of the physical lattice

$$\begin{aligned} \boldsymbol{d}_\mathrm{i} &= \frac{2a}{5}\left(\cos\frac{2\pi i}{5} - 1,\ \sin\frac{2\pi i}{5},\ 0,\ \cos\frac{6\pi i}{5} - 1,\ \sin\frac{6\pi i}{5}\right), \quad i = 1, \ldots, 4 \\ \boldsymbol{d}_5 &= a_5\,(0,\ 0,\ 1,\ 0,\ 0), \end{aligned} \tag{10.4}$$

with $a = 1/a^*$ and $a_5 = 1/a_5^*$. Figure 10.5 shows the projections $\boldsymbol{a}_\mathrm{i}^*$ of the basis vectors of the reciprocal lattice $\boldsymbol{d}_\mathrm{i}^*$ onto the physical space. The periodic direction $\boldsymbol{a}_5^*$ points along the $\boldsymbol{v}_3$ axis. The vectors $\boldsymbol{a}_1^*$ to $\boldsymbol{a}_4^*$ span the quasiperiodic plane. The diffraction pattern of an Al-15at%Ni-15at%Co single quasicrystal in Fig. 10.6 taken at the P2 pole (for this notation see [665]) with twofold symmetry shows the periodic sequence of the quasiperiodic planes.

The atomic structure of the stable phase d-Al-15at%Ni-15at%Co was studied by X-ray diffraction [667]. It has the lattice constants $a = 0.3794$ nm and $c = 0.40807$ nm. The structure consists of a periodic stacking of two quasiperiodic layers rotated against each other by 36°. The basic structural building elements corresponding to the clusters in icosahedral quasicrystals are columnar clusters parallel to the tenfold axis. The central pentagons of the stacked layers form pentagonal antiprismatic channels.

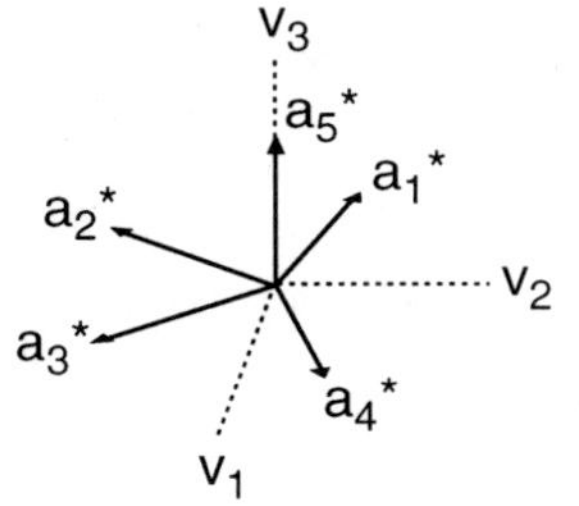

Fig. 10.5. Projection of the five-dimensional reciprocal basis vectors onto the physical space according to Steurer [664]

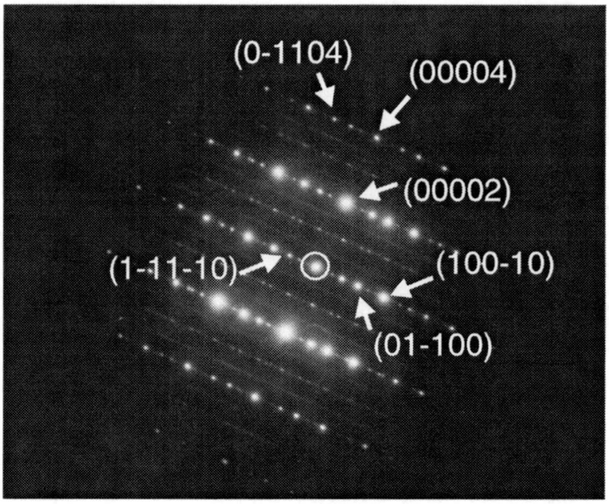

Fig. 10.6. Diffraction pattern of an Al-15at%Ni-15at%Co single quasicrystal taken at the P2 twofold pole. From [666]. Copyright Mater. Res. Soc. (2005)

10.2 Defects in Quasicrystals

Quasicrystals generally exhibit the same types of lattice defects as crystals, i.e., point defects like vacancies and interstitials [668–670], one-dimensional defects as the dislocations [671], and two-dimensional defects like grain boundaries [672], antiphase boundaries [673] and twins [674]. In addition, quasicrystals contain phason defects mentioned already earlier.

10.2.1 Vacancies

Measurements of diffusion coefficients of ^{54}Mn and ^{59}Fe in icosahedral Al–Pd–Mn reveal a behavior similar to that in metals, pointing at a vacancy mechanism of diffusion [668]. Thus, the thermal vacancy concentration can be described by an equation like (4.76). The formation and migration enthalpies of vacancies were determined by time-differential dilatometry to amount to $\Delta H_{\mathrm{f}} = 0.6\,\mathrm{eV}$ and $\Delta H_{\mathrm{m}} = 0.8\,\mathrm{eV}$ [675]. Thus, they are similar to those in pure aluminium. Positron annihilation spectroscopy reveals a neighborhood of the vacancies dominated by Al atoms [676]. However, Pd and Mn diffuse by several orders of magnitude slower than Al [677]. In addition to thermal vacancies, structural vacancies of a concentration of about 10^{-3} are observed in Al–Pd–Mn. Some papers (e.g., [678]) suggest that structural vacancies in the center of clusters of 12 Al atoms stabilize the quasicrystalline phase.

10.2.2 Phason Defects

If in physical space part of a quasicrystal is shifted with respect to the rest, there does not exist a three-dimensional displacement vector leading to an

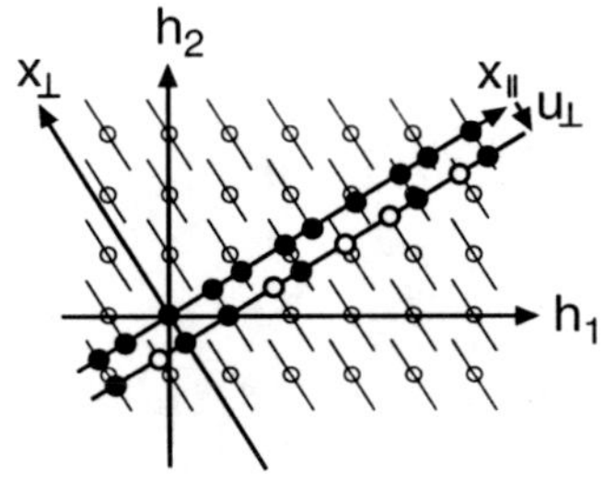

Fig. 10.7. Generation of phason faults by a displacement in the perpendicular space

undistorted lattice, because of the missing periodicity. Along the cutting plane, the matching rules are violated. The corresponding defects are called phasons. The process can be illustrated by the cut and projection procedure as shown in Fig. 10.7 for the generation of a one-dimensional quasiperiodic structure from a two-dimensional periodic lattice. Lattice defects are described by their displacement field in the higher-dimensional hyperspace. A general displacement $\boldsymbol{U}$ can then be split into components in both subspaces according to

$$\boldsymbol{U} = \boldsymbol{u}_{\parallel}(x_{\parallel}) + \boldsymbol{u}_{\perp}(x_{\parallel}), \tag{10.5}$$

which are called phonon and phason displacements. As all physical properties of quasicrystals depend only on the coordinates in physical (parallel) space, both displacements depend only on $x_{\parallel}$. The respective strains are the phonon and phason strains. If the shift vector $\boldsymbol{U}$ is a lattice vector in the hyperspace, there results an undisturbed quasicrystal. In other cases, the displacement vectors can be split according to (10.5). Shifts parallel to the $x_{\parallel}$ axis produce an elongation or contraction of the distances between the atoms, i.e., an elastic or phononic strain. However, displacements of the lattice points of the hyperlattice parallel to the $x_{\perp}$ axis, which, in Fig. 10.7, is identical with a shift $\boldsymbol{u}_{\perp}$ of the $x_{\parallel}$ axis into $x'_{\parallel}$, lead to a change in the sequence of lattice points since now other decorations of hyperlattice points intersect the $x_{\parallel}$ axis. In the new Fibonacci sequence, many atoms marked by full circles are in the same position as in the original sequence. Some, however, which are marked by open circles, do not fit the original quasiperiodic order and are called phason flips. The phason flips represent violations of the matching rules of the quasiperiodic lattice. Phason flips can appear individually as random phasons. Spatial variations of $\boldsymbol{u}_{\perp}$ result in collective flipping of unit cells in worms causing mismatches at the boundaries of the flipped areas. Since real quasicrystals are always alloys of two or more constituents, phason disorder is also connected with chemical disorder.

The properties of defects in quasicrystals were discussed in detail in [679]. Arrangements of atoms in a two-dimensional array were calculated using the density-wave method, which is an additional way to generate quasiperiodic lattices ([680] and other references in [679]), and were compared with results

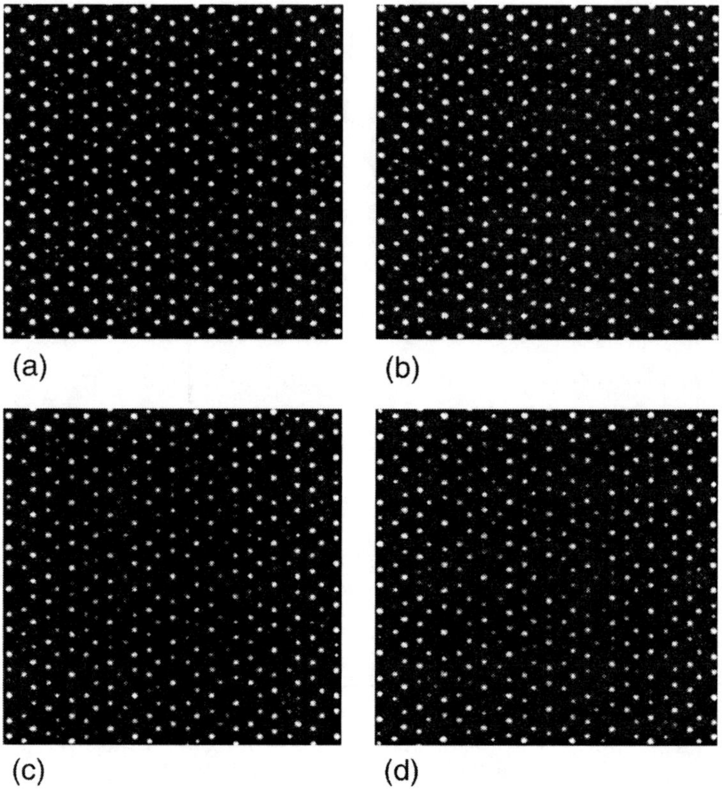

Fig. 10.8. Two-dimensional quasiperiodic structures generated by the density-wave method. (**a**) Perfect pattern. (**b**) Pattern of (**a**) distorted by a spatial variation of the phonon degree of freedom. (**c**) Pattern distorted by varying the phason degree of freedom. (**d**) Dislocation. Reprinted with permission from Socolar et al. [679]. Copyright (1986) by American Physical Society (http://link.aps.org/abstract/PRB/v34/p3345/y1986)

of the cut and projection procedure. The undisturbed arrangement of lattice points is demonstrated in Fig. 10.8a, showing different kinds of clusters forming the structure. Looking at this figure at a glazing angle reveals the different atom rows. In Fig. 10.8b, the same structure is distorted by spatially varying displacements of only phononic character. The originally straight rows of atoms are now curved, corresponding to elastic strains. The effect of varying phasonic displacements is demonstrated in Fig. 10.8c. The rows of atoms are straight again but they are jagged, i.e., they show lateral displacements. Similar images were revealed by high-resolution electron microscopy (e.g., [681]). Relaxation of the phason faults requires local diffusion of atoms. Therefore, these processes occur far more slowly than phonon relaxations. The dynamics of long-wavelength phason fluctuations was studied by changes in the speckle contrast in a coherent X-ray-scattering experiment [682]. According to that, phason fluctuations in i-Al–Pd–Mn are frozen-in below 500°C. At 650°C, the corresponding diffusion coefficient is $2.2 \times 10^{-18}\,\mathrm{m}^2\,\mathrm{s}^{-1}$ with an activation energy of 2.3(±1)eV. This diffusion coefficient is in the range of those of Pd and Mn. Phason flips were observed in situ at a high temperature in a decagonal Al–Cu–Co quasicrystal by high-resolution TEM [683].

10.2.3 Dislocations

As in crystals, plastic deformation is realized by the motion of dislocations. A recent review has been given by Bonneville, Caillard and Guyot [684]. Figure 10.9 shows the result of a computer simulation of the shearing and introduction of an edge dislocation into a two-dimensional decagonal quasicrystal from [685]. A Burgers circuit (see Sect. 3.1.1) is carried out around the dislocation in B and the corresponding reference circuit is shown in A. It has to be closed by the Burgers vector of the dislocation. Since the shearing was carried out in physical space, there remain a number of places with violations of the matching rules marked by grey shading, i.e., of phason defects. Therefore, the Volterra process to generate a dislocation should not be performed in the physical space but in the higher-dimensional hyperspace so that in icosahedral quasicrystals the Burgers vector is a six-dimensional vector [686]. The cutting procedure similar to Figs. 3.7 and 3.8 is carried out along a five-dimensional half-plane comprising the whole perpendicular space. The

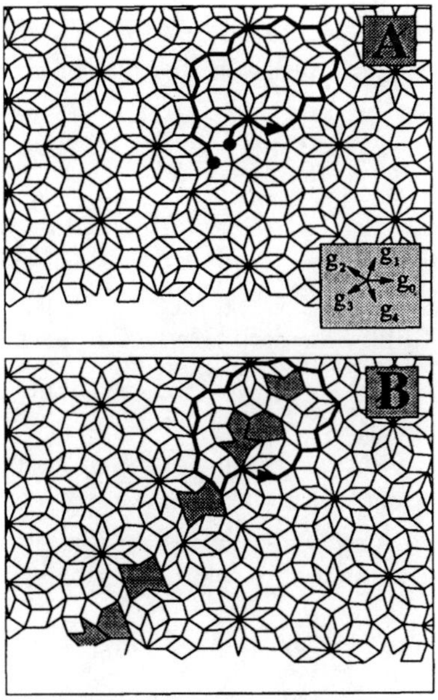

Fig. 10.9. Creation of phason strain by the introduction of a dislocation into a two-dimensional decagonal quasicrystal. From [685]. Copyright Taylor & Francis Ltd. (1995) (http://www.informaworld.com)

dislocation line is a four-dimensional manifold, a one-dimensional line in physical space plus the whole orthogonal space, since after (10.5), the strain field does not depend on the coordinates of the perpendicular space. The geometric properties of dislocations and their motion in the quasiperiodic lattice are described in detail in [687].

In analogy to (3.1), the Burgers vector $\boldsymbol{B}$ in quasicrystals can be defined by a line integral over the displacements $\boldsymbol{U}$ along a closed curve C around the dislocation in the physical space $E_{\|}$

$$\boldsymbol{B} = \oint_{\mathrm{C}} \mathrm{d}\boldsymbol{U} = \oint_{\mathrm{C}} \left(\mathrm{d}\boldsymbol{u}_{\|} + \mathrm{d}\boldsymbol{u}_{\perp}\right) = \boldsymbol{b}_{\|} + \boldsymbol{b}_{\perp}. \tag{10.6}$$

Thus, the integration over the phonon strain field yields the parallel part $\boldsymbol{b}_{\|}$ of the Burgers vector. It corresponds to the usual Burgers vector in crystals describing the continuous elastic strains around the dislocation line. For the edge dislocation, $\boldsymbol{b}_{\|}$ is perpendicular to the dislocation line, whereas for the screw dislocation it is parallel. In many processes, $\boldsymbol{b}_{\|}$ replaces the Burgers vector known from crystals. The perpendicular part of the Burgers vector $\boldsymbol{b}_{\perp}$ represents the phason strain field around the dislocation line. Both phenomena can be observed in the two-dimensional density wave image of an edge dislocation in Fig. 10.8d. The lattice rows are bent around the inserted half-row characterizing the elastic strain field of the dislocation. In addition, there appears a number of jags in the atom rows.

The full six-dimensional Burgers vector of (three-dimensional) quasicrystals can be determined by several methods. As shown by Wollgarten et al. [688], the contrast extinction criterion (2.9) of diffraction contrast TEM can be extended to yield

$$\boldsymbol{G} \cdot \boldsymbol{U} = \boldsymbol{g}_{\|} \cdot \boldsymbol{u}_{\|} + \boldsymbol{g}_{\perp} \cdot \boldsymbol{u}_{\perp} = 0, \tag{10.7}$$

with $\boldsymbol{G} = \boldsymbol{g}_{\|} + \boldsymbol{g}_{\perp}$. Since the cut in six-dimensional space has an irrational orientation, each $\boldsymbol{g}_{\|}$ belongs to a single $\boldsymbol{G}$ in the six-dimensional reciprocal lattice, and accordingly also to a single $\boldsymbol{g}_{\perp}$ in reciprocal space [654, 689]. The criterion (10.7) can be fulfilled in two ways. The first one is by making both terms zero, i.e.,

$$\boldsymbol{g}_{\|} \cdot \boldsymbol{u}_{\|} = \boldsymbol{g}_{\perp} \cdot \boldsymbol{u}_{\perp} = 0. \tag{10.8}$$

This is called a "strong" contrast extinction condition. Then, the dislocation contrast disappears for a whole systematic row of reflections through the origin of the diffraction pattern. Two such conditions have to be found to determine the direction of the Burgers vector in both the physical and orthogonal space. In the second case,

$$\boldsymbol{g}_{\|} \cdot \boldsymbol{u}_{\|} = -\boldsymbol{g}_{\perp} \cdot \boldsymbol{u}_{\perp}. \tag{10.9}$$

This is the "weak" contrast extinction condition, which is fulfilled only for periodically spaced reflections within a systematic row but not for other reflections of the same row. The determination of the length of the Burgers vector requires a quantitative analysis of the dislocation contrasts as in crystals.

Wang and Dai [690] were the first to use convergent beam electron diffraction (CBED) to determine the six components of the Burgers vector in icosahedral quasicrystals. In this technique, the dislocation image is superimposed with the defocus CBED pattern where zero-order Laue zone (ZOLZ) lines split n times when a dislocation line with $|\boldsymbol{G} \cdot \boldsymbol{B}| = n$ is crossed. $\boldsymbol{B}$ is determined completely by six independent equations $\boldsymbol{G} \cdot \boldsymbol{B} = n$. A further method employs high-resolution TEM by observing lattice fringe images taken in two different orientations of the specimen [691].

The elastic properties of dislocations in quasicrystals are treated in the framework of an extended elasticity theory (e.g., [692], for an introduction see, e.g., [693, 694]). Considering only short-time elastic responses, icosahedral quasicrystals behave isotropically so that their elastic properties can be described by two elastic constants (e.g., the Lamé constants λ and μ). They can be measured in the usual way by ultrasound transmission. In the generalized Hooke's law, there appear also two phason elastic constants K_1 and K_2, which are determined indirectly by diffuse X-ray or neutron scattering (e.g., [695]), and a phonon–phason coupling term R. Like for strains, there occur stress components in physical space and orthogonal space. The stresses in real space depend on the elastic strains via the Lamé constants, and on the phason strains via the coupling constant. On the other hand, the stresses in perpendicular space depend on the phason strain via the phason elastic constants, and on the phonon strains via the coupling constant. A set of elastic constants for i-Al–Pd–Mn used in another terminology in [692] comprises $\lambda = 85$, $\mu = 65$, $K_1 = 0.084$, $K_2 = 0.036$, and $R = 0.41$ (all in GPa). The elastic energy of a quasicrystal is composed of three terms, one from the phonon part, one from the phason part, and a coupling term.

Not only the phononic strains of dislocations but also the phasonic ones decrease with $1/r$, where r is the distance from the dislocation line. Only pure screw and edge dislocations along a twofold orientation have simple displacement fields with displacements only parallel to the line direction for the screws, and perpendicular to it for edges. In all the other cases, the properties are not so simple. The equilibrium positions of parallel dislocations with respect to glide were calculated in [696]. This is based on a generalized Peach–Koehler force, in analogy to (3.22),

$$\boldsymbol{F}_{\mathrm{PK}} = \left(\boldsymbol{b}_{\|}\sigma_{\|} + \boldsymbol{b}_{\perp}\sigma_{\perp}\right) \times \boldsymbol{\xi}, \tag{10.10}$$

where $\sigma_{\perp}$ is the stress tensor in perpendicular space. In [697], the formula is quoted for one-dimensional hexagonal quasicrystals.

In terms of the extended elasticity theory, the energy of a screw dislocation in the same quasicrystal is given by [697]

$$E_{\mathrm{d}} = \left(c_{44}b_{\|3}^2 + K_2 b_{\perp 3}^2 + 2R_3 b_{\|3} b_{\perp 3}\right) \frac{\ln\left(R/r_0\right)}{4\pi},$$

with R and r_0 being the outer and inner cut-off radii as introduced in Sect. 3.2.2. Thus, the dislocation energy depends on the squares of both

Burgers vector components $b_\parallel$ and $b_\perp$. Assuming that the phason part of the dislocation energy is proportional to the number of matching rule violations, there should arise a linear dependence on $b_\perp$ [679, 698]. Though the necessity of the extension of the elasticity theory for quasicrystals is quite obvious, the consequences have widely been ignored so far in interpreting the experimental results on dislocation motion and plastic deformation.

The microprocesses of dislocation motion depend on the temperature. If the temperature is high enough for a rapid phason rearrangement, the equilibrium state of the phason stress field around the dislocation can move together with the phonon field. The phason flips accompanied with the motion occur also far from the dislocation itself as illustrated by a simulation of the dislocation motion in a decagonal quasicrystal [699]. However, if the temperature is not so high, the dislocation may decompose [700] and the phononic part of it will move alone as it is outlined highly schematically in Fig. 10.10. In Fig. 10.10a, the dislocation of Burgers vector $\boldsymbol{b}_\parallel + \boldsymbol{b}_\perp$ sketched as a full circle is surrounded by its phason cloud marked by dashed circles. Caillard and coworkers called such a dislocation a perfect dislocation (e.g., in [684, 701]). In Fig. 10.10b, a dislocation with only the parallel part of the Burgers vector $\boldsymbol{b}_\parallel$ has moved from A to B owing to an external stress. It may be called an imperfect dislocation. In its wake, the imperfect dislocation created a layer of high

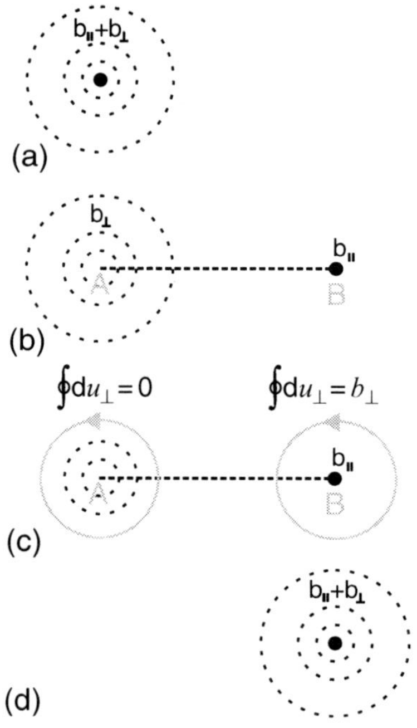

Fig. 10.10. Schematic of the motion of a perfect dislocation. (**a**) Starting dislocation with phonon and phason stress fields. (**b**) Intermediate state with separation of phonon and phason stress fields. (**c**) Burgers circuits around dislocations A and B. (**d**) Final state

phason disorder called a phason wall and indicated by the thick dashed line. At the original site A, there remains a dislocation with the Burgers vector $\boldsymbol{b}_\perp$ representing the phasonic strain field of the original dislocation. The situation is illuminated by the two grey Burgers circuits around A and B in Fig. 10.10c. While the dislocation in real space at B has only the Burgers vector $\boldsymbol{b}_\parallel$, the Burgers circuit in orthogonal space around B yields the Burgers vector $\boldsymbol{b}_\perp$ when crossing the concentrated phason strain of the phason wall. The Burgers circuit around A crosses the phason wall in opposite direction resulting in a contribution of $-\boldsymbol{b}_\perp$ from the phason wall, which cancels the contribution of the phason strain from the original dislocation. The whole system is not in thermal equilibrium. The equilibrium configuration with a perfect dislocation at B as in Fig. 10.10d can only be attained by diffusional retiling, where the phason field at A and the phason wall dissolve. Computer simulations on a two-dimensional decagonal quasicrystal estimate the stress necessary to create the phason wall by glide at zero temperature to amount to $\tau_{\text{ph}} \approx (1/27)\,\mu$ [685]. This would be a very high stress. Therefore, quasicrystals were considered brittle at low temperatures. On the other hand, measurements from the width of pairs of dislocations moving together by climb yield very low values [702].

Since the imperfect dislocations have only a phononic Burgers vector $\boldsymbol{b}_\parallel$, they fulfill only the strong extinction condition (10.8) with $\boldsymbol{g}_\parallel \cdot \boldsymbol{b}_\parallel = 0$. In the orientation where the corresponding perfect dislocations obey the weak extinction condition, a residual contrast remains. The phason walls have a displacement vector which can be defined either by $\boldsymbol{r}_\parallel = \boldsymbol{b}_\parallel$ or by $\boldsymbol{r}_\perp = \boldsymbol{b}_\perp$. The shift vectors of phason walls are determined in a way similar to that of stacking faults in crystals. Since $\boldsymbol{g}_\parallel \cdot \boldsymbol{r}_\parallel$ is irrational in quasicrystals, extinction occurs only for $\boldsymbol{g}_\parallel \cdot \boldsymbol{r}_\parallel = 0$, but not for $\boldsymbol{g}_\parallel \cdot \boldsymbol{r}_\parallel =$ integer. A TEM diffraction contrast analysis of phason faults is published in [703].

If the temperature is not too low, the distribution of the phason defects in the solid can be rearranged by spreading of the phasons of the wall and by annealing to form the equilibrium phason distribution around A and the new cloud around the dislocation at B. The process is called phason dispersion or retiling. All the existing phasons certainly do not diffuse to the dislocation at B. Instead, the retiling occurs by local diffusion processes which, however, include chemical diffusion over distances larger than the atomic distances. Thus, the atomic mobility has to be high enough. At the end, the new equilibrium configuration of phonon and phason strain fields may be re-established (Fig. 10.10d). During rapid deformation at lower temperatures, dislocations are certainly generated as imperfect dislocations. If the Burgers vector in hyperspace is a fraction of a lattice vector, the dislocation is a partial dislocation.

An important issue of dislocation motion in quasicrystals was the question whether the dislocations move by glide or climb. Although the first in situ straining experiments in a TEM on i-Al–Pd–Mn single quasicrystals were performed at a relatively high temperature (at $0.88\,T_{\text{m}}$, with T_{m} being the melting temperature) [647], they immediately showed that dislocation motion in this material is of crystallographic character, i.e., many dislocations showed straight crystallographically oriented segments, and they moved on crystallographic high-symmetry planes. Therefore, for several years it had been assumed that the dislocations move by glide without questioning this hypothesis. To prove the mode of dislocation motion, it is necessary to determine both the direction of the parallel component of the Burgers vector and the plane of motion of always the same dislocations. This is only possible by in situ straining experiments in a TEM. Only recently, Caillard and coworkers succeeded in performing such experiments, at first during in situ heating where the dislocations moved under internal stresses [704], and then also during in situ straining [705]. Both experiments showed dislocations moving by pure climb. Thus, climb is certainly a main mode of dislocation motion in quasicrystals. The following microstructural results have to be considered keeping this in mind.

10.3 Microscopic Observations of Dislocations

While information on dislocation mechanisms of quasicrystal deformation has been gained by TEM on a number of materials of either icosahedral or decagonal symmetry, for other materials like Al–Cu–Fe the evidence of dislocation motion is still missing, in spite of a wealth of macroscopic deformation data available. The following sections are restricted to two single quasicrystalline materials, icosahedral Al-21at%Pd-8.5at%Mn and decagonal Al-15at%Ni-15at%Co. These materials may be considered prototypes of quasicrystals. Their dislocation microstructures have most frequently been studied, also by the author of this book and his coworkers.

10.3.1 i-Al–Pd–Mn

As it was mentioned earlier, the dislocation properties strongly depend on the temperature. This influences also the macroscopic deformation parameters, which will be described in detail in Sect. 10.4. At high temperatures where diffusion is rapid, quasicrystals can be deformed up to high strains. After a yield drop effect, there develops a range of almost steady state deformation. Most deformation experiments and microstructure studies were carried out in this temperature range. Only later, experiments succeeded at lower temperatures where deformation at low rates is accompanied with very strong work-hardening. The following results consider both ranges. In situ straining experiments, however, were successful only in the high-temperature range where the flow stresses are not so high.

Transmission Electron Microscopy of Deformed Specimens

The first experiment which proved a dislocation mechanism of the plastic deformation of quasicrystals was the observation by Wollgarten et al. [646] of an increase of the dislocation density after deformation. Later on, the Burgers vectors of deformation-induced dislocations were analyzed by the CBED technique [706]. 60 dislocations were chosen from specimens deformed at 800°C up to different plastic strains, and at different temperatures down to 732°C at a fixed strain, all in the high-temperature range. Fifty-two of the 60 dislocations had parallel parts $\boldsymbol{b}_{\|}$ of the Burgers vectors in twofold directions, 32 of which with a length of $|\boldsymbol{b}_{\|}| = 0.183\,\mathrm{nm}$. Five dislocations had parallel Burgers vectors in the fivefold direction. They are super-partial dislocations. To characterize the contribution of phason strain, the so-called strain accommodation parameter $\zeta = |\boldsymbol{b}_{\perp}|/|\boldsymbol{b}_{\|}|$, i.e., the ratio between the lengths of the perpendicular and parallel parts of the Burgers vectors, was calculated. In deformed samples, the values varied between $\tau^7 \approx 29.0$ and $\tau^3 \approx 4.2$ with a maximum of $\tau^5 \approx 11.1$. Higher values of the ratio occurred more frequently with increasing plastic strain. In the in situ and post-mortem TEM studies of the group of Caillard [702, 704, 705] performed at lower temperatures, only dislocations with longer parallel parts of the Burgers vectors $\boldsymbol{b}_{\|}$ were analyzed, i.e., of 0.257, 0.296, 0.348, 0.456, 0.513, and 0.563 nm.

In addition to the Burgers vectors, also the line directions were determined of 25 dislocations with twofold Burgers vectors [706]. Forty percent of the lines were oriented along twofold directions. The character β of the dislocations ranged between 7 and 90°. Thus, pure screw dislocations have not been observed. The parallel parts of the Burgers vectors and the line directions span the slip planes of the dislocations, and the dislocations were believed to glide on these planes, although there were already hints that this may be wrong.

The dislocation microstructure of deformed samples was also imaged by diffraction contrast in the HVEM allowing thick specimens to be investigated so that the three-dimensional character of the dislocation structures could be studied [707]. After deformation, some of the samples were cooled under load to better preserve the microstructure. Some micrographs reveal dislocations being arranged in bands, but inside the bands, they form a homogeneous three-dimensional network at high temperatures, as shown in Fig. 10.11. Diffraction contrast analysis using strong contrast extinctions at two different $\boldsymbol{g}$ vectors indicates that dislocations with different Burgers vectors form the nodes of the network. There exist groups of dislocations with their parallel component of the Burgers vector being parallel to the compression axis, i.e., these dislocations experience a climb force but no glide force. At 800°C, the average link length in the network is of the order of magnitude of $L = 0.5\,\mu\mathrm{m}$. Many dislocation segments are quite straight and oriented along crystallographic directions. This is remarkable for the high temperature.

At the lowest temperature with steady state deformation, many dislocations form narrow bands. In Fig. 10.12, taken near a twofold pole parallel to the compression axis, two bands cut each other. Dislocations are arranged on two steeply inclined planes, most probably planes perpendicular to threefold directions. The bands include elongated dislocations as well as strongly bowed-out

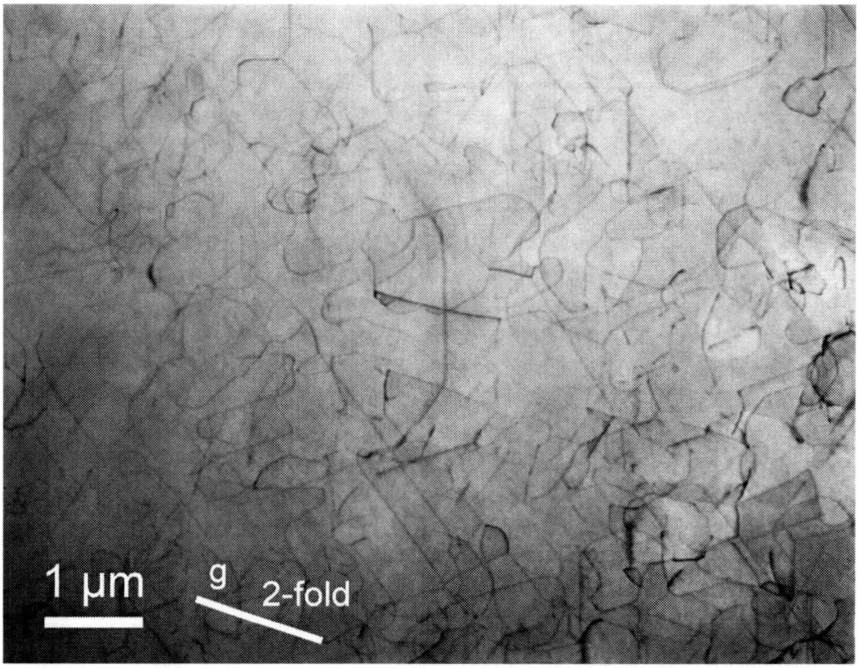

Fig. 10.11. Dislocation structure in an i-Al–Pd–Mn single quasicrystal deformed at 770, 800, and 818°C along a fivefold axis. Image taken near this pole. From [707]. Copyright Elsevier (2004)

dislocation segments and many small dislocation loops, as shown in Fig. 10.13. The radii of curvature of the bowings and the loops are as small as 70 nm. The dislocation density inside the bands is approximately $1.2 \times 10^{14}\,\mathrm{m}^{-2}$. Figures 10.12b to d exemplify the analysis of the parallel components of the Burgers vector of the band running downwards from the upper right corner of the figure. The hole in the specimen may serve as a marker. Because of the bending of the specimen, a field-limiting aperture of 2 µm diameter was placed near the center for taking the diffraction patterns shown in the insets. Thus, the respective diffraction conditions are fulfilled only near the centers of the images. The discussed dislocation band is extinguished in Figs. 10.12c and d. Accordingly, the parallel component of the Burgers vector is perpendicular to the extension of the band and to the compression axis. Thus, these dislocations are near the edge orientation and experience neither a glide force nor a climb force from the applied stress.

At a lower temperature of 532°C, where plastic deformation is achieved in stress relaxation tests in the strong work-hardening range, many dislocation segments are straight and oriented crystallographically but other segments very strongly bow out between cusps indicating pinning agents, as marked by an arrow in Fig. 10.14. The distances between the cusps (obstacle distances)

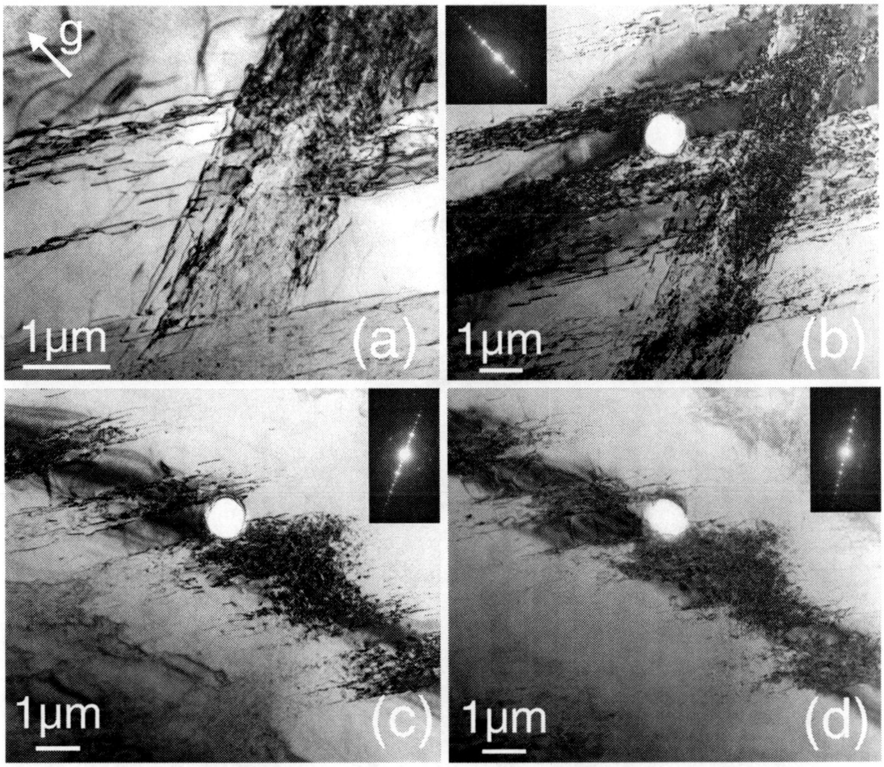

Fig. 10.12. Slip bands after steady state deformation by $\varepsilon = 2\%$ at 580°C at $10^{-6}\,\mathrm{s}^{-1}$ along a twofold axis with analysis of the parallel component of the Burgers vector of the band running approximately vertically. The deformation curve is shown in Fig. 10.32b. Compression axis and image normal [0/0 0/0 0/2] in (a, b), $\boldsymbol{g}$ vectors ($\bar{1}$/0 0/1 0/0) (a, b), (1/0 0/1 0/0) (c) and ($\bar{1}$/0 $\bar{1}$/$\bar{1}$ 0/1) (d). The insets show the respective diffraction patterns. From [708]

are of the order of magnitude of 100 nm and the radii of curvature of the bowed-out segments are about $\delta = 50\,\mathrm{nm}$. At the lowest temperatures of 487 and 482°C, where plastic deformation was obtained in conventional compression tests, different dislocation structures are observed. Some specimen regions again exhibit the straight crystallographically oriented dislocation segments. However, Fig. 10.15a presents a totally different dislocation structure. The dislocations are arranged in strongly curved loops, but the cusps indicating pinning agents at 532°C are missing now. Frequently, the area inside the loops exhibits a homogeneous area contrast, depending on the diffraction conditions applied. Stereo pairs show that the loops belong to different families of planes but that the individual loops have segments which bow out onto planes that differ from those of the main parts of the loops. In another specimen shown in Fig. 10.15b, two ordinary dislocation bands are crossing. In addition to the

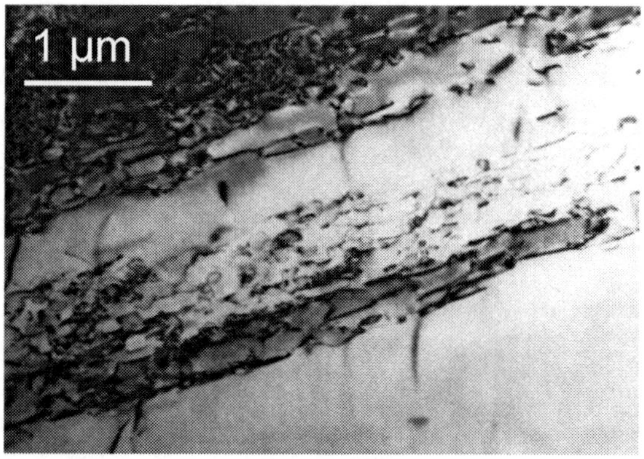

Fig. 10.13. Dislocations inside a band in i-Al–Pd–Mn from the same specimen as Fig. 10.12. Image taken near twofold pole. From [707]. Copyright Elsevier (2004)

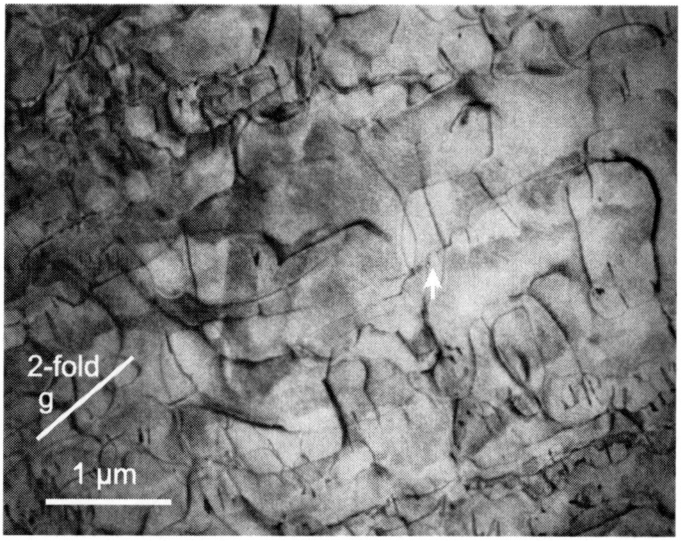

Fig. 10.14. Dislocation structure after deformation at 532°C in the range of high work-hardening. From [707]. Copyright Elsevier (2004)

dislocations, contrast fringes are imaged parallel to the extension of the bands (arrows), particularly to the left of the image at the upper edges of the bands. These fringes are visible only under strong excitation conditions. They can be imaged over most parts of the dislocation bands. Two types of pronounced fringes were observed in one experiment at 610°C, as illustrated in Fig. 10.16. Strong fringes marked A extend between two dislocations whereas weak fringes B "decorate" the whole path of the dislocations. The fringes belong to planar faults. In the HVEM studies, it had not been possible to characterize these

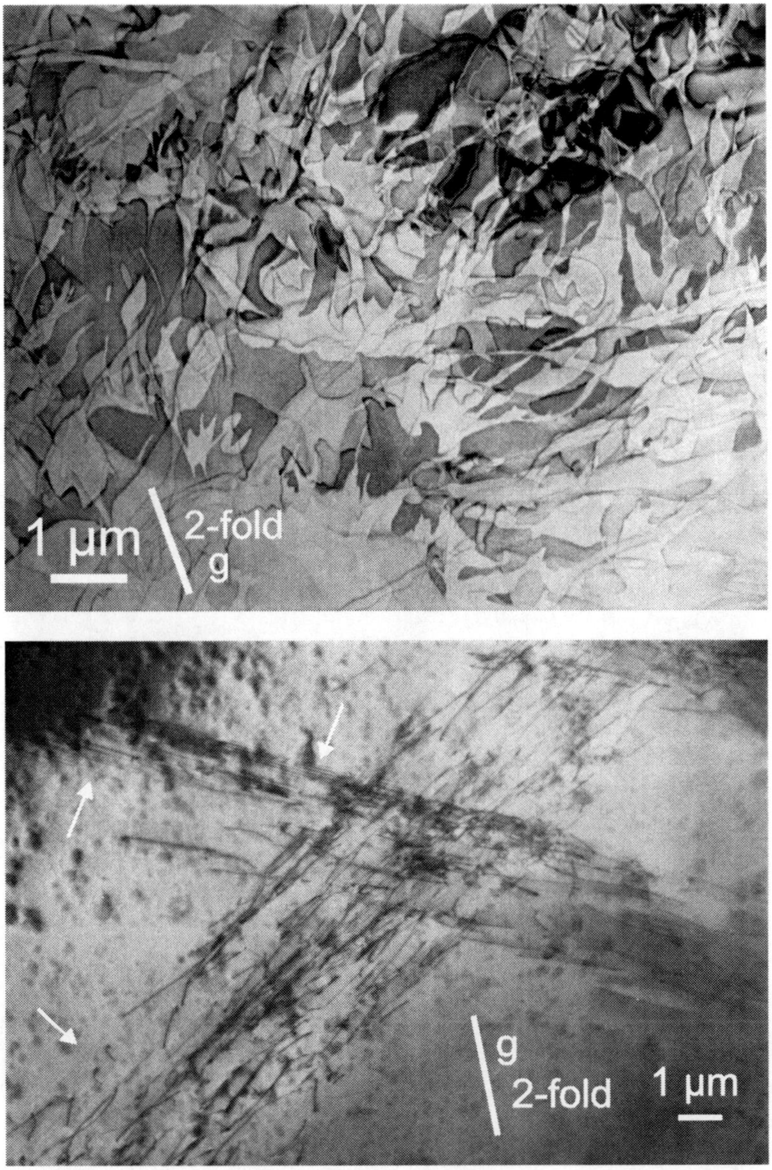

Fig. 10.15. Dislocation structure after deformation at the lowest temperatures in the range of high work-hardening. (**a**) 482°C, fivefold compression axis. (**b**) 487°C, twofold compression axis. Images taken near the poles of the compression axes. From [707] and [709]. Copyright Elsevier (2004) and Wiley-VCH (2003)

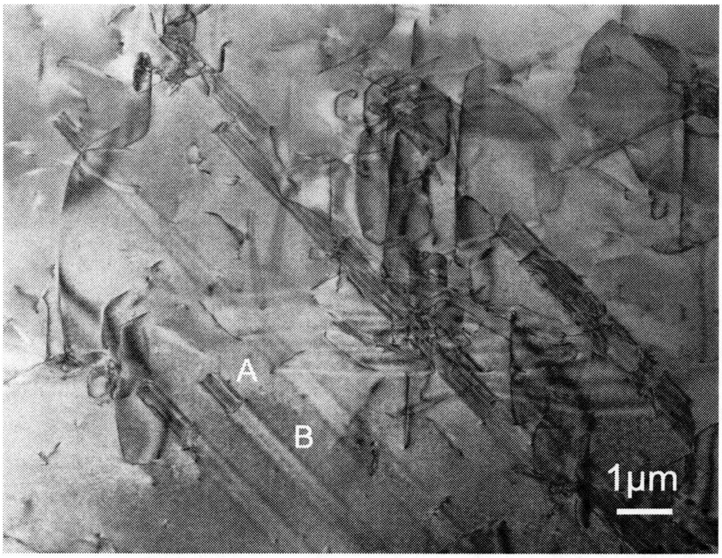

Fig. 10.16. Microstructure in i-Al–Pd–Mn after deformation at 610°C. From [709]. Copyright Wiley-VCH (2003)

fringes in more detail. In post-mortem TEM of deformed specimens, they have never been observed above 610°C.

Deformation at even lower temperatures can be achieved by applying confining hydrostatic pressure as described for ZrO_2 in Sect. 7.3. Such experiments have been performed in a multianvil apparatus at 300°C, a pressure of 7.5 GPa and compression along a fivefold axis [702]. Pairs of superpartial dislocations with Burgers vectors parallel to the compression axis extended on planes perpendicular to the compression axis and had dragged phason faults on these planes. Other dislocations with Burgers vectors in twofold directions were extended on planes perpendicular to their Burgers vectors. It is assumed that these dislocations had moved on the mentioned planes, i.e., they had moved by (mostly pure) climb. Part of the dislocations experience a high climb stress and another part climbs under zero mechanical stress. The dislocations drag phason faults and are all curved. Similar experiments [710], also including room temperature, showed dislocations concentrated in narrow bands. From the directions of the Burgers vectors and the planes which the dislocations were extending on, it was concluded that climb and glide contributed to the deformation.

Split dislocations have also been observed by HVEM in specimens deformed at lower temperatures. One example is presented in Fig. 10.17.

In the high-temperature range, the dependence of the dislocation density on the strain and temperature was studied in detail in [338]. The dislocation

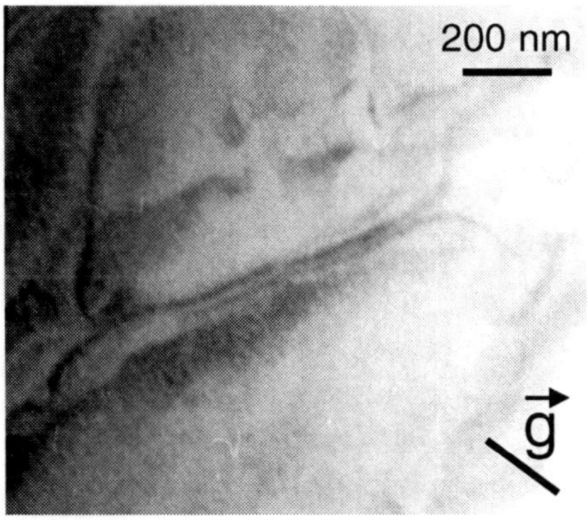

Fig. 10.17. Split dislocation in an i-Al–Pd–Mn single quasicrystal deformed at 487°C along a twofold direction. From [708]

density increases strongly during initial loading, reaching a maximum at about 5% plastic strain before it decreases. The temperature dependence in the steady state range of the flow stress was presented in Fig. 5.29b together with a few data measured from HVEM micrographs of deformed samples. The figure also includes values from the deformation associated with strong work-hardening at lower temperatures. During annealing at a deformation temperature of 730°C, about 60% of the dislocation density anneals out within one hour with a half-time of 10 min. The dislocation density data will be discussed in Sect. 10.5.7.

In summary, at high temperatures dislocations of different Burgers vectors in bands form a three-dimensional network. At low temperatures, the dislocations are concentrated in narrow bands. They show straight oriented segments particularly at high temperatures, whereas many dislocations are curved at low ones. Some segments strongly bow out between pinning agents near 550°C. At 482°C, also curved dislocation loops are formed, which do not show pinning agents. At 610°C and below, dislocations may trail fringe contrasts. There are sets of dislocations with either parallel components of their Burgers vectors parallel to the deformation axis, which should experience only a climb force from the external stress, or with Burgers vectors perpendicular to the loading axis, which experience no mechanical force at all. In the following, these sets are termed A and B.

In situ Straining Experiments in an High-Voltage Electron Microscope

In situ straining experiments on i-Al–Pd–Mn were successful only in a relatively narrow temperature window between 625 and 765°C, i.e., in the high-temperature range, where steady state deformation is possible at usual strain rates. Details of the experiments performed at the Halle HVEM facility are described in [711].

During in situ straining, the first dislocation bands are frequently formed at inhomogeneities, mostly near the perforation of the specimen, like the primary band PB in Fig. 10.18. Dislocations are emitted from the primary band into secondary bands SB. The dislocations imaged in dark contrast trail traces, which mark the planes of their motion. The latter were identified by simulating the respective geometry. The planes observed are listed in [711]. They are perpendicular to twofold, threefold, fivefold, and pseudo-twofold orientations. Often, the primary bands are almost edge-on with a trace perpendicular to the tensile direction as in Fig. 10.18. These bands have only a small orientation factor for glide but a maximum one for climb, therefore belonging to set A. Many dislocations consist of straight segments in preferred orientations like those in the conventional HVEM micrographs from the high-temperature range described earlier. Well identified are the three twofold directions on planes perpendicular to the threefold directions as of the segments marked by the small arrow. Twofold orientations are generally the dominating directions of the dislocation lines in macroscopic tests as well as in in situ experiments [704].

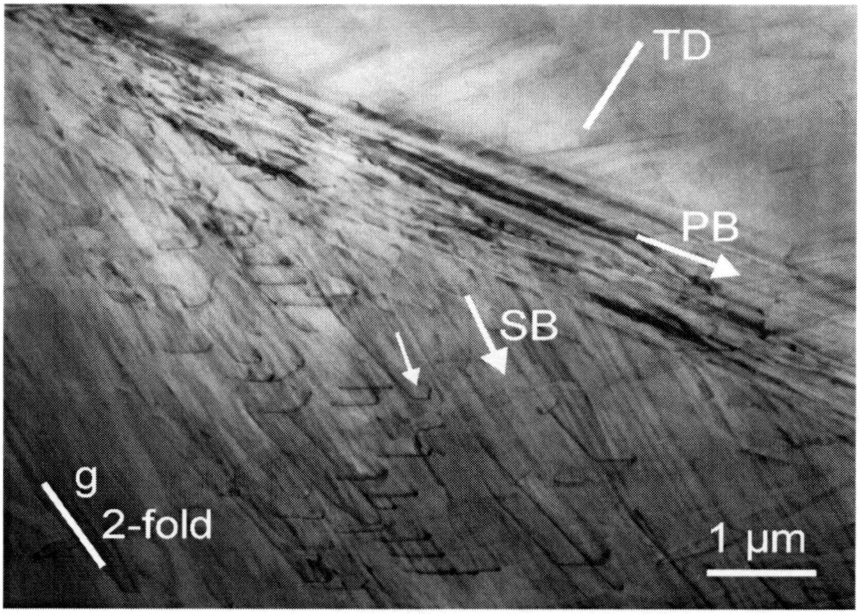

Fig. 10.18. Dislocation structure during in situ deformation of i-Al–Pd–Mn at 750°C in an HVEM. Threefold image plane. Twofold tensile direction (TD). From [709]. Copyright Wiley-VCH (2003)

The dynamic dislocation behavior is illustrated in the following video sequences.

Video 10.1. *Dislocation motion in* i-Al–Pd–Mn *single quasicrystals at* 750, 720 *and* 685 °C*:* Two of the three of these videos are from the series of in situ straining experiments performed in Halle which, for the first time, yielded direct evidence that plastic deformation in quasicrystals occurs by the motion of dislocations [647]. The most important observation is that the dislocations move in a smooth viscous way on different crystallographic planes. As described above, many dislocations move on planes perpendicular to the tensile direction (TD) indicated at the beginning of the video, i.e., they move under a high climb force (set A). In the second sequence at C, the dislocation moving upwards interacts with a dislocation on another plane. As a consequence, the dislocation changes its plane of motion to a more steeply inclined one without changing its direction of motion. Such a change is not possible by slip and cross slip alone and also not by climb alone. The other dislocation continues moving without any changes. In the third sequence, a dislocation moves on a plane parallel to the tensile axis. On this plane, there neither acts a glide nor a climb force, except residual forces originating from the inhomogeneous stress field in the in situ specimen. This dislocation belongs to set B.

These observations confirm the conclusion from the post-mortem TEM analysis that dislocations move under the action of pure climb forces or even without a force component from the applied stress. In the following video clip, the shapes of moving dislocations can well be observed.

Video 10.2. *Dislocation motion in an* i-Al–Pd–Mn *single quasicrystal at* 710 °C*:* The dislocations exhibit almost straight crystallographically oriented segments, as illustrated in the secondary band SB in Fig. 10.18. In the first sequence, relatively long dislocations assume three preferred orientations. In the second one, long dislocations move on a steeply inclined plane. The dislocations keep their oriented shape during the motion.

Both the oriented shape and the fact that the dislocations do not take the shortest length in the thin foil indicate the action of a high lattice resistance. The next video sequences are recorded at the relatively high temperature of 750°C.

Video 10.3. *Dislocation motion in an* i-Al–Pd–Mn *single quasicrystal at* 750°C*:* The video clip consists of five sequences. They all show dislocation motion on different planes. Many of the planes are oriented so that they do not correspond to very high orientation factors for climb. Also at this high temperature, many dislocations are straight and oriented along crystallographic directions. Dislocations with an angular shape with quite an acute angle are visible in the third sequence. Some dislocations change the width of their traces of motion without changing the orientation, especially in the first (marked by C) and fifth sequences. The example of the first sequence is displayed also in Fig. 10.19. In the fourth sequence at R, two dislocations moving on different planes react with each other, the result is not entirely clear. Figure 10.18 belongs to this sequence.

All the videos demonstrate the crystallographic character of the dislocation motion in i-Al–Pd–Mn at high temperatures. The orientations of the planes of

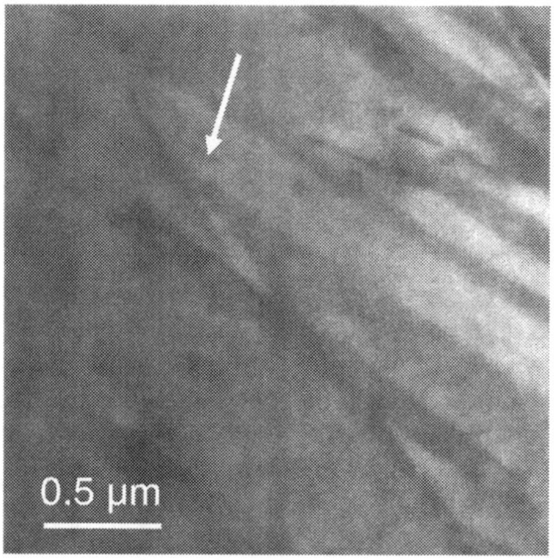

Fig. 10.19. Section of a frame of the first sequence of Video 10.3 showing a slip trace changing its width without changing its orientation (*arrow*). From the work in [647]

motion indicate partly high climb forces without glide forces from the external stress, and partly no mechanical force at all. The mode of motion, i.e., glide or climb, could not be decided from the videos shown here since the Burgers vectors had not been determined. This decision was made by the work of Caillard and coworkers by their combined analysis of the Burgers vectors and the planes of motion during in situ straining [701, 704]. In the cases analyzed, the dislocations always moved by pure climb. During in situ straining, sometimes dislocations of opposite sign moving across each other annihilate. In [705], two edge dislocations are shown approaching each other on slightly displaced fivefold planes forming a dipole which can annihilate only by glide on a twofold plane. At the high temperature of 740°C, this dipole is stable over more than 30 min indicating that, under these conditions, glide is, by more than three orders of magnitude, slower than climb. In conclusion, at high temperatures straight dislocations in icosahedral Al–Pd–Mn oriented mainly along twofold directions primarily move by climb along high-symmetry planes.

The velocities of dislocations moving on certain traces were analyzed by plotting the position of the dislocations versus time followed by differentiation of these curves. Histograms of the velocities indicate that in some traces dislocations move at two well-defined different velocities. In a special experiment at 625°C, the velocity v_d of a single dislocation was followed under decreasing external load F, exerted by the in situ stage. The slope of a plot of $\ln v_\mathrm{d}$ versus $\ln F$ in Fig. 10.20 yields the effective stress exponent of the motion of an individual dislocation (4.12) $m' = \mathrm{d}\ln v_\mathrm{d}/\mathrm{d}\ln \tau = \mathrm{d}\ln v_\mathrm{d}/\mathrm{d}\ln F$. The result is $m' \approx 6$ [711]. This value can be compared with the macroscopic one discussed in Sect. 10.4.

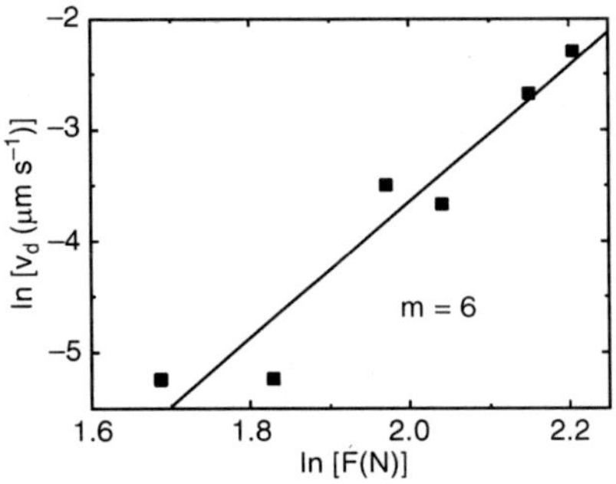

Fig. 10.20. Dependence of the velocity of a single dislocation on the force acting on the specimen during an in situ straining experiment on an i-Al–Pd–Mn single quasicrystal at 625°C. Data from [711]

Video 10.4 presents two sequences of dislocations with particular contrasts.

Video 10.4. *Special contrasts at the motion of dislocations in* i-Al–Pd–Mn *single quasicrystals:* The two sequences were taken at 675 and 685°C. The first sequence was recorded at a twofold specimen normal and a twofold tensile axis. The dislocation clearly shows a double contrast. Thus, these dislocations move in the split configuration. The dislocation in the second sequence trails a fringe contrast which should result from the phason fault dragged behind. The contrast becomes weak about two or three micrometers behind the dislocation.

Split dislocations were observed earlier in macroscopic specimens (Fig. 10.17). Special core configurations may control the dislocation mobility. Dislocations D trailing phason walls are also illustrated in Fig. 10.21.

Almost all dislocations shown in the videos above did not drag phason defects in their wake. Since the parallel part of the displacement vector of the phason faults ought to equal the parallel part of the Burgers vector, the faults should have been visible under the same imaging conditions as the dislocations themselves. Since this is not the case, the dislocations should be perfect dislocations above about 700°C, i.e., in the high-temperature range of deformation. In the border region as in the above video, the faults evaporate after a short time so that they are absent a short distance behind the dislocations. Similar observations are described in [704]. Phason faults generated by the movement of grown-in dislocations anneal out at high temperatures. During in situ heating, the fringe contrasts between the dislocations in the upper part of Fig. 10.22 anneal out between 610 and 690°C and have completely disappeared at 700°C (lower part of the figure). According to [712], phason walls at grown-in dislocations disappear at 560°C within hours, whereas those behind moving dislocations vanish at 750°C within seconds. An Arrhenius plot of the reciprocal time to disappearance yields an activation energy of 4.3 eV. This value is unreasonably high with respect to the atomic diffusion energies (2.32 eV for Pd [670], 1.99 eV for Mn [713] and an even lower value for Al [714])

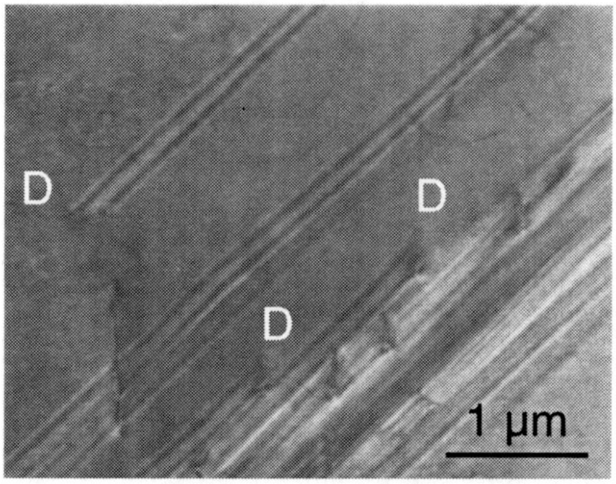

Fig. 10.21. Dislocations D trailing phason walls in i-Al–Pd–Mn during in situ deformation in an HVEM at 695°C. From the work in [647]

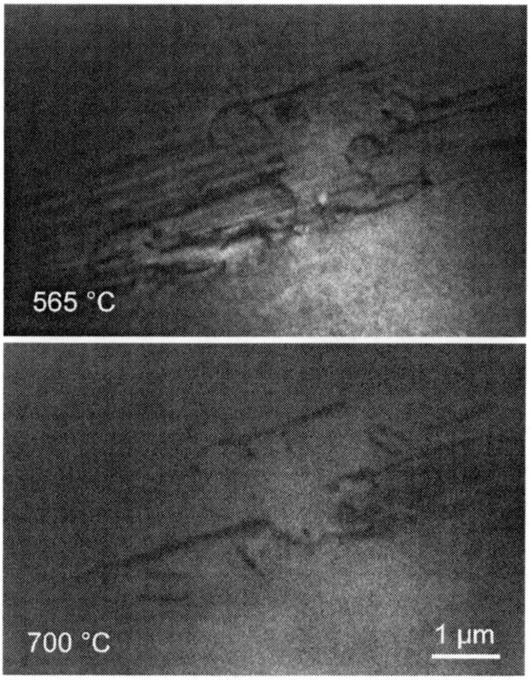

Fig. 10.22. In situ annealing of grown-in phason faults in an i-Al–Pd–Mn single quasicrystal

and to the results on phason relaxation studied by X-ray scattering mentioned in Sect. 10.2.2. Owing to longer times available for annealing in macroscopic deformation experiments, fringe contrasts of phason walls have never been observed there above 610°C. On the other hand, plastic deformation in the high-temperature range (760°C) causes linear phason strains which can be observed by a shift of the diffraction spots at selected area electron diffraction in regions between the dislocations [715]. The spots are displaced out of their systematic rows with the shift increasing with increasing absolute value of the component of the reciprocal vector $\boldsymbol{g}_{\perp}$ in perpendicular space. The orientation of the respective phason stress field varies spatially. It is argued that the phason strain is not attributed to the dislocations but to the material itself.

An important question is the origin of the dislocations created during plastic deformation. In crystals, dislocations frequently multiply by the double-cross slip mechanism (Sect. 5.1.1). In situ experiments often show this process with a characteristic intermediate stage similar to the letter α as highlighted in Fig. 5.5c. Such configurations are also observed during the in situ deformation of Al–Pd–Mn quasicrystals, as marked by arrows in Fig. 10.23.

In contrast to crystals, dislocations in quasicrystals move on well-defined planes, mostly by climb. After a change from the primary plane of motion PP onto a secondary one SP, the dislocations usually do not return onto a plane parallel to the primary one but spread on the secondary plane as outlined in Fig. 10.24. If only part of the dislocation changes onto another plane as in position 2, the point connecting the two segments can move only along the intersection line between the planes. Thus, it acts as an obstacle to the forward motion so that dislocation segments bow out on both planes (stage 3). Accordingly, the two ends of the α-like configurations in Fig. 10.23 extend on

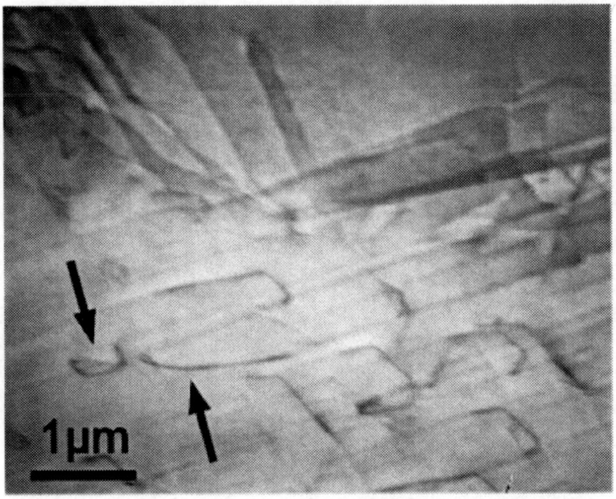

Fig. 10.23. α-shape configurations during multiplication of a dislocation in an i-Al–Pd–Mn single quasicrystal. From the work in [711]

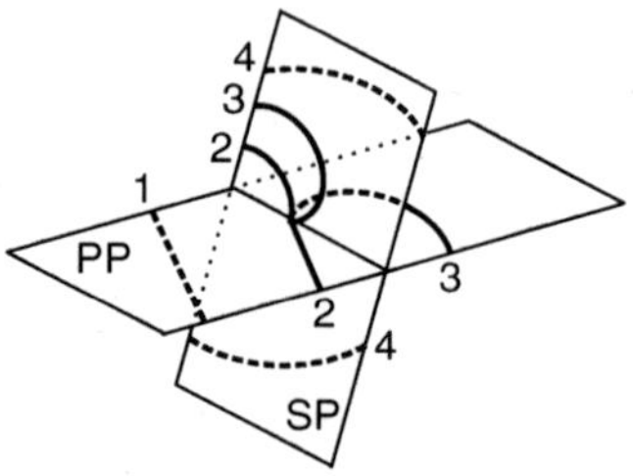

Fig. 10.24. Schematic of the multiplication of dislocations in Al–Pd–Mn after a change from the primary plane of motion PP onto a secondary one SP. After [711]

different planes. After full bow-out, two dislocations move on the secondary plane in opposite directions (stage 4). Thus, dislocation motion spreads from one plane onto other planes as discussed in connection with Fig. 10.18, leading to the dislocation networks formed at high temperatures. However, at low temperatures, the dislocations seem to return to planes parallel to the original ones similar to the multiplication in crystals as indicated by the narrow dislocation bands observed (Fig. 10.15b).

The following video clip illustrates dislocation generation, always connected with some change of the plane of motion.

Video 10.5. *Dislocation multiplication in* i-Al–Pd–Mn *single quasicrystals:* The three sequences were taken at 720, 685, and 675°C. The parent dislocations are marked by M. In the first sequence, the dislocation changes its plane of motion to a new one as described above. Two new dislocations emerge and move in opposite directions on a plane with its trace perpendicular to the tensile direction. In the second sequence, the multiplication event is connected with a small parallel displacement of the plane of motion. One dislocation quickly moves backwards, followed by two dislocations moving in forward direction. In the third sequence, the dislocation undergoes a shift onto a parallel plane. At A, one dislocation moving back annihilates with another one on the same plane. At R, two dislocations moving on planes with parallel traces of different inclination in the foil react with each other. As a result, a dislocation with double contrast moves to the right.

The in situ study in [704] also revealed dislocation multiplication.

10.3.2 d-Al–Ni–Co

As classified in [716,717], the decagonal phase of Al–Ni–Co may have different modifications depending on the exact composition. In the electron diffraction patterns, they show particular superstructure reflections, being first observed in [718]. The different modifications correspond to different states of structural and chemical order. At the Co and Ni-rich ends of the decagonal phase field, the decagonal structures do not show superlattice reflections. These structures are termed 'basic Co-rich' and 'basic Ni-rich' variants. The results following below are confined to the composition Al-15at%Ni-15at%Co with the type I superstructure showing the so-called S1 and S2 superlattice reflections [718].

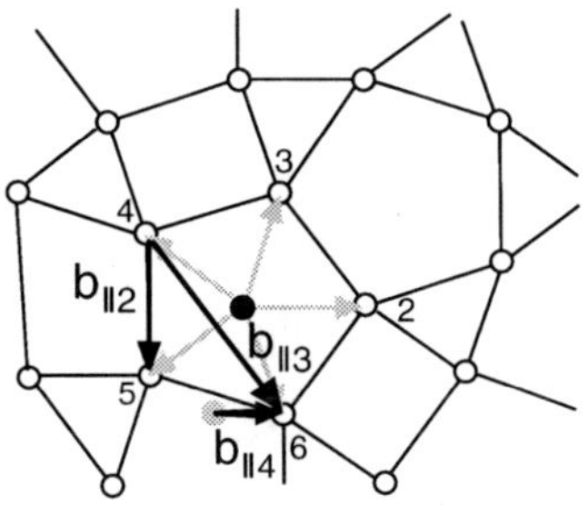

Fig. 10.25. Parallel components of the Burgers vectors in d-Al–Ni–Co single quasicrystals as determined in [720]. $b_{\|1}$ is directed perpendicular to the quasiperiodic image plane

The Burgers vectors of dislocations in decagonal quasicrystals were first studied by Zhang and Urban [719] via contrast extinction, and later on by other authors also by applying the CBED method (e.g., [719,720]). The parallel components of the observed Burgers vectors are outlined in Fig. 10.25. The grey arrows marked by 2 to 6 are the basis vectors in the quasicrystalline plane. They are longer by a factor of τ than those introduced by Steurer et al. [667]. The Burgers vectors observed are the following: the periodic Burgers vector directed along the periodic axis of the length of the c lattice constant in parallel space, in the full five-dimensional notation of Steurer et al. [667] written as $B_1 = [00001]$, the Burgers vectors in the quasiperiodic plane $B_2 = [\bar{1}0010]$ and $B_3 = [\bar{1}\bar{1}010]$ of the lengths of 0.29 nm and 0.47 nm in parallel space, and a mixed Burgers vector $B_4 = [\bar{1}/5, \bar{1}/5, \bar{1}/5, \bar{1}/5, \bar{1}/2]$ of the length of 0.25 nm in parallel space. The mixed Burgers vector is not a lattice vector in hyperspace so that the respective dislocations are partial dislocations. The strain accommodation parameters ζ for the quasiperiodic Burgers vectors are τ^{-1} and τ^{-3}. Compared to i-Al–Pd–Mn, in d-Al–Ni–Co these values indicate a lower tendency of the dislocations to be accompanied with phason defects.

The dislocations induced by plastic deformation show very strong preferential orientations parallel or perpendicular to the periodic axis. The dislocations with periodic Burgers vectors assume both these line directions whereas those with quasiperiodic or mixed Burgers vectors are arranged only parallel to the periodic axis.

While, because of the high symmetry of icosahedral Al–Pd–Mn, its plastic behavior is very isotropic, decagonal Al–Ni–Co has a restricted number of Burgers vectors of different character and only a few dislocation line directions resulting in a remarkable plastic anisotropy. It was first revealed on single Al–Ni–Co quasicrystals in [721]. Mostly, stress–strain curves were measured with loading axes parallel to the periodic axis, perpendicular to it, and at an angle of 45°, which may be denoted $A_{\|}$, $A_{\perp}$, and A_{45}. The highest yield stress always has the orientation $A_{\perp}$, whereas the lowest one has A_{45}. The amount of the intermediate value of $A_{\|}$ depends on the composition, it is slightly higher than that of A_{45} for Al-15at%Ni-15at%Co and slightly less than that of $A_{\perp}$ for the basic Co-rich material.

A characteristic deformation microstructure produced during in situ deformation along the $A_{\parallel}$ axis is shown in Fig. 10.26a. The figure was taken with a systematic row of reflections parallel to the periodic axis, i.e., containing the $\boldsymbol{G} = (00002)$ reflection, with a higher reflection excited (high-order bright-field technique). This direction is also the tensile direction TD. Under these conditions, dislocations with quasiperiodic Burgers vectors are extinguished. The arrows in the figure point to dislocations with a periodic Burgers vector, which appear as short dark lines that are attached to extended gray lines, which are the traces of the dislocation motion. The periodic dislocations, which are seen almost end-on, move along quasiperiodic planes perpendicular to the periodic axis. Many other traces in the same direction as of dislocations running out of the image area are visible as well. The dark double line in the center of the figure is the trace of many dislocations. The gray lines parallel to the tensile direction, marked by M, are dislocations which, according to [720], have mixed Burgers vectors.

Figure 10.26b refers to a different specimen region, taken with a systematic row containing the quasiperiodic reflection $\boldsymbol{G} = (\bar{1}0010)$. Dislocations with mixed and quasiperiodic Burgers vectors are visible but those with the periodic Burgers vector are now extinguished. The entire image shows a background structure with a striation contrast parallel to the tenfold axis. This contrast possibly originates from phason boundaries resulting from phason disorder [722]. It appears under all diffraction vectors that exhibit any component in the quasiperiodic direction. It makes the observation of dislocations difficult.

In macroscopic deformation experiments, all specimens were deformed within the high-temperature range. In the $A_{\parallel}$ orientation, the dislocations form networks between dislocations with periodic and mixed Burgers vectors as illustrated in Fig. 10.27a. Similar structures are observed in Al-15at%Ni-15at%Co. Burgers vector analysis by the CBED method [720] showed that the networks are formed by dislocation reactions with the periodic component of the Burgers vector of the mixed dislocation being opposite to the direction of the Burgers vector of the periodic dislocations, as outlined in Fig. 10.28. The dislocations in Fig. 10.28a move across each other, forming a junction in (b). The reaction is connected with a gain in energy. In (c), the junction has rearranged. According to the definition in Sect. 10.3.1, the dislocations with periodic Burgers vectors running horizontally in Fig. 10.27a are equivalent to set A experiencing a high climb force but no glide force.

In the A_{45} orientation, periodic dislocations arrange along the screw and edge orientations with the screws dominating as illustrated in Fig. 10.27b. The image plane is parallel to planes marked by slip steps on the specimen surface. The figure also shows a dislocation loop. If it is assumed that the loop plane is also the plane of motion, the loop has been formed by glide. It may therefore be concluded that periodic dislocations in d-Al–Ni–Co can

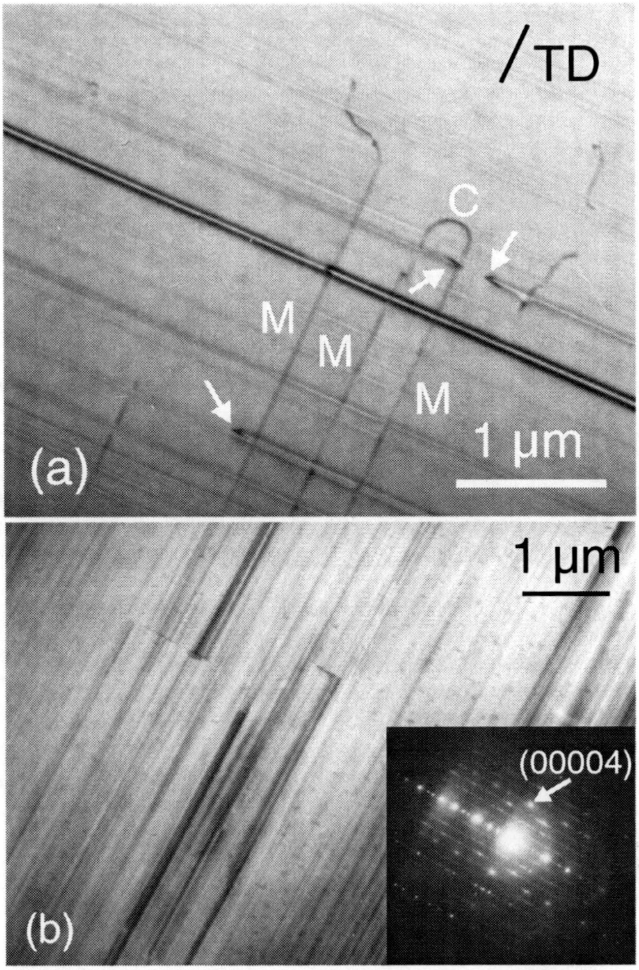

Fig. 10.26. Deformation microstructure during in situ straining of a d-Al–15at%Ni–15at%Co single quasicrystal along the $A_{\parallel}$ axis in an HVEM at 730 °C, taken at the P2 twofold pole of the foil normal. (**a**) $\boldsymbol{g}$ vector in the direction of the periodic axis. (**b**) Quasiperiodic $\boldsymbol{g}$ vector. From [666]. Copyright Mater. Res. Soc. (2005)

glide. In addition, also dislocations with mixed Burgers vectors contribute to the deformation. At the $A_{\perp}$ deformation axis, dislocations with quasiperiodic and mixed Burgers vectors dominate. They are arranged along the periodic axis.

Some dislocation density data of deformed Al–Ni–Co single quasicrystals are compiled in Table 10.1, revealing a slight dominance of periodic dislocations in the A_{45} orientation, where slip of periodic dislocations is possible, with a comparatively very low value of the total dislocation density. All

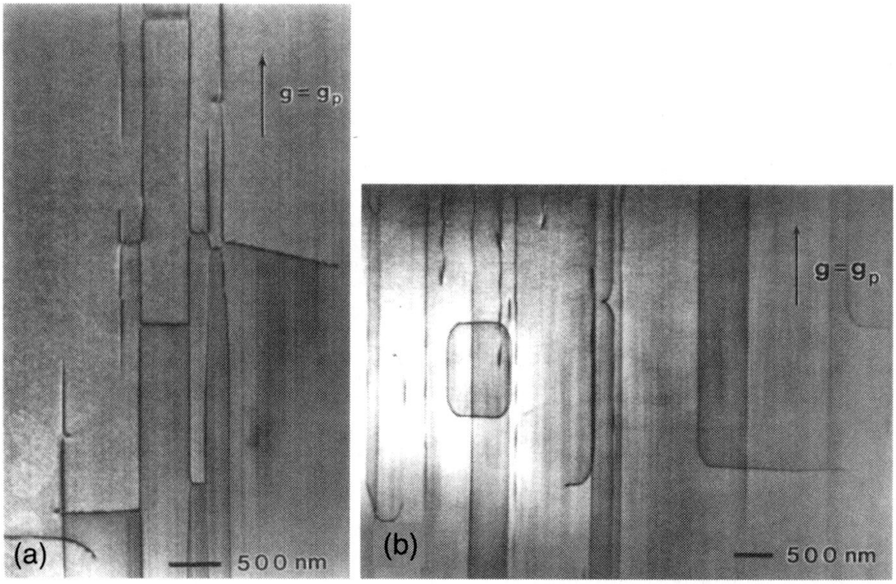

Fig. 10.27. Dislocation structures in decagonal basic Co-rich Al–Ni–Co single quasicrystals deformed in macroscopic compression. (**a**) $A_{\|}$ compression axis, 890°C, image plane containing the tenfold $\boldsymbol{g}$ vector g_{p}. (**b**) A_{45} compression axis, 790°C, image plane parallel to slip plane. $\boldsymbol{g}$ vector parallel to tenfold direction. From [723], courtesy of Peter Schall

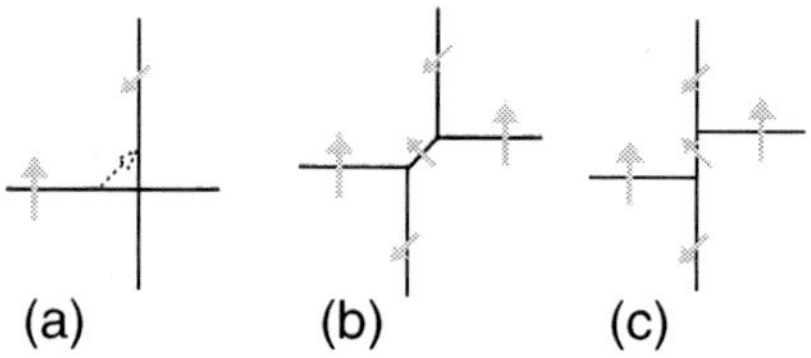

Fig. 10.28. Schematic of the reaction after [720, 723] between dislocations with periodic and mixed Burgers vectors in d-Al–Ni–Co to form a dislocation network. The Burgers vectors are indicated by grey arrows

Table 10.1. Densities of dislocations with periodic ϱ_{p}, mixed ϱ_{m}, and quasiperiodic Burgers vectors ϱ_{qp} in deformed Al–Ni–Co single quasicrystals in units of $10^{12}\ \mathrm{m}^{-2}$. ϱ_{ps} and ϱ_{pe} are densities of periodic screw and edge dislocations. Data from [723]

material, axis	T (°C)	ε (%)	ϱ_{p}	ϱ_{ps}	ϱ_{pe}	ϱ_{m}	ϱ_{qp}
Basic Co, $A_{\|}$	860	6	4.2			14.5	
Basic Co, A_{45}	860	6		0.6	0.2	0.5	
Al-15Ni-15Co, A_{45}	790	4.5		0.1	0.6	0.6	
Basic Co, $A_{\perp}$	860	6	1.2			17	19

dislocation densities are remarkably higher than those in undeformed but annealed reference specimens. They decrease with increasing temperature.

In situ straining experiments in an HVEM [666] yielded some characteristics of dislocation motion in the $A_{\|}$ orientation at 730°C. This temperature is relatively low compared with the temperatures of the conventional TEM experiments described above. Most dislocations observed moved on planes perpendicular to the periodic axis, like those indicated by arrows in Fig. 10.26a. These dislocations consist of two straight segments with a sharp knee between them. They move by pure climb since their Burgers vector is perpendicular to the plane of motion. As the tensile direction TD is also perpendicular to the plane of motion, the climb force acting on the dislocations from the applied load is maximum, but there is no glide force. These dislocations may be called periodic dislocations on the climb system and are equivalent to the dislocations of set A in i-Al–Pd–Mn. The following video illustrates their motion.

Video 10.6. *Motion of dislocations with periodic Burgers vectors in a* d-Al–Ni–Co *single quasicrystal:* The dominating feature is the continuous viscous motion of the dislocations in d-Al–Ni–Co. In some cases, dislocations move on planes with traces on the specimen surface parallel to those in Fig. 10.26a, but which have a different width, i.e., they are inclined with respect to the climb plane. In the first sequence, the plane of motion of the dislocation marked by a yellow circle changes continuously accompanied with a reduction of the width of the trace. The motion should then occur by a combination of climb and glide. In the second sequence, two dislocations of opposite sign annihilate each other. In the third sequence, periodic dislocations cross those of mixed Burgers vectors. The two types of dislocations do not seem to interact with each other. As described above, the reaction occurs only if the periodic Burgers vector component of the mixed dislocations is opposite to the Burgers vector of the periodic dislocations.

Figure 10.29 demonstrates that dislocations with periodic Burgers vectors may change between the climb plane, labeled c, and a plane with a trace inclined by about 18° with respect to the trace of the climb plane, labeled i. The places of the transition are marked by black arrows. In Fig. 10.29, these inclined planes are also imaged edge-on. With reference to the diffraction pattern in Fig. 10.6 (Fig. 10.29 is slightly rotated against Fig. 10.6), these planes may be indexed $(0\bar{1}104)$. The mode of motion on these planes is a combination of climb and glide. The inclined trace i in the middle of the figure changes into the direction of the Burgers vector or the tensile direction, marked by g, and back. The corresponding plane of motion is imaged edge-on, too. It may be indexed $(01\bar{1}00)$. This plane is a glide plane of these dislocations.

Similar processes are illustrated in the first sequence of the next video with another dislocation iteratively changing its plane of motion.

Video 10.7. *Climb and glide motion of a dislocation in* d-Al–Ni–Co*:* In the first part, the dislocation D extends approximately perpendicular to the periodic axis. The traces directly connected with the dislocation are parallel to the periodic direction. Afterwards, the dislocation shifts to a plane with inclined traces, which have the

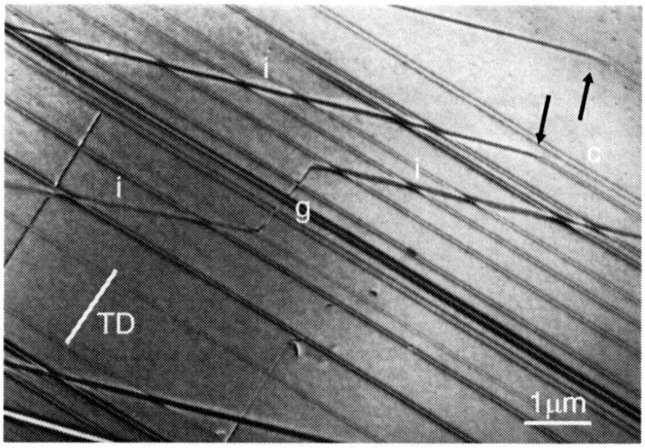

Fig. 10.29. Micrograph showing traces of periodic dislocations moving on different planes. From [666]. Copyright Mater. Res. Soc. (2005)

same orientation on the surface as the traces i in Fig. 10.29. Later on, it moves again on the same plane as at the beginning of the recording. In contrast to Fig. 10.29, the planes with the inclined trace and with the trace in the direction of the periodic axis are not imaged edge-on. Considering the ratio between the widths of the two traces, the corresponding planes of motion may be indexed $(0\bar{1}002)$ for the inclined trace and (01000) for the trace in the direction of the periodic axis. In principle, the planes of motion of a dislocation can change by glide and cross glide. The Burgers vector would then be parallel to the intersection line between the two planes, which is parallel to a quasiperiodic direction. These Burgers vectors are, however, extinguished under the present imaging conditions with a $\boldsymbol{g}$ vector in the periodic direction. Therefore, it is supposed that the dislocation also has a periodic Burgers vector, since, according to [720], only these dislocations assume line directions perpendicular to the periodic axis. Thus, the dislocation is a pure edge dislocation. Consequently, the mode of motion is a combination of climb and glide for the inclined traces, and it is a pure glide motion for the other one. The occurrence of the glide motion is surprising, since in the current geometry, there should be no glide force in a specimen with a homogeneous cross section. A glide force acting on the dislocation could be due to the inhomogeneous shape of the microtensile specimen. The second part of the video shows essentially the same. Now the dislocation changes between a plane which is probably inclined with respect to the tenfold axis and the one normal to it.

The dislocation velocities in the different modes of motion are quite different. Measurements from the video recording yield 1.1 and $7\,\mathrm{nm\,s^{-1}}$ for glide, and 325 and $285\,\mathrm{nm\,s^{-1}}$ for the combination of climb and glide. In the same specimen area, a dislocation moved on the climb plane at a velocity of $460\,\mathrm{nm\,s^{-1}}$. This shows that even for the periodic dislocations the glide velocity may be much lower than the climb velocity. However, because of the very different forces acting, the relation between the velocities does certainly not describe the relation of the dislocation mobilities on the different planes.

Both the macroscopic experiments in the $A_{\|}$ orientation and the in situ experiments show another set of dislocations with mixed Burgers vectors marked by M in Fig. 10.26a. They move on planes containing the periodic axis. The dislocation MCM expands by the curved segment C moving upwards thus trailing the long straight dislocations labeled M. Accordingly, the long straight dislocations parallel to the periodic axis do not represent the mobile parts of the mixed dislocations. The mode of motion can be determined only by indexing both the plane of motion and the Burgers vector. The determination of the Burgers vector was not possible during the in situ experiments because of the unsuitable imaging conditions with most $\boldsymbol{g}$ vectors (see above, Fig. 10.26b). As the plane of motion contains the tensile direction, neither a glide nor a climb force from the applied load acts on the dislocations in a macroscopic sense. Thus, these dislocations are equivalent to set B in i-Al–Pd–Mn. Dislocation velocities were measured from a video recording of a moving mixed dislocation segment as well as periodic dislocations on the climb system. At the same applied stress, the dislocations on the climb system are 20 times faster than the mixed dislocation.

As described above (Fig. 10.28), dislocations with periodic Burgers vectors can interact with dislocations with mixed Burgers vectors. In most cases, however, the moving periodic dislocations cut the mixed ones without a visible sign of reaction (third sequence in Video 10.6). However, sometimes the reaction takes place as in the following video.

Video 10.8. *Reaction between a dislocation with a periodic Burgers vector and one with a mixed Burgers vector in a* d-Al–Ni–Co *single quasicrystal:* The event is marked by a yellow dot. The periodic dislocation exhibits a sharp knee, like the dislocations marked by arrows in Fig. 10.26a. It moves to the left towards the mixed dislocation. The latter shows a similar curved shape as that in Fig. 10.26a. Both dislocations attract each other and join to form a segment of the product dislocation. At the same time, the climbing periodic dislocation segment is shifted onto a parallel climb plane (downwards in the video) due to the mutual elastic interaction. The latter segment is fixed at the junction with the product dislocation and starts to extend on the new climb plane (E). Later on, the segment indicated by the straight yellow line moves back (upwards in the figure) and removes the product dislocation. As a result, the mixed dislocation is restored and moves slowly upwards while the periodic dislocation resumes its motion on the new climb plane. The dark contrasts mark the traces of the dislocation motion.

In other cases, the same dislocation reaction results in the formation of a dislocation source, as demonstrated in the video.

Video 10.9. *Dislocation reaction and formation of a dislocation source in a* d-Al–Ni–Co *single quasicrystal:* Several dislocations with a periodic Burgers vector move on the climb plane from right to left close to the bowed part of the dislocation M with a mixed Burgers vector. After the mixed dislocation has slowly proceeded upwards, the periodic dislocation following P reacts with the mixed one to form a junction. Now, however, the segment of the periodic dislocation remains attached to the junction and draws a long dislocation towards the left side. After the new

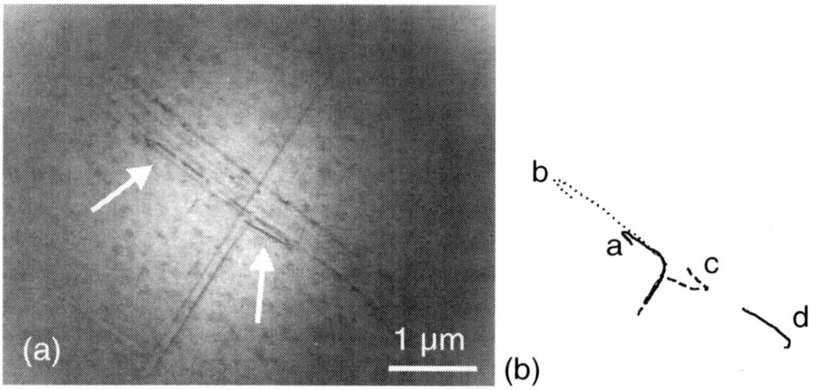

Fig. 10.30. Video frame of a dislocation climb source in d-Al–Ni–Co (**a**) and schematic of several stages of the source (**b**). From [666]. Copyright Mater. Res. Soc. (2005)

long segment of the periodic dislocation parallel to the specimen surface escaped through the surface, the remaining new dislocation moves out of the image area and the segment attached to the junction rotates to the right side of the mixed dislocation and expands there. Finally, the source has emitted a dislocation moving to the left whereas the rotating arm of the source has already turned to the right side (arrows), with the source being restored, as illustrated in Fig. 10.30a. Some stages of the operation of the source are schematically drawn in Fig. 10.30b. Afterwards, the source continues its revolutions. After each revolution, the rotating segment shifts slightly down onto a parallel plane. In the second part of the video, this spiral character of the source is clearly visible.

Dislocation sources operating with climbing dislocations are called Bardeen-Herring sources [724]. Several sources of this kind have been observed, emitting tens of dislocations each. Owing to the spiral character, the generated deformation bands grow in width.

During the observation of one source, the load was changed to study the dynamic dislocation behavior. From the video recordings, the frequency of the revolutions of the source f_{source} was determined as a function of the load F. In Fig. 10.31a, the results are plotted in a double-logarithmic scale. Each symbol corresponds to one revolution of the source. During the first part of the experiment, about 30 dislocations were emitted from the source. The frequency of the source decreased only slightly while the load remained almost constant (downward open triangles). Afterwards the specimen was partly unloaded until the source almost stopped to revolve. In the plot of Fig. 10.31a, this phase of the experiment is characterized by a linear decrease of the frequency of the source with a load or stress exponent $m'_{\text{sc ul}} = \mathrm{d}\ln(f_{\text{source}})/\mathrm{d}\ln F = \mathrm{d}\ln(f_{\text{source}})/\mathrm{d}\ln\sigma = 7.1$. Afterwards, the specimen was reloaded until fracture (full upward triangles). The figure shows that unloading and reloading are reversible over the main part of the load range. The reloading curve is linear, too, yielding a stress exponent of $m'_{\text{sc l}} = 4.4$.

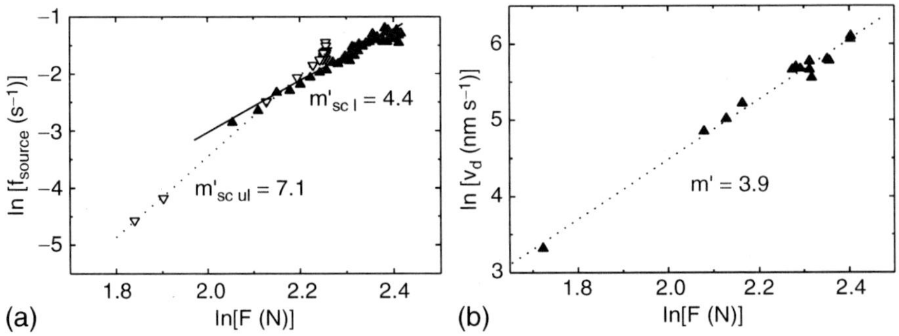

Fig. 10.31. Load dependence of the frequency of the operation of a dislocation source (**a**) and velocity of dislocations with periodic Burgers vectors moving away from the source (**b**). *Open downward triangles*: unloading; *full upward triangles*: re-loading. Data from [666]

For higher loads, the unloading and reloading curves do not coincide. In addition, the velocity v_d of the dislocations emitted to the left was determined from digitized video frames and plotted in Fig. 10.31b also as a function of the load. This curve is linear, too, yielding a load or stress exponent of the dislocation velocity of $m' = \mathrm{d}\ln v_\mathrm{d}/\mathrm{d}\ln F = \mathrm{d}\ln v_\mathrm{d}/\mathrm{d}\ln\sigma = 3.9$.

In summarizing this section, the following may be stated.

- At high temperatures, dislocations in icosahedral Al–Pd–Mn quasicrystals are mostly straight and oriented along crystallographic directions. They move on crystallographic planes.
- The motion is smooth in a viscous way with a stress exponent of about 4–6.
- Many dislocations move on planes where they experience a climb force but no glide force (set A), or even no force at all from the external load (set B). The determination of both the Burgers vectors and the plane of motion showed that climb is the dominating mode of dislocation motion in icosahedral quasicrystals.
- The dislocations drag phason walls which disappear at high temperatures by retiling.
- At low temperatures, the dislocations are curved.
- In decagonal quasicrystals, periodic dislocations may climb and glide. The mode of motion of the quasiperiodic and mixed dislocations is not yet clear.

10.4 Macroscopic Deformation Parameters

The whole methodology, which for more than 50 years had been designed to study the plastic deformation of crystals, was applied to quasicrystals within a few years. As diffusion is involved in the plastic deformation of quasicrystals, respective data, usual deformation experiments at ambient pressure are restricted to relatively high temperatures. The low-temperature experiments performed under confining hydrostatic pressure [702, 710] did not allow the measurement of detailed deformation data.

10.4.1 i-Al–Pd–Mn

Abundant deformation data are available on icosahedral quasicrystals including Al–Cu–Fe and Zn–Mg–Dy (e.g., [725, 726]) as well as on Al–Pd–Mn (e.g., [727–730]). The first intrinsic deformation data on icosahedral single quasicrystals were those in [727]. At present, the most comprehensive set of data of an icosahedral single quasicrystal is certainly that of i-Al-Pd-Mn in [16, 337, 709, 731]. The following description is based on these data. Of the low-temperature experiments performed under confining hydrostatic pressure [702, 710, 732], only the measurements on icosahedral Mg–Zn–Y in [732] yielded flow stress data. Accordingly, the yield stress is possibly independent of temperature near room temperature but decreases rapidly above about 100°C.

At ambient pressure and a standard strain rate of $\dot{\varepsilon}_{\mathrm{t}} = 10^{-5}\,\mathrm{s}^{-1}$, there is a brittle-to-ductile transition (BDT), that means most specimens break below about 630°C. The BDT temperatures for two other strain rates are indicated in Fig. 5.29. Typical stress–strain curves of icosahedral Al-21at%Pd-8.5at%Mn are shown in Fig. 10.32. Figure 10.32a presents an example of 487°C, which is almost the lowest temperature achieved without fracture. Since the specimens are very brittle at these low temperatures, low total strain rates were applied, mostly $10^{-6}\,\mathrm{s}^{-1}$. There appears a knee in the deformation curve, which marks a transition from almost elastic to elastic-plastic deformation with a very high rate of work-hardening. This transition is called the elastic limit and labeled EL in Fig. 10.32. It corresponds to an instantaneous increase in the plastic strain rate. As shown in Fig. 10.32a, the elastic limit is determined as the intersection between tangents to the stress–strain curve before the transition point (dotted line) and after it (dashed line). Below 550°C, the experiments were stopped at stresses around 1.5 GPa to prevent fracture. For measuring the strain rate sensitivity of the flow stress, stress relaxation tests R_{n} were carried out by keeping the total strain $\dot{\varepsilon}_{\mathrm{t}}$ constant. These tests increase the plastic strain at low strain rates. Additional plastic deformation can also be achieved under creep conditions at constant load (L).

Figure 10.32b shows the stress–strain curve for deformation at 580°C. With increasing strain, the curve bends down, passes a maximum, i.e., the upper yield point (UYP), and ends up in a range of almost steady state deformation. During steady state deformation, the work-hardening is balanced by recovery. The temperature in Fig. 10.32b is the lowest one with the upper yield point being overcome before sample failure. The stress at the elastic limit is much lower than the steady state flow stress. During the stress relaxation R_2 performed after the upper yield point, the stress relaxes more strongly than during the relaxation R_1 in the initial work-hardening range. Both relaxations

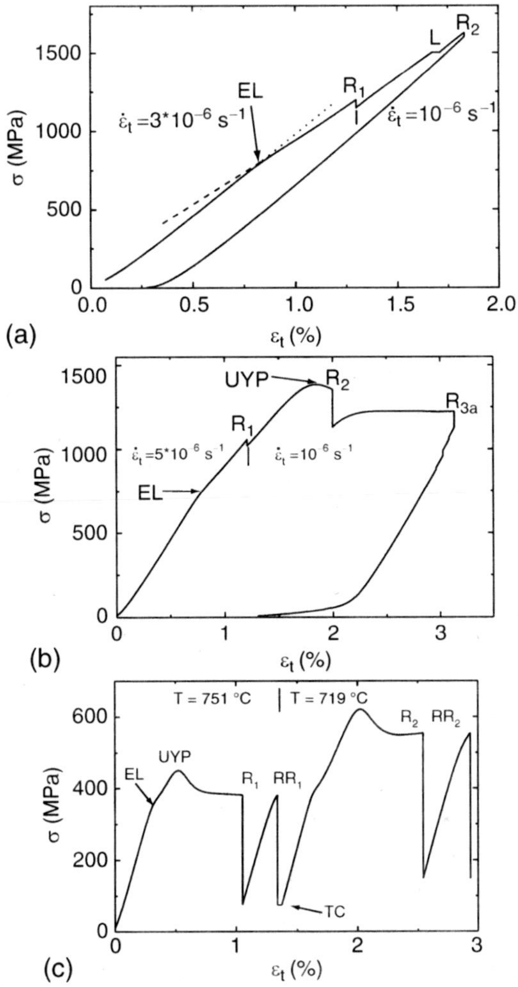

Fig. 10.32. Stress–strain curves of i-Al–Pd–Mn at different temperatures including stress relaxation curves R_n. (**a**) $T = 487°\mathrm{C}$, $\dot{\varepsilon}_t = 3 \times 10^{-6}\,\mathrm{s}^{-1}$ and change to $10^{-6}\,\mathrm{s}^{-1}$ as well as some deformation under constant load (L). (**b**) $T = 580°\mathrm{C}$, $\dot{\varepsilon}_t = 5 \times 10^{-6}\,\mathrm{s}^{-1}$ and change to $10^{-6}\,\mathrm{s}^{-1}$. (**c**) $T = 751°\mathrm{C}$ and change to $T = 719°\mathrm{C}$, $\dot{\varepsilon}_t = 10^{-4}\,\mathrm{s}^{-1}$. Data from [337]

were stopped at about $\ln[\dot{\sigma}(\mathrm{MPa\,s^{-1}})] = -3$, corresponding to a plastic strain rate of about $5 \times 10^{-7}\,\mathrm{s}^{-1}$.

Turning to higher temperatures, the steady state flow stress decreases drastically, so that it becomes only slightly higher than the elastic limit, as demonstrated in Fig. 10.32c for 751°C at a strain rate of $\dot{\varepsilon}_t = 10^{-4}\,\mathrm{s}^{-1}$. In this experiment, a temperature change experiment TC was performed down to the lower temperature of 719°C. This is connected with an increase in the flow stress. During the second deformation into the steady state, a yield drop effect again occurs. Apart from these changes of the deformation conditions, the flow stress reaches a steady state value after passing the yield drop, which is characteristic of the deformation at high temperatures. Such deformation curves are described in most papers on quasicrystal deformation. At large strains, the true flow stress may decrease (e.g., [728–730]). In many papers, this softening is considered a characteristic feature of quasicrystal deformation. Not all experiments exhibit a distinct initial range of high work-hardening. It is pronounced in specimens with a twofold compression axis.

Figure 10.33 summarizes the temperature dependence of the elastic limit and of the steady state flow stress. The latter is chosen since the interpretation is more simple for data measured during steady state. The flow stresses are quite isotropic, i.e., there is practically no difference between deformation along the twofold and fivefold compression axes. Both the elastic limit and the steady state flow stress decrease with increasing temperature. As mentioned above, the steady state flow stress is only slightly higher than the stress at the elastic limit for temperatures above about 650°C and a strain rate of $10^{-5}\,\mathrm{s}^{-1}$. For lower temperatures, however, the difference between both increases with decreasing temperature. At a total strain rate of $10^{-6}\,\mathrm{s}^{-1}$, the elastic limit becomes equal to the steady state flow stress at 700°C and is not observed above this temperature. The small open symbols below 570°C, where steady

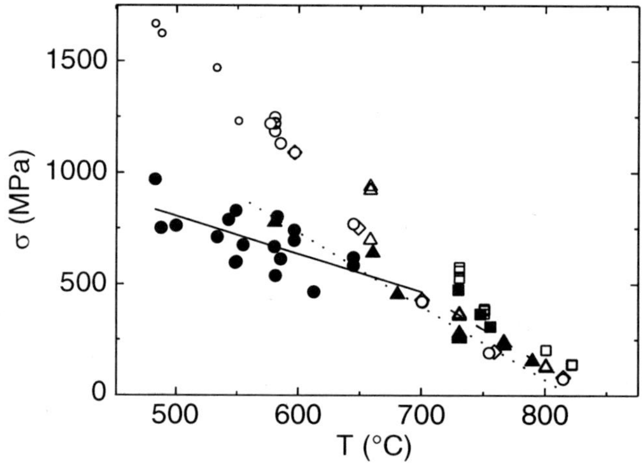

Fig. 10.33. Temperature dependence of the steady state flow stress (*open symbols*) and the elastic limit (*full symbols*). The four small open symbols at low temperatures correspond to the maximum stresses reached in the experiments where steady state flow was not achieved. The total strain rates $\dot{\varepsilon}_t$ are $10^{-6}\,\mathrm{s}^{-1}$ (*circles*), $2 \times 10^{-6}\,\mathrm{s}^{-1}$ (*diamonds*), $10^{-5}\,\mathrm{s}^{-1}$ (*triangles*), and $10^{-4}\,\mathrm{s}^{-1}$ (*squares*). The straight lines represent linear regression curves through the elastic limits measured at $10^{-6}\,\mathrm{s}^{-1}$ (*full line*) and $10^{-5}\,\mathrm{s}^{-1}$ (*dotted line*) as well as the steady state flow stresses at higher temperatures at $10^{-5}\,\mathrm{s}^{-1}$ (*dashed line*). Data from [337]

state deformation was not achieved because of the brittleness of the specimens, correspond to the maximum stresses reached in these deformation tests still before the upper yield point.

In the initial range of high work-hardening, the work-hardening coefficient $\theta = \mathrm{d}\sigma/\mathrm{d}\varepsilon$ is in the range of the elastic stiffness modulus S of the specimen and the parts of the deformation machine inside the strain measurement circuit. θ increases with decreasing temperature. As described in Sect. 2.1, S is measured from the slope of the unloading curves at the end of the experiments. The stiffness values increase with decreasing temperature between about 80 and 120 GPa. Using these values results in $\theta \approx S$ at high temperatures, and $\theta \approx 3S$ at low ones. These are very high values of work-hardening. They will be interpreted in Sect. 10.5.8.

As indicated in Fig. 10.32, stress relaxation tests were performed during all experiments to study the strain rate sensitivity r of the flow stress. Figure 10.34a presents three relaxation curves measured above the elastic limit within the work-hardening range at the low temperature of 547°C. These

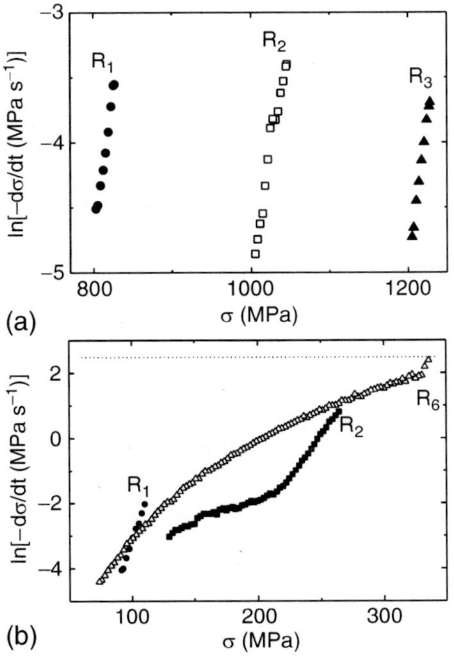

Fig. 10.34. Stress relaxation curves of i-Al–Pd–Mn taken in different stages of deformation and at different temperatures. (**a**) $T = 547°\mathrm{C}$. Relaxations taken above the elastic limit at different plastic strains. The total strain rate $\dot{\varepsilon}_t$ before the relaxations corresponds to $\ln(-\dot{\sigma}) = -2.23$. (**b**) $T = 760°\mathrm{C}$. R_1 relaxation taken below the elastic limit, R_2 taken above the elastic limit within the work-hardening range, R_6 first relaxation in the steady state range of deformation. The dotted horizontal line corresponds to the total strain rate $\dot{\varepsilon}_t$ before the relaxations. Data from [337]

curves are almost straight and shifted along the stress axis corresponding to the work-hardening during deformation. There is no sign of recovery, which is confirmed by the creep test L under constant load in Fig. 10.32a, where the creep rate continuously decreases. Figure 10.34b shows respective curves taken from an experiment in the high-temperature range at 760°C. The first relaxation R_1 is performed below the elastic limit. It is steep, like the curves at low temperatures, but demonstrates that at high temperatures the specimens deform plastically at a low rate even in the "elastic" range. At low temperatures, practically no relaxation occurs at much higher stresses. Above the elastic limit, in the work-hardening range, the relaxation curve R_2 still starts with a steep range before it bends to a second range of low slope. This transition causing an inverse curvature of the relaxation curves is due to recovery becoming important at low relaxation rates. Curve R_6 is the first one measured in the range of steady state deformation. The steep initial range has almost disappeared. Further curves exhibit only the flat part. Strain rate sensitivity data were always measured from the slope at the beginning of the curves, corresponding to the strain rate before the relaxations.

Because of the very high values of the work-hardening coefficient in the work-hardening range of the deformation curves, the work-hardening during the relaxation tests pronouncedly decreases the relaxation rate, which does not represent the stress dependence of the activation parameters of the thermally activated process. Therefore, the strain rate sensitivity data have to be corrected according to

$$r = r_{\text{ex}} \left(1 + \frac{\theta}{S}\right) . \tag{10.11}$$

With the values of θ and S quoted above, this correction is essential. In the high-temperature range, the situation is different. It was described in Sect. 10.3.1 that part of the deformation-induced dislocations recovers in the time scale of usual experiments. Thus, the microstructure may change in the opposite sign during the stress relaxation tests at high temperatures. To study these changes, after the original stress relaxation curves labeled R_{n} in Fig. 10.32c, "repeated" stress relaxation curves were measured by loading the sample after an original relaxation just up to or below the starting stress of the original relaxation without reaching the steady state deformation again. Such repeated relaxation tests are labeled RR_{n}. As shown in Fig. 2.3 in Sect. 2.1, the repeated curves are below the original ones, which indicates the microstructural changes during the original relaxation tests. These effects underline the importance of the evolution of the microstructure, i.e., essentially of the dislocation density. The changes in the microstructure have to be taken into account in evaluating the activation volume via (4.9) by introducing a correction factor

$$r \approx \frac{r_{\text{ex}}}{1 - C_{\text{r}}} , \tag{10.12}$$

where

$$C_{\text{r}} = \Delta \ln(-\dot{\sigma})|_{\text{r}} / \Delta \ln(-\dot{\sigma})|_{\text{t}} ,$$

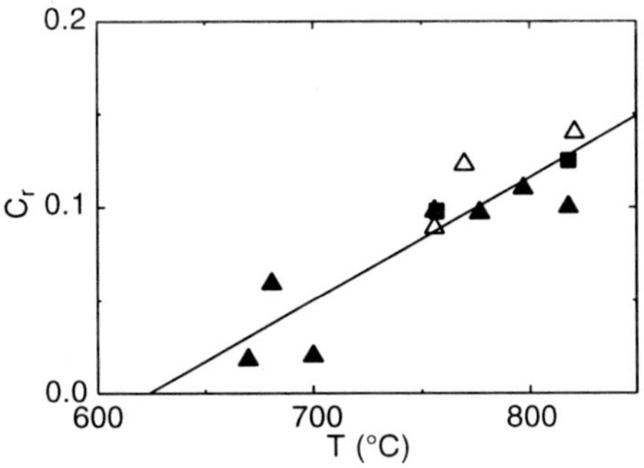

Fig. 10.35. Dependence of the recovery correction factor C_r on the temperature. $\dot{\varepsilon}_t = 10^{-5}\,\mathrm{s}^{-1}$ (*triangles*), $10^{-4}\,\mathrm{s}^{-1}$ (*squares*), fivefold compression axis (*full symbols*), twofold compression axis (*open symbols*). Data from [16]

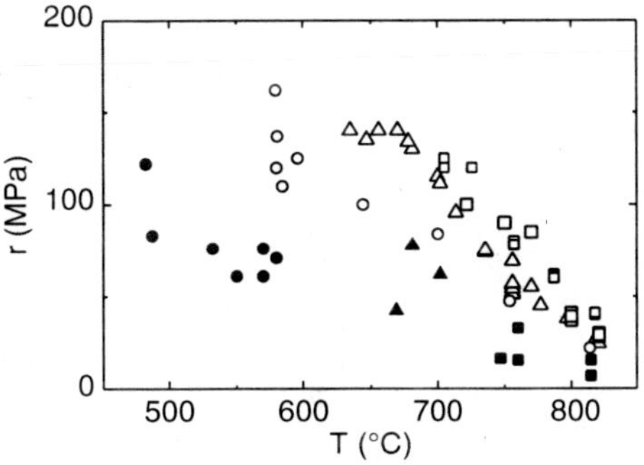

Fig. 10.36. Strain rate sensitivity of the flow stress versus temperature. *Full symbols*: stress relaxation tests during the (initial) work-hardening range, the experimental data are corrected for work-hardening; *open symbols*: relaxation tests during steady state deformation, the data are corrected for recovery transients. Strain rate symbols as in Fig. 10.33. Data from [16, 337]

with the respective quantities being defined in Fig. 2.3. As presented in Fig. 10.35, this correction factor decreases with decreasing temperature and (for $\dot{\epsilon} = 10^{-5}\,\mathrm{s}^{-1}$) becomes zero at about 625°C, i.e., at the temperature of the brittle-to-ductile transition. The shift $\Delta\sigma|_r$ of the beginning of the relaxation curves along the stress axis in Fig. 2.3 is almost equal to the yield point appearing during the transient to the steady state deformation at reloading. The effect of the evolution of the microstructure on the flow stress and the determination of the activation parameters at high temperatures are discussed in detail in [16, 731].

All strain rate sensitivity data are compiled in Fig. 10.36. The data from the steady state deformation (open symbols) increase with decreasing temperature and increasing total strain rate. Those obtained from the work-hardening range (full symbols) show the same tendency but are lower than the

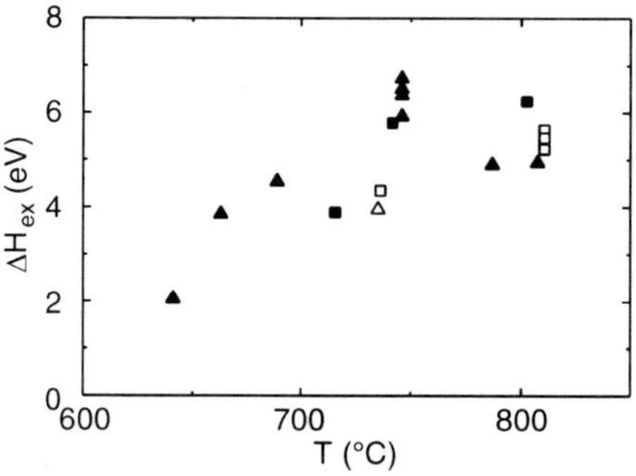

Fig. 10.37. Activation enthalpy of the deformation of i-Al–Pd–Mn single quasicrystals in the high-temperature range. Symbols as in Fig. 10.33. Data from [731]

steady state ones. This is partly due to the much lower strain rates but mainly to different mechanisms in the work-hardening and steady state ranges. The strain rate sensitivity can also be expressed in terms of the stress exponent $m' = \sigma/r$ (4.12). It ranges from about 3 at the high temperature of 800°C to about 6 at the brittle-to-ductile transition at 630°C, and to more than 10 in the low-temperature range without steady state deformation at 550°C.

Activation energies were determined by temperature change tests and applying (4.14). In the high-temperature range with steady state deformation, the stress increments were read after steady state was reached again. Both the temperature sensitivity and the strain rate sensitivity were not corrected for transient effects since they partly cancel each other [731]. The resulting activation enthalpy data are plotted in Fig. 10.37. The values are very high, particularly at high temperatures, which would lead to very low activation rates when inserted into an Arrhenius equation. As it will be discussed later, the Gibbs free energies of activation are much lower.

A few temperature change experiments were also performed in the range of high work-hardening above the elastic limit. The stress increments were read between relaxation curves at a fixed relaxation rate. They were corrected for work-hardening. At 564°C, a nominal strain rate of $10^{-6}\,s^{-1}$ and a stress of $\sigma = 918\,\text{MPa}$, the activation enthalpy amounts to $\Delta H \approx 3\,\text{eV}$. This is much higher than the values of steady state deformation in Fig. 10.37 at even higher temperatures.

10.4.2 d-Al–Ni–Co

In contrast to icosahedral Al–Pd–Mn which exhibits isotropic plastic behavior, decagonal quasicrystals show a pronounced plastic anisotropy as demonstrated first in d-Al–Ni–Co in [721] and afterwards also in d-Al–Cu–Co [733], and d-Al–Ni–Co again [734, 735]. The quasicrystals were deformed in the $A_{\|}$, A_{45},

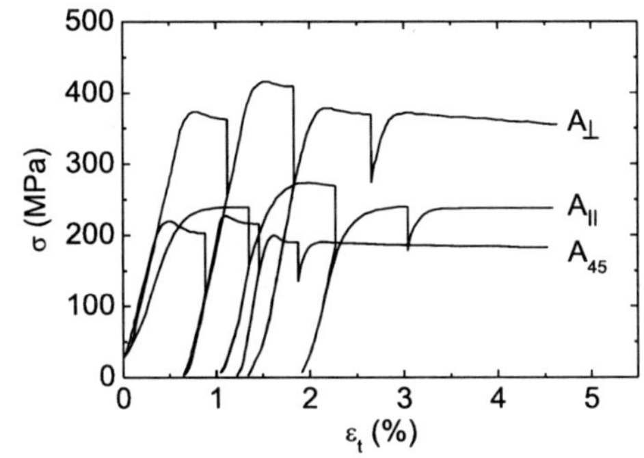

Fig. 10.38. Stress–strain curves of d-Al–15at%Ni–15at%Co at 790° C and a nominal strain rate of $10^{-5}\,s^{-1}$ with the $A_{\|}$, A_{45}, and $A_{\perp}$ compression axes. The deformation curves are interrupted by stress relaxation and temperature change experiments. Data from [735]

and $A_{\perp}$ orientations defined in Sect. 10.3.2. The $A_{\perp}$ axis is parallel to the P2 quasicrystal orientation. The following data are from d-Al–15at%Ni–15at%Co [735], the same material as that used for the in situ deformation study in an HVEM described above.

Figure 10.38 presents the stress–strain curves of the three orientations measured at 790°C. Comparing these curves with those of i-Al–Pd–Mn indicates that the deformation was performed in the high-temperature range in which upper and lower yield points clearly develop. The lowest flow stress occurs in the A_{45} orientation, and the highest one along $A_{\perp}$. The exact relations between the three orientations depend on the temperature and the crystal structure within the decagonal phase.

The strain rate sensitivity of the flow stress was determined by stress relaxation tests. Assuming an orientation factor of 0.5 for all orientations, the results can be represented by a hyperbolic dependence of the experimental activation volume on the flow stress, according to $V = 185\,\text{nm}^3\,\text{MPa}\,/\,\sigma$ or by an experimental stress exponent m' decreasing from about 6 at 700°C down to 4.5 at 850°C, both independent of the orientation of the loading axis.

The temperature dependence of the lower yield points is presented in Fig. 10.39 for the three orientations. These data are comparable to those of the steady state flow stresses for i-Al–Pd–Mn. The curves for the $A_{\|}$ and A_{45} compression axes do not essentially differ. All three curves have approximately the same slope except of an anomaly at 730°C in the curve for the $A_{\|}$ orientation. The activation enthalpies resulting from temperature change tests and the strain rate sensitivity data are quite independent of the temperature. For $A_{\|}$, the ΔH value of 3.9 eV is considerably lower than that of 5.8 eV for A_{45}. Unreasonably high values > 10 eV were obtained for temperatures above 790°C.

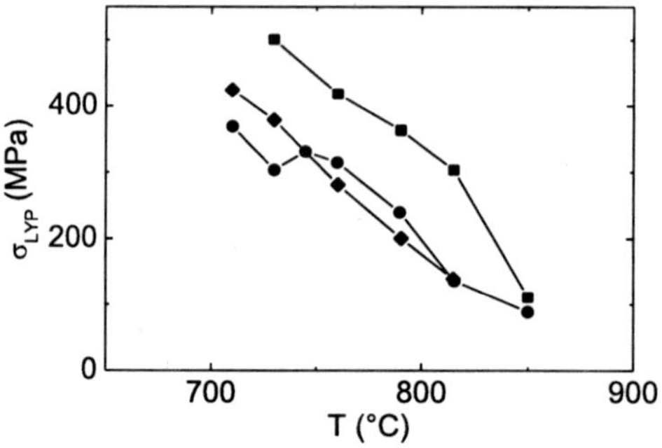

Fig. 10.39. Temperature dependence of the lower yield points of d-Al–15at%Ni–15at%Co single crystals. $A_{\parallel}$ (*circles*), A_{45} (*diamonds*), and $A_{\perp}$ orientation (*squares*). Data from [735]

In conclusion, the following may be said.

- Quasicrystals can plastically be deformed at temperatures above about 75% of the melting temperature.
- Then, the stress–strain curves exhibit an upper and a lower yield point followed by a range of almost steady state deformation, sometimes also of work-softening.
- At lower temperatures, icosahedral Al–Pd–Mn exhibits a transition from elastic to plastic deformation with a very high work-hardening coefficient.
- The plastic behavior is isotropic for icosahedral quasicrystals but anisotropic for decagonal ones. Markedly higher flow stresses appear for compression perpendicular to the periodic axis.
- The strain rate sensitivity of the flow stress can be characterized by intermediate values of the stress exponent m' in the range between about 3 and 10, with the higher values belonging to lower temperatures.
- At high temperatures, the activation enthalpies are high (6 eV), but they decrease with decreasing temperature. In d-Al–Ni–Co, the enthalpy for deformation along the periodic axis is lower than that along an axis inclined by 45° to it.

10.5 Mechanisms of Dislocation Motion and Plastic Deformation

Although abundant data have been accumulated about dislocation properties and plastic deformation of quasicrystals, certain information is still missing. For example, the stress dependence of the velocity of individual dislocations has not been measured yet. Besides, dislocations probably move in high super- or subsaturations of point defects causing a chemical stress, which is not proved experimentally. This makes it difficult to separate the effects of the dynamics of dislocation motion from those of the development of the microstructure in which the dislocations move. Accordingly, dislocation properties and macroscopic deformation data will be discussed in parallel.

10.5.1 Glide or Climb Motion of Dislocations

It's amazing how long it can take to see the obvious. But of course it's only obvious now.

- Ian Stewart, Mathematician

While the deformation curves of different quasicrystals measured in the high-temperature range are quite similar, irrespective of the particular chemical composition and quasicrystal structure, the ranges of the temperature and flow stresses scatter widely. However, Takeuchi showed that in a plot of the upper yield stress versus the temperature, the curves of the different icosahedral quasicrystals approach each other, if the axes are suitably normalized [736]. The scaling relations are $\sigma'_{\mathrm{y}} = \sigma_{\mathrm{y}}/E$ and $T' = T/(E\bar{a}^3)$, where E is the bulk modulus, and $\bar{a}$, an average of the Goldschmidt diameters of the atoms constituting the quasicrystal. The possibility of normalizing the temperature dependence of the flow stress suggests that similar mechanisms control the plastic deformation in different quasicrystalline materials.

Since the first direct observation of dislocation motion in an icosahedral quasicrystal in 1995 in [647], it had been assumed that the main mode of dislocation motion is glide, although there had been hints that this may be wrong. As described in Sect. 10.3.1, in situ straining experiments by the group of Caillard have clearly shown that the motion of dislocations with quasiperiodic Burgers vectors is dominated by climb [701, 704]. In decagonal Al–Ni–Co quasicrystals deformed along the tenfold axis, dislocations with a periodic Burgers vector move also by pure climb (Sect. 10.3.2).

Nevertheless, some combination of climb and glide seems to be possible, too. In the Videos 10.1 and 10.3, dislocations change the projected width of their planes of motion without changing the orientation of the traces of the planes, as also illustrated in Fig. 10.19. For geometrical reasons, this is only possible by a combination of climb and glide. In the post-mortem TEM study of samples deformed at 300°C, some dislocations had moved in a mixed mode [702]. Thus, some contribution of glide to the deformation should also be possible.

While many dislocations have zero orientation factors for glide and maximum ones for climb (set A), there also exist dislocation bands with zero orientation factors for both glide and climb (set B) like those presented in Fig. 10.12. As it will be discussed in Sect. 10.5.8, the latter dislocations may climb under a chemical stress originating from sub- or supersaturations of point defects, which are generated by the dislocations climbing under the external stress.

The slip plane of a dislocation is a selected plane of conservative motion with respect to the volume of the system, i.e., $\Delta V = 0$ in (3.3). The question arises: what makes the climb plane a preferred plane of dislocation motion in quasicrystals? This problem is discussed in [687]. The dislocation motion is considered to occur in a first step by the motion of the phonon field, characterized by $\boldsymbol{b}_{\parallel}$, followed by retiling in a second step as illustrated in Fig. 10.10. The quasicrystalline vectors of forward motion of the dislocation $\boldsymbol{x}_{\parallel}$ and the line direction $\boldsymbol{\xi}_{\parallel}$ in parallel space have single images $\boldsymbol{x}_{\perp}$ and $\boldsymbol{\xi}_{\perp}$ in orthogonal space. To respect the tiling matching without overlaps and empty spaces, a four-dimensional lattice embedded in a subspace $(x_{\parallel}, x_{\perp}, \xi_{\parallel}, \xi_{\perp})$ has to be kept invariant during the shift of the dislocation by $\boldsymbol{b}_{\parallel}$. This is achieved for $\boldsymbol{b}_{\perp} \perp \boldsymbol{x}_{\perp}$ and $\boldsymbol{b}_{\perp} \perp \boldsymbol{\xi}_{\perp}$, implying also that $\boldsymbol{b}_{\parallel} \perp \boldsymbol{x}_{\parallel}$ and $\boldsymbol{b}_{\parallel} \perp \boldsymbol{\xi}_{\parallel}$. Thus, the mode of dislocation motion maintaining the internal tiling coherent is pure climb of a pure edge dislocation. This result is illustrated in Fig. 10.40, based on the simulation of the glide motion in a two-dimensional decagonal quasicrystal [685]. For glide in the lower part of the figure, a high density of phason faults are generated in the wake of the dislocation, which cannot be retiled by the tiles used for building the regular structure. For climb in the upper part of the figure, the quasicrystal halves can be glued together without overlaps or open spaces by removing or adding so-called worms indicated by the grey band. Although the climb motion by shifts of $\boldsymbol{b}_{\parallel}$ leaves the lattice coherent, a phason wall is formed, too, which requires diffusional processes to be annihilated by retiling. Thus, climb should be the easier mode of dislocation motion as long as the dislocation does not move as a perfect dislocation in equilibrium with its phason field. The preference of crystallographic dislocation orientations and planes of motion is possibly supported by particular core configurations, e.g., the splitting observed in Fig. 10.17.

In decagonal Al–Ni–Co quasicrystals, the activation enthalpy at compression along $A_{\parallel}$ is by almost 2 eV lower than that along A_{45} (Sect. 10.4.2). In the $A_{\parallel}$ orientation, the deformation undoubtedly is carried by climb of

Fig. 10.40. Lattice mismatch produced by climb and glide. The tiling pattern with the phason wall generated by glide is from [685], copyright Taylor & Francis Ltd. (1995) (http://www.informaworld.com)

dislocations with periodic Burgers vectors. In the A_{45} orientation, periodic screw dislocations dominate, which should have moved preferentially by glide (Sect. 10.3.2). Although the flow stress is lower for glide, particularly if the different orientation factors for the two loading axes are considered (1 for climb in $A_{\|}$ and 0.5 for glide in A_{45}), the activation enthalpy is higher (for calculating the activation enthalpy, the orientation factor cancels in (4.14)). Thus, even for periodic dislocations which do not drag phason faults, the activation energy of glide is higher than that of climb.

The confirmation of climb as the main mode of dislocation motion makes it necessary to re-formulate all models of dislocation dynamics and plastic deformation of quasicrystals.

10.5.2 Components of the Flow Stress

For quasicrystals, in addition to the thermal and athermal components of the flow stress in (5.15) further contributions have to be considered

$$\sigma = \sigma_{\mathrm{ph}} + \sigma_{\mathrm{i}} + \sigma^{*} + \sigma_{\theta}. \tag{10.13}$$

This equation is written in terms of compressive or tensile stresses since the appropriate orientation factors are not always clear. The orientation factor m_{s} for glide may assume a maximum value of 0.5. In earlier papers of the author and his coworkers interpreting the plastic behavior of icosahedral quasicrystals by glide, e.g., [337,731], a value of $m_{\mathrm{s}} = 0.4$ was applied. For determining the climb force, the extended Peach–Koehler formula (10.10) has to be applied. Considering only normal stress components, an orientation factor for climb m_{c} may be introduced, which for dislocations with $\boldsymbol{b}_{\|}$ parallel to the loading axis moving on a plane perpendicular to it, as it has been observed experimentally, is $m_{\mathrm{c}} = 1$. Since also in the mixed mode of dislocation motion climb is dominating, $m_{\mathrm{c}} = 0.8$ may be taken as a suitable average value.

In (10.13), the first term results from the formation of phason walls trailed behind imperfect dislocations. The second term is the athermal contribution of long-range dislocation interactions. The third one represents the thermal stress component which originates from the thermally activated processes controlling the dislocation mobility. The last term describes the stress contribution causing the very strong work-hardening observed at low temperatures above the elastic limit, e.g., in Figs. 10.32a and b.

10.5.3 Formation of Phason Faults

As outlined in connection with Fig. 10.10, dislocation motion in quasicrystals is accompanied with the rearrangement of the phason disorder. In computer simulation experiments, the trailing of phason walls by dislocations gliding at low temperatures was studied in [699,737]. A recent simulation of glide at high temperatures in a three-dimensional quasicrystal approximant demonstrates that the phason wall is relatively sharp directly behind the dislocation but broadens afterwards [738]. Up to now, there is no theoretical model of the dislocation motion by climb. The energy per area γ_{ph} necessary to create the phason wall causes a back stress on the moving dislocation given by

$$\sigma_{\mathrm{ph}} = \gamma_{\mathrm{ph}} / \left(b_{\|}\, m_{\mathrm{c}}\right) .$$

As mentioned in Sect. 10.2.3, the shear stress to trail the phason wall by glide is very high ($\tau_{\mathrm{ph}} \approx (1/27)\,\mu$) [685], in accordance with the strongly defective lattice structure behind the moving dislocation. On the other hand, values of γ_{ph} of the order of magnitude of $25\,\mathrm{mJ\,m^{-2}}$ were concluded from splitting widths of dislocations by climb at 300°C (by analogy with the determination of the stacking fault energy (3.51)) yielding a back stress of the phason faults of only about $\sigma_{\mathrm{ph}} = 30\,\mathrm{MPa}/m_{\mathrm{c}}$ [702]. Such a stress would be important above about 750°C, but is negligible at low temperatures.

The experimental situation about the occurrence of phason walls behind the moving dislocations is slightly controversial (Sect. 10.3.1). In post-mortem TEM studies of macroscopically deformed specimens, phason faults have never been observed above 610°C, whereas during in situ straining in a TEM, they have sometimes been detected up to about 700°C (Fig. 10.21) or even 750°C [712]. From X-ray scattering experiments, the activation energy of the annealing of phason fluctuations of 2.3 eV is of the order of magnitude of the diffusion energies of Pd and Mn [682], which seems to be reasonable since the process is connected with diffusion. This enables effective phason annealing at 650°C, which roughly agrees with the in situ annealing study of Fig. 10.22. On the other hand, the time the phason walls need to disappear during in situ annealing in the TEM can be described by an activation energy of about 4.3 eV [712], which is unreasonably high. It is argued in [684] that a dislocation may be considered to move in the perfect state if the length of the trailed phason wall $l = v_{\mathrm{d}}\, t_{\mathrm{l}}$ is shorter than about 1 nm. Here, v_{d} is the dislocation velocity, and t_{l}, the lifetime of the phason wall. Using the data from [712], this occurs above 800°C for strain rates $\geq 5 \times 10^{-5}\,\mathrm{s}^{-1}$. Thus, there may be different situations depending on the true activation energy of phason dispersion. Either phason walls are dragged effectively only below some temperature between about 600 and 650°C, where the respective stress component is unimportant, or the creation of phason walls has to be considered up to quite high temperatures with the dragging stress being remarkable. In the first case, the onset of (Pd and Mn) diffusion and of phason mobility may cause the transition from brittle to ductile behavior around about 630°C.

The question is not clear whether the dragging of a phason wall is a thermally activated process, or not. Since there is some similarity to the creation of a stacking fault, the formation of a phason wall may be considered of athermal character. σ_{ph} is then a constant independent of temperature and strain rate.

10.5.4 Long-Range Dislocation Interactions

The model of Taylor hardening (5.11) in Sect. 5.2.1 to estimate the long-range internal stress contribution σ_i to the flow stress from the dislocation density ϱ has to be modified if the parallel dislocations bypass each other by climb instead of glide, which is usually considered. With the parallel part $\boldsymbol{b}_{\parallel}$ of the Burgers vectors of the edge dislocations parallel to the x axis and with the dislocation lines parallel to z as in Fig. 3.8, the climbing dislocation now moves in y direction on a yz plane with $x = x_0$. The respective stress component of the dislocation fixed to the origin is the normal stress σ_{11} from the first line of (3.11). Applying an equivalent normalization as in Fig. 3.14, namely y/x_0 for the coordinate of motion and $2\pi(1-\nu)x_0 f_{cy}/(\mu b^2)$ for the force, yields the dependence of the climb force on y/x_0 plotted in Fig. 10.41. The force is zero at $y = 0$ and has extrema at $y \approx \pm 0.681\, x_0$. The maximum values are

$$f_{cym} = \pm 1.1 \frac{\mu b_{\parallel}^2}{2\pi(1-\nu)y_0} \tag{10.14}$$

instead of (3.26) for the glide of screw dislocations. Using again (5.10), the Taylor equation (5.11) may then read

$$\sigma_i = \frac{1.1 \times 2\pi}{1-\nu} \frac{\mu b_{\parallel}}{2\pi m_c} \sqrt{\varrho} = \alpha_c \frac{\mu b_{\parallel}}{2\pi m_c} \sqrt{\varrho}. \tag{10.15}$$

A proper set of parameters may be $b_{\parallel} = 0.3\,\mathrm{nm}$ as in [739], $m_c = 0.8$ as suggested above, and $\alpha_c = 2\pi$, which may be justified by considering that $\alpha_c/\alpha_p = 1.1 \times 2/(1-\nu)$ following from comparing (5.11) with (10.15). Thus, the mutual interaction between the climbing edge dislocations is much stronger than that for gliding screws. The temperature dependence of the shear modulus measured in [740] can be represented by

$$\mu(T) = 77.745\,\mathrm{GPa} - 0.014\,\mathrm{GPa\,K^{-1}}\,T - 1.3 \times 10^{-5}\,\mathrm{GPa\,K^{-2}}\,T^2, \tag{10.16}$$

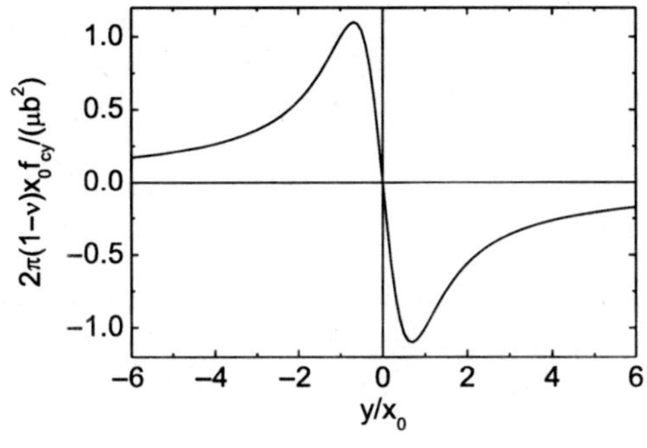

Fig. 10.41. Normalized interaction climb force acting on an edge dislocation bypassing another one at distance x_0 by climb

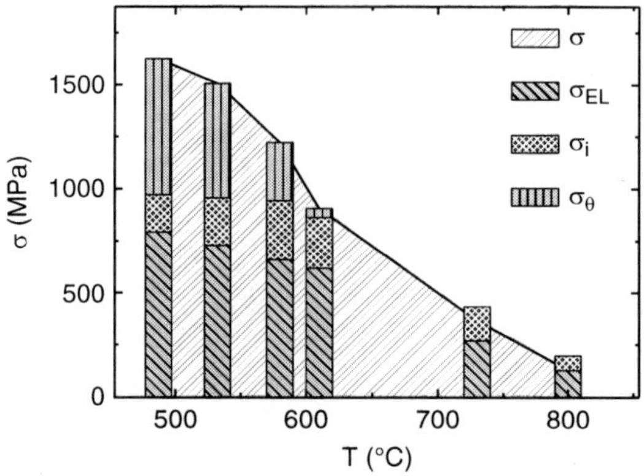

Fig. 10.42. Comparison of different components of the flow stress of i-Al–Pd–Mn single quasicrystals. σ total stress, σ_{EL} elastic limit, σ_{i} athermal stress, σ_θ stress component causing strong work-hardening. Data from [337]

where T is measured in K. The above data set yields almost the same values of σ_{i} as those estimated in [337]. The values based on the dislocation density data of Fig. 5.29 are plotted in Fig. 10.42 as dotted parts of the columns. At high temperatures, averages of the data from [338] and [337] are used. This athermal stress component is important at high temperatures in contrast to the estimation in [739] based on $\alpha_{\mathrm{c}} = 0.7\,\pi$.

In the high-temperature range, the dislocations form a three-dimensional network with segment lengths of $L \approx 0.5\,\mu\mathrm{m}$ (Sect. 10.3.1). For macroscopic deformation, the segments have to bow out beyond the half-circle (Frank–Read) configuration in Fig. 3.20c with the radius $r = L/2$. In analogy to (3.46), the necessary stress is given by $\sigma_{\mathrm{i}} = 2\Gamma/(Lb\,m_{\mathrm{c}})$. For the prismatic edge dislocation loops, the line tension Γ can be replaced by the line energy of (3.46). With a shear modulus of 48 GPa for 800°C, $r_0 = b_{\|}$, $\nu = 0.254$, and the constant in the logarithmic factor (3.46) $C = -0.59$ for the semihexagon, it follows that $\sigma_{\mathrm{i}} = 53\,\mathrm{MPa}$. This is of the same order of magnitude as the high-temperature value of $\sigma_{\mathrm{i}} = 70\,\mathrm{MPa}$ in Fig. 10.42. This confirms that the athermal component of the flow stress from long-range dislocation interactions should not be neglected at high temperatures. Possibly, also the stress σ_{ph} necessary to create the phason walls has to be taken into account in the athermal stress part.

10.5.5 Activation Parameters of Plastic Deformation

Before discussing possible mechanisms controlling the dislocation mobility and the thermal component of the flow stress in quasicrystals, the activation parameters of the deformation, i.e., the activation volume and the Gibbs free energy of activation, will be presented. In analogy with (4.9), the experimental activation volume is calculated from the strain rate sensitivity r by

$$V_{\mathrm{ex}} = \frac{kT}{r}.$$

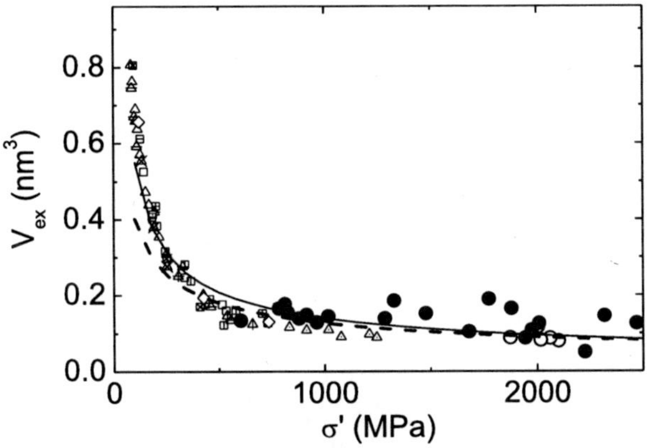

Fig. 10.43. Dependence of the experimental activation volume V_{ex} on the reduced stress σ'. Data from stress relaxations starting from steady state deformation at different strain rates between $10^{-5}\,\mathrm{s}^{-1}$ and $10^{-3}\,\mathrm{s}^{-1}$ (*small open symbols*) and $10^{-6}\,\mathrm{s}^{-1}$ (*large open circles*) and from the stage of high work-hardening at low temperatures at $10^{-6}\,\mathrm{s}^{-1}$ (*full large circles*). Data from [337,731]. Curves: activation volume according to the Peierls mechanism (10.18) (*full line*) and the second term in (10.18) (*broken line*)

For comparing the values with the predictions of the models, an orientation factor has to be considered which, however, is close to unity as discussed in Sect. 10.5.2. In Fig. 10.43, activation volume data of i-Al–Pd–Mn are plotted versus the applied stress, which are based on the same experiments as Fig. 10.36. The strain rate sensitivity was measured in stress relaxation tests under very different experimental conditions: different temperatures, different strains, different starting strain rates of the relaxation tests, and data taken at different stages along the relaxation curves. The data of the high-temperature range were corrected for recovery using (10.12), and those of the strong work-hardening range for hardening using (10.11). To correlate the data measured at different temperatures, a reduced stress should be applied

$$\sigma' = \frac{\mu_0 \sigma}{\mu(T)},$$

where μ_0 is the shear modulus at zero temperature, and $\mu(T)$ that at temperature T. The temperature dependence of the shear modulus was taken from [741], which is similar to (10.16). Except data points measured in the range of strong work-hardening at low temperatures, all the other data of the plot in Fig. 10.43 form a single curve. This is quite striking since the conditions of the measurements were very different. Part of the data was obtained near steady state deformation, and the other part far from steady state during relaxation tests. Thus, there exists an unequivocal relation between the stress and the activation volume. It will be shown in Sect. 10.5.8 that this relation describes the dislocation mobility. A typical value of the experimental activation volume in the high-temperature range is $V_{ex} \approx 0.2\,\mathrm{nm}^3 \approx 13\,\Omega$, where $\Omega = 0.0157\,\mathrm{nm}^3$ is the average atomic volume of i-Al–Pd–Mn [714].

The strain rate sensitivity of the flow stress can also be expressed by the stress exponent m', according to (4.12). It assumes values between 3 at the highest temperatures, and up to 10 at low ones in macroscopic tests, and between about 4 and 7 during the observation of single dislocations in in situ straining experiments.

The activation enthalpies ΔH presented in Fig. 10.37 are very high, in particular for high temperatures. However, the rate of the thermally activated processes during dislocation motion described by the Arrhenius law (4.7) is not controlled by ΔH, but by the Gibbs free energy ΔG. For crystals, ΔG is usually calculated from ΔH by applying Schoeck's formula (4.15), assuming that the entropy results from the temperature dependence of the shear modulus only. For quasicrystals, there might be other contributions to the entropy, particularly for the motion of imperfect dislocations, e.g., a configurational entropy originating from the change in the phason disorder. However, no theoretical treatment is available up to now. Therefore, the data in Fig. 10.44 were calculated by the Schoeck formula and the shear modulus data in [741]. The differences between ΔG and ΔH are greater than they usually are for crystals because of the strong temperature dependence of the shear modulus. The values in Fig. 10.44 yield reasonable rates of thermal activation.

10.5.6 Friction Mechanisms of Dislocation Motion

This section describes possible mechanisms controlling the dislocation mobility and accordingly the thermal component of the flow stress. These models have been established for dislocation glide and now have to be modified to consider that the dislocations move predominantly by climb. At the turning point of the discussions, the Scripta Materialia View Point Set No. 30 was

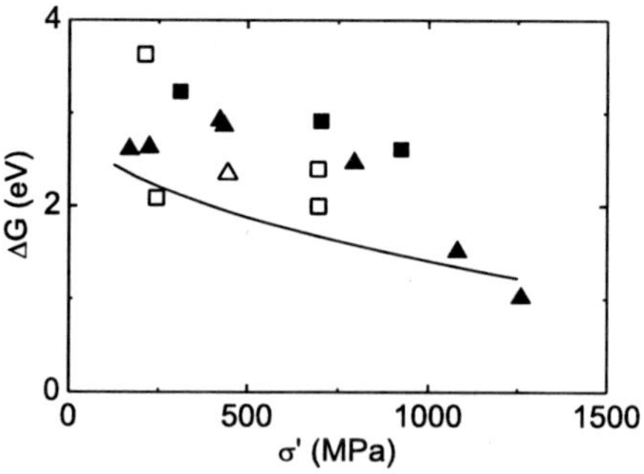

Fig. 10.44. Gibbs free energy of activation of the deformation of i-Al–Pd–Mn single quasicrystals in the high-temperature range. $\dot{\varepsilon}_t = 10^{-5}\,\mathrm{s}^{-1}$ (*triangles*), $10^{-4}\,\mathrm{s}^{-1}$ (*squares*), fivefold compression axis (*full symbols*), twofold compression axis (*open symbols*). Data from [742]. Theoretical curve from the activation energy in (10.17)

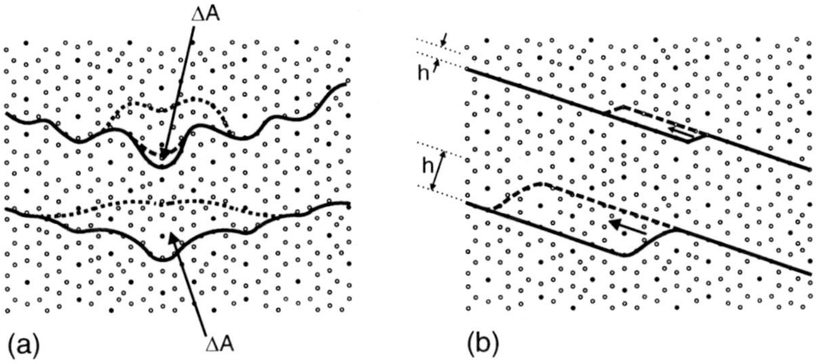

Fig. 10.45. Models of dislocation motion in quasicrystals. (**a**) Cluster friction model. Original version [747] (*upper part*). Refined version [748] (*lower part*). (**b**) Peierls mechanism on the atomic scale (*upper part*) and on the cluster scale (*lower part*). Atom plane after structure model in [656]

published with papers [743–746] on this topic. In this book, the dynamics of dislocation climb was treated in Sect. 4.10. However, the linearized equations of the climb velocity cannot be applied under the high stresses which cause rapid climb. The arguments are mainly that the activation volume of $V_{ex} \approx 13\,\Omega$ is too large for climb where $V_{ex} \approx \Omega$ is expected or, in other terms, that the stress exponents m' are too high, 3 to 10 instead of 1.

Two types of mechanisms were mainly discussed for glide, lattice friction mechanisms similar to the Peierls mechanism in crystals and the so-called cluster friction mechanism. The latter was originally suggested in [747] to explain the experimental values of the activation volume. The maximum experimental activation volumes plotted in Fig. 10.43 correspond to $50\,\Omega$, which is quite large for a simple Peierls model. If the orientation factor is considered for glide, the discrepancy is even larger. As outlined in the upper part of Fig. 10.45a, the original cluster friction model assumes that the strongly bound Mackay type clusters (rings of 10 atoms (open circles) around an Mn atom (full circles)) form obstacles to the dislocation motion like localized obstacles, e.g., small precipitates in crystals (Sect. 4.5).

The dislocation plotted as a full line bows out between these obstacles. After a single obstacle is overcome individually by thermal activation, the dislocation may be in the dotted position. The activation area $\Delta A = l\,\Delta d$ (4.4) of this process, i.e., the area between the stable equilibrium position before activation and the unstable equilibrium position after activation, may be the area between the full and dashed lines in the figure. Estimates show that with a diameter of the clusters and a distance between them of about 1 nm, ΔA is too small to account for the experimental activation volumes. The discrepancy can be avoided by a refined model [748] assuming that the clusters are not overcome individually but in a collective way, similarly to the Mott theory of solution hardening in crystals [210] described in Sect 4.5.2. In this theory, the clusters are not considered localized (point-like) obstacles, as above, but extended ones. In the field of these extended obstacles of high density, the dislocation is relatively stiff so that it only weakly bows out. As a consequence, the obstacles are not overcome individually but in a collective way so that during activation the dislocation position changes not only between the central obstacle and its neighbors but over a wider range, as illustrated in the lower part of Fig. 10.45a. As a consequence, both the experimental activation volume V and the activation energy ΔG are much larger than the values of the individual obstacles. Thus, in the refined model, the clusters are extended weak obstacles to the dislocation motion of high density. The model was confirmed in the three-dimensional computer simulation study of dislocation glide in [738].

An alternative to the cluster friction model is the Peierls model (Sect. 4.2). Originally, it was applied to quasicrystals on an atomic scale [744, 749, 750]. It is illustrated in the upper part of Fig. 10.45b. The dislocations form kinks of height h of the distances between atom rows on the slip plane. Unlike in crystals, in quasicrystals these distances are not equal but distributed quasiperiodically. The kinks spread sidewise as indicated by the arrow and the dashed succeeding position. In the elementary elastic theory of the Peierls mechanism in crystals [12], the formation energy of a kink ΔF_{fk} is given by (4.26), where W_{p} is the Peierls energy, E_{d} the line energy (or tension) of the dislocations, and τ_{p}, the Peierls stress at zero temperature. Using characteristic values of $h = 0.2\,\text{nm}$ and $\tau_{\text{p}} \approx 0.03\,\mu \approx 1.5\,\text{GPa}$ from the computer simulation in [751], it follows that $\Delta F_{\text{fk}} \approx 0.3\,\text{eV}$. In the Peierls model, the experimental activation energy equals the energy to form a kink pair, i.e., $2\Delta F_{\text{fk}}$, plus the energy of kink motion, which is mostly low. As a result, the energy obtained from the Peierls model on an atomic scale is too low to explain the experimental values in Fig. 10.44.

To remove the discrepancy it may be assumed that in quasicrystals, the kinks form on the cluster level as suggested in [742, 752] and illustrated in the lower part of Fig. 10.45b. In this case, the kink height is approximately equal to the cluster diameter, $h \approx 0.9\,\text{nm}$, so that $\Delta F_{\text{fk}} \approx 2.7\,\text{eV}$. The respective activation energy is now too high, but this may be due to general uncertainties of the models and the parameters applied.

The Peierls model can quite easily be adapted to the case of dislocation motion occurring predominantly by climb instead of glide. Instead of the shear stress τ, the resolved normal stress σ has to be used in the equations. The process of the formation of kink pairs is replaced by the formation of jog pairs, and the spreading of kinks along the dislocations, by the climb motion of jogs. In their comprehensive modeling of the deformation of icosahedral quasicrystals by climb, Mompiou and Caillard point out that the usual linearized equations for the climb velocity (4.90) or (4.92) cannot be applied to the high-temperature deformation of quasicrystals [739]. This is mainly a consequence of the high applied forces and is expressed, e.g., by the stress exponent m' being clearly higher than unity. They suggest to analogously apply the Peierls model to the situation of climb with (4.47) for the dislocation velocity v_{d}

with $\Delta F^*_{\mathrm{fdk}}$ from (4.40) for the formation energy of the jog pair. These are the equations for "short" dislocations of length λ_0, predicting straight dislocations arranged along the Peierls valleys, as it is observed at least at high temperatures. Since the spreading of kinks by climb involves self-diffusion, the diffusion energy ΔF_{sd} replaces the kink migration energy. The equation for the dislocation velocity now reads

$$v_{\mathrm{d}} = \frac{\sigma^* b_{\parallel} h^2 \lambda_0 \nu_0}{kT} \exp\left(- \frac{\Delta F_{\mathrm{sd}} + 2\Delta F_{\mathrm{fj}} - \sqrt{\sigma^* \mu b_{\parallel}^3 h^3/(2\pi)}}{kT} \right), \tag{10.17}$$

with the parameters from Sect. 4.2.2, h being now the height of the jogs, and ΔF_{fj}, their formation energy. The model should be able to account for the strain rate sensitivity experimentally observed. The activation volume is then (4.9)

$$V_{\mathrm{ex}} = kT \frac{\mathrm{d}\ln v_{\mathrm{d}}}{\mathrm{d}\sigma^*} = \frac{kT}{\sigma^*} + \sqrt{\frac{\mu b_{\parallel}^3 h^3}{8\pi\sigma^*}}. \tag{10.18}$$

In Fig. 10.43, the results of (10.18) are plotted for the approximation of $\sigma^* = \sigma'$, which neglects both σ_{ph} and σ_{i}. σ_{ph} may be small and the Cottrell–Stokes law may hold, i.e., σ_{i} is proportional to σ^* (Sect. 5.2.2). This would not alter the functional form of (10.18). However, as in Fig. 10.42 indicated by the decreasing values of σ_{i} at low temperatures, the Cottrell–Stokes law is violated at low temperatures. In (10.18), the first term stems from the linear stress dependence of the climb velocity of the jogs, and the second one, from the stress dependence of the formation energy of the jog pairs. The latter term is plotted as a dashed line, and the whole activation volume, as a solid one. The difference between both curves is the contribution of the climbing motion. It does not contain adjustable parameters (except the orientation factor assumed to be close to unity). The correlation between σ and T was obtained from the temperature dependence of the flow stress in Fig. 10.33. As the curves in Fig. 10.43 demonstrate, the contribution of climb to the activation volume is small except at low stresses corresponding to high temperatures or low strain rates. The parameters of the second term, $\mu = 50\,\mathrm{GPa}$, $b_{\parallel} = 0.3\,\mathrm{nm}$, and $h = 0.67\,\mathrm{nm}$, are adjusted to fit the high-stress data from steady state deformation. It turns out that the whole theoretical curve of the activation volume agrees quite well with the universal experimental curve. In view of neglecting σ_{i}, this is quite astonishing. An exception are the measuring points from the range of strong work-hardening at low temperatures (full circles in Fig. 10.43).

The numerical values above can also be introduced into the expression of the activation energy in (10.17) to estimate its stress dependence. Assuming that the low values of ΔG at the higher stresses are correct, the curve in Fig. 10.44 is plotted with $\Delta F_{\mathrm{sd}} + 2\Delta F_{\mathrm{fj}} = 3\,\mathrm{eV}$ to force agreement there. This is certainly a lower limit. A value of $4\,\mathrm{eV}$ may also be realistic. With decreasing stress, the experimental values of ΔG increase more strongly than predicted by the model, which may be due to further processes becoming relevant at high temperatures. The energies ΔF_{sd} and ΔF_{fj} cannot be separated experimentally. The climb velocity should be determined by the diffusion of the species diffusing more slowly, which is Pd or Mn with activation energies of 2.32 eV [670] and 1.99 eV [713]. Lower diffusion energies may result if pipe diffusion plays a role in the climb process. The jog height $h = 0.67$ nm adjusted to the experimental activation volumes agrees with the Peierls model better on the cluster scale than on the atomic one. This conclusion had already been drawn for glide in [742], and for climb in [739]. The elastic energy of the jog pair of this height should be in the range of several eV as estimated above for kinks. This would not leave much space for the diffusion energy, in accordance with the results on the activation volume. Unfortunately, more precise data cannot be given at present.

Takeuchi [753] compared the Peierls model for glide and climb and argued that in quasicrystals jog pairs can form more easily than kink pairs. However, the main reason for the predominance of climb in quasicrystals rather is the different structure of the defect walls created by the imperfect dislocations than the difference in the elastic properties of kink and jog pairs. A role may also play the dissociation of the dislocation cores on the climb plane. Split dislocations have been observed in Fig. 10.17 and were analyzed in [702]. In many intermetallic alloys, dislocations have complex cores which extend on planes other than the glide plane, leading to a low glide mobility and to the influence of non-glide components of the stress field (e.g., [508, 754]). In particular, the dislocations can undergo a climb dissociation, where glide is strongly impeded so that climb may be facilitated. Thus, dislocation motion in quasicrystals may show similarities to that in other intermetallic alloys.

As to the friction models, numerical estimates indicate that the refined cluster friction model predicts a stress dependence of the activation parameters similar to that of the Peierls model if the same parameters are applied. That means that the refined cluster friction model and the Peierls model on a cluster scale are equivalent with respect to the prediction of the macroscopic deformation data. The cluster friction model is supported by two-dimensional [751] and three-dimensional molecular dynamics simulations of glide [738, 755], where the dislocations move jerkily between strong spots in the quasicrystal structure. On the other hand, the Peierls model fits better the straight, crystallographically oriented shape of many dislocations (Fig. 10.18) as well as the viscous type of dislocation motion. Thus, this model will be preferred in the following. However, note that coming from low temperatures, at about 530°C at least part of the dislocations strongly bow out between pinning agents (Fig. 10.14), which is not in agreement with the Peierls model at these temperatures. The obstacles are most probably kinks in the climbing dislocation lines. The radius of curvature of $\delta = 50$ nm, mentioned in Sect. 10.3.1, corresponds to a local effective stress of $\sigma^* = \Gamma/(\delta b) \approx 240$ MPa. In [739], the change from curved dislocations at lower temperatures to straight ones at higher temperatures is explained by a transition in the Peierls model from the case of "long" dislocations (4.46) to that of "short" ones (4.47). However, the transition may also be connected with special core configurations at high temperatures as in intermetallic alloys.

In decagonal quasicrystals, the activation parameters have values similar to those in icosahedral ones, the dislocations are also straight and move in a viscous way preferentially by climb, even those with periodic Burgers vectors. Thus, the dislocation motion seems to be controlled by the same mechanisms.

10.5.7 Dislocation Kinetics in the High-Temperature Range

Since at the elastic limit the dislocation density is still low, the athermal stress component σ_i in (10.13) is low. Besides, work-hardening has not yet started ($\sigma_\theta = 0$) so that the elastic limit is given by

$$\sigma_{\mathrm{EL}} = \sigma_{\mathrm{ph}} + \sigma^*. \tag{10.19}$$

The elastic limit is included in the diagram of the stress components in Fig. 10.42. It shows that in the high-temperature range, say above about 650°C, the total steady state flow stress is given by

$$\sigma = \sigma_{\mathrm{EL}} + \sigma_\mathrm{i}.$$

As discussed in Sect. 10.5.6, the amount of σ_{ph} is not clear, especially at high temperatures. If it can be neglected here, the flow stress at high temperatures will be

$$\sigma = \sigma^* + \sigma_\mathrm{i},$$

as in many crystals. The kinetics of dislocation generation and annihilation as well as of the resulting flow stress was modeled by a set of constitutive equations in Sect. 5.2.4 applying a power law for the dependence of the dislocation velocity on the stress. A similar model for quasicrystals was first established by Guyot and Canova [756], and later on also by Feuerbacher et al. [757]. To consider work-softening as a specific feature of quasicrystal deformation, an order parameter was introduced, which is reduced by plastic strain and recovers in time. The modeling in Sect. 5.2.4 is primarily thought to describe the (quasi) steady state deformation near the lower yield point. It does not invoke any process specific to quasicrystals. The results were presented in Fig. 5.29. The model predicts quite correctly the temperature and the strain rate dependencies of the steady state flow stress as well as the temperature dependence of the dislocation density down to about 700°C. At low temperatures, the real dislocation densities are much lower than those predicted by the model. This agrees with the difference between the ratio σ_i/σ following from the model in Fig. 10.46 and the respective value of σ_i calculated from the measured dislocation densities in Fig. 10.42. Nevertheless, at high temperatures, the plastic properties of icosahedral Al–Pd–Mn can well be interpreted without taking into account anything specific to quasicrystals.

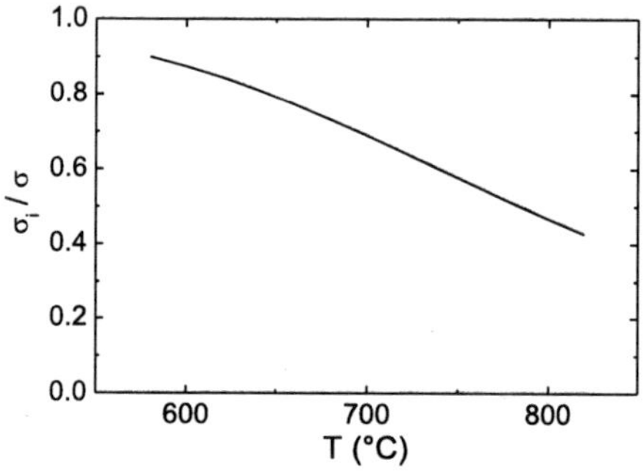

Fig. 10.46. Temperature dependence of σ_i/σ at $\dot{\varepsilon} = 10^{-5}\,\mathrm{s}^{-1}$ after the modeling in Sect. 5.2.4. Data from [336]

The equations used for the modeling are effectively independent of the mode of dislocation motion, i.e., glide or climb. Likewise, the kinetics of dislocation generation in crystals and icosahedral quasicrystals seems to be similar. Like the double-cross slip sources dominating in crystals, the Bardeen–Herring climb sources, too, do not seem to be localized. A necessary ingredient of the steady state deformation is the recovery of the dislocation structure, particularly the dislocation density, which in [338] is observed experimentally during annealing. Besides, the formation of a three-dimensional dislocation network at high temperatures, as in Fig. 10.11, hints at the action of recovery. It is also indicated by the inverse curvature of the stress relaxation curve R_2 in Fig. 10.34b. This is the typical recovery-type curve as observed for ZrO_2 single crystals in Fig. 7.34b and interpreted in Sect. 7.3.5. The steep range at the beginning of the relaxation tests corresponds to the dislocation dynamics whereas the flat part of the curve is mainly governed by recovery.

The physical processes controlling the dislocation annihilation are still not clear. If the dislocations move preferentially by climb, the mutual annihilation of dislocations of opposite sign has to take place by glide, which occurs very slowly as demonstrated in [705]. It was discussed in Sect. 10.5.1 that strong structural disorder is generated by glide. To reduce the disorder, extensive atomic rearrangements are necessary, which certainly involve diffusional steps. At present, it is not yet clear which process of what activation energy controls the recovery. To describe the strong decrease of the steady state flow stress with increasing temperature, high activation energies of both dislocation mobility ΔG_0 and recovery ΔG_{ann} had to be chosen. They are higher than the diffusion energies of 2–2.3 eV for Pd and Mn [670, 713]. The value of $\Delta G_0 = 3\,\mathrm{eV}$ agrees with the activation parameters evaluated by the modified Peierls model in the preceding section. The supposed stress exponent of the dislocation mobility of $m = 4$ is consistent with the exponents observed for single dislocations during in situ straining (Figs. 10.20 and 10.31) as well as with those from macroscopic stress relaxation tests ($m' = 3 \ldots 6$ in the high-temperature range, Sect. 10.4.1). However, the macroscopic stress exponents can conveniently be interpreted by recovery where similar stress exponents are expected (Sect. 5.2.4). There seems to be a coincidence of the stress exponents of dislocation mobility with recovery-controlled deformation.

10.5.8 The Climb-Exchange Model

While at high temperatures, the sum of the long-range internal stress component σ_i and the friction stress σ^* of moving dislocations is sufficient to interpret the macroscopic flow stress of i-Al–Pd–Mn, Fig. 10.42 indicates that a further stress contribution σ_θ is required to explain the strong work-hardening at low temperatures. It was suggested in [337] that the hardening results from the accumulation of phason defects, which cannot anneal out at low temperatures and which cause hardening instead of softening, which was assumed in other models.

The strong hardening may also originate from the chemical force due to the point defect super- or subsaturations, generated by the climbing dislocations. As observed in both icosahedral Al–Pd–Mn and decagonal Al–Ni–Co, there exist sets of dislocations termed A in Sects. 10.3.1 and 10.3.2, which climb under high climb forces arising from the external load. Other sets B experience zero or small mechanical forces. As illustrated in Fig. 10.47 for the case of compression, the outward motion of the dislocations of set A occurs by absorbing vacancies causing a subsaturation, which induces the dislocations of set B to climb by producing vacancies under a chemical force. The flux of vacancies is indicated by the open squares and grey arrows. This idea was first put forward by Schall [723] for the motion of periodic and mixed dislocations in d-Al–Ni–Co and later on by Mompiou et al. [702] and Ledig [708] for i-Al–Pd–Mn. It may be called the climb-exchange model.

Recently, the model was formulated in a quantitative way by Mompiou and Caillard [739]. The super- or subsaturation of vacancies results in a chemical

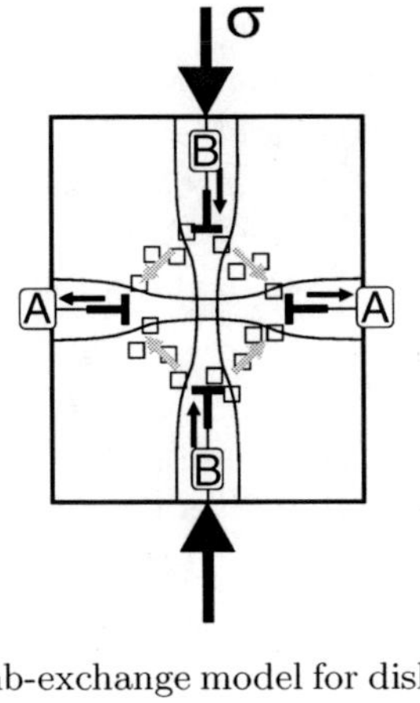

Fig. 10.47. Schematic of climb-exchange model for dislocation motion in icosahedral quasicrystals

climb stress, which is given by the osmotic force f_{os} from (4.84)

$$\sigma_{\mathrm{ch}} = \frac{f_{\mathrm{os}}}{b \sin \beta} = \frac{kT}{\Omega} \ln \frac{c_{\mathrm{v}}}{c_{\mathrm{v0}}},$$

where β is the character of the dislocation, Ω, the average atomic volume, c_{v}, the actual vacancy concentration, and c_{v0}, the equilibrium concentration after (4.76). The chemical stress counteracts the applied stress resulting in chemical hardening with a strain-hardening coefficient

$$\theta_{\mathrm{ch}} = \frac{\Delta \sigma_{\mathrm{ch}}}{\Delta \varepsilon},$$

with $\Delta\varepsilon$ being the increment in plastic strain. The latter is given by the number of absorbed vacancies and, for pure climb, by the area swept by the dislocations, i.e.,

$$\Delta \varepsilon = c_{\mathrm{v}} - c_{\mathrm{v0}}.$$

Note that $\Delta\varepsilon$ is negative for compression. Thus, the chemical hardening coefficient becomes

$$\theta_{\mathrm{ch}} = \frac{kT}{\Omega} \frac{\ln(c_{\mathrm{v}}/c_{\mathrm{v0}})}{c_{\mathrm{v}} - c_{\mathrm{v0}}}.$$

Near the elastic limit where $c_{\mathrm{v}} \approx c_{\mathrm{v0}}$, it may be approximated by

$$\theta_{\mathrm{ch}} \approx \frac{kT}{\Omega c_{\mathrm{v0}}}.$$

These hardening coefficients are very high, higher than the shear modulus. In [739], they are estimated to agree well with the temperature-depending initial hardening coefficients observed by Ledig et al. [337] in the range of strong work-hardening. During in situ straining experiments in a TEM, the near surfaces can take the role of the dislocation family B so that high super- or subsaturations of point defects do not develop.

The chemical stress σ_{ch} impedes the climb motion of dislocations of the family A. It is thus identical with the hardening term σ_θ in (10.13). In addition, it causes the climb motion of the dislocations of set B, which do not experience a stress component from the external load. In compressive deformation, these dislocations emit vacancies which reduce the subsaturation leading the chemical hardening coefficient to decrease. Finally, a steady state develops with $\theta_{\mathrm{ch}} = 0$ when equal numbers of vacancies are absorbed and emitted. This is achieved when

$$\varrho_{\mathrm{A}} b_{\mathrm{A}} v_{\mathrm{A}} = \varrho_{\mathrm{B}} b_{\mathrm{B}} v_{\mathrm{B}}. \tag{10.20}$$

The dislocations of sets A and B with densities ϱ_{A} and ϱ_{B} move at velocities $v_{\mathrm{A}}(\sigma - \sigma_{\mathrm{ch}})$ and $v_{\mathrm{B}}(\sigma_{\mathrm{ch}})$. In [739], it is argued that for equal average Burgers vectors $b_{\mathrm{A}} = b_{\mathrm{B}}$ during steady state also $\varrho_{\mathrm{A}} = \varrho_{\mathrm{B}}$ holds. Then, it follows from (10.20) that $v_{\mathrm{A}} = v_{\mathrm{B}}$, which is fulfilled for

$$\sigma_{\mathrm{ch}} \approx \sigma_{\mathrm{ss}}/2. \tag{10.21}$$

$\sigma_{\rm ss}$ is the steady state flow stress. According to these considerations, the deformation behavior of quasicrystals at low temperatures can be interpreted in the following way. At the elastic limit, only dislocations of set A are moving. As shown above (10.19)

$$\sigma_{\rm EL} = \sigma_{\rm ph} + \sigma^* . \tag{10.22}$$

σ^* is the friction stress of climbing dislocations discussed above in terms of the modified Peierls model (10.17).

For the steady state stress, from (10.13), (10.21), and (10.22) and by neglecting $\sigma_{\rm i}$ it follows that

$$\begin{aligned} \sigma_{\rm ss} &= \sigma_{\rm EL} + \sigma_{\rm ch} = 2\sigma_{\rm EL} \\ &= 2(\sigma^* + \sigma_{\rm ph}) . \end{aligned} \tag{10.23}$$

Thus, the climb-exchange model predicts that the steady state flow stress equals twice the elastic limit. Neglecting the different strain rates in Fig. 10.33, Mompiou and Caillard claim that this relation is fulfilled over the whole temperature range of this figure (Fig. 4 in their paper [739]). Indeed, it is reasonably well fulfilled at the lowest temperature of 580°C, where steady state deformation was achieved at a strain rate of $\dot{\varepsilon} = 10^{-6}\,{\rm s}^{-1}$. The flow stress values at lower temperatures (small open circles in Fig. 10.33) belong to the range of strong work-hardening but not to the steady state. At higher temperatures, the ratio between $\sigma_{\rm ss}$ and $\sigma_{\rm EL}$ decreases so that $\sigma_{\rm ss} < 2\sigma_{\rm EL}$. At 700°C, $\sigma_{\rm ss}$ and $\sigma_{\rm EL}$ become equal, and above that temperature no elastic limit is observed at $\dot{\varepsilon} = 10^{-6}\,{\rm s}^{-1}$. At a strain rate of $10^{-5}\,{\rm s}^{-1}$, the high-temperature data represented by the dotted regression curve for the elastic limit and the dashed one for the steady state flow stress differ by a small constant value. Only at the lowest temperature of 658°C, the steady state flow stress values are considerably higher than the elastic limit. It may therefore be concluded that relation (10.23) is approximately valid at the lowest temperatures where steady state deformation was achieved but not in the high-temperature range.

The climb-exchange model may also help to interpret the measurements of the strain rate sensitivity by means of stress relaxation tests. If these tests are discussed in terms of the power law (4.10) for the strain rate $\dot{\varepsilon}$, the relaxation equation (2.7) can be written as

$$-\dot{\sigma} = S\,\dot{\varepsilon} = \dot{\sigma}_0\,\sigma^{*m} ,$$

where all constants are lumped together into $\dot{\sigma}_0$. The experimental strain rate sensitivity $r_{\rm ex}$ from (2.8) is then given by

$$\frac{1}{r_{\rm ex}} = \frac{\partial \ln \dot{\varepsilon}}{\partial \sigma} = \frac{\partial \ln(-\dot{\sigma})}{\partial \sigma} = m\,\frac{\partial (\ln \sigma^*)}{\partial \sigma} = m\,\frac{\partial \sigma^*}{\partial \sigma}\,\frac{1}{\sigma^*} .$$

Accordingly, the experimental strain rate sensitivity depends on $\partial\sigma^*/\partial\sigma$, which deviates from unity if $\sigma_{\rm i}$ and/or $\sigma_{\rm ch}$ change during the relaxation. It is usually argued that $\sigma_{\rm i}$ changes only slowly at the beginning of relaxation tests. Mompiou and Caillard [739] estimate the change in the stress acting on the dislocation family B. They conclude that it is zero near the steady state resulting in $\partial\sigma^*/\partial\sigma = 1$, or

$$r_{\mathrm{ex}} = \frac{\sigma^*}{m} = r.$$

Thus, considering (4.11), the initial experimental strain rate sensitivity of relaxation tests starting from steady state deformation represents that of the dislocation mobility expressed by σ^* and m.

After long relaxation times, it is assumed that the recovery of the chemical stress reduces the stresses acting on both dislocation sets A and B at the same rate yielding $\partial\sigma^*/\partial\sigma = 1/2$, or

$$r_{\mathrm{ex}} = \frac{2\sigma^*}{m} = 2r. \tag{10.24}$$

Consequently, the experimental strain rate sensitivity increases by a factor of two after long relaxation times. This behavior is indeed observed, e.g., in the two-stage relaxation curves R_2 and R_6 in Fig. 10.34b. Similar considerations show that in the range of strong work hardening the correction after (10.11) has to be applied to obtain the strain rate sensitivity of the dislocation mobility from r_{ex}. In conclusion, after the appropriate corrections, the initial slope of the relaxation curves should always represent the strain rate sensitivity of the dislocation mobility. At a fixed temperature, Fig. 10.36 exhibits very different strain rate sensitivities for measurements in the hardening and steady state ranges. If these data are plotted as the activation volume versus the temperature-corrected stress as in Fig. 10.43, they fit quite well a single curve describing the processes of dislocation mobility.

The climb-exchange model is an attractive option to explain the strong work-hardening at low temperatures without invoking other unknown stress contributions. It was derived in [739] supposing that the internal stress σ_{i} is small. There is little experimental evidence of the true amount of σ_{i}. The estimation in Sect. 10.5.4 via (10.15) and the resulting values in Fig. 10.42 are based on a large factor α_{c}. Nevertheless, this is consistent with the stress necessary to bow out segments of the dislocation network. On this basis, σ_{i} cannot be neglected. This will not basically change the general conclusions from the model but it will lead to quantitative alterations, e.g., smaller differences between σ_{ss} and σ_{EL} at higher temperatures.

Experimentally, the recovery of the chemical stress causing the two-stage relaxation curves R_2 and R_6 in Fig. 10.34b cannot be distinguished from the recovery of the dislocation structure, which was modeled for the high-temperature deformation in Sects. 5.2.4 and 10.5.7, and which also results in the two-stage relaxation curves. On the other hand, there is no sign of recovery in the range of high work-hardening at low temperatures, where the climb-exchange model agrees best with the experimental flow stress data. The relaxation curves in Fig. 10.34a do not exhibit an inverse curvature towards a recovery stage. They are simply shifted to higher stresses indicating the hardening occurring between the relaxation curves. This is confirmed by the creep curve included in the experiment of Fig. 10.32a, which exhibits a continuously decreasing creep rate.

Besides, the climb-exchange model ignores the different dislocation structures formed at low and high temperatures. The recovery-type network structure (Fig. 10.11) forms only at high temperatures. At low ones, the dislocations are arranged in typical low-temperature structures of narrow bands (Figs. 10.12 and 10.15b). Part of the dislocations strongly bows out to radii of about 50 nm. From the line tension model analogous to (3.38), a shear modulus at 550°C of 57.4 GPa, $b_{\|} = 0.3\,\text{nm}$, and an orientation factor of 0.8, there follows a back stress of $\sigma_{\text{i}} = 315\,\text{MPa}$. This is at least part of the difference between the elastic limit and the steady state flow stress, though the obstacles causing the dislocations to strongly bow out are not yet clear at all. Thus, the climb exchange model appears as an attractive suggestion to explain the strong hardening of i-Al–Pd–Mn at low temperatures although many questions still remain unanswered.

Mompiou and Caillard argue that the climb-exchange model is not restricted to quasicrystals but may operate also at high temperatures in crystalline materials like TiAl and γ/γ' superalloys [758]. Indeed, rapid climb as a mode of dislocation motion has been observed also in some intermetallics described above (TiAl, FeAl). As the stress-driven climb motion of one dislocation causes non-equilibrium concentrations of point defects, other dislocations can climb under the action of the resulting chemical force. Thus, the climb-exchange model may be of a more general character at high-temperature deformation.

10.6 Conclusions on Quasicrystals

The following conclusions can be drawn on the dislocation motion and plastic deformation of quasicrystals.

- Quasicrystals are intermetallic solids with long-range order but without periodicity (translational symmetry). In contrast to crystals, which are built by periodically arranging a single unit cell, at least two "unit cells" called tiles are necessary to form a quasicrystalline structure, in a one-dimensional quasicrystal they are long and short distances of an irrational ratio of the Golden Mean. The arrangement of the atoms can be described in terms of hyperspaces of more than three dimensions, six dimensions for quasicrystals with icosahedral symmetry, and five dimensions for those with decagonal symmetry. Three axes form the real space, which is also called the physical or parallel space. The remaining axes form the perpendicular or orthogonal space. The hyperspace exhibits translational symmetry. The positions of atoms are obtained by projecting the lattice sites of the hyperspace onto the axes of the parallel space.

- In addition to the usual defects in crystals, the quasicrystalline order of the two "unit cells" may be disturbed leading to violations of the so-called matching rules. Such defects are called phasonic defects or phasons. Diffusion is necessary to anneal out phason defects. This process is called retiling.
- Dislocations in quasicrystals carry a long-range elastic strain field as in crystals. In addition, they disturb the quasicrystalline order, i.e., they induce a field of phason defects. The Burgers vector $\boldsymbol{B}$ of dislocations can be defined in the higher-dimensional hyperspace. It can be separated into a so-called phonon component $\boldsymbol{b}_{\|}$ in real space, and a phason component $\boldsymbol{b}_{\perp}$ in perpendicular space. $\boldsymbol{b}_{\|}$ replaces the Burgers vector in crystals and describes the elastic strain field, whereas $\boldsymbol{b}_{\perp}$ describes the field of phason defects in the regular quasiperiodic structure around the dislocation lines.
- In situ straining experiments in a TEM at high temperatures have shown that dislocations in quasicrystals move in a viscous way on well-defined crystallographic planes. The determination of both the Burgers vectors and the planes of motion on the same dislocations revealed that the dominating mode of motion is climb. Only during climb, the tiling matching is respected without overlaps and empty spaces in the quasicrystalline structure. A combination of climb and glide seems to be possible, too. Dislocations with periodic Burgers vectors in decagonal quasicrystals can both glide and climb. The mobility of the dislocations can best be described by a double-kink model on the cluster scale adapted to climb. For dislocation annihilation, glide is necessary.
- At high temperatures, dislocations may move in their perfect state, i.e., they carry their phonon and phason fields. At low temperatures, imperfect dislocations having only the phonon field drag phason walls. The formation energy of phason walls is probably very high for glide, and low for climb. The transition temperature between the motion in the perfect and imperfect states is not clear. It may be between about 610 and 650°C, but it may also be substantially higher.
- During plastic deformation, dislocation systems are activated with high orientation factors for climb, and low ones for glide. In addition, dislocations move on systems with zero or near-zero orientation factors for climb and glide. At high-temperature deformation, three-dimensional networks of mostly straight, crystallographically oriented dislocations are formed. At low temperatures, curved dislocations may form narrow bands.
- At high-temperature deformation, the dislocation density increases strongly during the initial deformation and the upper yield point. Afterwards, it reaches a steady state value or decreases slightly at higher strains. After the deformation has stopped, the dislocation density recovers.
- Beyond the elastic range, the stress–strain curves of quasicrystals exhibit a transition to a range of very high work-hardening. The transition point may be called the elastic limit. At low temperatures (and usual strain rates), the specimens break at a stress of about 1.6 GPa. At high

temperatures, the curves show an upper and a lower yield point followed by a range of steady state deformation and work-softening at high plastic strains. The brittle-to-ductile transition is connected with the possibility to overcome the upper yield point. It depends strongly on the strain rate.

- The rapid work-hardening in the first stage of deformation can be explained by a chemical stress arising from the point defect super- or subsaturation resulting from the dislocations climbing under the action of the external stress. In the climb-exchange model, the chemical stress owing to the excess or deficiency concentrations of point defects causes dislocations to climb which have zero or near-zero orientation factors for glide and climb. The model predicts that the steady state flow stress is twice the elastic limit. This condition holds true at low temperatures. High-temperature deformation can be interpreted only by the long-range internal stress between dislocations and the effective stress due to the dislocation mobility.

11 Conclusion

Dislocation motion is quite a complex process spanning several orders of magnitude in length scales. The motion is realized by breaking and re-establishing bonds on the atomic level. These problems are tackled by molecular dynamics simulations using semi-heuristic methods. Ab initio calculations are still rare and, for more complex problems, they suffer from the still low number of atoms involved. There is still poor microscopic experimental verification of these dynamic processes. The static core configurations of dislocations are studied by high-resolution electron microscopy. The spatial resolution of the latter techniques has drastically improved by the new aberration-corrected microscopes. However, the informative value of the micrographs is limited by the geometrical requirements to the specimens. Simple interpretations are possible only for edge dislocations arranged end-on in the specimen. Up to now, there are no TEM in situ loading studies under high-resolution conditions revealing changes of the core structure of relatively compact cores under load. This is due partly to experimental difficulties and partly to surface effects from the low specimen thicknesses, which influence the processes to be studied. Thus, in understanding the atomic processes of dislocation motion, there are still some challenges to be respected.

In the materials with low Peierls stress, these atomistic processes do not control the dislocation dynamics. In addition to the disturbance of the local coordination of atoms, dislocations carry an elastic stress field, which gives rise to the line contrast of dislocations in diffraction contrast TEM. Besides, the stress field comprises the main part of the energy of the dislocation. As, in a zero-order approximation, the energy is proportional to the length of the dislocation, many processes can well be described within the framework of the line tension model. This model and the elastic interactions between dislocations and other crystal defects as well as with other dislocations form the bulk of the basic understanding of the dislocation motion. In the 1950s and 60s, many models were formulated on an intuitive way, which are now partly proven by diffraction contrast TEM and other methods. Frequently, these processes take place on a scale between about 10 nm and a few micrometers. They constitute the bulk of the present book.

In the last decades, research has turned to more complex materials like the intermetallic alloys where the dislocation mobility depends sensitively on the structure of the dislocation cores. In addition to the nonplanar spreading of the cores, which directly influences the glide resistance, nonconservative processes within the cores seem to play a role in several intermetallics, leading to a similar macroscopic behavior controlled by diffusion, irrespective of the particular crystal structure. However, there are still no methods for monitoring the small diffusion fluxes in the dislocation cores to prove these models.

In the new field of dislocation motion in quasicrystals, the application of the whole set of methods developed for crystals for over 50 years provided good insight within the short time of about 20 years, only. Nevertheless, the present models of climb controlled by the formation of jog pairs on a cluster level require more direct experimental evidence. Here, it should be possible to prove the dynamic nonequilibrium point defect concentrations by respective in situ methods.

Thus, for a number of years, the basic mechanisms controlling the dislocation dynamics have been understood quite well. However, to determine their actual parameters in the more complex materials requires the combined application of macroscopic and microscopic methods as the atomistic interactions, for example, between dislocations and intrinsic or extrinsic point defects, cannot be observed directly, and so the interpretation of deformation data on the basis of microscopic models is frequently still of speculative nature.

It is certainly one aim of the study of dislocation dynamics to establish phenomenological laws to be incorporated into theories of the collective behavior of many dislocations, which governs macroscopic plasticity. First steps are here the formulation of constitutive laws combining dislocation dynamics with the kinetics of dislocation generation and immobilization and annihilation, as well as the statistical models of plastic instabilities. Other approaches are the computer simulation of the motion of a greater ensemble of dislocations or the insertion of dislocation mobility laws into finite element calculations. Though these methods have brought some progress, there is, on the other hand, still no good method to describe the effect of long-range elastic fields on the motion of individual dislocations with input parameters that can easily be measured. In this connection, very little is known about the role of the athermal stress component in the deformation of intermetallic alloys. Besides, the interplay is not well understood between the effective stress, which influences the dislocation multiplication, and the development of the dislocation density and the resulting athermal stress component. Thus, we may conclude that the basic knowledge of dislocation dynamics has reached some completion. It appears, however, challenging and requiring more efforts to uncover and to design the details of the continuously developing novel materials, and to implement physics-based dislocation mobility laws into the theories of collective dislocation behavior.

List of Abbreviations and Symbols

a	area, swept area, lattice constant
A	area, cross section
APB	antiphase boundary
b	absolute value of Burgers vector
$\boldsymbol{b}$	Burgers vector
B	viscous drag coefficient
$\boldsymbol{B}$	Burgers vector in higher-dimensional hyperspace
b_{e}	edge component of Burgers vector
b_{s}	screw component of Burgers vector
$\boldsymbol{b}_{\parallel}$	Burgers vector component in physical (parallel) space
$\boldsymbol{b}_{\perp}$	Burgers vector component in perpendicular (orthogonal) space
b.c.c.	body centered cubic
c	concentration (of defects), crack length
c_0	(thermal) equilibrium concentration
c_+, c_-	concentration of positive and negative kinks
c_{j}	concentration of jogs
c_{k}	concentration of kinks
c_{t}	transverse sound velocity
CBED	convergent beam electron diffraction
CSF	complex stacking fault
d	width of decomposed dislocations
D	diffusion coefficient, diameter of particles (extended obstacles), total density of cross slip events
D_{k}	diffusion coefficient of kinks
D_{sd}	self-diffusion coefficient
E	elastic (Young's) modulus, elastic energy
E_{c}	core energy per unit length of dislocation
E_{d}	energy per unit length of dislocation
E_{e}	energy of edge dislocation
E_{m}	dislocation energy per length in matrix

E_p	dislocation energy per length in particle
E_s	energy of screw dislocation
E_{de}	energy per unit length of edge dislocation
E_{ds}	energy per unit length of screw dislocation
E_{0e}	prelogarithmic factor of energy per unit length of edge dislocation
E_{0s}	prelogarithmic factor of energy per unit length of screw dislocation
f	normalized (dimensionless) force acting on obstacle, volume fraction of precipitates
f_0	normalized (dimensionless) obstacle strength
f.c.c.	face centered cubic
f_d	force per unit length acting on dislocation
f_c	climb force per unit length resulting from external stress
f_{cl}	total climb force per unit length acting on dislocation
f_{gl}	glide force per unit length acting on dislocation
f_{os}	osmotic force per unit length due to point defect sub- of supersaturation
F	applied external force, force on a dislocation segment, force acting on obstacles
F_0	obstacle strength
F_s	scaling factor in Mott–Labusch statistics
F_{int}	interaction force between kinks of double-kink
$\boldsymbol{g}$	diffraction vector
$\boldsymbol{G}$	diffraction vector in higher-dimensional reciprocal space
G_c	crack extension force, energy release rate
$\boldsymbol{g}_{\parallel}$	parallel component of diffraction vector
$\boldsymbol{g}_{\perp}$	perpendicular component of diffraction vector
h	distance between Peierls valleys, cross slip height
h_c	critical height of dislocation dipole
$\bar{h}$	average cross slip height
$H(h)$	frequency of cross slip of height h
h.c.p.	hexagonal close packed
HVEM	high-voltage electron microscope/microscopy
J	(diffusion) flux (of kinks)
J_c	J integral
J_{dk}	double-kink nucleation rate
k	Boltzmann's constant
K	stress intensity factor
K_c	critical stress intensity factor
K_s, K_e	energy factors of screw and edge dislocations
K_x, K_z	resistance force at particle border
K_1, K_2	phason elastic constants
l	specimen length, length of dislocation segments
l_{sq}	square lattice distance
L	length of dislocation, segment length, width of double-kink
L^*	critical width of double-kink

LYP	lower yield point
m	(true) stress exponent
m'	effective stress exponent
m_c	orientation factor for climb
m_d	effective mass of dislocation per unit length
m_s	orientation (Schmid) factor for glide
n	coordination number, creep exponent, ageing exponent, reciprocal exponent of strain rate in equation for yield stress of semiconductors, strain-softening exponent
$\boldsymbol{n}$	normal vector of specimen, pole near direction of electron beam
p	hydrostatic pressure
p_{ext}	external pressure
p_{int}	internal pressure
P_{ul}	probability of unlocking
q	dislocation immobilization or annihilation coefficient
Q_e	experimental activation energy
s	slip distance
S	elastic stiffness
SISF	superlattice intrinsic stacking fault
r	radius vector, radius of curvature of dislocation segment, strain rate sensitivity
r_0	inner cut-off radius
r_{ex}	experimental strain rate sensitivity
R	outer cut-off radius
R, R_3	phonon–phason coupling constant
$\boldsymbol{R}$, $\boldsymbol{R}_F$	fault or displacement vector of planar faults
$\boldsymbol{r}_{\parallel}$	parallel component of displacement (shift) vector
t	time
t^*	lifetime of kinks
t_l	lifetime
t_w	waiting time
T	absolute temperature
T_m	absolute melting temperature
TEM	transmission electron microscopy
UYP	upper yield point
$\boldsymbol{U}$	displacement in higher-dimensional hyperspace
u_i	elastic displacements
$\boldsymbol{u}_{\parallel}$	elastic displacement in physical (parallel) space
$\boldsymbol{u}_{\perp}$	elastic displacement in perpendicular (orthogonal) space
v_d	dislocation velocity
v_{d0}	preexponential factor of dislocation velocity
v_j	velocity of jog along dislocation
v_k	kink velocity
V	volume, activation volume
V_{ex}	experimental activation volume

w	dislocation multiplication coefficient
W	mechanical energy
W_{el}	elastic energy
W_{fdk}	formation energy of double-kink
W_{fk}	formation energy of kink
W_{int}	interaction energy between kinks of double-kink
W_{p}	peierls energy
W_{t}	total mechanical energy
W_{pot}	potential energy
x, y, z	coordinates (x in forward direction of motion, z along dislocation line)
$x_{\mathrm{h}}, x_{\mathrm{s}}$	area fractions of hard and soft regions in composite model
x_0	width of localized obstacles in x direction
z	number of atoms in unit cell, atomic number
β	angle between Burgers vector and dislocation line direction, size effect interaction constant between dislocations and spherical inclusions
ΔA	activation area
Δd	activation distance
ΔF	Helmholtz free energy of activation
ΔF_{cb}	Helmholtz free energy of formation of critical bulge
ΔF_{fdk}	Helmholtz free energy of formation of double-kink
ΔF_{fj}	Helmholtz free energy of formation of jog
ΔF_{fk}	Helmholtz free energy of formation of kink
ΔF_{mk}	Helmholtz free energy of motion of kinks
ΔG	Gibbs free energy of activation
ΔG_{ann}	Gibbs free energy of dislocation annihilation
ΔG_{fi}	Gibbs free energy of formation of an interstitial
ΔG_{fj}	Gibbs free energy of formation of a jog
ΔG_{fv}	Gibbs free energy of formation of a vacancy
ΔG_{sd}	Gibbs free energy of self-diffusion
ΔH	activation enthalpy
ΔS	activation entropy
ΔV_{ext}	external volume change during formation of point defect
ΔW	work term
ε	plastic (normal) strain
$\varepsilon_{\mathrm{el}}$	elastic strain
ε_{t}	total strain, elastic plus plastic
ε_{ij}	components of elastic strain tensor
$\dot{\varepsilon}$	plastic (normal) strain rate
$\dot{\varepsilon}_{\mathrm{el}}$	elastic strain rate
$\dot{\varepsilon}_{\mathrm{t}}$	total strain rate, elastic plus plastic
$\Delta\epsilon$	difference between longitudinal and transversal elastic strain
γ	(plastic) shear strain, energy of planar faults
γ_{APB}	antiphase boundary energy

γ_s	surface energy
γ_{sf}	stacking fault energy
$\dot{\gamma}$	plastic shear strain rate
$\dot{\gamma}_0$	preexponential factor of shear strain rate
Γ	line tension
Θ	work-hardening coefficient
λ	displacement of dislocation, sweeping distance of kinks
μ	shear modulus, chemical potential
ν	Poisson's ratio
ν_D	Debye frequency
ν_0	attempt frequency
ξ	line vector of dislocation
ξ_0	normalized width of localized obstacles in x (ξ) direction
ϱ	dislocation density
ϱ_0	mass density of material
ϱ_f	forest dislocation density
ϱ_m	mobile dislocation density
σ	normal (engineering) stress
σ^*	effective stress in quasicrystals
σ_i	internal stress in quasicrystals
σ_{ij}	components of stress tensor
σ_{ph}	stress contribution due to the creation of phason walls
σ_y	yield stress or strength
$\sigma_{\parallel}$	stress components in parallel space
$\sigma_{\perp}$	stress components in perpendicular space
Σ	stress tensor
τ	shear stress, value of golden mean
τ^*	effective shear stress
τ'_0	normalized athermal flow stress of random obstacle array
τ'	normalized (dimensionless) stress
τ_f	friction stress inside particles
τ_{FR}	critical stress of Frank–Read source
τ_h, τ_s	shear strength in hard and soft regions in composite model
τ_i	(long-range) internal stress
τ_{id}	theoretical shear strength of an ideal crystal
τ_{loc}	locally acting shear stress
τ_{ly}, τ_{uy}	lower and upper yield stress at yield drop
τ_{OR}	Orowan stress
τ_p	Peierls stress
τ_{self}	self-stress of dislocation
τ_{th}	threshold stress, related to internal stress component τ_i
ω	jump frequency of kinks, of point defects
Ω	atomic volume, elementary strain if all dislocations overcome an obstacle
$\Delta\Omega$	atomic volume difference

List of Video Clips

Most of the video sequences were produced within the research of the Plasticity Group in the former Institute of Solid State Physics of the Academy of Sciences of the GDR and the present Max Planck Institute of Microstructure Physics, both in Halle (Saale). The main authors are Ulrich Messerschmidt and Martin Bartsch. Exceptions are the Videos 8.2 and 9.1, the authors of which, their affiliations, as well as further authors are listed together with the particular materials.

6 Semiconductors

- Polycrystalline Silicon
  Detlef Hoehl and Martina Werner
  6.1 Dislocation motion in polycrystalline silicon at about 500 and 550°C
  6.2 Dislocation motion and generation in polycrystalline silicon at about 650°C
- Ge and Ge-Si Alloys
  Martina Werner and Ichiro Yonenaga
  6.3 Several slip bands in Ge at about 430°C
  6.4 Motion of dislocations in a dislocation network in Ge at about 610°C
  6.5 Dislocation band in a network in Ge-4 at% Si at about 655°C
  6.6 Dislocation bands in Si-5 at% Ge at about 745°C
  6.7 Dislocation band in Si-5 at% Ge at about 835°C

7 Ceramic Single Crystals

- Magnesium Oxide
  Fritz Appel
  7.1 Dislocation motion and generation in MgO single crystals at room temperature
  7.2 Motion of a dislocation in MgO at room temperature

- Zirconia Single Crystals
  Dietmar Baither, Bernd Baufeld, and Manfred Rühle

7.3 Dislocation motion in c-ZrO_2 on a $\{001\}$ plane during deformation along $[\bar{1}1\bar{2}]$ at 1150°C
7.4 Dislocation processes in c-ZrO_2 on a (211) plane during in situ straining along $[\bar{1}1\bar{2}]$ at 1150°C
7.5 Dislocation motion in c-ZrO_2 on a $\{10\bar{1}\}$ plane during deformation along [100] at 1150°C
7.6 Domain switching in t′-ZrO_2 during ferroelastic deformation at 1150°C
7.7 Domain switching in partially stabilized ZrO_2 during ferroelastic deformation at 1150°C
7.8 Dislocation motion in partially stabilized ZrO_2 after ferroelastic deformation at 1150°C

8 Metallic Alloys

- Al-1 at % Ag
  Rainer Hattenhauer, Peter J. Wilbrandt, and Peter Haasen
  8.1 Dislocation motion in an Al-Ag single crystal at room temperature
- Computer Simulation of Dislocation Motion in a Material Hardened by Precipitates of an Intermetallic Phase
  Video by Volker Mohles and Eckhard Nembach, Institute of Materials Research of Münster University
  8.2 Simulation of the motion of a dislocation pair in Nimonic PE16
- Dislocation Generation in Al-1 at% Ag
  Rainer Hattenhauer, Peter J. Wilbrandt, and Peter Haasen
  8.3 Loop generation and opening in an Al-Ag single crystal
- Dislocation Generation in Ti-6 wt% Al
  T. Neeraj and Michael J. Mills
  8.4 Creation and opening of dislocation loops in air-cooled Ti-6 wt% Al at room temperature
  8.5 Dislocation generation at a jog in step-annealed Ti-6 wt% Al at room temperature
  8.6 Localized (Frank–Read) sources in step-annealed Ti-6 wt% Al at room temperature
- Dislocation Generation in Duplex Steel
  Witold Zielinski
  8.7 Dislocation multiplication in a ferrite grain during the deformation of duplex steel at room temperature
  8.8 Frank–Read source in an austenite grain during the deformation of duplex steel at room temperature
- Oxide Dispersion Strengthened (ODS) Material INCOLOY MA956
  Anna Wasilkowska and Aleksandra Czyrska-Filemonowicz
  8.9 Dislocation motion in INCOLOY MA956 at room temperature
  8.10 Dislocation motion in MA956 at 700°C
  8.11 Dislocation motion in annealed MA956 at 700°C
  8.12 Dislocation motion in MA956 at 1010°C
  8.13 Collective dislocation motion in MA956 at 1010°C

- Oxide Dispersion Strengthened (ODS) Material INCONEL MA754
Dietrich Häussler and Bernd Reppich
8.14 Dislocation motion in the ODS alloy INCONEL MA754 at 655, 745, and 790°C
- Plastic Deformation During Fracture of Al_2O_3/Nb Sandwich Specimens
Christian Dietzsch, Christina Scheu, Wolfgang Kurtz, and Manfred Rühle
8.15 Dislocation motion in the plastic zone of cracks in Al_2O_3/Nb sandwiches

9 Intermetallic Alloys

- Ni_3Al
Video by G. Molénat and Daniel Caillard, Centre d'Elaboration de Materiaux et d'Etudes Structurales, CNRS, Toulouse
9.1 APB jumps in Ni_3Al-0.25at%Hf at room temperature
- γ-TiAl
Dietrich Häussler, Mark Aindow, and Ian Jones
9.2 Motion of ordinary dislocations in γ-TiAl at room temperature
9.3 Motion of ordinary dislocations in slip bands in γ-TiAl at room temperature
9.4 Motion of $\langle 101 \rangle$ superdislocations in γ-TiAl at room temperature
9.5 Motion of twinning dislocations in γ-TiAl at room temperature
9.6 Motion of ordinary dislocations in γ-TiAl at 655°C
- NiAl
Ralf Haushälter and Dietrich Häussler
9.7 Motion of dislocations with $\langle 100 \rangle$ Burgers vectors in an NiAl single crystal at room temperature
9.8 Motion of dislocations with $\langle 100 \rangle$ Burgers vectors in NiAl-0.2 at% Ta at room temperature
9.9 Motion of dislocations with $\langle 100 \rangle$ Burgers vectors in an NiAl single crystal at 475°C
9.10 Motion of dislocations with $\langle 100 \rangle$ Burgers vectors in an NiAl single crystal at 565°C
9.11 Motion of dislocations with $\langle 110 \rangle$ Burgers vectors on $\{110\}$ planes in NiAl-0.2 at% Ta at 475°C
- FeAl
Christian Dietzsch and Easo P. George
9.12 Motion of dislocations with $\langle 111 \rangle$ Burgers vectors in Fe-43 at% Al at room temperature
9.13 Motion of dislocations with $\langle 111 \rangle$ Burgers vectors in Fe-43 at% Al at 530°C
9.14 Motion of dislocations with $\langle 100 \rangle$ Burgers vectors in Fe-43 at% Al at 447 and 530°C
9.15 Dynamic formation and annihilation of dislocations in Fe-43 at% Al at 700°C

References

1. A. Smekal, Handb. Phys. **24**, 2 (1933)
2. J. Frenkel, Z. Phys. **37**, 572 (1926)
3. E. Orowan, Z. Phys. **89**, 634 (1934)
4. M. Polanyi, Z. Phys. **89**, 660 (1934)
5. G.I. Taylor, Proc. Roy. Soc. A **145**, 362 (1934)
6. J.M. Burgers, Proc. R. Neth. Akad. Sci. **42**, 293, 378 (1939)
7. V. Volterra, Ann. Ecole Norm. Super. **24**, 400 (1907)
8. U. Messerschmidt, M. Bartsch, Ch. Dietzsch, Intermetallics **14**, 607 (2006)
9. U.F. Kocks, A.S. Argon, M.F. Ashby, in *Progress in Materials Science*, vol. 19, ed. by B. Chalmers, J.W. Christian, T.B. Massalski (Pergamon, Oxford, 1975)
10. E. Nadgornyi, in *Progress in Materials Science*, vol. 31, ed. by B. Chalmers, J.W. Christian, T.B. Massalski (Pergamon, Oxford, 1988)
11. D. Caillard, J.-L. Martin, *Thermally Activated Mechanisms in Crystal Plasticity* (Pergamon, Amsterdam, 2003)
12. J.P. Hirth, J. Lothe, *Theory of Dislocations*, 2nd edn. (Wiley, New York, 1982)
13. E. Schmid, W. Boas, *Kristallplastizität* (Springer, Berlin, 1935)
14. M. Bartsch, Z.-F. Zhang, Ch. Scheu, M. Rühle, U. Messerschmidt, Z. Metallkd. **95**, 9 (2004)
15. Ch. Dietzsch, M. Bartsch, U. Messerschmidt, Phys. Status Solidi A **202**, 2249 (2005)
16. B. Geyer, M. Bartsch, M. Feuerbacher, K. Urban, U. Messerschmidt, Philos. Mag. A **80**, 1151 (2000)
17. A.G. Evans, R.D. Rawlings, Phys. Status Solidi **34**, 9 (1969)
18. J.J. Gilman, W.G. Johnston, in *Dislocations and Mechanical Properties of Crystals* (Wiley, New York, 1957), p. 116
19. W.G. Johnston, J.J. Gilman, J. Appl. Phys. **30**, 129 (1959)
20. D.L. Stein, J.R. Low, J. Appl. Phys. **31**, 362 (1960)
21. B.Ya. Farber, S. Chiarelli, A.H. Heuer, Philos. Mag. **72**, 59 (1995)
22. P.B. Hirsch, A. Howie, P.B. Nicholson, D.W. Pashley, M.J.Whelan, *Electron Microscopy of Thin Crystals* (Butterworths, London, 1965)
23. M. von Heimendahl, *Electron Microscopy of Materials: An Introduction* (Academic, New York, 1980)

24. J.W. Edington, *Practical Electron Microscopy in Materials Science*, Repr. of original of 1976 (TechBooks, Herndon, Va., 1991)
25. L.E. Murr, *Electron and Ion Microscopy and Microanalysis: Principles and Applications*, 2nd edn. (Dekker, New York, 1991)
26. S. Amelinckx, *Electron Microscopy: Principles and Fundamentals* (VCH, Weinheim, 1997)
27. A. Couret, J. Crestou, S. Farenc, G. Molenat, N. Clement, A. Coujou. D. Caillard, Microsc. Microanal. Microstruct. **4**, 153 (1993)
28. A. Nohara, N. Yonezawa, T. Chikada, T. Imura, in *Proc. 8th Int. Conf. on High Voltage Electron Microscopy* (Canberra, 1974), p. 570
29. U. Valdré, in *Proc. 7th Int. Conf. on Electron Microscopy* Part I (Grenoble, 1970), p. 131
30. H. Saka, K. Noda, T. Imura, Y. Kashikuwa, S. Katagiri, in *Proc. 5th Europ. Congr. on Electron Microscopy* (Manchester, 1972), p. 354
31. H. Saka, T. Imura, N. Yukawa, Jpn. J. Appl. Phys. **10**, 1 (1971)
32. R. Valle, J.-L. Martin, in *Proc. 8th Int. Conf. on High Voltage Electron Microscopy* (Canberra, 1974), p. 180
33. U. Messerschmidt, F. Appel, Ultramicroscopy **1**, 223 (1976)
34. U. Messerschmidt, M. Bartsch, Ultramicroscopy **56**, 163 (1994)
35. H. Fujita, K. Komatsu, in *Proc. 7th Int. Conf. on High Voltage Electron Microscopy* (Berkeley, 1983), p. 371
36. B. Baufeld, U. Messerschmidt, D. Baither, M. Bartsch, A. Foitzik, M. Rühle, in *Plastic Deformation of Ceramics*, ed. by R.C. Bradt et al. (Plenum, New York, 1995), p. 43
37. A. Coujou, P. Lours, N.A. Roy, D. Caillard, N. Clement, Acta Metall. Mater. **38**, 825 (1990)
38. J.L. Martin, L.P. Kubin, Ultramicroscopy **3**, 215 (1978)
39. J.L. Martin, L.P. Kubin, Phys. Status Solidi A **56**, 487 (1979)
40. B. Baufeld, D. Baither, U. Messerschmidt, M. Bartsch, I. Merkel, J. Am. Ceram. Soc. **76**, 3163 (1993)
41. B. Raghothamachar, G. Dhanaraj, J. Bau, M. Dudley, Microsc. Res. Tech. **69**, 343 (2006)
42. K. Sumino, H. Harada, Philos. Mag. A **44**, 1319 (1981)
43. F. Vallino, A. Jaques, A. George, Phys. Status Solidi B **222**, 51 (2000)
44. U. Messerschmidt, Y. Nishino, T. Imura, H. Saka, Phys. Status Solidi A **76**, 277 (1983)
45. A. Seeger, J. Diehl, S. Mader, H. Rebstock, Philos. Mag. **2**, 323 (1957)
46. C. Schwink, H. Neuhäuser, Phys. Status Solidi **6**, 679 (1964)
47. H. Neuhäuser, in *Dislocations in Solids*, vol. 6, ed. by F.R.N. Nabarro (North Holland, Amsterdam, 1983), p. 319
48. U. Hoffmann, A. Hampel, H. Neuhäuser, J. Microscopy (Oxford) **203**, 108 (2001)
49. L. Casarotto, R. Tutsch, R. Ritter, H. Dierke, F. Klose, H. Neuhäuser, Comput. Mater. Sci. **32**, 316 (2005)
50. G.A. Bassett, Philos. Mag. **3**, 1042 (1958)
51. H. Bethge, Phys. Status Solidi **2**, 3 and 775 (1962)
52. D.E. Kramer, M.F. Savage, L.E. Levine, Acta Mater. **53**, 4655 (2005)
53. J. Fikar, C. Coupeau, T. Kruml, J. Bonneville, Mater. Sci. Eng. A **387**, 926 (2004)

54. R. De Batist, *Internal Friction of Structural Defects in Crystalline Solids* (North-Holland, Amsterdam, American Elsevier Publishing, New York, 1972)
55. R. Schaller, G. Fantozzi, G. Gremaud (eds.), *Mechanical Spectroscopy 2001 with Applications to Materials Science*, Mater. Sci. Forum **366–368** (2001)
56. M. Weller, Mater. Sci. Forum **366–368**, 95 (2001)
57. J.L. Snoek, Physica **6**, 161, 321, 591, 797 (1939)
58. J.L. Snoek, Physica **8**, 711 (1941)
59. M. Weller, Mater. Sci. Eng. **442**, 21 (2006)
60. P.G. Bordoni, J. Acoust. Soc. Am. **26**, 495 (1954)
61. G. Alefeld, in *Lattice Defects and Their Interaction*, ed. by R. Hasiguti (Gordon and Breach, New York, 1967)
62. J.S. Koehler, in *Proc. Symp. on Imperfections in Nearly Perfect Crystals, Poconon Manor PA, 1950*, ed. by W. Shockley, J.H. Holomon, R. Maurer, F. Seitz (Wiley, New York 1952), p. 197
63. A.V. Granato, K. Lücke, J. Appl. Phys. **25**, 583 (1956)
64. V.V. Zoninashvili, I.A. Naskidashvili, U. Messerschmidt, V.A. Melik-Shakhnazarov, N.M. Yasterbova, L.M. Kolshnikova, Phys. Status Solidi A **75**, K205 (1983)
65. B.K. Kardashev, S.B. Kustov, A.B. Lebedev, G.V. Berezhkova, P.P. Perstnev, F. Appel, U. Messerschmidt, Phys. Status Solidi A **91**, 79 (1985)
66. Y. Estrin, K. Lücke, Scripta Metall. **14**, 341 (1980)
67. A.B. Lebedev, J. Alloys Compd. **211/212**, 177 (1994)
68. A.B. Lebedev, Phys. Solid State **35**, 1141 (1993)
69. S.L. Archbutt, Trans. Faraday Soc. **17**, 22 (1921)
70. H. Blaha, B. Langenecker, Naturwissenschaften **42**, 556 (1955)
71. J.Th.M. DeHosson, O. Kanert, A.W. Sleeswyk, in *Dislocations in Solids*, vol. 6, ed. by F.R.N. Nabarro (North Holland, Amsterdam, 1983), p. 441
72. O. Kanert, Phys. Status Solidi **30**, 127 (1968)
73. W.H.M. Alsem, J.Th.M. DeHosson, Philos. Mag. **46**, 327 (1982)
74. A. Christian, O. Kanert, J.Th.M. DeHosson, Acta Metall. Mater. **38**, 2479 (1990)
75. K.L. Murty, K. Detemple, O. Kanert, G. Peters, J.Th.M. DeHosson, J. Mater. Sci. **31**, 3289 (1996)
76. F.C. Frank, Phil. Mag. **42**, 809 (1951)
77. E. Orowan, Proc. Phys. Soc., **52**, 8 (1940)
78. S. Timoshenko, J.N. Goodier, *Theory of Elasticity* (McGraw-Hill, New York, 1951), p. 116
79. P. Haasen, U. Messerschmidt, W. Skrotzki, Mater. Sci. Eng. **81**, 493 (1986)
80. G. Schoeck, R. Frydman, Phys. Status Solidi B **53**, 661 (1972)
81. W.M. Lomer, Philos. Mag. **42**, 1327 (1951
82. A.H. Cottrell, Philos. Mag. **43**, 645 (1952)
83. J.P. Hirth, J. Appl. Phys. **32**, 700 (1961)
84. M. Peach, J.S. Koehler, Phys. Rev. **80**, 436 (1950)
85. F.R.N. Nabarro, Advanc. Phys. **1**, 269 (1952)
86. D. Kuhlmann-Wilsdorf, in *Strength of Metals and Alloys*, ed. by P.O. Kettunen, T.K. Lepistö, M.E. Lehtonen (Pergamom, Oxford, 1988), p. 221
87. A. Seeger, Z. Naturforschg. A **11**, 985 (1956)
88. W.T. Read, *Dislocations in Crystals* (McGraw-Hill, New York, 1953)
89. C.S. Hartley, J.P. Hirth, Acta Metall. **13**, 79 (1965)

90. R. Bullough, J.V. Sharp, Philos. Mag. A **11**, 605 (1965)
91. P. Haasen, in *Dislocations in Solids*, vol. 4, ed. by F.R.N. Nabarro (North Holland, Amsterdam, 1979), chap. 15
92. H. Neuhäuser, C. Schwink, in *Materials Science and Technology*, vol. 6, ed. by H. Mughrabi (VCH, Weinheim, 1993), chap. 5
93. R.L. Fleischer, J. Appl. Phys. **33**, 3504 (1962)
94. G.B. Gibbs, Philos. Mag. **20**, 611 (1969)
95. D.M. Barnett, W.D. Nix, Acta Metall. **21**, 1157 (1973)
96. V.M. Chernov, M.M. Savin, Phys. Status Solidi A **55**, 113 (1979)
97. G. DeWitt, J.S. Koehler, Phys. Rev. **116**, 1113 (1959)
98. D.J. Bacon, R.O. Scattergood, J. Phys. F: Metal Phys. **4**, 2126 (1974)
99. U. Essmann, Phys. Status Solidi **12**, 723 (1965)
100. V.K. Pare, J.C. Crump, J. Appl. Phys. **40**,790 (1969)
101. U. Messerschmidt, F. Appel, Krist. Tech. **14**, 1331 (1979)
102. U. Messerschmidt, F. Appel, in *Strength of Metal and Alloys*, ed. by P. Haasen, V. Gerold, G. Kostorz (Pergamon, Toronto, 1979), p. 1305
103. H. Schmid, H.E. Exner, Z. Metallkunde **72**, 412 (1981)
104. U. Messerschmidt, F. Appel, H. Schmid, Philos. Mag. **51**, 781 (1985)
105. R.O. Scattergood, D.J. Bacon, Philos. Mag. **31**, 179 (1975)
106. L.M. Brown, Philos. Mag. **15**, 363 (1967)
107. V.L. Indenbom, S.S. Orlov, Soviet Phys. JETP Lett. **6**, 274 (1967)
108. V.L. Indenbom, S.S. Orlov, Soviet Phys. Crystallogr. **12**, 849 (1968)
109. H. Schmid, H.O.K. Kirchner, Philos. Mag. A **58**, 905 (1988)
110. F. Prinz, H.O.K. Kirchner, G. Schoeck, Philos. Mag. A **38**, 321 (1978)
111. F. Prinz, A. Korner, H.O.K. Kirchner, Philos. Mag. A **47**, 441 (1983)
112. J. Lothe, J.P. Hirth, Phys. Status Solidi B **242**, 836 (2005)
113. U. Messerschmidt, F. Appel, M. Bartsch, in *Deformation of Ceramics II*, ed. by R.E. Tressler, R.C. Bradt (Plenum, New York, 1984), p. 223
114. M.P. Puls, M.J. Norgett, J. Appl. Phys. **47**, 466 (1976)
115. U. Messerschmidt, Z. Metallkunde **84**, 391 (1993)
116. F.C. Frank, W.T. Read, Phys. Rev. **79**, 722 (1950)
117. F.C. Frank, Physica **15**, 131 (1949)
118. R.D. Heidenreich, W. Shockley, in *Report on a Conference on Strength of Solids* (Physical Society, London, 1948), p. 57
119. M. Bartsch, U. Messerschmidt, A.D. Vasilev, Phys. Status Solidi A **146**, 173 (1994)
120. F.C. Frank, Proc. Phys. Soc. **62A**, 202 (1949)
121. J. Eyring, J. Chem. Phys. **4**, 283 (1936)
122. G. Schoeck, Phys. Status Solidi **8**, 499 (1965)
123. J. Friedel, *Dislocations* (Pergamon, Oxford, 1964)
124. A.V. Granato, K. Lücke, J. Schlipf, L.J. Teutonico, J. Appl. Phys. **35**, 2732 (1964)
125. U. Dehlinger, A. Kochendörfer, Z. Phys. **116**, 576 (1940)
126. J. Frenkel, T. Kontorova, Phys. Z. Sowj. **13**, 1 (1938)
127. R.E. Peierls, Proc. Phys. Soc. **52**, 23 (1940)
128. F.R.N. Nabarro, Proc. Phys. Soc. **59**, 256 (1947)
129. D. Eshelby, Philos. Mag. **40**, 903 (1949)
130. J.P. Hirth, J. Lothe, *Theory of Dislocations*, 1st edn. (McGraw-Hill, New York, 1968), pp. 217–220

131. A.J.E. Foreman, M.A. Jawson, J.K. Wood, Proc. Phys. Soc. **64A**, 156 (1951)
132. F. Granzer, G. Wagner, J. Eisenblätter, Phys. Status Solidi **30**, 587 (1968)
133. C.H. Woo, M.P. Puls, Philos. Mag. **35**, 727 (1977)
134. M.S. Daw, S.M. Foiles, M.I. Baskes, Mater. Sci. Reports **9**, 251 (1993)
135. V. Vitek, Intermetallics **6**, 579 (1998)
136. V. Vitek, A. Mrovec, R. Groger, J.L. Bassani, V. Racherla, L. Yin, Mater. Sci. Eng. A **387**, 138 (2004)
137. H. Koizumi, H.O.K. Kirchner, T. Suzuki, Philos. Mag. **86**, 3835 (2006)
138. A. Seeger, P. Schiller, in *Physical Acoustics III A*, ed. by W.P. Mason (Academic, New York, 1966), p. 361
139. P.G. Shewmon, *Diffusion in Solids* (McGrw-Hill, New York, 1963)
140. P. Heitjans, *Diffusion in Condensed Matter: Methods, Materials, Models* (Berlin, Springer, 2005)
141. L. Farkas, Z. Phys. Chem. **125**, 236 (1927)
142. J.B. Zeldovich, Acta Physicochem. USSR **18**, 1 (1943)
143. J. Frenkel, *Kinetic Theory of Liquids* (Oxford University Press, Fair Lawn, 1946)
144. J.P. Hirth, G.M. Pound, *Condensation and Evaporation* (Pergamon, New York, 1963)
145. J. Feder, K.C. Russell, J. Lothe, G.M. Pound, Adv. Phys. **15**, 111 (1966)
146. M. Volmer, *Kinetik der Phasenbildung* (Steinkopff, Dresden, 1939)
147. A. Seeger, P. Schiller, Acta Metall. **10**, 348 (1962)
148. A. Seeger, in *Dislocations, Int. Colloq. CNRS*, ed. by P. Veyssiere, L. Kubin, J. Castaing (CNRS, 1984), p. 141
149. D. Brunner, J. Diehl, A. Seeger, in *The Structure and Properties of Crystal Defects*, Mater. Sci. Monographs, vol. 20, ed. by V. Paidar, L. Lejcek (Elsevier, Amsterdam, 1984), p. 175
150. J. Lothe, J.P. Hirth, Phys. Rev. **115**, 543 (1959)
151. V. Celli, M. Kabler, T. Ninomiya, R. Thomson, Phys. Rev. **131**, 58 (1963)
152. J.E. Dorn, S. Rajnak, Trans. Metall. Soc. AIME **230**, 1052 (1964)
153. B.V. Petukhov, V.L. Pokrovskii, JETP **36**, 336 (1973)
154. V.L. Indenbom, B.V. Petukhov, J. Lothe, in *Elastic Strain Fields and Dislocation Mobility*, ed. by V.L. Indenbom, J. Lothe (Elsevier Science Publisher, 1992), p. 489
155. I.V. Gektina, F.F. Lavrentev, V.D. Natsik, JETP **79**, 1927 (1980)
156. P. Haasen, C. Barthel, T. Suzuki, in *Dislocations in Solids, Proc. Yamada Conference IX*, ed. by H. Suzuki, T. Ninomiya, K. Sumino, S. Takeuchi (University of Tokyo Press, Tokyo, 1985), p. 165
157. B.V. Petukhov, in *Crystal Lattice Defects and Dislocation Dynamics*, ed. by R.A. Vardanian (Nova Science Publishers Inc., Huntington, 2000)
158. B.V. Petukhov. Soviet Phys. - Solid State **13**, 1204 (1971)
159. J.-P. Bouchaud, A. Georges, Phys. Rep. **195**, 127 (1990)
160. V. Vitek, Philos. Mag. **84**, 415 (2004)
161. K. Edagawa, T. Suzuki, S. Takeuchi, Phys. Rev. B **55**, 6180 (1997)
162. V. Bulatov, Wei Cai, *Computer Simulations of Dislocations* (Oxford University Press, Oxford, 2006)
163. M. Werner, M. Bartsch, U. Messerschmidt, D. Baither, Phys. Status Solidi A **146**, 133 (1994)
164. U. Messerschmidt, F. Appel, Mater. Sci. Eng. A **113**, 409 (1989)

165. R.L. Fleischer, Acta Metall. **7**, 134 (1959)
166. G. Schoeck, A. Seeger, *Rep. Conf. on Defects in Crystalline Solids* (Physical Society, London, 1955)
167. B. Escaig, J. Phys. (France) **29**, 225 (1968)
168. J. Friedel, *Dislocations and Mechanical Properties of Crystals* (Wiley, New York, 1957), p. 330
169. A.N. Stroh, Proc. Phys. Soc. **B67**, 427 (1954)
170. W. Püschl, Prog. Mater. Sci. **47**, 415 (2002)
171. S. Rao, T.A. Parthasarathy, C. Woodward, Philos. Mag. **79**, 1167 (1999)
172. T. Rasmussen, in *Multiscale Phenomena in Plasticity*, ed. by J. Lépinoux et al. (Kluwer, London, 2000), p. 281
173. D. Mordehai, I. Kelson, G. Makov, Mater. Sci. Eng. A **400–401**, 37 (2005)
174. G. Schöck, Philos. Mag. Lett. **89**, 505 (2009)
175. J. Bonneville, B. Escaig, Acta Metall. **27**, 1477 (1979)
176. J. Bonneville, B. Escaig, J.L. Martin, Acta Metall. **36**, 1989 (1988)
177. F. Appel, U. Messerschmidt, V. Schmidt, O.V. Klyavin, A.V. Nikiforov, Mater. Sci. Eng. **56**, 211 (1982)
178. F. Appel, U. Messerschmidt, Phys. Status Solidi **35**, 1003 (1969)
179. F. Appel, H. Grube, U. Messerschmidt, B.I. Smirnov, Cryst. Lattice Defects **7**, 65 (1982)
180. H. Wiedersich, J. Appl. Phys. **33**, 854 (1962)
181. F. Appel, U. Messerschmidt, in *Dislocations in Solids, Proc. Yamada Conference IX*, ed. by H. Suzuki, T.Ninomiya, K. Sumino, S. Takeuchi (University of Tokyo Press, Tokyo, 1985), p. 463
182. P. Haasen, in *Physical Metallurgy*, 2nd edn., ed. by R. Cahn (North-Holland, Amsterdam, 1970), chap. 17
183. B. Escaig, J. de Physique **35**, C7, 151 (1974)
184. V. Vitek, Phys. Status Solidi **18**, 687 (1966)
185. A. Couret, D. Caillard, Philos. Mag. A **59**, 783 (1989)
186. D. Caillard, Philos. Mag. Lett. **89**, 517 (2009)
187. R.B. Schwarz, R. Labusch, J. Appl Phys. **49**, 5174 (1978)
188. J. Friedel, *Les Dislocations* (Gautier-Villars, Paris, 1956), p. 205
189. N.F. Mott, F.R.N. Nabarro, in *Rep. Conf. on the Strength of Solids* (Physical Society, London, 1948), p. 1
190. U. Messerschmidt, F. Appel, in *Electron Microscopy in Solid State Physics*, ed. by H. Bethge, J. Heydenreich (Elsevier, Amsterdam, 1987), p. 325
191. G.B. Gibbs, Mater. Sci. Eng. **4**, 313 (1969)
192. A.J.E. Foreman, M.J. Makin, Canad. J. Phys. **45**, 511 (1967)
193. J.W. Morris, D.H. Klahn, J. Appl. Phys. **44**, 4882 (1973)
194. J.W. Morris, D.H. Klahn, J. Appl. Phys. **45**, 2027 (1974)
195. S.I. Zaitsev, E.M. Nadgornyi, Sov. Phys. Solid State **15**, 1777 (1974)
196. S.I. Zaitsev, E.M. Nadgornyi, Nuclear Metallurgy **20**, 707 (1976)
197. S.I. Zaitsev, E.M. Nadgornyi, Nuclear Metallurgy **20**, 816 (1976)
198. K. Hanson, J.W. Morris, J. Appl. Phys. **46**, 983 (1975)
199. R. Labusch, J. Appl. Phys. **48**, 4550 (1977)
200. S. Altintas, Ph. D. Thesis, Univ. of California, Berkeley 1978
201. A.I. Landau, Phys. Status Solidi A **61**, 555 (1980), **65**, 119 (1981), **65**, 415 (1981)
202. A.I. Landau, Phys. Status Solidi A **30**, 659 (1975)

203. K. Hanson, J.W. Morris, J. Appl. Phys. **46**, 2378 (1975)
204. U. Messerschmidt, Doctoral Thesis B, Martin Luther University, Halle-Wittenberg (1986)
205. V.N. Vydashenko, A.I. Landau, Fiz. Nizkikh. Temp. **5**, 794 (1979)
206. U.F. Kocks, in *Physics of Strength and Plasticity*, ed. by A.S. Argon (MIT, 1969), p. 143
207. G. Schoeck, Phys. Status Solidi A **87**, 571 (1985)
208. R.C. Picu, R.G. Li, Z.J. Xu, Mater. Sci. Eng. A **502**, 164 (2009)
209. B.M. Strunin, A.B. Popov, Phys. Status Solidi A **34**, 761 (1976)
210. R. Labusch, R.B. Schwarz, in *Strength of Metals and Alloys*, ed. by D.G. Brandon, R. Chaim, A. Rosen (Freund Publishing, London, 1991), p. 47
211. B.V. Petukhov. Phys. Metall. Metall. **56**, 123 (1983)
212. B.V. Petukhov, M. Bartsch, U. Messerschmidt, Eur. Phys. J. AP **9**, 89 (2000)
213. B. Baufeld, B.V. Petukhov, M. Bartsch, U. Messerschmidt, Acta Mater. **46**, 3077 (1998)
214. P. Teracher, H. Garem, J. Rabier, in *Strength of Metals and Alloys*, ed. by D.G. Brandon, R. Chaim, A. Rosen (Freund Publishing, London, 1991), p. 217
215. B.V. Petukhov, Crystallography Reports, **52**, 112 (2007)
216. A. Melander, Scripta Metall. **11**, 393 (1977)
217. U. Messerschmidt, Mater. Sci. Eng. **100**, 101 (1988)
218. B. Reppich, Acta Metall. **23**, 1055 (1975)
219. E. Orowan, in *Symp. on Internal Stresses in Metals and Alloys* (The Institute of Metals, London, 1948), p. 451
220. D.J. Bacon, U.F. Kocks, R.O. Scattergood, Philos. Mag. **26**, 1241 (1973)
221. M.F. Ashby, in *Physics of Strength and Plasticity* (MIT, 1969), p. 113
222. R.O. Scattergood, E.S.P. Das, in *Conf. on Computer Simulation for Materials Applications* (Gaithersburg Maryland, 1976)
223. V. Gerold, in *Dislocations in Solids*, vol. 4, *Dislocations in Metallurgy*, ed. by F.R.N. Nabarro (North-Holland, Amsterdam, 1979), p. 219
224. A.J. Ardell, Metall. Trans. A **16**, 2131 (1985)
225. E. Nembach, *Particle Strengthening of Metals and Alloys* (Wiley, New York, 1997)
226. U. Messerschmidt, F. Appel, M. Bartsch, R. Gerlach, Phys. Status Solidi A **78**, 93 (1983)
227. W. Püschl, Philos. Mag. Lett. **80**, 199 (2000)
228. F. Appel, Philos. Mag. A **63**, 71 (1991)
229. V.I. Alshits, V.L. Indenbom, in *Dislocations in Solids*, vol. 7, ed. by F.R.N. Nabarro (North-Holland, Amsterdam, 1986), p. 43
230. P. Gumbsch, H. Gao, Science **283**, 965 (1999)
231. G. Leibfried, Z. Phys. **127**, 344 (1950)
232. F.R.N. Nabarro, Proc. Roy. Soc. A **209**, 278 (1951)
233. B. Tittmann, H. Bommel, Phys. Rev. Lett. **14**, 296 (1965)
234. E. Bitzek, P. Gumbsch, Mater. Sci. Eng. A **387**, 11 (2004)
235. H. Kojima, T. Suzuki, Phys. Rev. Lett. **21**, 898 (1968)
236. V.V. Pustovalov, V.I. Startsev, V.S. Fomenko, Preprint Fiz. Tekh. Inst. Nizk. Temp. Akad. Nauk Ukr. SSR, Kharkov (1968)
237. V.P. Soldatov, V.I. Startsev, T.I. Vainblat, J. Low Temp. Phys. **2**, 641 (1970)
238. G. Kostorz, Phys. Status Solidi B **58**, 9 (1973)
239. V.V. Pustovalov, V.S. Fomenko, Low Temperature Phys. **32**, 1 (2006)

240. V.D. Natsik, A.I. Osetskii, V.P. Soldatov, V.I. Startsev, Phys. Status Solidi B **54**, 99 (1972)
241. H. Koizumi, H.O.K. Kirchner, T. Suzuki, Phys. Rev. B **65**, 214104 (2002)
242. J.F. Stohr, J.P. Poirier, Philos. Mag. **25**, 1313 (1972)
243. G. Edelin, J.P. Poirier, Philos. Mag. **28**, 1203 (1973)
244. F.R.N. Nabarro, in *Report on the Conference on the Strength of Solids* (Physical Society, London, 1947), p. 75
245. C. Herring, J. Appl. Phys. **21**, 437 (1950)
246. D.N. Seidman, R.W. Baluffi, Phys. Status Solidi **17**, 531 (1966)
247. R.W. Baluffi, Phys. Status Solidi **31**, 443 (1969)
248. N.F. Mott, Proc. Phys. Soc. B **64**, 729 (1951)
249. N. Louat, C.A. Johnson, Philos. Mag. **7**, 2051 (1962)
250. U. Messerschmidt, Phys. Status Solidi **26**, 701 (1968)
251. U. Messerschmidt, Phys. Status Solidi **41**, 549 (1970), **48**, 781 (1971)
252. N.F. Mott, Philos. Mag. **43**, 1151 (1952)
253. F. Seitz, Phys. Rev. **88**, 722 (1952)
254. H. Saka, T. Kondo, T. Imura, Philos. Mag. A **47**, 859 (1983)
255. H. Suzuki, Sci. Repts. Tohoku Univ. A **4**, 455 (1952)
256. H. Suzuki, J. Phys. Soc. Japan **17**, 322 (1962)
257. H. Suzuki, in *Dislocations in Solids*, vol. 4, *Dislocations in Metallurgy*, ed. by F.R.N. Nabarro (North-Holland, Amsterdam, 1979), p. 191
258. A.H. Cottrell, M.A. Jawson, Proc. Roy. Soc. A **199**, 104 (1949)
259. A.H. Cottrell, Philos. Mag. **74**, 829 (1953)
260. B.V. Petukhov, Crystallography Reports **54**, 82 2009
261. G. Schoeck, A. Seeger, Acta Metall. **7**, 469 (1959)
262. M. Imai, K. Sumino, Philos. Mag. A **47**, 599 (1983)
263. J.J. Gilman, Australian J. Phys. **13**, 327 (1960)
264. J.J. Gilman, J. Appl. Phys. **36**, 3195 (1965)
265. Wei Cai, V.V. Bulatov, Mater. Sci. Eng. A **387–389**, 277 (2004)
266. D. Lorenz, A. Zeckzer, U. Hilpert, P. Grau, H. Johansen, H.S. Leipner, Phys. Rev. B **67**, 172101 (2003)
267. L. Johnson, M.F. Ashby, Acta Metall. **16**, 219 (1968)
268. J.B. Bilde-Sörensen, Acta Metall. **21**, 1495 (1973)
269. S. Saimoto, H. Saka, T. Imura, Scripta Metall. **11**, 903 (1977)
270. C.B. Carter, Philos. Mag. **35**, 75 (1977)
271. A.R. Rosenfield, G.T. Hahn, in *Dislocation Dynamics*, ed. by A.R. Rosenfield, G.T. Hahn, A.L. Bement Jr., R.I. Jaffee (McGraw Hall, New York, 1968), p. 255
272. S.I. Zaitsev, E.M. Nadgornyi, Phys. Status Solidi A **9**, 353 (1971)
273. S.I. Zaitsev, E.M. Nadgornyi, Fiz. Tverd. Tela **12**, 1846 (1970)
274. J.S. Koehler, Phys. Rev. **86**, 52 (1952)
275. E. Orowan, in *Dislocations in Metals* (Am. Inst. Mining and Metallurg. Eng., New York, 1954), p. 103
276. W.G. Johnston, J.J. Gilman, J. Appl. Phys. **31**, 632 (1960)
277. P.P. Gillis, J.J. Gilman, J. Appl. Phys. **36**, 3370 (1965)
278. P. Haasen, Z. Phys. **167**, 461 (1962)
279. H. Alexander, P. Haasen, in *Solid State Physics*, vol. 22, ed. by F. Seitz, D. Turnbull, H. Ehrenreich (Academic, New York, 1968), p. 27
280. U. Messerschmidt, M. Bartsch, Mater. Chem. Phys. **81**, 518 (2003)

281. F. Appel, H. Bethge, U. Messerschmidt, Phys. Status Solidi A **42**, 61 (1977)
282. D. Baither, B. Baufeld, U. Messerschmidt, M. Bartsch, Mater. Sci. Eng. A **233**, 75 (1997)
283. R. Hattenhauer, M. Bartsch, U. Messerschmidt, P. Haasen, P.-J. Wilbrandt, Philos. Mag. A **70**, 447 (1994)
284. U. Messerschmidt, M. Bartsch, Mater. Sci. Eng. A **164**, 332 (1993)
285. M. Bartsch, T. Neeraj, M. Mills, unpublished results
286. D. Häussler, M. Bartsch, M. Aindow, I.P. Jones, U. Messerschmidt, Philos. Mag. A **79**, 1045 (1999)
287. U. Messerschmidt, R. Haushälter, M. Bartsch, Mater. Sci. Eng. A **234–236**, 822 (1997)
288. S. Guder, M. Bartsch, M. Yamaguchi, U. Messerschmidt, Mater. Sci. Eng. A **261**, 139 (1999)
289. A. Dominguez-Rodriguez, D.-S. Cheong, A.H. Heuer, Philos. Mag. A **64**, 923 (1993)
290. A.S. Tetelman, Acta Metall. **10**, 813 (1962)
291. H.S. Chen, A.K. Head, J.J. Gilman, J. Appl. Phys. **35**, 2502 (1964)
292. P. Neumann, Acta Metall. **19**, 1233 (1971)
293. P.P. Gillis, J.J. Gilman, J. Appl. Phys. **36**, 3380 (1965)
294. D.N. Seidman, R.W. Baluffi, Philos. Mag. **13**, 649 (1966)
295. K.P. Lagerlöf, T.E. Mitchell, A.H. Heuer, J. Am. Ceram. Soc. **72**, 2159 (1989)
296. B. Devincre, T. Hoc, L.P. Kubin, Mater. Sci. Eng. A **400–401**, 182 (2005)
297. F.J. Basinski, Z.S. Basinski, in *Dislocations in Solids*, vol. 4, ed. by F.R.N. Nabarro (North-Holland, Amsterdam, 1979), p. 261
298. B. Devincre, L.P. Kubin, Modelling Simul. Mater. Sci. Eng. **2**, 559 (1994)
299. B. Devincre, L.P. Kubin, Phil. Trans. R. Soc. Lond. A **355**, 2003 (1997)
300. B. Devincre, T. Hoc, L. Kubin, Science **320**, 1745 (2008)
301. B. Neuhaus, C. Schwink, Philos. Mag. A **65**, 1463 (1992)
302. F. Appel, Phys. Status Solidi A **61**, 477 (1980)
303. D. Kuhlmann-Wilsdorf, Trans. AIME **224**, 1047 (1962)
304. K. Mecke, U. Messerschmidt, Krist. Tech. **14**, 1319 (1979)
305. H. Mughrabi, Acta Metall. **31**, 1367 (1983)
306. C. Holste, G.K. Schmidt, R. Torber, Phys. Status Solidi A **54**, 305 (1979)
307. H. Mughrabi, Philos. Mag. **86**, 4037 (2006)
308. H. Mughrabi, Mater. Sci. Eng. A **319–321**, 139 (2001)
309. H. Mughrabi, Mater. Sci. Eng. A **309–310**, 237 (2001)
310. J.J. Gilman, J. Appl. Phys. **33**, 2703 (1962)
311. F. Kroupa, Acta Metall. **14**, 60 (1966)
312. M. Beneš, J. Kratochvíl, J. Křištan, V. Minarik, and P. Pauš, Eur. Phys. J. Special Topics **177**, 177 (2009)
313. A. Seeger, Philos. Mag. **46**, 1194 (1955)
314. H.S. Chen, J.J. Gilman, A.K. Head, J. Appl. Phys. **35**, 2502 (1964)
315. J.C.M. Li, in *Dislocation Dynamics*, ed. by A.R. Rosenfield, G.T. Hahn, A.L. Bement Jr., R.I. Jaffee (McGraw Hall, New York, 1968), p. 87
316. W.D. Nix, W.A. Coghlan, C.R. Barrett, Mater. Sci. Eng. **4**, 98 (1969)
317. M.I. Baskes, A.W. Thompson, J. Appl. Phys. **44**, 892 (1973)
318. A.H. Cottrell, R.J. Stokes, Proc. Roy. Soc. A **233**, 17 (1955).
319. Y.J. Li, J. Mueller, H.W. Höppel, M. Göken, W. Blum, Acta Mater. **55**, 5708 (2007)

320. E. Kröner, *Kontinuumstheorie der Versetzungen und Eigenspannungen*, Ergebnisse der Angewandten Mathematik, vol. 5 (Springer, Berlin, 1958)
321. C.S. Hartley, Philos. Mag. **83**, 3783 (2001)
322. G.B. Gibbs, Philos. Mag. **13**, 317 (1966)
323. T. Kruml, O. Coddet, G. Saada, J.L. Martin, Philos. Mag. Lett. **83**, 651 (2003)
324. B. Baufeld, M. Bartsch, U. Messerschmidt, D. Baither, Acta Metall. Mater. **43**, 1925 (1995)
325. S. Straub, W. Blum, H.J. Maier, T. Ungar, A. Borbely, H. Renner, Acta Mater. **44**, 4337 (1996)
326. M. Hecker, E. Thiele, C. Holste, Acta Mater. **50**, 2357 (2002)
327. G. Schoeck, J. Appl. Phys. **33**, 1745 (1962)
328. J. Bailey, P.B. Hirsch, Philos. Mag. **5**, 485 (1960)
329. A.S. Keh, J. Appl. Phys. **31**, 1501 (1960)
330. R.K. Ham, Philos. Mag. **6**, 1183 (1961)
331. T. Ungar, H. Mughrabi, M. Wilkens, Acta Metall. **30**, 1861 (1982)
332. T. Ungar, A. Borbely, Appl. Phys. Lett. **69**, 3173 (1996)
333. T. Ungar, I. Dragomir, A. Revesz, A. Borbely, J. Appl. Cryst. **32**, 992 (1999)
334. F. Appel, U. Messerschmidt, Phys. Status Solidi A **131**, 289 (1992)
335. D. Brunner, J. Diehl, Phys. Status Solidi A **160**, 355 (1997)
336. U. Messerschmidt, B.V. Petukhov, M. Bartsch, Ch. Dietzsch, B. Geyer, D. Häussler, L. Ledig, M. Feuerbacher, P. Schall, K. Urban, Mater. Sci. Eng. A **319–321**, 107 (2001)
337. L. Ledig, M. Bartsch, U. Messerschmidt, Mater. Sci. Eng. A **429**, 79 (2006)
338. P. Schall, M. Feuerbacher, M. Bartsch, U. Messerschmidt, K. Urban, Philos. Mag. Lett. **79**, 785 (1999)
339. S. Takeuchi, A.S. Argon, J. Mater. Sci. **11**, 1542 (1976)
340. W.R. Cannon, T.G. Langdon, J. Mater. Sci. **18**, 1 (1983)
341. W. Blum, in *Materials Science and Technology*, vol. 6, ed. by R.W. Cahn, P. Haasen, E.J. Kramer (VCH, Weinheim, 1993), p. 359
342. P. Eisenlohr, W. Blum, Mater. Sci. Eng. A **400–401**, 175 (2005)
343. B. Burton, Mater. Sci. Technol. **5**, 1005 (1989)
344. M. Bartsch, A. Tikhonovsky, U. Messerschmidt, Phys. Status Solidi A **201**, 46 (2004)
345. Y. Estrin, L.P. Kubin, Res Mechanica **23**, 197 (1988)
346. M.A. LeChatelier, Rev. Metall. **6**, 914 (1909)
347. A. Portevin, F. LeChatelier, Comp. Rend. Sci. Paris **176**, 507 (1923)
348. L.P. Kubin, Y. Estrin, J. Phys. (Paris) **47**, 497 (1986)
349. L.P. Kubin, K. Chihab, Y. Estrin, Acta Metall. **36**, 2707 (1988)
350. L.P. Kubin, Y. Estrin, Acta Metall. Mater. **38**, 697 (1990)
351. M. Zaiser, P. Hähner, Phys. Status Solidi B **199**, 267 (1997)
352. C. Fressengeas, A.J. Beaudoin, M. Lebyodkin, L.P. Kubin, Y. Estrin, Mater. Sci. Eng. A **400–401**, 226 (2005)
353. A. van den Beukel, Phys. Status Solidi A **30**, 197 (1975)
354. P.G. McCormick, Acta Metall. **36**, 3061 (1988)
355. T. Wutzke, Ch. Schwinck, Phys. Status Solidi A **137**, 337 (1993)
356. N. Louat, Scripta Metall. **15**, 1167 (1981)
357. H. Alexander, in *Dislocations in Solids*, vol. 7, ed. by F.R.N. Nabarro (North-Holland, Amsterdam, 1986), p. 113
358. K. Sumino, in *Handbook of Semiconductors*, vol. 3, ed. by S. Mahajan (Elsevier, 1994), p. 73

359. I. Yonenaga, Mater. Trans. **46**, 1979 (2005)
360. I.L.F. Ray, D.J.H. Cockayne, Proc. Roy. Soc. A **325**, 543 (1971)
361. G. Packeiser, P. Haasen, Philos. Mag. **35**, 821 (1977)
362. H. Gottschalk, G. Patzer, H. Alexander, Phys. Status Solidi A **14**, 177 (1978)
363. M. Sato, K. Hiraga, K. Sumino, Jpn. J. Appl. Phys. **19**, L155 (1980)
364. D. Gerthsen, F.A. Ponce, G.B. Anderson, Philos. Mag. A **59**, 45 (1989)
365. M. Sato, K. Sumino, Krist. Tech. **14**, 1351 (1979)
366. H. Gottschalk, J. Phys. **40**, C-6-177 (1979)
367. K. Wessel, H. Alexander, Philos. Mag. **35**, 1523 (1977)
368. I. Yonenaga, M. Werner, M. Bartsch, U. Messerschmidt, E.R. Weber, Phys. Status Solidi A **171**, 35 (1999)
369. G. Packeiser, Philos. Mag. A **41**, 459 (1980)
370. R. Meingast, H. Alexander, Phys. Status Solidi A **17**, 229 (1973)
371. T. Suzuki, H. Kojima, Acta Metall. **14**, 913 (1966)
372. A. George, C. Escavarage, G. Champier, W. Schröter, Phys. Status Solidi B **53**, 483 (1972)
373. B.Ya. Farber, Yu.L. Iunin, V.I. Nikitenko, Phys. Status Solidi A **97**, 469 (1986)
374. I. Yonenaga, J. Mater. Sci. (Materials in Electronics) **10**, 329 (1999)
375. F. Louchet, Philos. Mag. A **43**, 1289 (1981)
376. F. Louchet, D. Cochet Muchy, Y. Brechet, J. Pelissier, Philos. Mag. A, **57**, 327 (1988)
377. C. Levade, J.J. Couderc, D. Caillard, A. Couret, in *Electron Microscopy in Plasticity and Fracture Research of Materials*, ed. by U. Messerschmidt, F. Appel, J. Heydenreich, V. Schmidt. Physical Research, vol. 14 (Akademie-Verlag, Berlin, 1990), p. 199
378. F. Louchet, J. Pelissier, D. Caillard, J.P. Peyrade, C. Levade, G. Vanderschaeve, Microscopy, Microanalysis, Microstructures **4**, 199 (1993)
379. Y.M. Huang, J.C.H. Spence, O.F. Sankey, Phys. Rev. Lett. **74**, 3392 (1995)
380. R. Krause-Rehberg, M. Brohl, H.S. Leipner, Th. Brost, A. Polity, U. Beyer, H. Alexander, Phys. Rev. B **47**, 13266 (1993)
381. M. Werner, E.R. Weber, M. Bartsch, U. Messerschmidt, Phys. Status Solidi A **150**, 337 (1995)
382. K. Maeda, Y. Yamashita, N. Maeda, S. Takeuchi, MRS Symp. Proc. **184**, 69 (1990)
383. K. Maeda, S. Takeuchi, J. Physique **44**, C-375 (1983)
384. K. Maeda, S. Takeuchi, in *Dislocations in Solids*, chap. 54, ed. by F.R.N. Nabarro, M.S. Duesbery (Elsevier Science publishers B.V., Amsterdam, 1996), p. 443
385. I. Yonenaga, K. Sumino, Appl. Phys. Lett. **69**, 1264 (1996)
386. I. Yonenaga, K. Sumino, Phys. Status Solidi A **50**, 685 (1978)
387. T.E. Mitchell, A.H. Heuer, in *Dislocations in Solids*, , vol. 12, chap. 68, ed. by F.R.N. Nabarro, J.P. Hirth (Elsevier, Amsterdam, 2004), p. 339
388. M.T. Sprackling, *The Plastic Deformation of Simple Ionic Crystals* (Academic, London, 1976)
389. E.Yu. Gutmanas, E.M. Nadgornyi, A.W. Stepanov, Fiz. Tverd. Tela **5**, 1021 (1963)
390. E.Yu. Gutmanas, E.M. Nadgornyi, Fiz. Tverd. Tela **10**, 2284 (1968), Soviet Physics Solid State **10**, 1799 (1969)
391. P. Haasen, Mater. Sci. Technol. **1**, 1013 (1985)

392. W. Skrotzki, P. Haasen, J. Phys. (France) **42**, C3, 119 (1981)
393. W. Skrotzki, T. Suzuki, Radiation Effects **74**, 315 (1983)
394. J.J. Gilman, J. Appl. Phys. **44**, 982 (1973)
395. W. Frank, Phys. Status Solidi **29**, 391 and 767 (1968)
396. F. Appel, Phys. Status Solidi A **25**, 607 (1974)
397. F. Appel, Phys. Status Solidi A **31**, 615 (1975)
398. F. Appel, Mater. Sci. Eng. **50**, 199 (1981)
399. J. Hesse, Physica Status Solidi **9**, 209 (1965)
400. L. Kempter, H. Strunk, Phys. Status Solidi A **40**, 385 (1977)
401. C.Y. Cheng, Scripta Metall. **4**, 623 (1970)
402. M. Srinivasan, T.G. Stoebe, J. Mater. Sci. **8**, 1567 (1973)
403. R.N. Singh, R.L. Coble, J. Appl. Phys. **45**, 981, 990, and 5129 (1974)
404. K. Ogawa, Philos. Mag. **13**, 1097, **14**, 519 (1966)
405. F. Appel, H. Bethge, U. Messerschmidt, Phys. Status Solidi A **38**, 103 (1976)
406. F. Appel, U. Messerschmidt, E.M. Nadgornyi, S.I. Zaitsev, Mater. Sci. Eng. **52**, 69 (1982)
407. A.I. Landau, V.N. Vydashenko, personal communication
408. F. Appel, U. Messerschmidt, M. Kuna, Phys. Status Solidi A **55**, 529 (1979)
409. U. Messerschmidt, F. Appel, in *Strength of Metals and Alloys*, ed. by P.O. Kettunen, T.K. Lepistö, M.E. Lehtonen (Pergamon, Oxford, 1988), p. 139
410. V.N. Vydashenko, A.I. Landau, S.V. Lubenets, E.I. Ostapcuk, Ukrainskii Fiz. Zh. **26**, 177 (1981)
411. B.V. Petukhov, S.I. Zaitsev, private communication
412. S.N. Valkovskii, E.M. Nadgornyi, Fiz. Tverd. Tela **17**, 2603 (1975); Soviet Phys. Solid State **17**, 1733 (1975) and Fiz. Tverd. Tela **17**, 2603 (1975); Soviet Phys. Solid State **17**, 1733 (1975)
413. F. Appel, M. Bartsch, U. Messerschmidt, E.M. Nadgornyi, S.N. Valkovskii, Phys. Status Solidi A **83**, 179 (1984)
414. M. Bartsch, unpublished results
415. F. Appel, B. Wielke, Mater. Sci. Eng. **73**, 97 (1985)
416. H. Knoch, B. Reppich, Acta Metall. **23**, 1061 (1975)
417. A.A. Griffith, Phil. Trans. Roy. Soc. London A **221**, 163 (1920)
418. G.R. Irwin, *Encyclopedia of Physics*, vol. 6 (Springer, Heidelberg, 1958)
419. A.S. Tetelman, A.J. McEvily Jr., *Fracture of Structural Materials* (Wiley, New York, 1967)
420. D.S. Dugdale, J. Mech. Phys. Solids **8**, 100 (1960)
421. J.N. Sneddon, Proc. Roy. Soc. London A **187**, 229 (1946)
422. S.M. Ohr, Mater. Sci. Eng. **72**, 1 (1985)
423. A.H. Heuer, J. Am. Ceram. Soc. **70**, 689 (1987)
424. D.J. Green, R.H.J. Hannink, M.W. Swain, *Transformation Toughening of Ceramics* (CRC, Boca Raton Fla, 1989)
425. A. Dominguez-Rodriguez, V. Lanteri, A.H. Heuer, J. Am. Ceram. Soc. **69**, 285 (1986)
426. A.H. Heuer, V. Lanteri, A. Dominguez-Rodriguez, Acta Metall. **37**, 559 (1989)
427. A. Virkar, R. Matsumoto, J. Am. Ceram. Soc. **69**, C224 (1986)
428. C.-J. Chan, F.F. Lange, M. Rühle, J.-F. Jue, A.V. Virkar, J. Am. Ceram. Soc. **74**, 807 (1991)
429. A.H. Heuer, R. Chaim, V. Lanteri, in *Science and Technology of Zirconia III*, Advances in Ceramics, vol. 24, ed. by S. Somiya, N. Yamamoto, H. Hanagida (American Ceramic Society, Westerville OH, 1988), p. 3

430. J. Martinez-Fernandes, M. Jimenez-Melendo, A. Dominguez-Rodriguez, Acta Metall. Mater. **43**, 593 (1995)
431. D. Gomez-Garcia, J. Martinez-Fernandez, A. Dominguez-Rodriguez, K. Westmacott, J. Am. Ceram. Soc. **79**, 487 (1996)
432. U. Messerschmidt, B. Baufeld, D. Baither, in *Zirconia Engineering Materials*, ed. by E. Kisi (Trans Tech, Uetikon-Zürich, 1998), Key Eng. Mater. **153–154**, 143 (1998)
433. U. Messerschmidt, D. Baither, B. Baufeld, M. Bartsch, Mater. Sci. Eng. A **233**, 61 (1997)
434. B. Baufeld, D. Baither, M. Bartsch, U. Messerschmidt, Phys. Status Solidi A **166**, 127 (1998)
435. A. Tikhonovsky, M. Bartsch, U. Messerschmidt, Phys. Status Solidi A **201**, 26 (2004)
436. D. Gomez-Garcia, J. Martinez-Fernandez, A. Dominguez-Rodriguez, P. Eveno, K. Westmacott, Acta Mater. **44**, 991 (1996)
437. D.S. Cheong, A. Dominguez-Rodriguez, A.H. Heuer, Philos. Mag. A **63**, 377 (1991)
438. U. Messerschmidt, B. Baufeld, K.J. McClellan, A.H. Heuer, Acta Metall. Mater. **43**, 1917 (1995)
439. A. Dominguez-Rodriguez, K.P.D. Lagerlöf, A.H. Heuer, J. Am. Ceram. Soc. **69**, 281 (1986)
440. A. Lakki, R. Herzog, M. Weller, H. Schubert, C. Reetz, O. Görke, M. Kilo, G. Borchardt, J. Europ. Ceram. Soc. **20**, 285 (2000)
441. D. Gomez-Garcia, J. Martinez-Fernandes, A. Dominguez-Rodriguez, and K.H. Westmacott, Philos. Mag. A **79**, 1839 (1999)
442. Y. Oishi, K. Ando, Y. Sakka, in *Advances in Ceramics*, vol. 7, ed. by M.F. Yan, A.H. Heuer (American Ceramic Society, Westerville OH, 1983), p. 208
443. F.R. Chien, A.H. Heuer, Philos. Mag. A **73**, 681 (1996)
444. U.F. Kocks, J. Eng. Mater. Technol. **98**, 76 (1976)
445. U.F. Kocks, *Physical Basis for Non-Elastic Constitutive Relations*, Discussion Paper at Symp. Adv. in Metal Deformation (Cornell University, Ithaca NY, 1976)
446. J. Martinez-Fernandez, M. Jimenez-Melendo, A. Dominguez-Rodriguez, A.H. Heuer, J. Am. Ceram. Soc. **73**, 2452 (1990)
447. M. Kilo, G. Borchardt, S. Weber, S. Scherrer, K. Tinschert, Ber. Bunsenges. Phys. Chem. **101**, 1361 (1997)
448. D. Baither, B. Baufeld, U. Messerschmidt, A. Foitzik, M. Rühle, J. Am. Ceram. Soc. **80**, 1691 (1997)
449. D. Baither, M. Bartsch, B. Baufeld, A. Tikhonovski, A. Foitzik, M. Rühle, U. Messerschmidt, J. Am. Ceram. Soc. **84**, 1755 (2001)
450. B. Baufeld, D. Baither, U. Messerschmidt, M. Bartsch, A. Foitzik, M. Rühle, J. Am. Ceram Soc. **80**, 1699 (1997)
451. D. Baither, B. Baufeld, U. Messerschmidt, Phys. Status Solidi A **137**, 569 (1993)
452. K.H. Dünkeloh, G. Kralik, V. Gerold, Z. Metallk. **65**, 291 (1994)
453. A. Zahra, C.Y. Zahra, M. Laffitte, W. Lacom, H.P. Degischer, Z. Metallk. **70**, 172 (1979)
454. F. Ernst, P. Haasen, Phys. Status Solidi A **104**, 403 (1987)
455. F. Rabe, Diploma Thesis, Göttingen University (1987)

456. P.B. Hirsch, A. Kelly, Philos. Mag. **12**, 881 (1965)
457. E. Nembach, Scripta Metall. **20**, 763 (1986)
458. R. Baur, V. Gerold, Acta Metall. **10**, 637 (1962)
459. T. Al-Kassab, P. Haasen, Z. Metallk. **84**, 248 (1993)
460. H. Gleiter, E. Hornbogen, Phys. Status Solidi **12**, 235 (1965)
461. J.L. Castagné, J. de Physique **27**, C3-233 (1966)
462. R.K. Ham, Trans. Jpn. Inst. Metals **9**, Suppl. 52 (1968)
463. Y. Miura, A. Matsui, M. Furukawa, M. Nemoto, in *Aluminium-Lithium III*, ed. by C. Baker, P.J. Gregson, S.J. Harris, C.J. Peel (Institute of Metals, London, 1986), p. 427
464. J. Glazer, J.W. Morris Jr., Philos. Mag. A **56**, 107 (1987)
465. A.J. Ardell, J.C. Huang, Philos. Mag. Lett. **58**, 189 (1988)
466. H.-J. Gudladt, J. Lendvai, J. Schneider, Acta Metall. **37**, 3327 (1989)
467. E. Nembach, U. Messerschmidt, M. Kolbe, C. Schlesier, M. Bartsch, J. Pesicka, in *Strength of Metals and Alloys*, ed. by D.G. Brandon, R. Chaim, A. Rosen (Freund Publishing, London, 1991), p. 1087
468. J.Th.M. DeHosson, A. Huisin't Veld, H. Tamler, O. Kanert, Acta Metall. **32**, 1205 (1984)
469. V. Mohles, Philos. Mag. A **81**, 971 (2001)
470. B. Noble, S.J. Harris, K. Dinsdale, J. Mater. Sci. **17**, 461 (1982)
471. T. Neeraj, D.-H. Hou, G.S. Daehn, M. Mills, Acta Mater. **48**, 1225 (2000)
472. T. Neeraj, M. Mills, Mater. Sci. Eng. A **319–321**, 415 (2001)
473. J. Rösler, E. Arzt, Acta Metall. **36**, 1043 (1988)
474. E. Arzt, D.S. Wilkinson, Acta Metall. **34**, 1893 (1986)
475. D.J. Srolovitz, R. Petkovic-Luton, M.J. Luton, Scripta Metall. **16**, 1401 (1982)
476. D.J. Srolovitz, M.J. Luton, R. Petkovic-Luton, D.M. Barnett, W.D. Nix, Acta Metall. **32**, 1079 (1984)
477. E. Arzt, J. Rösler, Acta Metall. **36**, 1053 (1988)
478. J. Rösler, E. Arzt, Acta Metall. **38**, 671 (1990)
479. Y. Xiang, D.J. Srolovitz, Philos. Mag. **86**, 3937 (2006)
480. J.H. Schroeder and E. Arzt, Scripta Metall. **19**, 1129 (1985)
481. R. Herzog, A. Wasilkowska, A. Czyrska-Filemonowicz, H. Schuster, in *IXth Conf. on Eletron Microscopy of Solids* (Krakow-Zakopane, 1996), p. 379
482. M. Bartsch, A.Wasilkowska, A. Czyrska-Filemonowicz, U. Messerschmidt, Mater. Sci. Eng. A. **272**, 152 (1999)
483. D. Häussler, M. Bartsch, U. Messerschmidt, B. Reppich, Acta Mater. **49**, 3647 (2001)
484. W. Hermann, Ph.D. thesis, University of Erlangen-Nürnberg, 1995
485. A. Wasilkowska, PhD thesis, University of Mining and Metallurgy, Krakow, 1998
486. A. Czyrska-Filemonowicz, B. Dubiel, J. Mater. Proc. Technol. **64**, 53 (1997)
487. B. Reppich, M. Heilmaier, J. Wunder, H. Baumeister, T. Huber, in *Microstructure and Mechanical Properties of Metallic High-Temperature Materials*, ed. by H. Mughrabi, G. Gottstein, H. Mecking, H. Riedel, J. Tobolski (Wiley-VCH, Weinheim, 1999), p. 509
488. R. Herzog, PhD thesis, RWTH Aachen, 1997
489. L.P. Kubin, Y. Estrin, J. Phys. Paris III **1**, 929 (1991)
490. M. Heilmeier, B. Reppich, in *Proc. 5th Int. Conf. on Creep and Fracture of Eng. Mater. and Struct.*, ed. by B. Wilshire, R.W. Evans (Institute of Materials, London, 1993), p. 231

491. F. Ernst, Mater. Sci. Eng. R **14**, 97 (1995)
492. J.R. Rice, R. Thomson, Philos. Mag. **29**, 73 (1974)
493. J. Rice, Z. Suo, J.S. Wang, in *Metal-Ceramic Interfaces*, vol. 4, ed. by M. Rühle, A.G. Evans, M.F. Ashby, J.P. Hirth, Acta Scripta Metall. Proc. Ser., (Pergamon, Oxford, 1990), p. 269
494. G.E. Beltz, J.R. Rice, Acta Metall. Mater. **40**, S321 (1992)
495. D. Korn, G. Elssner, H.F. Fischmeister, M. Rühle, Acta Metall. Mater. **40**, S355 (1992)
496. D. Korn, G. Elssner, R.M. Cannon, M. Rühle, Acta Mater. **50**, 3881 (2002)
497. R.M. Cannon, D. Korn, G. Elssner, M. Rühle, Acta Mater. **50**, 3903 (2002)
498. U. Messerschmidt, M. Bartsch, C. Dietzsch, W. Kurtz, C. Scheu, M. Rühle, Z. Metallkd. **95**, 785 (2004)
499. M.S. Duesbery, R.A. Foxall, Phil. Mag. **20**, 719 (1969)
500. A.G. Atkins, Y.-W. Mai, *Elastic and Plastic Fracture* (Ellis Horwood, Chichester, 1985)
501. G. Sauthoff, *Intermetallics* (VCH, Weinheim, 1995)
502. J.H. Westbrook, R.L. Fleischer (ed.), *Intermetallic Compounds, Principles and Practice* (Wiley, Chichester, 1995)
503. M. Yamaguchi, Y. Umakoshi, Prog. Mater. Sci. **34**, 1 (1990)
504. C.T. Liu, D.P. Pope, in *Intermetallic Compounds, Principles and Practice*, ed. by J.H. Westbrook, R.L. Fleischer (Wiley, Chichester, 1995), p. 17
505. M.J. Marcinkowski, N. Brown, R.M. Fisher, Acta Metall. **9**, 124 (1961)
506. V. Vitek, Cryst. Latt. Def. **5**, 1 (1974)
507. M. Yamaguchi, in *Mechanical Properties of BCC Metals*, ed. by M. Meshii (TMS-AIME, Warrendale, 1981), p. 31
508. V. Vitek, V. Paidar, in *Dislocations in Solids*, vol. 14, ed. by J.P. Hirth (Elsevier, Amsterdam, 2008), p. 439
509. D.L. Anton, in *Intermetallic Compounds, Principles and Practice*, vol. 2, ed. by J.H. Westbrook, R.L. Fleischer (Wiley, Chichester, 1995), p. 3
510. P. Veyssiere, J. Douin, in *Intermetallic Compounds, Principles and Practice*, vol. 1, ed. by J.H. Westbrook, R.L. Fleischer (Wiley, Chichester et al., 1995), p. 519
511. H.P. Karnthaler, E.Th. Mühlbacher and C. Rentenberger, Acta Mater. **44**, 547 (1996)
512. G. Molénat, D. Caillard, Philos. Mag. A **64**, 1291 (1991)
513. G. Molénat, D. Caillard, Philos. Mag. A **69**, 939 (1994)
514. A. Couret, Y.Q. Sun, P.B. Hirsch, Philos. Mag. A **67**, 29 (1993)
515. E.M. Nadgorny, Yu.L. Iunin, in *High-Temperature Ordered Intermetallic Alloys VI*, Mater. Res. Soc. Symp. Proc. **364**, 707 (1995)
516. B.H. Kear, H.G.F. Wilsdorf, Trans. Metal. Soc. AIME **224**, 382 (1962)
517. S. Takeuchi, E. Kuramoto, Acta Metall. **21**, 415 (1973)
518. V. Paidar, D.P. Pope, V. Vitek, Acta Metall. **32**, 435 (1984)
519. D. Caillard, V. Paidar, Acta Mater. **44**, 2759 (1996)
520. D. Caillard, Acta Mater. **44**, 2774 (1996)
521. G. Saada, P. Veyssiere, Philos. Mag. A **66**, 1081 (1992) and **70**, 925 (1994)
522. C.T. Chou, P.B. Hirsch, Philos. Mag. A **68**, 1097 (1993)
523. M.J. Mills, D.C. Chrzan. Acta Metall. Mater. **40**, 3051 (1992)
524. D.C. Chrzan, C.K. Erdonmez, Mater. Sci. Eng. A **319–321**, 31 (2001)
525. B.V. Petukhov, Mater. Sci. Eng. A **309–310**, 345 (2001)

526. S.C. Huang, J.C. Chesnutt, in *Intermetallic Compounds, Principles and Practice*, vol. 2, ed. by J.H. Westbrook, R.L. Fleischer (Wiley, Chichester, 1995), p. 73
527. M. Yamaguchi, H. Inui, K. Kishida, M. Matsumuro, Y. Shirai, in *High-Temperature Ordered Intermetallic Alloys VI*, Mater. Res. Soc. Symp. Proc. **364**, 3 (1995)
528. F. Appel, M. Oehring, in *Titanium and Titanium Alloys*, ed. by C. Leyens, M. Peters (Wiley-VCH, Weinheim 2003), p. 89
529. M. Legros, A. Couret, D. Caillard, Philos. Mag. A **73**, 61 and 81 (1996)
530. M. Legros, Y. Minonishi, D. Caillard, Philos. Mag. A **76**, 995 (1997)
531. D. Häussler, U. Messerschmidt, M. Bartsch, F. Appel, R. Wagner, Mater. Sci. Eng. A **233**, 15 (1997)
532. S. Farenc, A. Coujou, A. Couret, Philos. Mag. A **67**, 127 (1993)
533. D. Caillard, in *High-Temperature Ordered Intermetallic Alloys VI*, Mater. Res. Soc. Symp. Proc., **364**, 17 (1995)
534. M.H. Yoo, private communication (1994)
535. U. Messerschmidt, M. Bartsch, S. Guder, D. Häussler, in *High-Temperature Ordered Intermetallic Alloys VIII*, Mater. Res. Soc. Symp. Proc., **552**, KK10.9.1 (1999)
536. H. Inui, M. Matsumoto, D.-H. Wu, M. Yamaguchi, Philos. Mag. A **75**, 395 (1997)
537. S.A. Court, V.K. Vasudevan, H.L. Fraser, Philos. Mag. A **61**, 141 (1990)
538. S. Farenc, A. Couret, 1993, in *High-Temperature Ordered Intermetallic Alloys V*, Mat. Res. Soc. Proc. **288**, 465 (1993)
539. A. Girshick, V. Vitek, in *High-Temperature Ordered Intermetallic Alloys VI*, Mater. Res. Soc. Symp. Proc., **364**, 145 (1995)
540. J. Panova, D. Farkas, in *High-Temperature Ordered Intermetallic Alloys VI*, Mater. Res. Soc. Symp. Proc., **364**, 151 (1995)
541. J.P. Simmons, S.I. Rao, D.M. Dimiduk, Philos. Mag. A **75**, 1299 (1997)
542. J.P. Simmons, M.J. Mills, S.I. Rao, in *High-Temperature Ordered Intermetallic Alloys VI*, Mater. Res. Soc. Symp. Proc., **364**, 137 (1995)
543. B.A. Greenberg, O.V. Antonova, V.N. Indenbaum, L.E. Karkina, A.B. Notkin, M.V. Ponomarev, L.V. Smirnov, Acta Metall. Mater. **39**, 233 (1991)
544. M.A. Morris, Intermetallics **4**, 417 (1996)
545. M. Aindow, K. Chaudhuri, S. Das, H.L. Fraser, Scripta Metall. Mater. **24**, 1105 (1990)
546. B.M. Kad, H.L. Fraser, Philos. Mag. Lett. **70**, 211 (1994)
547. F. Louchet, B. Viguier, Philos. Mag. A **71**, 1313 (1995)
548. B. Viguier, K.J. Hemker, J. Bonneville, F. Louchet, J.-L. Martin, Philos. Mag. A **71**, 1295 (1995)
549. S. Sriram, D.M. Dimiduk, P.M. Hazzledine, V.K. Vasudevan, Philos. Mag. A **76**, 965 (1997)
550. V. Vitek, Prog. Mater. Sci. **36**, 1 (1992)
551. G. Hug, A. Loiseau, P. Veyssiere, Philos. Mag. A **57**, 499 (1988)
552. G. Molénat, A. Couret, D. Caillard, Mater. Sci. Eng. A **234–236**, 660 (1997)
553. J. Malaplate, D. Caillard, A. Couret, Philos. Mag. **84**, 3671 (2004)
554. F. Appel, U. Lorenz, U. Sparka, R. Wagner, in *Strength of Materials*, ed. by H. Oikawa, M. Maruyama, S. Takeuchi, M. Yamaguchi (Japan Institute of Metals, Sendai, 1994), p. 341

555. U. Brossmann, R. Würschum, K. Badura, H.-E. Schaefer, Phys. Rev. B **49**, 6457 (1994)
556. C. Herzig, T. Przeorski, Y. Mishin, Intermetallics **7**, 389 (1999)
557. D. Caillard, Mater. Sci. Eng. A **319–321**, 74 (2001)
558. U. Fröbel, F. Appel, Intermetallics **14**, 1187 (2006)
559. F. Appel, U. Lorenz, M. Oehring, U. Sparka, R. Wagner, Mater. Sci. Eng. A **233**, 1 (1997)
560. H.-E. Schaefer, K. Badura-Gergen, Defects and Diffusion Forum **143–147**, 193 (1996)
561. Z. Dong, W. Liming, Z. Dunxu, Y. Mei, C. Qigong, J. Mater. Sci. Lett. **11**, 1026 (1992)
562. D.B. Miracle, Acta Metall. Mater. **41**, 649 (1993)
563. G. Sauthoff, Z. Metallkunde **80**, 337 (1989)
564. G. Sauthoff, Intermetallics **8**, 1101 (2000)
565. U. Messerschmidt, D. Häussler, M. Bartsch, G. Sauthoff, Intermetallics **7**, 455 (1999)
566. D. Baither, F. Ernst, T. Wagner, M. Rühle, M. Bartsch, U. Messerschmidt, Intermetallics **7**, 479 (1999)
567. N. Rusovic, H. Warlimont, Phys. Status Solidi A **44**, 609 (1977)
568. M.H. Loretto, R.J. Wasilewski, Philos. Mag. **23**, 1311 (1971)
569. A. Ball, R.E. Smallman, Acta Metall. **14**, 1517 (1966)
570. R.T. Pascoe, C.W.A. Newey, Metal. Sci. J. **2**, 138 (1968)
571. R. Srinivasan, M.S. Daw, R.D. Noebe, M.J. Mills, Philos. Mag. **83**, 1111 (2003)
572. M.J. Mills, D.B. Miracle, Acta Metall. Mater. **41**, 85 (1993)
573. U. Messerschmidt, S. Guder, D. Häussler, M. Bartsch, in *Proc. XVIth Physical Metallurgy and Materials Science Conference on Advanced Materials and Technologies, AMT'2001*, Inzynieria Materialowa **5**, 627 (2001)
574. D. Goldberg, G. Sauthoff, Intermetallics **4**, 253 (1996)
575. Y. Sun, G. Taylor, R. Dariola, P.M. Hazzledine, in *High-Temperature Ordered Intermetallic Alloys VI*, Mater. Res. Soc. Symp. Proc., **364**, 261 (1995)
576. H.L. Fraser, M.H. Loretto, R.E. Smallman, Philos. Mag. A **28**, 667 (1973)
577. J.T. Kim, PhD Thesis, Univ. of Michigan (1990)
578. R.D. Field, D.F. Lahrman, R. Dariola, Acta Metall. Mater. **39**, 2951 (1991)
579. J.T. Kim, R. Gibala, in *High-Temperature Ordered Intermetallic Alloys IV*, Mater. Res. Soc. Symp. Proc. **213**, 261 (1991)
580. C.L. Fu, M.H. Yoo, Acta Metall. Mater. **40**, 703 (1992)
581. R. Srinivasan, J. Brown, M.S. Daw, R.D. Noebe, M.J. Mills, Philos. Mag. A **80**, 2841 (2000)
582. J. Brown, R. Srinivasan, M.J. Mills, M.S. Daw, Philos. Mag. A **80**, 2855 (2000)
583. M.J. Mills, R. Srinivasan, M.S. Daw, Philos. Mag. A **77**, 801 (1998)
584. H. Saka, Y.M. Zhu, Philos. Mag. A **51**, 629 (1985)
585. Y. Umakoshi, M. Yamaguchi, Philos. Mag. A **41**, 573 (1980)
586. I. Baker, D.J. Gaydosh, Mater. Sci. Eng. **96**, 147 (1987)
587. H. Xiao, I. Baker, Scripta Metall. Mater. **28**, 1411 (1993)
588. B.K. Kad, J.A. Horton, Mater. Sci. Eng. A **239–240**, 118 (1997)
589. K. Yoshimi, M.H. Yoo, S. Hanada, Acta Mater. **46**, 5769 (1998)
590. M.H. Yoo, K. Yoshimi, S. Hanada, Acta Mater. **47**, 3579 (1999)
591. H.J. Leamy, E.D. Gibson, F.X. Kayser, Acta Metall. **15**, 1827 (1967)
592. I. Baker, J.A. Horton, Philos. Mag. A **67**, 479 (1993)

593. E.P. George, I. Baker, Philos. Mag. A **77**, 737 (1998)
594. Y. Umakoshi, M. Yamaguchi, Y. Namba, M. Murakami, Acta Metall. **24**, 89 (1976)
595. K. Yoshimi, S. Hanada, in *Structural Intermetallics*, ed. by R. Dariola, J.J. Lewandowski, C.T. Liu, P.L. Martin, D.B. Miracle, M.V. Nathal (Metallurgical Society of AIME, Warrendale PA, 1993), p. 475
596. D.G. Morris, Philos. Mag. A **71**, 1281 (1995)
597. R.L. Carleton, E.P. George, R.H. Zee, Intermetallics **3**, 433 (1995)
598. U. Reimann, G. Sauthoff, Intermetallics **7**, 437 (1999)
599. E.P. George, I. Baker, Intermetallics **6**, 759 (1998)
600. D.G. Morris, C.T. Liu, E.P. George, Intermetallics **7**, 1059 (1999)
601. D. Wu, I. Baker, P.R. Munroe, E.P. George, Intermetallics **15**, 103 (2007)
602. H.-E. Schaefer, B. Damson, M. Weller, E. Arzt, E.P. George, Phys. Status Solidi A **160**, 531 (1997)
603. J. Friedel, *Dislocations* (Pergamon, Oxford, 1964), p. 364
604. M.H. Yoo, T. Takasugi, S. Hanada, O. Izumi, Mater. Trans. JIM **31**, 435 (1990)
605. H.-E. Schaefer, K. Badura-Gergen, Defect and Diffusion Forum **143–147**, 193 (1997)
606. H. Neuhuser, C. Schwink, in *Plastic Deformation and Fracture of Materials*, Materials Science and Technology, Vol. 6, ed. by H. Mughrabi (VCH, Weinheim, 1993), p. 191
607. D.G. Morris, J.C. Joyce, M. Lebeouf, Philos. Mag. A **69**, 961 (1994)
608. K. Yoshimi, S. Hanada, M.H. Yoo, Acta Metall. Mater. **43**, 4141 (1995)
609. M. Eggersmann, H. Mehrer, Philos. Mag. A **80**, 1219 (2000)
610. C. Vailhe, D. Farkas, Acta Mater. **45**, 4463 (1997)
611. C. Vailhe, D. Farkas, Philos. Mag. A **79**, 921 (1999)
612. A.K. Vasudevan, J.J. Petrovic, Mat. Sci. Eng. A **155**, 1 (1992)
613. Y. Umakoshi, T. Sakagami, T. Hirano, T. Yamane, Acta Metall. Mater. **38**, 909 (1990)
614. S.A. Maloy, A.H. Heuer, J.J. Lewandowski, T.E. Mitchell, Acta Metall. Mater. **40**, 3159 (1992)
615. S.A. Maloy, T.E. Mitchell, J.J. Lewandowski, A.H. Heuer, Philos. Mag. Lett. **67**, 313 (1993)
616. S.A. Maloy, T.E. Mitchell, A.H. Heuer, Acta Metall. Mater. **43**, 657 (1995)
617. K. Ito, H. Inui, S. Shirai, M. Yamaguchi, Philos. Mag. A **72**, 1075 (1995)
618. U. Messerschmidt, M. Bartsch, S. Guder, D. Häussler, R. Haushälter, M. Yamaguchi, Intermetallics **6**, 729 (1998)
619. S. Guder, PhD Thesis, Martin Luther University, Halle-Wittenberg (2000)
620. U. Messerschmidt, S. Guder, L. Junker, M. Bartsch, M. Yamaguchi, Mater. Sci. Eng. A **319–321**, 342 (2001)
621. U. Messerschmidt, S. Guder, D. Häussler, M. Bartsch, in *High-Temperature Ordered Intermetallic Alloys IX.*, Mater. Res. Soc. Symp. Proc. **652**, Y11.6.1 (2001)
622. L. Junker, M. Bartsch, U. Messerschmidt, Mater. Sci. Eng. A **328**, 181 (2001)
623. Y. Umakoshi, T. Sakagami, T. Yamane, Philos. Mag. Lett. **59**, 159 (1989)
624. B.K. Kad, K.S. Vecchio, R.J. Asaro, B.P. Bewlay, Philos. Mag. A **72**, 1 (1995)
625. K. Ito, T. Nakamoto, H. Inui, M. Yamaguchi, in *High-Temperature Ordered Intermetallic Alloys VII*, Mater. Res. Soc. Symp. Proc. **460**, 599 (1997)

626. T.E. Mitchell, S.A. Maloy, in *Critical Issues in the Development of High Temperature Structural Materials*, ed. by N.S. Stoloff, D.J. Duquette, A.F. Giamei (The Mineral, Metals and Materials Society, Warrendale 1993), p. 279
627. S. Guder, M. Bartsch, U. Messerschmidt, Philos. Mag. A **82**, 2737 (2002)
628. K. Ito, K. Matsuda, Y. Shirai, H. Inui, M. Yamaguchi, Mater. Sci. Eng. A **261**, 99 (1999)
629. D.J. Evans, S.A. Court, P.M. Hazzledine, H.L. Fraser, Philos. Mag. Lett. **67**, 331 (1993)
630. M.I. Baskes, Mater. Sci. Eng. A **261**, 165 (1999)
631. T.E. Mitchell, M.I. Baskes, R.G. Hoagland, A. Misra, Intermetallics **9**, 849 (2001)
632. M. Salamon, H. Mehrer, Defect and Diffusion Forum **216–217**, 161 (2003)
633. M. Salamon, A. Strohm, T. Voss, P. Laitinen, I. Riimimäki, S. Divinski, W. Frank, J. Räisänen, H. Mehrer, Philos. Mag. **84**, 737 (2004)
634. R. Schroll, P. Gumbsch, Phys. Status Solidi A **166**, 475 (1998)
635. D. Shechtman, I. Blech, D. Gratias, J.W. Cahn, Phys. Rev. Lett. **53**, 1951 (1984)
636. D. Levine, P.J. Steinhardt, Phys. Rev. Lett. **53**, 2477 (1984)
637. R. Penrose, Bull. Inst. Math. Appl. **10**, 266 (1974)
638. C. Janot, *Quasicrystals - A Primer* (Clarendon, Oxford, 1992)
639. K. Hiraga, M. Hirabayashi, Jpn. J. Appl. Phys. L155 (1987)
640. T. Shibuya, T. Hashimoto, S. Takeuchi, Jpn. J. Appl. Phys. **29**, 349 (1990)
641. L. Bresson, D. Gratias, J. Non-Crystalline Solids **153–154**, 468 (1993)
642. S. Takeuchi, X. Hashimoto, Jpn. J. Appl. Phys. **30**, 561 (1993)
643. J.E. Shield, M.J. Kramer, R.W. McCallum, J. Mater. Res. **9**, 343 (1994)
644. A.P. Tsai, A. Inoue, T. Masumoto, Mater. Trans. Japan Inst. Metals **30**, 666 (1989)
645. Y. Yokoyama, T. Miura, A.P. Tsai, A. Inoue, T. Masumoto, Mater. Trans. Japan Inst. Metals **33**, 97 (1992)
646. M. Wollgarten, M. Beyss, K. Urban, H. Liebertz, U. Köster, Phys. Rev. Lett. **71**, 549 (1993)
647. M. Wollgarten, M. Bartsch, U. Messerschmidt, M. Feuerbacher, R. Rosenfeld, M. Beyss, K. Urban, Philos. Mag. Lett. **71**, 99 (1995)
648. J.M. Dubois, S.S. Kang, Y. Massiani, J. Non-Crystalline Solids **153–154**, 443 (1993)
649. K.F. Kelton, in *Quasicrystals*, Mater. Res. Soc. Proc. **553**, 471 (1999)
650. A. Inoue, H.M. Kimura, in *Quasicrystals*, Mater. Res. Soc. Proc. **553**, 495 (1999)
651. D.H. Bae, S.H. Kim, W.T. Kim, D.H. Kim, in *Proc. 4th Pacific Rim Int. Conf. on Advanced Materials and Processing*, ed. by S. Hanada, Z. Zhong, S.W. Nam, R.N. Wright (2001), p. 99
652. A. Janner, T. Janssen, Phys. Rev. B **15**, 643 (1977)
653. M. Duneau, A. Katz, Phys. Rev. Lett. **54**, 2688 (1985)
654. V. Elser, Acta Cryst. A **42**, 36 (1986)
655. J.W. Cahn, D. Shechtman, D. Gratias, J. Mater. Res. **1**, 13 (1986)
656. M. Boudard, M. de Boissieu, C. Janot, G. Heeger, C. Beeli, H.-U. Nissen, H. Vincent, R. Ibberson, M. Audier, J.M. Dubois, J. Phys. Condens. Matter **4**, 10149 (1992)
657. A. Yamamoto, H. Takakura, A.P. Tsai, Phys. Rev. B **68**, 094201 (2003)

658. M. Quiquandon, D. Gratias, Phys. Rev. B **74**, 214205 (2006)
659. T. Kupke, H.R. Trebin, S. Hoffmann, J. Non-Crystalline Solids **153**, 196 (1993)
660. W. Yang, M. Feuerbacher, K. Urban, J. Alloys Compd. **342**, 164 (2002)
661. L.A. Bendersky, Phys. Rev. Lett. **55**, 1461 (1985)
662. K. Chattopadhyay, S. Lele, S. Ranganathan, G.N. Subbanna, N. Thangaraj, Current Science **54**, 895 (1985)
663. A.P. Tsai, A. Inoue, T. Masumoto, Mater. Trans. **30**, 463 (1989)
664. W. Steurer, Acta Cryst. B **45**, 534 (1989)
665. K.K. Fung, C.Y. Yang, Y.Q. Zhou, J.G. Zhao, W.S. Zhan, B.G. Shen, Phys. Rev. Lett. **56**, 2060 (1986)
666. M. Bartsch, P. Schall, M. Feuerbacher, U. Messerschmidt, J. Mater. Res. **20**, 1814 (2005)
667. W. Steurer, T. Haibach, B. Zhang, S. Kek, R. Lück, Acta Cryst. B **49**, 661 (1993)
668. Th. Zumkley, H. Mehrer, K. Freitag, M. Wollgarten, N. Tamura, K. Urban, Phys. Rev. B **54**, 6815 (1996)
669. W. Sprengel, T.A. Lograsso, H. Nakajima, Defects and Diffusion Forum **143–147**, 849 (1997)
670. R. Blüher, P. Scharwächter, W. Frank, H. Kronmüller, Phys. Rev. Lett. **80**, 1014 (1998)
671. Z. Zhang, M. Wollgarten, K. Urban, Philos. Mag. Lett. **61**, 125 (1989)
672. Y. Yan, R. Wang, Philos. Mag. Lett. **66**, 253 (1992)
673. J. Devaud-Rzepski, A. Quivy, Y. Calvayrac, M. Cornier-Quiquandon, D, Gratias, Philos. Mag. B **60**, 855 (1989)
674. M.X. Dai, K. Urban, Philos. Mag. Lett. **67**, 67 (1993)
675. F. Baier, M.A. Müller, B. Grushko, H.-E. Schaefer, Mater. Sci. Eng. A **294–296**, 650 (2000)
676. F. Baier, H.-E. Schaefer, J. Alloys Compd. **345**, 318 (2002)
677. H. Mehrer, R. Galler, J. Alloys Compd. **345**, 296 (2002)
678. K. Kimura, H. Matsuda, R. Tamamura, M. Fujimori, R. Schmechel, H. Werheit, in *Proc. 5th Int. Conf. on Quasicrystals*, ed. by C. Janot, R. Mosseri (World Scientific, Singapore, 1995), p. 730
679. J.E.S. Socolar, T.C. Lubensky, P.J. Steinhardt, Phys. Rev. B **34**, 3345 (1986)
680. P.A. Kalugin, A. Kataev, L. Levitov, Pisma Zh Eksp. Teor. Fiz. **41**, 100 (1985), JETP Lett. **41**, 120 (1985)
681. K. Hiraga, M. Hirabayashi, A. Inoue, T. Masumoto, Sci. Rep. Res. Inst. Tohoku Univ. A **32**, 309 (1985)
682. S. Francoual, F. Livet, M. de Boissieu, F. Yakhou, F. Bley, A. Letoublon, R. Caudron, J. Gastaldi, R. Currat, Philos. Mag. **86**, 1029 (2006)
683. K. Edagawa, K. Suzuki, S. Takeuchi, J. Alloys Compd. **342**, 271 (2002)
684. J. Bonneville, D. Caillard, P. Guyot, in *Dislocations in Solids*, vol. 14, ed. by J.P. Hirth (Elsevier, Amsterdam, 2008), p. 251
685. R. Mikulla, J. Roth, H.-R. Trebin, Philos. Mag. B **71**, 981 (1995)
686. D. Levine, T.C. Lubensky, S. Ostlund, P.J. Steinhardt, J. Toner, Phys. Rev. Lett. **54**, 1520 (1985)
687. D. Gratias, J.T. Beauchesne, F. Mompiou, D. Caillard, Philos. Mag. **86**, 4139 (2006)
688. M. Wollgarten, D. Gratias, Z. Zhang, K. Urban, Philos. Mag. A **64**, 819 (1991)
689. A. Katz, M. Duneau, J. Physique **47**, 181 (1986)

690. R. Wang, M.X. Dai, Phys. Rev. B **47**, 15326 (1993)
691. W.G. Yang, R.H. Wang, M. Feuerbacher, P. Schall, K. Urban, Philos. Mag. Lett. **80**, 281 (2000)
692. M. Ricker, J. Bachteler, H.-R. Trebin, Eur. Phys. J. B **23**, 351 (2001)
693. T.Y. Fan, Y.W. Mail, Appl. Mech. Rev. **57**, 325 (2004)
694. K. Edagawa, S. Takeuchi, in *Dislocations in Solids*, vol. 13, ed. by F.R.N. Nabarro, J.P. Hirth (Elsevier, Amsterdam, 2007), p. 365
695. A. Letoublon, M. de Boissieu, M. Boudard, L. Mancini, J. Gastaldi, B. Hennion, R. Caudron, R. Bellissent, Philos. Mag. Lett. **81**, 273 (2001)
696. M. Ricker, Diploma Thesis, Stuttgart University (2000)
697. X.F. Li, T.Y. Fan, Phys. Status Solidi B **212**, 19 (1999)
698. M. Wollgarten, C. Metzmacher, R. Rosenfeld, M. Feuerbacher, Philos. Mag. A **76**, 455 (1997)
699. S. Takeuchi, K. Shinoda, K. Edagawa, Philos. Mag. A **79**, 317 (1999)
700. I.A. Ovid'ko, Mater. Sci. Eng. A **154**, 29 (1992)
701. F. Mompiou, D. Caillard, Mater. Sci. Eng. A **400–401**, 283 (2005)
702. F. Mompiou, L. Bresson, P. Cordier, D. Caillard, Philos. Mag. **83**, 3133 (2003)
703. D. Caillard, G. Vanderschaeve, L. Bresson, D. Gratias, Philos. Mag. A **80**, 237 (2000)
704. F. Mompiou, D. Caillard, M. Feuerbacher, Philos. Mag. **84**, 2777 (2004)
705. F. Mompiou, D. Caillard, Philos. Mag. Lett. **84**, 555 (2004)
706. R. Rosenfeld, M. Feuerbacher, B. Baufeld, M. Bartsch, M. Wollgarten, G. Hanke, M. Beyss, U. Messerschmidt, K. Urban, Philos. Mag. Lett. **72**, 375 (1995)
707. U. Messerschmidt, L. Ledig, M. Bartsch, J. Non-Crystalline Solids **334–335**, 436 (2004)
708. L. Ledig, PhD Thesis, Martin Luther University, Halle-Wittenberg (2003)
709. U. Messerschmidt, M. Bartsch, B. Geyer, L. Ledig, in *Quasicrystals*, ed. by H.-R. Trebin (Wiley-VCH, Weinheim, 2003), p. 462
710. M. Texier, A. Proult, J. Bonneville, J. Rabier, Mater. Sci. Eng. A **387–389**, 1023 (2004)
711. U. Messerschmidt, M. Bartsch, B. Geyer, L. Ledig, M. Feuerbacher, M. Wollgarten, K. Urban, in *Quasicrystals - Preparation, Properties and Applications*, Mater. Res. Soc. Proc. **643**, K6.5.1 (2001)
712. M. Feuerbacher, D. Caillard, Acta Mater. **54**, 3233 (2006)
713. T. Zumkley, M. Wollgarten, M. Feuerbacher, K. Freytag, H. Mehrer, Defect and Diffusion Forum **143–147**, 843 (1997)
714. R. Galler, H. Mehrer, Mater. Sci. Eng. A **294–296**, 693 (2000)
715. V. Franz, M. Feuerbacher, M. Wollgarten, K. Urban, Philos. Mag. Lett. **79**, 333 (1999)
716. S. Ritsch, PhD Thesis, ETH Zürich, Nr. 11730 (1996)
717. S. Ritsch, C. Beeli, H.-U. Nissen, T. Gödecke, M. Scheffer, R. Lück, Philos. Mag. Lett. **78**, 67 (1998)
718. K. Edagawa, M. Ichihara, K. Suzuki, S. Takeuchi, Philos. Mag. Lett. **66**, 19 (1992)
719. Z. Zhang, K. Urban, Philos. Mag. Lett. **60**, 97 (1989)
720. P. Schall, M. Feuerbacher, K. Urban, Phys. Rev. B **69**, 134105 (2004)
721. M. Feuerbacher, M. Bartsch, B. Grushko, U. Messerschmidt, K. Urban, Philos. Mag. Lett. **76**, 369 (1997)

722. N.K. Mukhopadhyay, G.C. Weatherly, G.V.S. Sastry, Mater. Sci. Eng. A **294–296**, 135 (2000)
723. P. Schall, PhD Thesis, RWTH Aachen (2002), Reports of Jülich Research Centre, Nr. 3974
724. J. Bardeen, C. Herring, in *Imperfections in Nearly Perfect Crystals*, ed. by W. Shockley et al. (Wiley, New York, 1952), p. 261
725. E. Giacometti, P. Guyot, N. Baluc, J. Bonneville, Mater. Sci. Eng. A **319–321**, 429 (2001)
726. M. Heggen, M. Feuerbacher, P. Schall, H. Klein, I.R. Fisher, P.C. Canfield, K. Urban, Philos. Mag. Lett., **80**, 129 (2000)
727. M. Feuerbacher, B. Baufeld, R. Rosenfeld, M. Bartsch, G. Hanke, M. Beyss, M. Wollgarten, U. Messerschmidt, K. Urban, Philos. Mag. Lett. **71**, 91 (1995)
728. D. Brunner, D. Plachke, H.D. Carstanjen, Mater. Sci. Eng. A **234–236**, 310 (1997)
729. D. Brunner, D. Plachke, H.D. Carstanjen, Mater. Sci. Eng. A **294–296**, 773 (2000)
730. E. Kabutoya, K. Edagawa, R. Tamura, S. Takeuchi, J.Q. Guo, A.-P. Tsai, Philos. Mag. A **82**, 369 (2002)
731. U. Messerschmidt, M. Bartsch, B. Geyer, M. Feuerbacher, K. Urban, Philos. Mag. A **80**, 1165 (2000)
732. T. Saito, K. Miyaki, Y. Kamimura, K. Edagawa, S. Takeuchi, Mater. Trans. **46**, 369 (2005)
733. K. Edagawa, Y. Arai, T. Hashimoto, S. Takeuchi, Mater. Trans. Jap. Inst. Metals **39**, 863 (1998)
734. K. Edagawa, S. Ohta, S. Takeuchi, E. Kabutoya, J.Q. Guo, A.P. Tsai, Mater. Sci. Eng. A **294–296**, 748 (2000)
735. P. Schall, M. Feuerbacher, K. Urban, Philos. Mag. Lett. **81**, 339 (2001)
736. S. Takeuchi, Philos. Mag. **86**, 1007 (2006)
737. C. Dilger, R. Mikulla, J. Roth, H.-R. Trebin, Philos. Mag. A **75**, 425 (1997)
738. G.D. Schaaf, J. Roth, H.-R. Trebin, Philos. Mag. **83**, 2449 (2003)
739. F. Mompiou, D. Caillard, Acta Mater. **56**, 2262 (2008)
740. K. Tanaka, Y. Mitari, M. Koiwa, Philos. Mag. A **73**, 1715 (1996)
741. M. Feuerbacher, M. Weller, J. Diehl, K. Urban, Philos. Mag. Lett. **74**, 81 (1996)
742. U. Messerschmidt, D. Häussler, M. Bartsch, B. Geyer, M. Feuerbacher, K. Urban, Mater. Sci. Eng. A **294–296**, 757 (2000)
743. D. Caillard, F. Mompiou, L. Bresson, D. Gratias, Scripta Mater. **49**, 11 (2003)
744. S. Takeuchi, Scripta Mater. **49**, 19 (2003)
745. M. Feuerbacher, P. Schall, Scripta Mater. **49**, 25 (2003)
746. U. Messerschmidt, M. Bartsch, Scripta Mater. **49**, 33 (2003)
747. M. Feuerbacher, C. Metzmacher, M. Wollgarten, K. Urban, B. Baufeld, M. Bartsch, U. Messerschmidt, Mater. Sci. Eng. A **233**, 103 (1997)
748. U. Messerschmidt, M. Bartsch, M. Feuerbacher, B. Geyer, K. Urban, Philos. Mag. A **79**, 2123 (1999)
749. S. Takeuchi, T. Hashimoto, Jpn. J. Appl. Phys. **22**, 2063 (1993)
750. S. Takeuchi, R. Tamura, E. Kabutoya, Philos. Mag. A **82**, 379 (2002)
751. R. Mikulla, P. Gumbsch, H.-R. Trebin, Philos. Mag. Lett. **78**, 369 (1998)
752. P. Schall, Diploma Thesis, RWTH Aachen (1998)
753. S. Takeuchi, Mater. Sci. Eng. A **400–401** 306 (2005)

754. J.L. Bassani, K. Ito, V. Vitek, Mater. Sci. Eng. A **319–321**, 97 (2001)
755. G.D. Schaaf, R. Mikulla, J. Roth, H.-R. Trebin, Mater. Sci. Eng. A **294–296**, 799 (2000)
756. P. Guyot, G. Canova, in *Quasicrystals*, ed. by S. Takeuchi, T. Fujiwara (World Scientific, Singapore, 1998), p. 529
757. M. Feuerbacher, P. Schall, Y. Estrin, Y. Brechet, Philos. Mag. Lett. **81**, 473 (2001)
758. F. Mompiou, D. Caillard, Mater. Sci. Eng. A **483–484**, 143 (2008)

Index

Printed by Printforce, the Netherlands